International Conference on Climate Change (9th-ICCC)

Held online

E3S Web of Conferences Volume 467 (2023)

Surakarta, Indonesia and Online
27-28 September 2023

Editors:

Komariah
K. Hiramatsu

ISBN: 978-1-7138-8659-4

Printed from e-media with permission by:

Curran Associates, Inc.
57 Morehouse Lane
Red Hook, NY 12571

Some format issues inherent in the e-media version may also appear in this print version.

For additional information, please contact EDP Sciences – Web of Conferences at the address below.

EDP Sciences – Web of Conferences
17, Avenue du Hoggar
Parc d'Activité de Courtabœuf
BP 112
F-91944 Les Ulis Cedex A
France

Phone: +33 (0) 1 69 18 75 75
Fax: +33 (0) 1 69 28 84 91

contact-edps@webofconferences.org

Additional copies of this publication are available from:

Curran Associates, Inc.
57 Morehouse Lane
Red Hook, NY 12571 USA
Phone: 845-758-0400
Fax: 845-758-2633
Email: curran@proceedings.com
Web: www.proceedings.com

TABLE OF CONTENTS

IMPACT OF DEPLETION OR ENHANCE OF A CAPABILITY OF RESOURCES OF AIR, WATER, SOIL, AND VEGETATION

STRATEGY FOR ENVIRONMENTAL DISASTER REDUCTION RESEARCH

CLIMATE MODEL AND UNEVEN PRECIPITATION DISTRIBUTION

CARBON FOOTPRINT, GREENHOUSE GAS EMISSION, RECYCLE AND REUSE ENERGY RESEARCH

DIRECT AND INDIRECT RISKS TO WELLBEING

THE IMPLICATION OF CLIMATE ADAPTATION AND MITIGATION RESEARCH

INFRASTRUCTURES RISKS AND PLANNING ON CLIMATE ADAPTATION

MARINE ECOSYSTEM AFFECTED BY CLIMATE CHANGE

PREFACE

Climate change has resulted in an escalation of heat exposure and natural hazards caused by extreme weather events. This anthropogenic climate change phenomenon has adversely affected the most fundamental natural resources for living organisms: air, water, and soil. The air has become polluted and warmer due to human activities. The quantity of freshwater continues to diminish, with groundwater recharge being impacted by chemical and biological contamination. The sustainability of soil as the natural planting medium is under threat due to land use conversion and soil quality degradation. Water acidification and temperature increases have been proven to alter the salinity of seawater. These are the impacts of the increasing greenhouse gas emissions in the Earth's atmosphere.

The 9th International Conference on Climate Change (ICCC) aims to provide a platform for accommodating and discussing regional findings related to adaptation and mitigation strategies for global climate change. The increasing recognition of the significance of hosting the ICCC underscores its importance. This conference serves as a warning about the impending disaster, and the contributions of all participants and distinguished guests are expected to enhance our world and advance scientific knowledge. Sharing insights from recent research and cutting-edge technologies is expected to bolster our capabilities and may generate breakthrough ideas for sustainability in the face of climate change. The ICCC has the potential to be highly beneficial for each participant, offering a valuable opportunity to explore scientific solutions for our shared goals.

The 9th ICCC was organized by the Graduate School and Faculty of Agriculture, Sebelas Maret University, Surakarta, Indonesia in collaboration with The United Graduate School of Agricultural Sciences, Gifu University, Japan. The topic of the 9th ICCC is "*Linking Climate Change and Environmental Security*". The 9th ICCC was held at the Graduate School's auditorium, Sebelas Maret University, Surakarta, Indonesia from 27-28 September 2023. The 9th ICCC was conducted in the Hybrid method to provide options to the participants with limited budgets. The 9th ICCC 2022 delivers great appreciation to Prof. Dr. Sutarno, Dean of Graduate School and Prof. Dr. Samanhudi, Dean of Faculty of Agriculture, Sebelas Maret University, Indonesia; and Prof. Dr. Ken Hiramatsu, Dean of United Graduate School of Agricultural Science, Gifu University, Japan; for formal support to the conference.

Gratitude presented to the invited speakers: Dr. James MacGregor from Ecoplannet (Canada) and Mohammed VI Polytechnic University Morocco); Dr. Anurita Selvarajoo from the University of Nottingham Malaysia; Dr. Takashi Tanaka from Gifu University, Japan; and Dr. Sumani from Sebelas Maret University, Indonesia. The committee also acknowledged all the editorial boards and reviewers who have assisted us in directing the conference and reviewing the papers.

The 9th ICCC was attended by 164 people in physical attendance and 121 people in online attendance. The participants' geographical distribution includes Indonesia, Japan, Pakistan, Malaysia, Bangladesh, Morocco, Ethiopia, Nigeria, Netherlands, and Uzbekistan. The recent findings from the 9th ICCC highlighted the effects of climate change on the environment and delved into current strategies for both adapting to and mitigating these changes. The discourse has paved the way for potential future collaborations among participants, aiming to enhance research efforts collectively.

Committee of the 9th ICCC 2023

OPENING ADDRESS

The 9th International Conference on Climate Change (ICCC) 2023 is organized by Graduate School of Sebelas Maret University, Indonesia collaborated with The United Graduate School of Agricultural Science, Gifu University, Japan. The purpose of The 9th ICCC is to accommodate the new related inspiration and innovation about how to understand the climate change impact on immune modulation system for organis including human, crop and animal.

Climate change is a global problem promoted by global warming. It leads to uncertain climatic conditions such as floods, droughts, landslides, high waves and sea level rise. It also resulting into extreme changes in Biodiversity, including the emerge of new variety of microorganism, including viruses. It may influence the immune system of organism to survive. Hence, the 9th ICCC 2021 is held to understand how the climate change influence the immune system modulation in organism including human, animal and crops from the environmental perspective. Hopefully it will contribute to new ideas on how to increase the immune system more effective in the changing climate.

We acknowledge to all the invited speakers of The 9th ICCC 2023, i.e: Dr James MacGregor from Ecoplanet, Canada; Komariah, STP., M.Sc., Ph.D. from Sebelas Maret University; Dr. Anurita Selvarajoo from University of Nottingham, Malaysia; and Dr. Takashi Tanaka from Gifu University, Japan. I also express my highest gratitude to the steering and organizing committees for their contribution to the 9th ICCC 2023.

I would also like to thank all participants at the 9th International Climate Change Conference. We hope this conference will bring benefits to all of us.

Thank You.

Dean of Graduate School
Sebelas Maret University, Indonesia
Prof. Drs. Sutarno, M.Sc., Ph.D.

Impact of different soil tillage added bio-ameliorant on nutrient concentration and growth maize-soybean intercropping adaptive to climate change in dry land

Wahyu Astiko[1*], *Ni Made Laksmi* Ernawati[1], and *I Putu* Silawibawa[2]

[1]Agroecotechnology Study Program, Faculty of Agriculture University of Mataram, Jl. Majapahit 62 Mataram 83125, Lombok, Nusa Tenggara Barat, Indonesia
[2]Department of Soil Science Faculty of Agriculture, University of Mataram, Jl. Majapahit 62 Mataram 83125, Lombok, Nusa Tenggara Barat, Indonesia

Abstract. Climate change hurts the production of food crops, especially maize, and soybeans, on dry land. This climate change affects the concentration of nutrients and the growth of maize-soybean intercropping on dry land. This study aimed to determine the impact of different soil tillage added with bio-ameliorant on nutrient concentrations and the growth of climate change adaptive maize-soybean intercropping in the dry land of North Lombok, Indonesia. The research was conducted from May to July 2023 in West Pemenang Village, Pemenang District, and North Lombok Regency. The study used an experimental method in the field using a randomized block design by testing five tillage techniques i.e., O_0: without tillage, O_1: without tillage, only sprayed with pre-growth herbicides before planting, O_2: minimum tillage is only hoeing, O_3: simple tillage and leveling, O_4: intensive tillage by hoeing, loosening, and leveling. The results showed that the intensive processing technique, complete with the addition of bio-ameliorant, significantly affected the nutrient concentration and growth of the maize-soybean intercropping that was adaptive to climate change in dry land. Plants can be improved to be more adaptable to climate change through intensive tillage and adding bio-ameliorants.

1. Introduction

Tillage is a land preparation activity for plant cultivation. The tillage system is divided into intensive tillage and conservation tillage. Land preparation commonly used by farmers is intensive tillage. Farmers often use intensive tillage because it is more cost-effective, labor and time efficient. Intensive tillage is a system carried out twice to loosen the soil so that plant roots can grow well and the soil surface is clean without weeds or grass covering the soil surface [1, 2].

Global warming, on the other hand, has altered global, regional, and local climate conditions. This global climate change is produced, among other things, by an increase in

* Corresponding author: astiko@unram.ac.id

greenhouse gas emissions due to various activities that cause the earth's temperature to rise. Because climate is the most critical factor in plant metabolism, physiology, and ecosystem systems, global climate change will severely influence agricultural sustainability [3]. One of the consequences of global climate change is a drop in crop yield, which can impact food availability in Indonesia [4].

One strategy for increasing crop yield is to use heavy tillage with bio-ameliorant, a combination of biological agents like mycorrhizae, and soil additives, including manure, compost, and rice husk charcoal. Organic material ameliorants are essential in enhancing soil's physical, chemical, and biological qualities [5, 6]. The role of physical soil, among other things, as an adhesive material between soil particles to join into soil aggregates, enhance soil structure, increase soil porosity, and promote soil water retention [7]. Furthermore, adding indigenous mycorrhiza to the ameliorant can boost nutrient availability for plants, increasing the efficacy of plant roots to absorb nutrients by two to three times [8, 9]. This is one way to increase plant resistance to drought and climate change. This study aimed to determine the impact of different soil tillage added with bio-ameliorant on nutrient concentrations and the growth of climate change adaptive maize-soybean intercropping in the dry land of North Lombok, Indonesia.

2. Materials and method

2.1. Experimental design

The experimental method utilized was a randomized block design with five tillage technique treatments, notably O_0: without tillage, O_1: no without-tillage, only pre-growth herbicides applied before planting, O_2: only hoeing, O_3: simple tillage hoeing and leveling, O_4: intensive tillage with a hoe, loosened, and flattened. Each treatment was replicated four times.

2.2. Plants cultivation

The experiment was conducted in Telaga Wareng Hamlet, West Pemenang Village, Pemenang District, North Lombok Regency, Indonesia, from May to August 2023. Tillage removes weeds from the soil by using a hoe and spraying herbicides according to the tillage treatment. The area is then divided into 20 plots in 5 m by 5 m portions. The intercropping design employed is three rows of maize (NK212 variety) and three rows of soybeans (Dega 1 variety). With a maize spacing of 60 cm by 40 cm and a soybean spacing of 30 cm by 20 cm, each hole was filled with two seeds.

They give 40 g of bio-ameliorant per hole by placing it as deep as 5 cm and 7 cm away from the planting hole. At the same time, inorganic fertilizers are applied at 50% of the prescribed dosage in the form of Urea and Phonska (maize: 150 kg/ha Urea and 100 kg/ha Phonska, and soybeans: 30 kg/ha Urea and 60 kg/ha Phonska). Stitching is accomplished by replanting maize and soybean seeds seven days after planting (dap).

2.3. Observation of parameter

At 42 dap, soil nutrient concentrations and plant N and P uptake were measured, as were the weight of wet and dry stover roots and shoots and the number of spores and percentage of root infection.

3. Results and discussion

3.1. Wet and dry plant biomass weights

When the plants reached 42 DAP, the results of the analysis variance revealed that intense tillage by hoeing, loosening, and leveling (O_4) had a significant effect on the weight of wet and dry biomass of roots and plant shoots compared to the control (without-tillage) (Table 1).

Table 1. Average weights of wet and dry biomass of maize and soybean roots and shoots at some tillage, ages 42 DAP.

Soil Tillage	Maize		Soybean	
	Roots	Shoots	Roots	Shoots
Wet Biomass				
Oo : without tillage	5.53^{b}	47.35^{e}	1.62^{c}	3.37^{d}
O_1 : only herbicides	8.81$^{a.}$	53.18^{d}	1.77^{c}	5.59^{c}
O_2 : only hoeing	18.76^{ab}	148.48^{c}	1.89^{c}	1..75^{b}
O_3 : simple tillage	62.12^{ab}	192.80^{b}	3.51^{b}	13.83^{b}
O_4 : intensive tillage	65.32^{a}	368.57^{a}	4.52^{a}	32.60^{a}
HSD 5%	39.10	2.47	0.32	1.38
Dry Biomass				
Oo : without tillage	1.12^{d}	10.85^{e}	0.18^{d}	1.02^{e}
O_1 : only herbicides	2.72^{d}	18.28^{d}	0.22^{d}	1.94^{d}
O_2 : only hoeing	5.78^{c}	71.33^{c}	0.34^{c}	3.21^{c}
O_3 : simple tillage	32.43^{b}	78.55^{b}	1.65^{b}	4.66^{b}
O_4 : intensive tillage	44.58^{a}	189.14^{a}	2.68^{a}	8.26^{a}
HSD 5%	1.42	1.74	0.038	0.026

*Mean values in each column followed by the same letters are not significantly different between treatments.

Maize and soybean wet and dry weight yields were highest and most significant were obtained on intense tillage (O_4) compared to control, no-tillage. This is generated by intense tillage, which causes the soil to become loose, crumbly, and weed-free so that roots and plant shoots can grow and flourish. Maize and soybean plants require loose soil to grow and develop properly. The intensive tillage treatment (O_4) had the largest dry biomass weights of maize and soybean roots and shoots, substantially different from no-tillage (O_O). The low weight of wet and dry root and shoot biomass in the no-tillage treatment was attributable to plant-weed competition.

The wet and dry weight yields of maize and soybean gave the highest and most significant results in intensive tillage (O_4) compared to control without tillage. This is caused by intensive tillage, causing the soil to become loose and crumbly and clean of weeds to support the growth and development of roots and plant shoots [10, 11]. Maize and soybean plants require loose soil so that the roots grow and develop optimally. The highest dry biomass weights of maize and soybean roots and shoots were found in the intensive tillage treatment (O_4), significantly different from no-tillage (O_O). The low weight of wet and dry root and shoot biomass in the no-tillage treatment was attributable to early-growth competition between plants and weeds. Weeds are more fertilizer and water responsive than maize and soybean plants [12]. Furthermore, intense tillage affects the process of nutrient absorption and the availability of water for plants to provide water in the soil, which might alter the wet and dry weight of roots and plant shoots to better [13-15].

3.2. Soil Nutrient Concentration and Plant Nutrient Absorption

The HSD test results at the 5% level showed that intensive tillage with hoeing, loosening, and leveling (O_4) could increase N, P, and K concentrations, soil-organic C uptake, and plant nutrient uptake of N, P, K, and Ca at 42 DAP and is significantly different from the control (Table 2).

Table 2. Average N, P, K, C-organic, and pH contents in the rhizospheres of maize and soybean plants at 42 DAP soil treatments.

Treatment	N ($g\ kg^{-1}$)		P ($mg\ kg^{-1}$)		K ($cmol\ kg^{-1}$)		C-organic ($g\ kg^{-1}$)		pH	
	Maize	Soybear	Maize	Soybean	Maize	Soybean	Maiz	Soybear	Maize	oybean
Without tillage	1.21^{e}	1.18^{e}	12.16^{e}	12.76^{e}	0.64^{d}	0.62^{c}	7.27^{e}	6.64^{e}	6.37^{d}	6.71^{d}
Only herbicides	1.33^{d}	1.27^{d}	16.34^{d}	15.37^{d}	0.69^{c}	0.72^{bc}	7.61^{d}	7.23^{d}	6.63^{c}	6.76^{cd}
Only hoeing	1.42^{c}	1.33^{c}	20.62^{c}	17.29·	0.75^{b}	0.76^{b}	8.65^{c}	7.75^{c}	6.73^{b}	6.80^{c}
Simple tillage	1.61^{b}	1.42^{b}	23.72^{b}	20.2^{b}	0.78^{b}	0.83^{ab}	9.77^{b}	8.51^{b}	6.88^{a}	6.87^{b}
Intensive tillage	1.95^{a}	1.49^{a}	32.53^{a}	24.65^{a}	0.86^{a}	0.94^{a}	10.65^{a}	9.55^{a}	6.94^{a}	6.97^{a}
HSD 5%	0.050	0.027	2,010	0,861	0.026	0.084	0.044	0.193	0.044	0.040

*Mean values in each column followed by the same letters are not significantly different between treatments.

At 42 dap, the same thing happened to the plant nutrient absorption parameters N, P, K, and Ca (Table 3). Compared to the control (Oo), the intense tillage treatment by hoeing, loosening, and flattening (O4) produced the most significant and significantly different increase.

Table 3. Plant nutrient uptake (N, P, K, and Ca) concentration (mg/g) at 42 DAP aged soil treatment.

Treatment	Nutrient uptake (mg/g plant)							
	Maize				Soybean			
	N	P	K	Ca	N	P	K	Ca
Without tillage	28.2^{e}	1.87^{e}	25.41^{e}	2.38^{d}	35.25^{e}	1.91^{e}	17.54^{d}	5.68^{d}
Only herbicides	29.4^{d}	244^{d}	26.48^{d}	2.58^{cd}	36.73^{d}	2.26^{d}	22.80^{c}	6.62^{c}
Only hoeing	30.60^{c}	2.76^{c}	27.15^{c}	2.74^{bc}	40.47^{c}	2.80^{c}	23.65^{b}	6.78^{bc}
Simple tillage	31.81^{b}	3.32^{b}	27.73^{b}	2.83^{ab}	45.45^{b}	3.49^{b}	23.75^{b}	6.92^{b}
Intensive tillage	33.36^{a}	3.91^{a}	28.53^{a}	$2,98^{a}$	46.95^{a}	3.90^{a}	24.76^{a}	7.60^{a}
HSD 5%	0.65	0.13	0.12	0.15	0.25	0.12	0.21	0.17

*Mean values in each column followed by the same letters are not significantly different between treatments.

In the results of soil pH analysis on intensive tillage with additions of bio-ameliorant plus mycorrhiza (O_4), the highest pH was obtained, close to normal pH, and the lowest was in the treatment without tillage (O_O), which was 6.71. The addition bio-ameliorant plus mycorrhizae to the soil during intensive tillage can increase soil pH. The organic compounds released by bio-ameliorant organic matter can increase cations in the soil complex [16, 17].

A bio-ameliorant is a substance that can be applied to soil to enhance its physical, chemical, and biological qualities. Soils in North Lombok's dry land generally have a very low C-organic content. Cow dung, rice husk charcoal, compost, and mycorrhiza significantly boosted soil pH, N, P, K, and C-organic content, as well as N, P, K, and Ca nutrient uptake by maize and soybean plants grown on sandy soil [18]. Organic matter as an ameliorant can boost C-organic in the soil, allowing nutrients such as N, P, K, Ca, Mg, and S to be available for maize and soybean plants during growth and plant tissue formation [19]. pH is a key determinant in phosphorus availability in soil. Suspected the planting medium has increased accessible P levels. In that case, this is because the majority of phosphorus in the soil originates from weathering of natural mineral rocks, with the remainder coming from weathering of organic mineral components. The phosphorus ions in the soil are similarly impacted by soil pH [20].

3.3. Number of Spores and Root Colonization

The 5% HSD test findings revealed that intensive tillage using hoed, loosened, and flattened (O_4) was significantly different from the control treatment (O_O) on the parameters of spore count (per 100 g of soil) and root colonization (%) at 42 dap (Table 4).

Table 4. At 42 DAP, the average number of spores (per 100 g of soil) and root colonization (%) of maize and soybean plants.

Soil Tillage	Spores on Maize	Spores on Soybean	%-Maize Colonization	%-Soybean Colonization
Oo : without tillage	383.00^{d}	800.00^{c}	31.66^{c}	27.66^{d}
O_1 : only herbicides	628.00^{c}	1172.56^{b}	66.00^{b}	58.33^{c}
O_2 : only hoeing	641.00^{c}	1194.66^{b}	74.33^{b}	61.66^{c}
O_3 : simple tillage	1288.33^{b}	1220.00^{b}	84.33^{a}	78.33^{b}
O_4 : intensive tillage	1438.33^{a}	1381.66^{a}	89.33^{a}	86.66^{a}
HSD 5%	10.53	45.84	5.808	5.016

*Mean values in each column followed by the same letters are not significantly different between treatments.

Improving the fertility of sandy soil in the arid region of North Lombok can be accomplished through intense processing combined with bio-ameliorant additives containing mycorrhiza. Mycorrhizal spores and colonization percentage increased more in the heavy tillage treatment than in the non-tillage treatment. To carry out their metabolic functions and to grow and develop, mycorrhizal fungi require specific organic substances as a source of carbon and nutrients [21]. According to the total quantity of spores and mycorrhizal colonization results, heavy tillage appears to be associated with bio-ameliorant organic output of higher value than non-tillage. Intensive tillage with the addition of a bio-ameliorant can boost mycorrhizal effectiveness on plant root infection. Organic substrates provide energy to mycorrhizal fungi, whereas growth hormones, which are also found in organic matter, stimulate the growth and regeneration of plant roots and create root exudate, accelerating the symbiotic interaction between mycorrhizal fungi and plants [22].

4. Conclusion and suggestions

The results showed that the intensive processing technique, and the addition of a bio-ameliorant had a substantial effect on the nutrient content and growth of the climate-adaptive maize-soybean intercropping in a dry field. Plants can adapt more to climate change by using intense tillage and adding bio-ameliorants.

The authors thank the DRPM KEMENDIKBUDRISTEK DIKTI and the University of Mataram for the research grants.

References

1. S.W. Duiker, D.C. Reicosky, K. Koeller, T. Friedrich, K. Weiss. Soil and Tillage Research. **137**, 16 (2014)
2. A. Morugán-Coronado, P. Pérez-Rodríguez, E. Insolia, D. Soto-Gómez, D. Fernández-Calvino, R. Zornoza. Agriculture, Ecosystems & Environment **329**, 107867 (2022)
3. G.S. Malhi, M. Kaur, P. Kaushik. Sustainability, **13** (3), 1318 (2021)
4. Y. Syaukat. J. ISSAAS, **17** (1), 40 (2011)
5. A.R. Dexter. Soil and Tillage Research, **43** (1-2), 41 (1997)
6. L. Abdollahi, L.J. Munkholm. Soil Science Society of America Journal, **78** (1), 262

(2014)

7. K.V. Usharani, Roopashree, K. M., & Naik, D. Journal of Pharmacognosy and Phytochemistry, **8** (5), 1256-1267. (2019)
8. P. Jeffries, S. Gianinazzi, S. Perotto, K. Turnau, J.M. Barea. Biology and fertility of soils. **37**, 16 (2003)
9. W. Astiko, M. Isnaini, M.T. Fauzi, I. International Journal of Innovative Science and Research Technology, **7** (9), 322-329 (2022)
10. A. Wasaya, T.A. Yasir, M. Ijaz, S. Ahmad. Agronomic Crops:Volume 2: Management Practices, **73**-99 (2019)
11. N.E. Korres, J.K. Norsworthy, P. Tehranchian, T.K. Gitsopoulos, D.A. Loka, D.M. Oosterhuis, M. Palhano. Agronomy for Sustainable Development, **36**, 22 (2016)
12. Rühlemann, L., & Schmidtke, K. European Journal of Agronomy, **65**, 83-94. (2015)
13. S.Y. Sorokina, N.S. Sorokin, S.M. Sychev, O.A. Okorokova. In IOP Conference Series: Earth and Environmental Science Vol. 650, No. 1, p. 012084 (2021)
14. S.P. Long, X.G. Zhu, S.L. Naidu, D.R. Ort. Plant, cell & environment, **29** (3), 315 (2006)
15. N.K. Fageria, A. Moreira. Advances in agronomy, **110**, 251 (2011)
16. W. Astiko, M.T. Fauzi, I. Muthahanas. International Journal of Innovative Science and Research Technology, **7** (10), 168 (2022)
17. A. Herawati, J. Syamsiyah, M. Mujiyo, M. Rochmadtulloh, A.A. Susila, M.R. Romadhon. SAINS TANAH-Journal of Soil Science and Agroclimatology, **18** (1), 73 (2021)
18. R.A. Saputra, N.N. Sari. In IOP Conference Series: Earth and Environmental Science, Vol. 648, No. 1, p. 012183 (2021)
19. C.J. Penn, J.J. Camberato. (2019. Agriculture, **9** (6), 120 (2019)
20. N. Helber, K. Wippel, N. Sauer, S. Schaarschmidt, B. Hause, N. Requena. The Plant Cell, **23** (10), 3812 (2011)
21. A. Medina, R. Azcón. Journal of soil science and plant nutrition, **10**(3) 354 (2010)
22. R.D. Finlay. Journal of experimental botany, **59** (5), 1115-1126 (2008)

Enhancing chemical properties and maize yield through dolomite application on rock phosphate-amended oxisol

Antonius Kasno[1*], *Neneng* Nurida[1], *Adha Fatmah* Siregar[2], *Ahmad* Samsun[2], *Ladiyani Retno* Widowati[2], and *Husnain*[2]

[1]National Research and Innovation Agency (BRIN), Cibinong Science Center, Jl. Raya Jakarta-Bogor, KM. 46, Cibinong, Bogor, West Java, 16991, Indonesia
[2]Indonesian Agency for Agricultural Instrument Standardization, Jl. Tentara Pelajar No. 12, Ciwaringin, Bogor Tengah, Kota Bogor, Java Barat 16114, Indonesia

Abstract. Improving soil quality is vital so plants can adapt to climate change. Indonesia's acid soils cover about 147.5 million hectares, and are generally acidic, with low levels of calcium and magnesium, and high levels of aluminum. This study aims to study the effect of dolomite application on acid upland soils that have been applied with rock phosphate to improve soil chemical properties and increase maize yields. The research was conducted in Jati Agung using a randomized block design with 5 treatments and 5 replications. The treatment consisted of 5 application levels of dolomite: 0, 500, 1000, 1500, and 2000 kg ha^{-1}. The size of the experimental plot was 10 m x 10 m. Hybrid maize varieties were used as indicators. The results showed that applying dolomite increased soil pH, Ca, Mg, and K levels and decreased exchangeable and Al saturation. It also increases plant height, maize yield, and biomass. The implication of this study is that the application of dolomite 1000 kg ha^{-1} after rock phosphate is applied can improve soil chemical properties in acid soils.

1 Introduction

The main source of water for upland farming is rainwater. Climate change that occurs causes planning for planting corn to be not ideal. Acid upland soil is one of the potential suboptimal lands that can be optimized for agricultural development in Indonesia. Acid upland soils are formed in areas with high rainfall with acid parent material occupying around 109 million ha [1]; 5,30 million ha of them are available and suitable for agricultural activity in the upland areas. The acid upland soil is characterized by low soil pH (< 5,5), base saturation (<50%), and high Al saturation of >30% [2–4]. Moreover, acid upland soil was also noticed with low content of soil organic carbon, N, P, K, Ca, and Mg [5–7].

Generally, soil Ca and Mg concentrations in acid upland soil are categorized as low [8] and are limiting factors for plant growth and productivity. The application of soil conditioners such as dolomite can be used as a source of Ca and Mg which are needed in upland acid soils. The application of dolomite to acid upland soils in Lampung significantly increased soil pH

* Corresponding author: antkasno@gmail.com

and exchange of Ca and Mg [9]. A similar result showed that dolomite application in Ultisol in Hubei Province, China could increase soil pH [10].

Generally, lime or dolomite is applied at the beginning of planting with the aim of increasing soil pH and reducing the adverse effects of aluminum (Al). However, the effect of dolomite application in acid upland soil with previous rock phosphate application has not been investigated deeply. The aim of this research is to study the effect of the dolomite application on improving soil chemical properties and maize yield in acid upland soil which has been applied rock phosphate previously.

2 Methodology

This research was conducted at Karangrejo Village, Jati Agung Sub-district, South Lampung, Indonesia (-05° 14' 4.1" S, 105° 23' 28.8" E), with elevation at 60 m asl, two growing seasons from July 2021 to April 2022. This study was set up in a randomized complete block design with 5 treatments and 5 replications. The treatments consist of 5 rates of dolomite namely 0 (control), 500, 1000, 1500, and 2000 kg ha^{-1}. Urea and KCl dosages were 400 and 100 kg ha^{-1}, rock phosphate 1 t ha^{-1}, and manure 2 t ha^{-1} for each treatment respectively. The plot size was 10 m x 10 m, maize was planted in a zig-zag cropping pattern. Previous research showed that zig-zag planting could increase maize yield [11]. Hybrid maize varieties were used as indicator plants and were maintained until harvest.

Rock phosphate was given 7 days before dolomite application by spreading it over the surface of the plot and mixing it with the soil. Manure was given at the planting hole to cover the seeds after planting. Urea application was split three times at 10, 25, and 40 DAP as 120, 160, and 120 kg ha^{-1} respectively. KCl was applied twice at 10 and 25 DAP as 50 kg KCl ha^{-1} for each application. Urea and KCl were applied in a hole 3 cm away from plants and covered with soil.

Parameter observation consists of plant height, maize, and biomass yield at the first planting time and second planting time including soil chemical properties. Soil samples were collected before rock phosphate application, at 15, 30, and 90 DAP at PS1. Initial soil analysis parameters consist of soil texture, pH H_20 (1:5), C-organic (Waklkey dan Black), N-total (Kejldal), P_2O_5 and K_2O (HCl 25%), P-Bray 1, Ca, Mg, K, Na, and CEC (NH_4OAc 1 N pH 7), Al and H (KCl 1N) 3). Soil parameter analysis at 15, 30, and 90 DAP consist of soil pH H_2O (1:5), P-Bray 1, Ca, Mg, K, Na (NH_4OAc 1M pH 7), exchangeable Al and H (KCl 1N).

The total soil base cation is calculated with: Total cation = Ca + Mg + K + Na (1)
Aluminum saturation is calculated using the formula by [12] as below:

$$\text{Al saturation (\%)} = \frac{\text{Al}}{\text{Ca+Mg+K+Na+Al+H}} + 100\% \quad (2)$$

Maize harvesting is done when the corn looks ripe, the corn cobs are yellow. The size of the harvest plot is 6.3 m x 8 m, as the edge crops are two rows from the left and right plots. Harvesting activities consist of collecting corn, peeling, and shelling then weighing the wet and dry corn including wet and dry biomass.

The observation data of plant growth and yield were analyzed statistically using analysis of variance (ANOVA) at 95% and then followed with the Duncan Multiple Range Test (DMRT) at $\alpha = 5\%$. Descriptive analyses were also conducted for other parameters.

The soil in the site experiment was classified as Plinthic Hapludox, a loam soil texture, with low soil pH of 4.5 followed by low soil carbon content (< 1.5%). Previous research has proved that organic matter application could increase soil pH, P available, and CEC and decrease Al^{3+} concentration [13,14]. This site also showed low soil N-total, P, and K potential, exchangeable K, CEC, and base saturation, meanwhile, Al saturation (52%) was

 https://doi.org/10.1051/e3sconf/202346701002

categorized as medium. These soil chemical properties in acid upland soil limit maize productivity [15], and the minimum level of Al saturation for maize is 30%.

3 Result and Discussion

3.1 Maize Growth and Yield

Dolomite application at 1000 kg ha^{-1} could significantly increase plant height compared to the control **(Fig. 1)**. However, the application of dolomite at higher dosages (1500 and 2000 kg ha^{-1}) did not lead to a significant increase in plant height at the first planting season (PS-1). In contrast to PS-1, the dolomite application at a dosage of 2000 kg ha^{-1} at the second planting time (PS-2), significantly increases plant height compared to the control.

These results are similar to research in Timor Leste in 2016, where the application of 320 kg ha^{-1} of dolomite increased the number of pods, plant dry weight, 100 seed weight, and green bean seed dry weight [16]

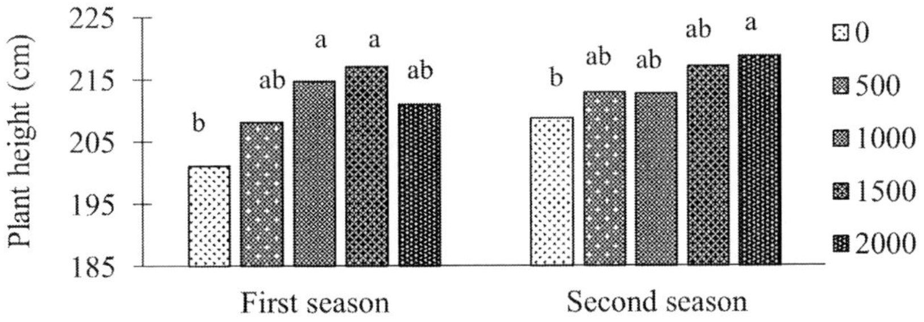

Fig. 1. Maize height which dolomite application at in the first and second seasons.

Applying dolomite as 500 kg ha^{-1} increased maize yield significantly compared to the control (**Fig. 2**). However, the application of dolomite of more than 500 kg ha^{-1} showed no significant difference in the yield. Dolomite application at < 1000 kg ha^{-1} showed an increase in dry maize weight, at >1000 kg ha^{-1} showed a flattened yield then tended to decrease at 1750 kg ha^{-1}. The maximum corn yield of 11.07 t ha^{-1} occurred in the application of dolomite 1550 kg ha^{-1}. Based on the correlation equation, showed that the optimum dosage of dolomite at acid upland soil that has been applied rock phosphate previously is around 1000 kg ha^{-1} (**Fig 3**).

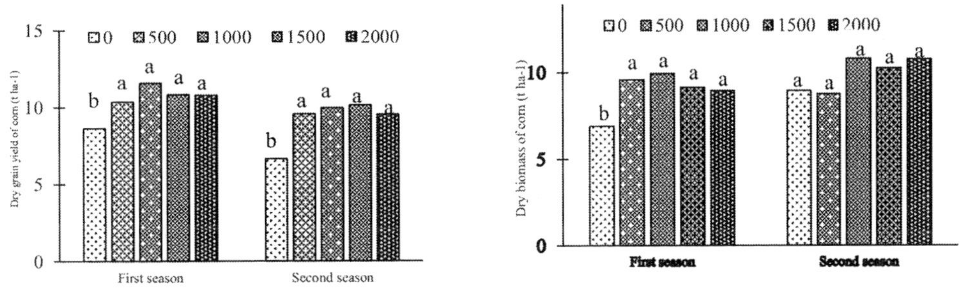

Fig. 2. Maize yield and biomass in the first and second seasons.

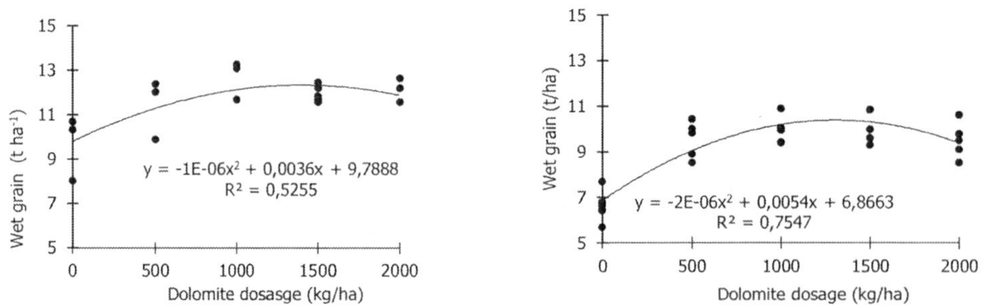

Fig. 3. Correlation between dolomite dosage and maize wet grain in the first and second seasons.

3.2 Nutrient Content Improvement

Dolomite as soil ameliorant contains Ca and Mg significantly increased soil pH at 15 DAT with 2000 kg ha^{-1}, at 30 and 90 DAP with 1500 kg ha^{-1} (Table 1). Soil pH increased to 0.47, 0.79, and 0.97 units at 15, 30, and 90 DAP respectively. This current research is in line with several studies, where the application of dolomite can increase soil pH [17–20]

Table 1. The effect of the dolomite application on soil pH at first planting time.

Dolomite dosage (kg ha^{-1})	15 DAP	30 DAP	90 DAP
0	4.56 b	4.53 c	4.45 c
500	4.50 b	4.53 c	4.33 c
1000	4.59 b	4.66 bc	4.40 c
1500	4.67 b	4.85 b	4.73 b
2000	5.03 a	5.32 a	5.42 a
CV (%)	4.49	6.25	5.42

This means values having the same letter in one column are not significantly different at 5% DMRT statistic analysis. DAP: days after planting.

Morocco rock phosphate contains 27.70% - 31.20 P_2O_5 [21,22]. Rock phosphate application will significantly affect soil with low pH (<5.5). To improve the effectiveness of rock phosphate, after one week application should be followed with the dolomite application. Current research showed that applying dolomite at varied dosages could not increase soil P concentration (Table 2).

Table 2. Soil Available P after dolomite application in the first planting time.

Dolomite dosage (kg ha^{-1})	15 DAP	%[1]	30 DAP	%[1]	90 DAP	%[1]
0	271.71 a		254.62 a		309.68 a	
500	319.32 a	17.5	282.36 a	10.9	273.29 a	-11.8
1000	327.67 a	20.6	271.89 a	6.8	315.08 a	1.7
1500	308.48 a	13.5	260.86 a	2.5	302.66 a	-2.3
2000	297.73 a	9.6	225.96 a	-11.3	268.49 a	-13.3
CV (%)	23.29		26.21		19.70	

This means values having the same letter in one column are not significantly different at 5% DMRT statistic analysis. DAP: days after planting. [1]increasing/decreasing to control.

Dolomite application as 2000 kg ha^{-1} significantly increased soil exchangeable Ca compared to the control. Dolomite application at 1500 kg ha^{-1} significantly increased soil exchangeable Mg compared to control and increased as the dosage of dolomite increased up to 2000 kg ha^{-1} (**Fig. 4**). Previous research showed a similar result, that dolomite application at acid soil in Centre Lampung could increase soil pH, increase soil Ca and mg and decrease soil Al^{3+} [23].

Dolomite application at 2000 kg ha^{-1} significantly increased total cation compared to 0 – 1500 kg ha^{-1} at 15 DAP (**Fig. 5**). The different pattern of total cation showed that 1500 kg dolomite ha^{-1} at 30 and 90 DAP increased total cation significantly compared to 1000 kg dolomite ha^{-1} treatment, the same pattern also occurred in 2000 compared to 1500 kg dolomite ha^{-1} treatment. A high concentration of Al in acid upland soil could fix P and not available for plant absorption, antagonistic reaction with K at soil site sorption which leads to decreasing of soil K.

The result showed that at 15 DAP, dolomite application as 2000 kg ha^{-1} significantly decreased Al saturity compared to others, meanwhile at 30 and 90 DAP, dolomite application at 1500 kg ha^{-1} gave significant effect on decreasing Al saturity compared to 0 and 500 kg dolomite ha^{-1} treatments. Dolomite at dosage 2000 kg ha^{-1} decreased Al saturiry significantly compared to 1500 kg dolomite ha^{-1} treatment (**Fig. 5**). A high concentration of Al could cover the root surface and interfere with nutrient uptake.

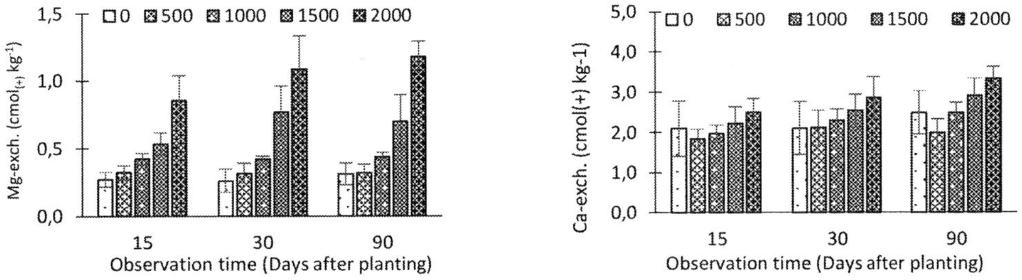

Fig. 4. Exchangeable Ca and Mg content after dolomite application in the first season.

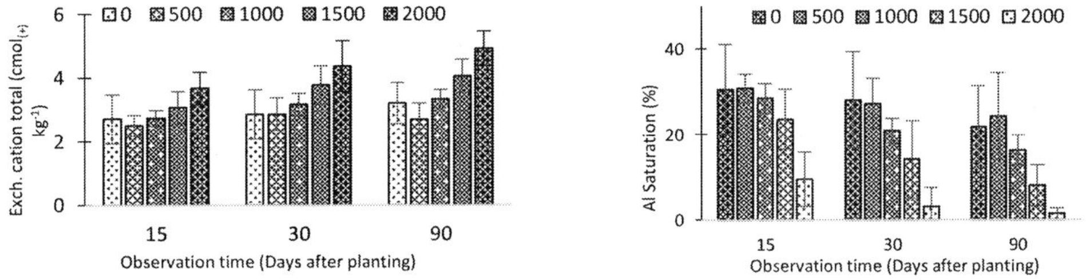

Fig. 5. Cation total and Al saturation after dolomite application in the first season.

Dolomite application at acid upland soil that has been applied rock phosphate as 1 t ha^{-1} could increase soil pH and decrease Al^{3+} concentration and saturation. Rock phosphate is a fertilizer that contains high Ca as 30.1%, 48.4 – 55.8% CaO [9]. Meanwhile dolomite as soil ameliorant contains 29.69% CaO and 21.17% MgO [23]. As a soil ameliorant, application of dolomite at acid upland soil that has been applied rock phosphate 1 t ha^{-1} could increase soil pH (**Fig. 5**), exchangeable Ca, and Mg.

4 Conclusions

Dolomite application on acid upland soils Plinthic Hapludox which was the rock phosphate application was able to

1. Increase plant height, corn yield, and biomass in both the first and second planting seasons. Soil pH, exchangeable Ca and Mg, and total cations increased while exchangeable Al and saturation decreased.
2. The optimal dosage of dolomite on acid upland soils, on which rock phosphate has previously been applied, is 1000 kg ha^{-1}.

These authors acknowledge the research grant from the Indonesian Agency of Agriculture Research and Development (IAARD), by the Ministry of Agriculture. We thank the technicians from the Indonesian Soil Research Institute (ISRI) for their valuable support in the fieldwork.

References

1. Mulyani Anny and Muhrizal Sarwani, Jurnal Sumberdaya Lahan **7**, 47 (2013)
2. N. L. Nurida, A. Dariah, and S. Sutono, Jurnal Tanah Dan Iklim **39**, 99 (2015)
3. Y. Sulaeman, Maswar, and Erfandi Deddy, Jurnal Pengkajian Dan Pengembangan Teknologi Pertanian **20**, 1 (2017)
4. Subiksa IGM, J. Agroecotania **1**, 1 (2018)
5. M. M. Hanafi, A. Hartinie, J. Shukor, and T. M. M. Mahmud, Pertanika J. Trop. Agric. Sci **32**, 225 (2009)
6. IPG, dan Wigena, and Andriati, Jurnal Sumberdaya Lahan **10**, 11 (2016)
7. Karnilawati, C. Mulia Sari, and Musfirah, Jurnal Sains Riset | **12**, 96 (2022)
8. D. N. Utami, Jurnal Sains Dan Teknologi Mitigasi Bencana **14**, (2019)
9. P. Cahyono, S. Loekito, D. Wiharso, Afandi, A. Rahmat, Komariah, N. Nishimura, and M. Senge, International Journal of GEOMATE **19**, 173 (2020)
10. M. Shaaban, Qi-an Peng, R. Hu, Y. Wu, S. Lin, and J. Zhao, Environmental Science and Pollution Research **22**, 19961 (2015)
11. W. Annisa, Husnain, and F. Djufry, in *IOP Conf Ser Earth Environ Sci* (Institute of Physics Publishing, 2020), pp. 1–7
12. M. Ikiriko, J. Omueti, and E. Thomas, Agri Res & Tech: Open Access J **1**, 1 (2016)
13. P. Siregar, Fauzi, and Supriadi, Jurnal Agroekoteknologi FP USU **5**, 256 (2017)
14. L. M. Septiana, H. Indhira, A. Afandi, and I. S. Banuwa, Jurnal Agrotek Tropika **9**, 251 (2021)
15. T. Dierolf T. H. Fairhurst E. W. Mutert, *Soil Fertility Kit: A Toolkit for Acid Upland Soil Fertility Management in Southeast Asia* (GTZ, FAO PPI, PPIC, 2000)
16. D. Ribeiro, N. Kartini, and G. Wijaya, AGROTROP **7**, 42 (2017)
17. H. S. Hasibuan, D. Sopandie, Trikoesoemaningtyas, and D. Wirnas, Jurnal Agron. Indonesia **46**, 175 (2018)
18. R. Domagoj, Z. Vladimir, and R. Mirta, in *13th Alps-Adria Scientific Workshop*, edited by B. K. Z. P. B. T. Á. BAKTI (Villach, Ossiacher See, Austria, 2014), pp. 193–196
19. M. Tállai, B. A. Kovács, Z. Á. Oláh, and J. Kátai, Natural Resources and Sustainable Development **9**, 174 (2019)
20. A. Maulana, Herviyanti, and T. Budi Prasetyo, Jurnal Tanah Dan Sumberdaya Lahan **7**, 209 (2020)
21. A. N. Flatian, S. Slamet, A. C. Pusat, A. Isotop, D. Radiasi, B. Tenaga, N. Nasional, J. Kuningan, and B. Raya, Jurnal Tanah Dan Iklim **42**, 83 (2018)
22. U. Ryszko, P. Rusek, and D. Kołodyńska, Materials **16**, (2023)
23. A. Krismawati and E. Latifah, in *International Conference on Tropical Agrifood, Feed and Fuel* (2022), pp. 5–20

Corn cob biochar as agricultural waste management for increasing soil resilient and crop yield

Neneng Laela Nurida[1*], *Selly* Salma[1] *and* Jubaedah[1]

[1]Research Centre for Food Crops, National Research and Innovation Agency, Bogor, Indonesia

Abstract. For the purpose of climate change mitigation and adaption, biochar is a solution for managing agricultural waste by maximizing its addition to soil. The conversion of corn cobs to biochar have advantages in terms of waste management and improving soil properties. This research aims to utilize corn cob biochar to enhance soil properties and corn yield. The study was conducted on September to December 2021 at Karangrejo village, South Lampung. The research arranged by randomized block design with 6 replications. The treatments tested were: 1) manure compost 2 tha^{-1}; 2) corn cob biochar (CCBC) 5 tha^{-1}+compost manure 2 tha^{-1}; 3) Co-compost CCBC-manure 7 tha^{-1} and 4) Co-compost CCBC-manure 5 tha^{-1}. Parameters observed were soil properties and dry grain yield. The research results showed that after one season, the application of CCBC, either as a single or in co-compost biochar-manure, increased soil porosity, soil pH to be 4.68 but no significantly differences in soil organic carbon content. The addition of co-compost CCBC-manure of 7 tha^{-1} or 5 $th^{-1}$was able to decrease the Al^{3+} content and increase dry grain as well. The conversion of corn cobs into biochar was not only increasing soil properties but also reduces carbon loss from agricultural waste.

1 Introduction

Corn is the main crop grown in the South Lampung Regency. The productivity in this region is about 5.9 t ha^{-1} and the harvested area about 134.9 hundred hectares [1]. Some studies reported that cobs accounted for 20% of the total corn yield [2], while another study found that corn harvest produces corn cobs that account for 21% of the dry cob weight [3]. These corn residues were not put to optimal use. Biochar is one of the solutions in agricultural waste processing, where thermal energy is also generated at the same time during the biochar production process [4,5]. From a climate change mitigation perspective, maximizing the use of biochar in soil is highly beneficial, as it is one of the carbon (C) sequestration effort using locally available materials or in situ. Most of the soil in South Lampung is acidic due to the nutrients leaching, leaving H^+ and Al^{3+} in the soil complex [6,7].

Biochar has the ability to reduce soil acidity in agricultural lands. It is rich in basic cations (Ca, Mg, K) and contributes to neutralize soil acidity and reduce levels of Al^{3+} and other heavy metals [8]. Biochar also contains essential nutrients for plants (N, P, K, Ca), and thus

* Corresponding author: lelanurida@yahoo.com

reduces nutrient leaching into the subsoil, especially in soils with low cation exchange capacity (CEC) [9]. Biochar benefit also include improving the water-holding capacity of soils with low available water content [10,11]. So, biochar is capable to improve soil resilience to various environmental. Corn cob based Biochar combined with compost has potential as a soil conditioner and an appropriate waste management effort as well [5].

The functions of compost and biochar in improving soil resilient have been widely proven [12–14]. The composting process can decrease nutrient content such as N and K, resulting in lower nutrient content compared to the raw materials [15–17][15-17]. On the other hand, biochar has the ability to retain NPK nutrients, making them less prone to loss or leaching. It can also create a favorable habitat for the development of symbiotic microorganisms like mycorrhizae due to its water and air retention capabilities, as well as its ability to create a neutral environment, particularly in acidic soils [18].

The combination of composting and converting agricultural waste (corn) into biochar is necessary to make the waste more productive for soil improvement. This research aims to utilize corn cob biochar to enhance soil resilient through improve soil properties and corn yield in upland acidic soil.

2 Research methods

The research was conducted in Karangrejo Village, Jati Agung District, South Lampung Regency, located at coordinates 05°13'45,5"S 105° 23' 19.2" E. The research activities in 2022 began with the production of biochar from corn cobs. Field activities started during the second planting season from October 2021 to January 2022, after the compost and biochar materials were ready for application. The research was carried out using a randomized complete block design with six replications. Four treatments tested were: Manure compost (MC) 2 t/ha, Corn Cob Biochar (CCBC) 5 t/ha + manure compost 2 t/ha, Co-compost CCBC-manure compost 7 t/ha (50:50 % w/w), and Co-compost CCBC-manure compost 5 t/ha (50:50 % w/w).

The raw material used for biochar production was corn cobs, produced by Kon Tiki method (digging a hole in the ground with a diameter of 150 cm and a height of 75 cm), with a temperature of approximately 250°C and a burning time of about 1 hour. The co-compost was prepared by mixing corn cob biochar with cow manure and then composting it. The plot size is 10 m x 2.5 m, planted with corn variety NK Jumbo. Corn cob biochar, cow manure, and cow manure-biochar co-compost were applied during the last (second) tillage by spreading and incorporating all ameliorants into the soil. Then the plots were incubated for 1 week before corn planting. Fertilization was done using a Rock phosphate package, consisting of 1 tha^{-1} Rock phosphate and 300 tha^{-1} of urea and 100 tha^{-1} KCl, while corn cob biochar and animal manure compost were applied according to the treatments.

The observations measured: 1) soil physical properties: porosity (aeration and water availability pores), 2) soil chemical properties: pH H_2O, organic carbon (Walkley and Black method), exchangeable cations, and cation exchange capacity (NH_4OAc 1 N, pH 7), as well as soil acidity (KCl 1 N), 3) dry biomass and dry corn grain. Soil samples were taken before planting (per block) and after harvest for each treatment plot.
The data analysis was conducted statistically on the observed variables using analysis of variance (ANOVA) or diversity test with a 95% confidence interval. To determine the significant differences caused by the treatments and their interactions, the Duncan's multiple range test (DMRT) was performed at a significance level of 95% (α = 5%).

3 Result and discussion

Table 1 showed the soil initial characteristics before treatment. The soil is classified as acidic, with pH H_2O ranging from 4.3 to 4.4, and the blocks were relatively homogeneous. The organic C content is low, ranging from 1.06% to 1.25%, and the total N content is also low, ranging from 0.10% to 0.11%, as were the other nutrient contents (Ca, Mg, and K), which are also low except for the potentially available P, which is classified as high. Soil quality at the research site was considered low or degraded, as indicated by the low cation exchange capacity (CEC) and nutrient contents [8,12]. The soil's aluminium levels range from 1.11 to 1.59 $cmol_+ kg^{-1}$, indicating relatively high acidity that may inhibit plant growth. The upland at the research site is classified as acidic, and a reduction in soil acidity is necessary to support optimal plant growth. The soil resilient at the research location is considered low, therefore soil improvement effort are needed to achieve normal yield.

Table 1. The initial of soil chemical properties of a six blocks

Parameter	Unit	Block 1	Block 2	Block 3	Block 4	Block 5	Block 6
pH H_2O		4.3	4.3	4.4	4.4	4.4	4.3
Organic C	%	1.06	1.08	1.18	1.16	1.25	1.18
N total	%	0.11	0.10	0.11	0.11	0.11	0.11
CEC	$cmol_+ kg^{-1}$	6.88	7.06	6.04	6.51	7.23	5.92
Al^{3+}	$cmol_+ kg^{-1}$	1.11	1.30	1.14	1.68	1.39	1.59
Ca	$cmol_+ kg^{-1}$	1.31	1.19	1.23	1.02	1.06	0.83
Mg	$cmol_+ kg^{-1}$	0.27	0.26	0.30	0.24	0.24	0.20
K	$cmol_+ kg^{-1}$	0.20	0.23	0.24	0.25	0.25	0.20
Potential P_2O_5	ppm	101.6	123.6	137.8	146.3	84.2	119.0
Available P_2O_5	mg 10^{-2}g	48	56	57	57	63	50
K_2O	mg 10^{-2}g	11	12	13	13	12	11

3.1 Soil physical and chemical properties

The application of CCBC+manure and co-compost CCBC+manure 7 tha^{-1} and 5tha^{-1} resulted in higher water available pore (13.53%Vol. and 13.72%Vol.) than others treatment after one planting season (Table 2). The lowest aeration porosity was obtained when co-compost CCBC+manure 5 tha^{-1} but there were no signifant differences to others CCBC applications. The decrease in aeration pores allows water to stay longer in the root zone without immediately moving to the lower layers of the soil. This is further supported by the increased availability of water pores when co-compost biochar from corn cobs is applied. The water retention ability of biochar leads to increased water availability in the root zone [10,15,19], particularly when co-compost CCBC was applied. The research site has sandy texture, making it crucial to enhance water availability in the root zone. Improving soil porosity could ensure that water will be more readily available to plants during the dry season. Increasing water availability, especially during the dry season, is critical as corn is a crop that is sensitive to water scarcity).

Table 2. Aeration pores and water available pores after maize harvested

Treatments	Aeration pores	Water available pores
	---------- % Vol. --------	
Manure compost 2 tha^{-1}	18.03 a	10.25 b
CCBC 5 tha^{-1} + manure compost 2 tha^{-1}	17.28 ab	11.78 b
Co compost BCCC- manure 7 tha^{-1}	17.50 ab	13.53 a
Co compost BCCC- manure 5 tha^{-1}	16.67 b	13.72 a

CCBC: Corn cob biochar

After treatment with CCBC, soil pH ncreased significantly compared to manure application (no CCBC). When treated with manure compost, the soil pH was only 4.58, whereas when CCBC was applied CCBC it increased to 4.68 in three tratments with biochar and co-compost (Table 3). This is a statistically significant increase, indicating that soil response improves as soil acidity decreases. The results of this research are consistent with some studies conducted on other acidic upland soils [7,20]. Application of CCBC+manure and co-compost CCBC did not result in a significant increase in soil organic carbon content over a short period of time (during corn planting). The increase in soil organic carbon is more affected by the presence of fertilizers (Table 3). The use of co-compost at a dosage of 7 tha^{-1}or 5 tha^{-1}could reduce the Al^{3+} content, but only the corn cob co-compost at a dosage of 7 tha^{-1}could lower the H^{+}concentration. Overall, the application of corn cob biochar co-compost at a dosage of 7 tha^{-1}after one growing season, had a significant impact on reducing soil acidity.

Table 3. pH, Organic C, Al^{3+} and H^{+} after maize harvested

Treatments	pH H_2O	Organic C	Al^{3+}	H^{+}
		%	cmol+ kg^{-1}	
Manure compost 2 tha^{-1}	4,58 b	1,07 a	1,04 a	0,47 a
CCBC 5 tha^{-1}+ manure compost 2 tha^{-1}	4,68 a	1,10 a	0,98 ab	0,43 b
Co compost BCCC- manure 7 tha^{-1}	4,68 a	1,08 a	0,94 b	0,44 b
Co compost BCCC-manure 5 tha^{-1}	4,62 ab	1,07 a	0,92 b	0,46 a

CCBC: Corn cob biochar

3.2 Corn yield

Figure 1 shows no differences in dry biomass between the treatments. The application of co-compost CCBC+manure resulted in significantly higher yields compared to the application of manure and CCBC+compost. The increase in dry grain yield with co-compost CCBC at dosages of 5 tha^{-1} and 7 tha^{-1} was 8% and 11%, respectively, which was statistically significant compared to the manure and CCBC+compost treatments. The increase in corn yield is associated with the improvement of soil characteristics, such as the availability of water pores and the decrease in soil acidity (Table 2 and 3). Crop yields obtained during one growing season did not yet reflect the effectiveness of adding CCBC, as its effects were more noticeable in the second and third growing seasons [6,7]. The residual effects of biochar will be highly beneficial in the subsequent growing seasons.

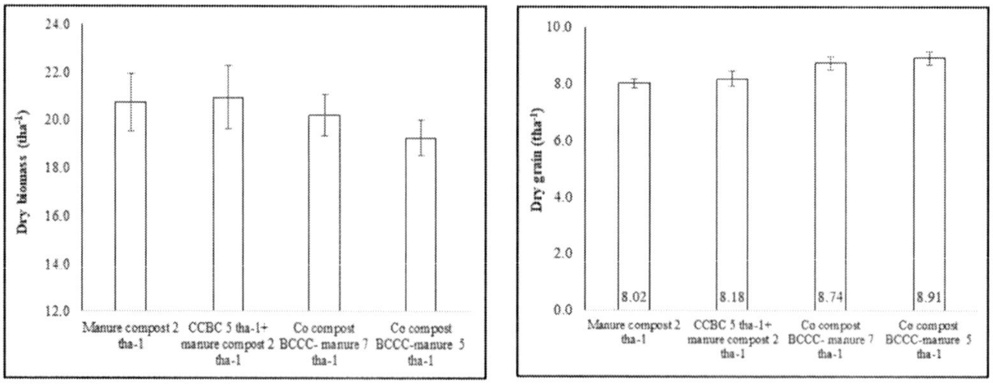

CCBC: Corn cob biochar

Figure 1. Biomas and dry grain after corn harvested

4 Conclusion

Application of CCBC, either as a single or in co-compost biochar-manure, increased the water available pore and soil pH to be 4.68, there was no significant differences in soil organic carbon content between all treatments. Improving soil porosity could ensure water more readily available to plants during the dry season. Addition of co-compost CCBC-manure oat a rate of 7 tha^{-1} or 5 th^{-1} had a significant impact on reducing soil acidity and increasing corn yield. Converting corn cobs to biochar not only improves soil properties of upland acidic soil but also increase soil resilience while reducing carbon loss from agricultural waste.

The authors acknowledge the research grant from the Indonesian Agency of Agriculture Research Development (IAARD), Ministry of Agriculture.

References

1. BPS Lampung Selatan, *Kabupaten Lampung Selatan Dalam Angka 2023* (2023)
2. Ł. Czajkowski, D. Wojcieszak, W. Olek, and J. Przybył, Constr. Build. Mater. **210**, 709 (2019)
3. Muhrizal, N. L. Nurida, and Fahmuddin Agus, (2013)
4. S. Menardo, G. Airoldi, V. Cacciatore, and P. Balsari, Biosyst. Eng. **129**, 352 (2015)
5. J. Y. Choi, J. Nam, B. Y. Yun, Y. U. Kim, and S. Kim, Ind. Crops Prod. **183**, 114931 (2022)
6. N. L. Nurida and . Jubaedah, J. Trop. Soils **24**, 33 (2019)
7. S. E. Hale, N. L. Nurida, Jubaedah, J. Mulder, E. Sørmo, L. Silvani, S. Abiven, S. Joseph, S. Taherymoosavi, and G. Cornelissen, Sci. Total Environ. **719**, (2020)
8. Z. Dai, X. Zhang, C. Tang, N. Muhammad, J. Wu, P. C. Brookes, and J. Xu, Sci. Total Environ. **581–582**, 601 (2017)
9. S. E. Hale, V. Alling, V. Martinsen, J. Mulder, G. D. Breedveld, and G. Cornelissen, Chemosphere **91**, 1612 (2013)
10. A. Obia, J. Mulder, S. E. Hale, N. L. Nurida, and G. Cornelissen, PLoS One **13**, (2018)
11. G. Cornelissen, Jubaedah, N. L. Nurida, S. E. Hale, V. Martinsen, L. Silvani, and J. Mulder, Sci. Total Environ. **634**, 561 (2018)
12. P. Liu, W. J. Liu, H. Jiang, J. J. Chen, W. W. Li, and H. Q. Yu, Bioresour. Technol. **121**, 235 (2012)
13. G. Agegnehu, M. I. Bird, P. N. Nelson, and A. M. Bass, Soil Res. **53**, 1 (2015)
14. I. S. Abujabhah, S. A. Bound, R. Doyle, and J. P. Bowman, Appl. Soil Ecol. **98**, 243 (2016)
15. L. Zhang and X. Sun, Bioresour. Technol. **171**, 274 (2014)
16. N. Hagemann, E. Subdiaga, S. Orsetti, J. M. de la Rosa, H. Knicker, H. P. Schmidt, A. Kappler, and S. Behrens, Sci. Total Environ. **613–614**, 20 (2018)
17. K. A. Frimpong, E. Abban-Baidoo, and B. Marschner, Heliyon **7**, e07089 (2021)
18. H. Schulz, G. Dunst, and B. Glaser, Agron. Sustain. Dev. **33**, 817 (2013)
19. M. K. Awasthi, Q. Wang, H. Chen, M. Wang, X. Ren, J. Zhao, J. Li, D. Guo, D. sheng Li, S. K. Awasthi, X. Sun, and Z. Zhang, Bioresour. Technol. **237**, 156 (2017)
20. N. R. Pandit, H. P. Schmidt, J. Mulder, S. E. Hale, O. Husson, and G. Cornelissen, Arch. Agron. Soil Sci. **66**, 250 (2020)

Effect of dolomite and micronutrients on the growth and yield of lowland rice on acid sulfate soil

Wiwik Hartatik[1*], *I Gusti Made* Subiksa[1], *Ladiyani* Retno Widowati[2] and *Dila* Aksani[1]

[1]National Research and Innovation Agency, Cibinong Science Centre, Jl. Raya Jakarta-Bogor, KM 46, Cibinong, Bogor, West Java 16991, Indonesia
[2]Indonesian Institute of Soil and Fertilizer Standardization, Cimanggu 12, Bogor, West Java, Indonesia

Abstract. Soil acidity, availability of P, base cations, micronutrients such as Cu and Zn, and climate change have become limiting factors in nutrient management on acid sulfate soils. Therefore, applying dolomite and micronutrient fertilizer is required to improve rice productivity and climate change adaptation. This study aims to determine the effect of dolomite, Cu, and Zn fertilizers on the growth and yield of rice in acid sulfate soils. The study was conducted in the greenhouse using a completely randomized design with seven treatments and four replications. The treatment consisted of control, 3 tons ha^{-1} dolomite application, and 4 levels of Cu and Zn fertilizer doses 0, 2, 4, and 6 g L^{-1}. NPK as basal fertilizer was applied according to the recommended dose for all treatments. Observations included soil pH, dissolved Fe, plant height, number of tillers, number and weight of panicle, dry shoots weight, 1000 grains weight, yield, and nutrient uptake (N, P, K, Fe, Cu, and Zn). The results showed that applying dolomite 3 tons ha^{-1} significantly increased soil pH, growth, and yield of rice, but applying Cu and Zn fertilizers without dolomite cannot support the growth of rice. Combination of dolomite 3 tons ha^{-1} with Cu and Zn fertilizers 4 ml L^{-1} increased nutrient uptake of N, P, K, Cu, and Zn in grains around 16% - 35% compared to without micronutrient fertilizers. Applying a combination of dolomite and micronutrient fertilizers can increase soil productivity on acid sulfate soils, nutrient uptake, and increase rice yield.

1 introduction

Indonesia has around 20.12 million ha of tidal swamps land that consists of 10.89 million ha of peatland, 4.23 million ha acid sulfate land, 2.07 million ha of potential land, and 0.44 million ha saline affected land [1]. The development of acid sulfate soils for agricultural land faces many constraints, including high soil acidity and low available P due to Al and Fe fixation. The low productivity of acid sulfate soil is not only caused by the high acidity that increases the solubility of toxic elements such as Fe, Al, and Mn but also because of low base saturation. Low pH increases the solubility of toxic elements and nutrient deficiency, leading

* Corresponding author: wiwik_hartatik@yahoo.com

to unproductive land [2,3]. These conditions cause rice plants to experience often iron toxicity [4]. Liming on acid sulfate soil is needed to improve the plant growing environment, and increase pH and P nutrient availability [5]. The novelty of this research is that the application of dolomite and micronutrients such as Cu and Zn can increase nutrient uptake and rice yields.

Climate change is one of the inhibiting factors for increased rice production. Climate changes such as drought cause increased pyrite oxidation, iron solubility, soil acidity, and decreased nutrient availability, therefore, requiring land management of acid sulfate soils through dolomite amelioration and balanced fertilization to support climate change adaptation. Dolomite amelioration will increase the pH and reduce dissolved Fe to reduce environmental pollution. Liming in acid sulfate soils is required to improve soil pH, toxic elements, and nutrient availability for better plant growth. [5] suggested that acid sulfate soils required dolomite at around 4-6 tons ha^{-1}. Liming at the recommended dose will not significantly affect the soil Fe concentration. Liming seems to have more influence on plant growth by providing nutrients and reducing the harmful effects of Fe toxicity.

The application of Cu and Zn micronutrient fertilizers on acid sulfate soils is expected to meet the needs of Cu and Zn for lowland rice. The micronutrient Zn is very important in several biochemical processes in plants such as synthesis of cytochromes, nucleotides, auxin metabolism, chlorophyll production, enzyme activation, and maintenance of membrane integrity. Zn plays a significant role in plant physiological and photosynthetic processes and increases plant growth and yield [6]. Zn deficiency causes necrotic spots on leaves, chlorosis, deformed and stunted leaves, and stunted plants. Previous studies reported that Zn deficiency can reduce up to 40% of yield without any leaf symptoms [7,8]. [7] reported that the application of 20-25 kg ha^{-1} $ZnSO_4$ can increase Zn levels in the acid soil. The application of Zn fertilizer by spraying onto the leaves or spreading directly into the soil is effective for overcoming Zn deficiency in soil and plants. Copper is a constituent of ascorbic acid, oxidase and phenolase enzymes, and plastocyanin. Ascorbic acid and these enzymes are regulatory factors in enzyme reactions (effectors, stabilizers, and inhibitors) and catalysts of oxidation reactions, which play an important role in the processes of N metabolism, proteins and hormones, photosynthesis and respiration, and pollen formation. The application of micronutrient fertilizers containing Cu_2O and ZnO active ingredients to meet the Cu and Zn nutrients requirement in plants is also expected to improve the quality of lowland rice yield. This study aimed to determine the effect of Dolomite and Cu and Zn fertilizer on growth, nutrient uptakes, and yield of lowland rice on acid sulfate soil in a greenhouse.

2 material and methods

This research was conducted in a greenhouse in Bogor, West Java, Indonesia. Soil bulk sample was taken out from Tanjung Lago, Banyuasin Regency, South Sumatra Province (-2°40'24" S, 104°44'42" E, 9.0 m asl). The actual acid sulfate soil samples were taken from paddy fields in tidal swamp land in Banyuasin Regency, South Sumatra. Bulk soil samples were taken within 0-40 cm depth. The bulk soil sample was prepared in a moist condition and mixed evenly before being weighed in about 10 kg into the plastic pot. The moisture content of the soil in the pot is maintained in flooded conditions. The initial soil sample was analyzed in the soil laboratory and the analysis result showed that acid sulfate soil was very acidic with a pH value lower than 4.5, high C-organic content (4.66%), low total N-content (0.13%), and very high C/N ratio (38). The low pH of acid sulfate soil affected the availability of nutrients in the soil, i.e. very low P-total (6 ppm), low P-available and K-total classified (12 and 16 mg $100g^{-1}$), low cation exchange capacity (13.44 cmolc kg^{-1}), very low exchangeable K (0.03 cmolc kg^{-1}), low exchangeable Ca (2.88 cmolc kg^{-1}), and moderate exchangeable Na (37 cmolc kg^{-1}) and Mg (1.28 cmolc kg^{-1}). Low base saturation (37%), high total-Fe (4.79%), high total-S (3.98%), and very high pyrite content (7.46%).

The experiment was arranged in a completely randomized design with 7 treatments and 4 replications, consisting of control, micronutrient fertilizer 2,0 g L^{-1}, micronutrient fertilizer 4,0 g L^{-1}, micronutrient fertilizer 6,0 g L^{-1}, dolomite 36 g pot^{-1}, dolomite 36 g pot^{-1} + micronutrient fertilizer 2,0 g L^{-1} and dolomite 36 g pot^{-1} + micronutrient fertilizer 4,0 g L^{-1}. NPK was applied according to the recommended dose as basal fertilizer for all treatments. Dolomite was applied by spreading it on the soil surface in the pot and then mixing it evenly into the flooded soil one week prior to rice transplanting. NPK with a dose of 350 kg ha^{-1} was applied as basal fertilizer 10 days after planting (DAP). Meanwhile, urea with a dose of 200 kg ha^{-1} was applied twice (21 DAP and 35 DAP). Fertilizers were applied by dressing around the plants with saturated soil moisture content. Micronutrient fertilizer containing Cu and Zn was applied thrice namely the first three days before planting applied by spraying to the soil surface, second (7 DAP), and third (21 DAP) applied by spraying it around the clump of rice plant according to the treatment. The plant indicator used Inpari 32 rice variety and their seeds were sown in nurseries. On the 14th day old seedlings were transplanted into the pots that have been prepared according to the treatment. Plant maintenance consists of weeding, watering every two days, and preventing pests, and diseases.

Soil pH was observed every four weeks in stagnant water. The observation of plant height and number of tillers were carried out at 4, 8, and 12 weeks after planting (WAP). Samples of biomass harvesting (grain, shoots, and roots) were carried out to analyze N, P, K, Fe, Cu, and Zn nutrients. Observed yield components were the number of panicles, the weight of the panicle, the weight of dry shoots, the weight of dry grain, and the weight of 1000 grains, and N, P, K, Fe, Cu, and Zn uptake by the plant were also calculated.

The collected data were subjected to the SPSS statistical analysis program to determine the effect of fertilizer application on the growth yield of rice. The Duncan Multiple Range Test (DMRT) was carried out at a 5% level of accuracy to determine the differences between treatments.

3 results and discussion

3.1. Effect of Dolomite and micronutrient fertilizers on soil pH and Fe hydrous oxide

The pH at 4, 8, and 12 WAP showed no significant difference between the application of micronutrients without dolomite (K0V2, K0V4, and K0V6) and the control treatment (K0V0). The treatment with dolomite, namely the K3V0, K3V2, and K3V4, was significantly different from K0V0 and the micronutrient fertilizer treatment without dolomite (K0V2, K0V4, K0V6). The treatment K3V0 at 8 WAP was significantly different from the K0V0, K0V2, K0V4, K0V6, and K3V2, but not significantly different from the K3V4. The use of dolomite as an ameliorant can increase the pH of acid sulfate soils (Fig. 1). According to [9], liming can improve the physical, biological, and chemical properties of acid soil, one of which is by increasing soil pH so that plants can grow better. Soil pH at 8 WAP increased in almost all treatments, except for the K3V0 treatment, which decreased. These findings are in line with the previous study[4] where the application of 3.69 tons ha^{-1} dolomite increased the pH of stagnant water up to 6.2 - 6.6 compared to control 5.3. The application of microfertilizers has no effect on soil pH. This was shown in the K3V2 and K3V4 treatments which were not significantly different from the K3V0.

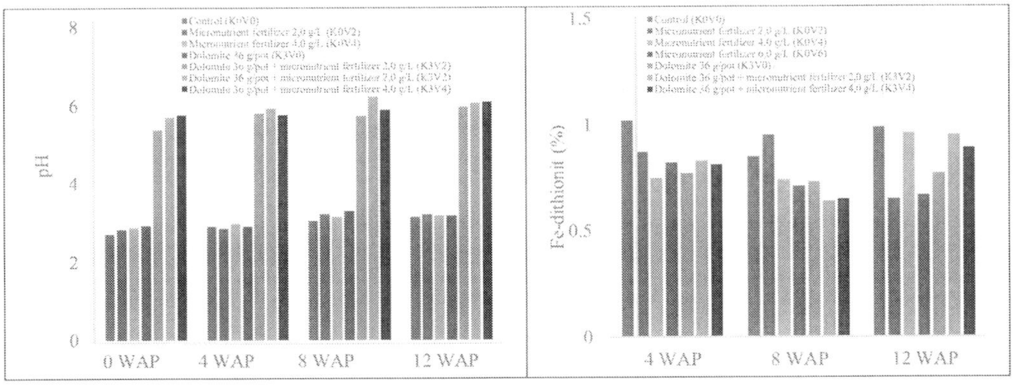

Figure 1. pH at 0, 4, 8, and 12 WAP and Fe dithionite treated with dolomite and micronutrient fertilizers.

Dithionite extract can dissolve hydrous iron – oxide. The K0V0 and K0V2 treatments gave higher hydrous iron oxide than the other treatments, this was because without liming it caused a low pH so that Fe hydrous - oxide dissolved more. According to[5] Fe levels will decrease due to calcification, and soil pH will increase. Increasing the dose of micronutrient fertilizers, both with dolomite and without dolomite, did not make a difference to Fe hydrous oxide. The application of micronutrient fertilizers was not able to reduce Fe hydrous-oxide, except for the K3V2 treatment at 12 WAP observations (Fig. 1).

3.2. Rice growth

Treatment of dolomite application and without dolomite on plant height at 4, 8, and 12 WAP showed significant differences. The treatments of K0V0, K0V2, and K0V6 at 4 WAP were not significantly different because rice plants could not grow due to very acidic soil pH. Liming seems to have more influence on plant growth by providing nutrients and reducing the harmful effects of Fe toxicity. Rice plants in the K0V4 treatment only lasted up to 4 WAP. According to [10], the application of dolomite can increase soil pH which provides soil conditions that support the growth of rice plants. Micronutrient fertilizer doses did not significantly affect plant height in the treatment with dolomite at 3 levels of micronutrient doses (K3V0, K3V2, and K3V4). The highest plants at 4, 8, and 12 WAP were achieved by K3V4, K3V0, and K3V2 treatments respectively. Plant height is influenced by nutrient availability and nutrient absorption by roots.

In line with the plant height, the treatment with dolomite and without dolomite showed significant differences in the number of tillers 4, 8, and 12 WAP. The application of micronutrient fertilizers was not significantly different in the number of tillers, both in the treatment with dolomite and without dolomite. In contrast, the treatments K3V2 and K3V4 were not significantly different. Plants with the highest number of tillers at 4, 8, and 12 WAP were achieved by treatments K3V0, K3V4, and K3V6 respectively. The number of productive tillers is affected by the nitrogen fertilizer content. According to [11], the most influential nutrients in the number of tillers are N and P (Fig. 2).

https://doi.org/10.1051/e3sconf/202346701004

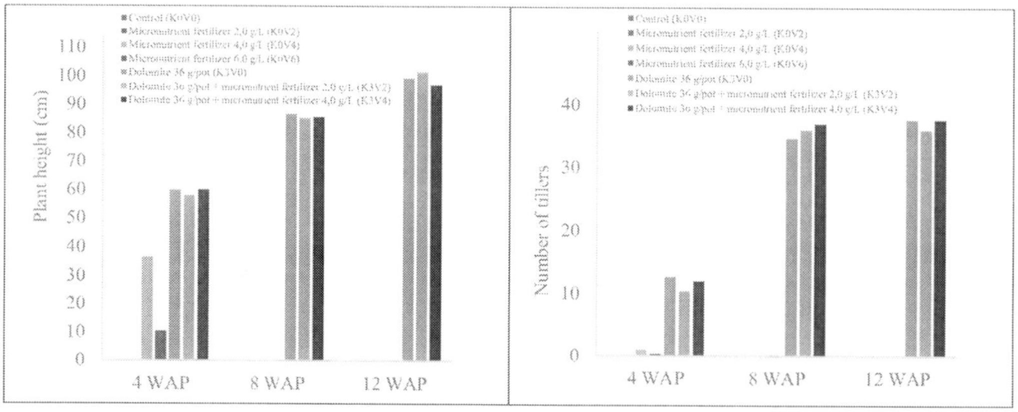

Figure 2. Plant height and number of tillers at 4, 8, and 12 WAP treated with dolomite and micronutrient fertilizers.

3.2. Rice yield

The dolomite treatment was significantly different from without dolomite treatment on the parameters of rice yield components. Control treatment showed that the plants could not grow properly and were unable to produce panicles and grain [12,13]. The application of micronutrient fertilizers affected shoot weight at a dose of 4 g L^{-1}. This was shown in the dry shoot weight of the K3V4 treatment which was significantly different from the treatment without micronutrient fertilizer (K3V0). The highest dry shoot weight was the K3V4 treatment with a dry weight value of 80.67 g pot^{-1}.

The highest number of panicles was 35 panicles pot^{-1} in the K3V4 treatment which according to IRRI criteria is very high. [14] divided the criteria for the number of panicles (productive tillers) into 4 levels, namely very low: < 5.0, low: 5 - 9, moderate: 10 - 19, high: 20 -25, and very high: > 25. The more panicles per hill formed, the greater the potential grain yield to be achieved. The application of micronutrient fertilizers was not significantly different from those without micronutrient fertilizers. The K3V0 treatment had the highest panicle weight of 88.17 g pot^{-1}. These dry grain weights were not significantly different from the treatment using micronutrients K3V2 and K3V4. The largest dry grain weight was achieved by the K3V0 treatment with a weight of 68.33 g pot^{-1}, this result was significantly different from the treatment without dolomite. Micronutrient fertilizers did not affect the weight of dry grain which was shown between the K3V0, K3V2, and K3V4 treatments which were not significantly different. A sufficient supply of N and P should be provided to rice plants for optimum yield [15] (Table 1).

Table 1. Number of panicles, panicle weight, dry shoots, and grain weight, and 1000 grain weight treated with dolomite and micronutrient fertilizers.

Treatment	Number of panicles	Weight of panicle	Dry shoots weight	Dry-grain weight	1000 grain weight
		.... g pot^{-1}			
K0V0	0,00a	0,00a	0,00a	0,00a	0,00a
K0V2	0,00a	0,00a	0,00a	0,00a	0,00a
K0V4	0,00a	0,00a	0,00a	0,00a	0,00a
K0V6	0,00a	0,00a	0,00a	0,00a	0,00a
K3V0	34,00b	88,17b	64,47b	68,33b	24,60b
K3V2	33,00b	79,30b	76,63bc	59,77b	24,80b
K3V4	35,00b	84,20b	80,67c	63,73b	24,87b

*) The numbers in the column followed by the same letter are not significantly different at the 5% level of the DMRT test.

3.3. Nutrient uptake

N-total uptake of rice plants in shoots > grain > root (Fig.3). The higher N in leaves/shoots compared to grain and roots because the process of photosynthesis occurs in the leaves and formation of plant tissue cells to support plant vegetative growth, whereas as the plant ages, N uptake decreases and is shifted to generative growth forming panicles and grain. Meanwhile [15] reported that nitrogen uptake strongly influences the number of productive tillers per square meter. N-shoots uptake ranged from 529 – 853 mg pot^{-1}, N-roots 290–430 mg pot^{-1} and N-grain ranged from 483 – 585 mg pot^{-1}. The application of micronutrient fertilizers in the dolomite treatment (K3V0, K3V2, and K3V4) on shoots, roots, and grain tended to increase N uptake (20-40%).

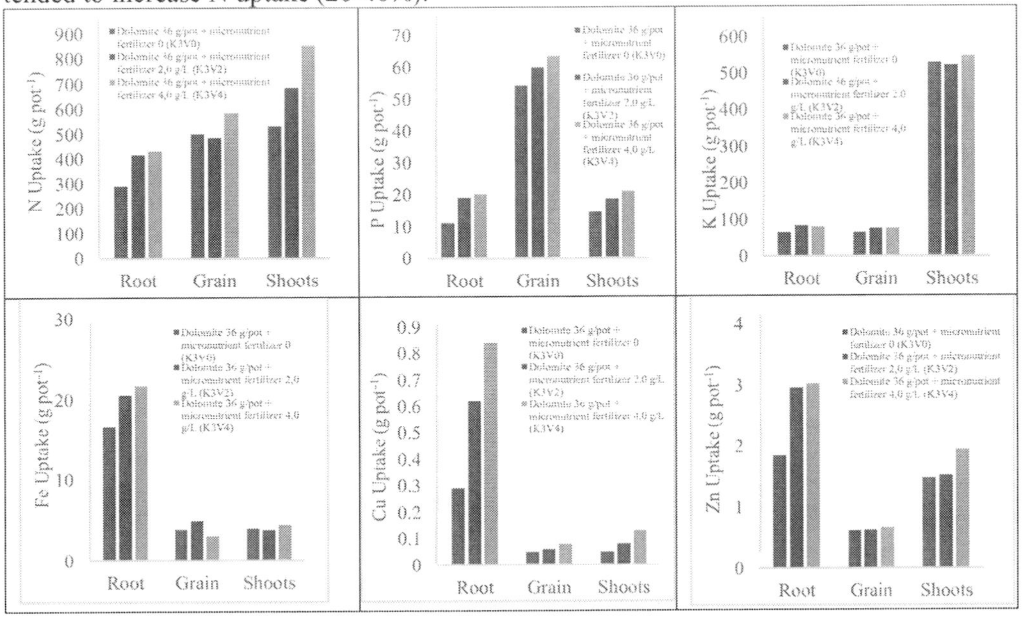

Figure 3. N, P, K Fe, Cu, and Zn uptake in shoots, roots, and grains of rice treated with dolomite and micronutrient fertilizers.

The total P uptake of rice plants at harvest showed a higher P-grain absorption than P-shoots > P-roots. The higher P in harvested grain (54 – 63 mg pot^{-1}) compared to P-shoots (14 – 21 mg pot^{-1}) and P-root (11 – 20 mg pot^{-1}) is due to the concentrated P nutrient which is a constituent of protein in grain yields. P is a component of several enzymes, proteins, and fruits. The application of dolomite and micronutrient fertilizer doses of 2 and 4 ml l^{-1} (K3V2 and K3V4) on rice roots, shoots, and grain tended to increase P uptake by around 16 – 42% (Fig. 3).

The total K uptake of rice plants in the shoots section was significantly higher than the K-grain and K-roots at harvest. The uptake of K-shoots at harvest ranged (522 – 547 mg pot^{-1}), K-grain (63 – 76 mg pot^{-1}), and K-root (64 – 84 mg pot^{-1}). Potassium acts as a regulator of plant physiological processes such as photosynthesis, accumulation, translocation, and carbohydrate transport, and K will be accumulated in shoots [11]. The high K in shoots is because about 80% of K nutrients are accumulated in the stem as a substance that strengthens the vigor of rice plants so that they are resistant to pests and disease attacks (Fig.3).

The high uptake of Fe in plant tissues is due to the high level of iron in the soil as a result of the oxidation of pyrite in acid sulfate soils. Iron in excess ferrous Fe^{3+} will be reduced to ferrous iron Fe^{2+} on the root surface of rice plants so that it settles on the root surface and inhibits the uptake of other nutrients from roots to leaves. As a result, plant growth will be disrupted and even rice plants will fail to grow. Fe uptake in the roots of rice plants growing on acid sulfate soils was very high at harvest. Fe-root absorption ranged from 16-21 mg pot^{-1}, Fe-grain 3-4 mg pot^{-1}, and Fe-shoots 3-4 mg pot^{-1}. Low soil pH without calcification causes higher dissolved Fe hydroxide and it will be absorbed by plants. The high concentration of Fe^{2+} in the soil solution was responded to by plant roots by releasing O_2 resulting in the oxidation of Fe^{2+} to Fe^{3+} forming FeOOH and $Fe(OH)_3$ which can precipitate on the roots' surface [16]. (Fig 3).

Cu and Zn uptake of rice plants was observed when the plants were harvested at 12 WAP presented in Figure 3. Cu and Zn are micronutrients needed by rice plants in small quantities but are beneficial for the quality of the grain produced. Uptake of Cu-root > Cu-shoots > Cu-grain successively ranged from Cu-root (0.29 – 0.84 mg pot^{-1}), Cu-shoots (0.05 – 0.13 mg pot^{-1}), and Cu-grain (0.05 – 0.08 mg pot^{-1}). High Zn uptake in grain is very beneficial for the health of stunting sufferers in infants. Zn-root uptake > Zn-shoots > Zn-grain ranged from Zn-root (1.84 – 3.02 mg pot^{-1}), Zn-shoots (1.48 - 1.94 mg pot^{-1}), and Zn-grain (0.62 – 0.66 mg pot^{-1}) respectively. The rhizosphere produces root exudates in the form of organic acids which form complex organic metal compounds. The application of dolomite and micronutrient fertilizer increased Cu and Zn uptake in grain by 16 and 31%, respectively. Zn fertilization is integrated with dolomite on tropical acid sulfate soils in Vietnam, significantly increasing the Zn content in grain up to 30% in unhulled rice and 27% in white rice compared to the control [17].

4 conclusion

The application of dolomite 3 tons ha^{-1} significantly increased the pH (5.96 - 6.09), growth, and yield of rice, but applying Cu and Zn fertilizers without dolomite cannot support the growth of rice. A combination of dolomite 3 tons ha^{-1} with Cu and Zn fertilizers 4 ml L^{-1} increased nutrient uptake of N, P, K, Cu, and Zn in grain around 16% - 35% compared to without micronutrient fertilizers. P uptake in grain>shoots>root, N, and K uptake in shoots>grain>root; Fe uptake in root>grain>shoots, Cu, and Zn uptake in root>shoots>grain. Applying a combination of dolomite and micronutrient fertilizers can increase soil and rice productivity on acid sulfate soils. Dolomite amelioration and balanced fertilization are land management that support climate change adaptation.

Deep appreciation to the Indonesian Soil Research Institute and PT Tritama Wirakarsa for providing financial assistance and support to micronutrient fertilizer. Wiwik Hartatik, I Gusti Made Subiksa, Ladiyani Retno Widowati, and Dila Aksani are the main contributors.

References

1. A. Mulyani, B. Mulyanto, B. Barus, D. R. Panuju, and Husnain, Land (Basel) **12**, (2023)
2. D. S. Fanning, M. C. Rabenhorst, and R. W. Fitzpatrick, Geoderma **308**, 191 (2017)
3. J. Shamshuddin, Q. Ali Panhwar, R. Shazana, and E. Azura Azman, *Improving the Productivity of Acid Sulfate Soils for Rice Cultivation Using Limestone, Basalt, Organic Fertilizer and/or Their Combinations* (2016)
4. J. Shamshuddin, S. Muhrizal, I. Fauziah, and E. Van Ranst, Commun Soil Sci Plant Anal **35**, 117 (2004)

5. A. Susilawati, A. Fahmi, P. Badan, L. Pertanian, D. Balai, P. Pertanian, L. Rawa, and J. K. Karet, *Dinamika Besi Pada Tanah Sulfat Masam Yang Ditanami Padi Dynamics of Iron in Acid Sulfate Soils Cultivated with Rice* (2013)
6. H. Barman, S. K. Das, and A. Roy, Univers J Agric Res **6**, 149 (2018)
7. L. Ozturk, M. A. Yazici, C. Yucel, A. Torun, C. Cekic, A. Bagci, H. Ozkan, H. J. Braun, Z. Sayers, and I. Cakmak, Physiol Plant **128**, 144 (2006)
8. A. Roy, S. K. Das, A. K. Tripathi, N. U. Singh, and H. K. Barman, Progressive Agriculture **15**, 182 (2015)
9. S. Kumar, R. Meena, G. Yadav, and A. Pandey, Int J Plant Soil Sci **14**, 1 (2017)
10. M. Suriani and M. Mahbub, *Impact of Rice Straw Compost on Ferro (Fe 2+) Solubility, Soil PH and The Growth of Ciherang Rice Plant in Acid Sulfate Soil Pengaruh Kompos Jerami Padi Terhadap Kelarutan Ferro (Fe 2+) Dan PH Tanah Serta Pertumbuhan Tanaman Padi Ciherang Di Tanah Sulfat Masam* (2020)
11. Achim. Dobermann, Thomas. Fairhurst, and International Rice Research Institute., *Rice : Nutrient Disorders & Nutrient Management* (Potash & Phosphate Institute, East & Southeast Asia Programs, 2000)
12. A. Audebert and K. L. Sahrawat, J Plant Nutr **23**, 1877 (2000)
13. Y. Sulaeman, E. Maftu'ah, M. Mukhlis, K. Anwar, V. Karolinoerita, N. Wakhid, M. Saleh, I. Khairullah, M. Agustiani, A. W. Anggara, P. Sasmita, I. A. Rumanti, Y. Sastro, M. Ahmad, and D. Wurjanto, Resources **11**, (2022)
14. (n.d.)
15. S. Kheroar, B. Patra, D. Halder, S. Barman, and K. Mandal, Current Journal of Applied Science and Technology **26**, 1 (2018)
16. T. Wang and J. H. Peverly, *DIVISION S-10-WETLAND SOILS Iron Oxidation States on Root Surfaces of a Wetland Plant (Phragmites Australis}* (n.d.)
17. S. M. Soltani, M. M. Hanafi, A. W. Samsuri, S. K. S. Muhammed, and M. A. Hakim, Chemical Speciation and Bioavailability **28**, 152 (2016)

Selection of M3 mutant strains from 200 gy gamma irradiation on Cempo Ireng

Nandariyah[1*], Sukaya[1], Sutarno[1], Riyatun[1] and *Yasmin Mutia* Widianingrum[2]

[1]Department of Agrotechnology, Faculty of Agriculture, Universitas Sebelas Maret, Surakarta, Indonesia
[2]Undergraduate Department of Agrotechnology, Faculty of Agriculture, Universitas Sebelas Maret, Surakarta, Indonesia

Abstract. Climate change is a serious issue that can disrupt crop production and pest resistance, especially in rice crops. To tackle climate change, plant breeding can be employed to develop new characteristics that are resistant, using physical mutagenesis like gamma irradiation. Cempo Ireng is one type of rice in Indonesia with plant properties such as tall plant, long harvesting periods, and susceptibility to brown planthopper attacks. The research aims to obtain information and select M3 mutant strains with properties of shorter plants, early harvesting, and brown planthopper resistance. The study was conducted from February-July 2022 in Klaten Regency, Central Java, Indonesia. The research material consisted of 3 mutant strains of black rice M3(GH8'), M3(GH51'), and M3(GH52'), with the control variety of Cempo Ireng. The experimental design used was a Randomized Complete Block Design without replication. Data were analyzed descriptively and used t-test, at a significance level of 5%. The results showed that M3 mutant black rice had better growth and yield components than the control plant. Brown planthopper attacks of M3 mutant strains decreased to 24% compared to the control.

1 Introduction

Climate change is one of the challenges in agriculture that has a serious impact on food production. The threat of climate change is the increasing overall temperature which causes changes in the natural environment, affecting plants and animals natural patterns. The lack of adapting plants to climate change and changes in pest behavior is a catastrophic scenario for food production. The inability of plants to adapt to climate change leads to a decrease in yield due to a mismatch in plant requirements and conditions of plant growth. Also, changes in pest distribution can promote extensive pesticide application, threatening food safety[1–4].

Rice is the most widely consumed staple food commodity in Indonesia. There are several types of rice, including black rice. Black rice is usually chosen as an alternative to white rice due to superior nutritional content as anthocyanins [5]. However, Cempo Ireng black rice has not been extensively cultivated because it has some drawbacks that are considered to result

* Corresponding author: nandariyah@staff.uns.ac.id

in low yield potentials such as long harvest period, tall plant height, relatively low yield potential, and low resistance to pest attacks [6].

Plant breeding help develop plant varieties for sustainable food production by generating new plant characteristics. One of the breeding techniques is gamma-ray irradiation. Gamma-ray induction is a physical mutagen that causes changes in DNA structure resulting in the emergence of improved characteristics of plants [7]. Gamma-ray mutations change genotype structure, enhance species variation, and increase plant adaptability which is advantageous for plant breeding [8]. The breeding program for Cempo Ireng black rice by gamma-ray irradiation helps emerges new traits in plants to cope with climate change, especially pest attacks. This study aims to obtain phenotype and pest resistance in shorter plant height and early flowering of M3 black rice for individual selection.

2 Methodology

This study was conducted in Jogonalan, Klaten District, Central Java, Indonesia from February to July 2022. The materials are three mutant numbers of M3 rice strains with an irradiation dose of 200 Gy. Plant code of M3(GH8') are M3(GH8')-7, M3(GH8')-8, M3(GH8')-14, M3(GH8')-25. Plant code of M3(GH51') are M3(GH51')-1, M3(GH51')-4, M3(GH51')-6, M3(GH51')-7, M3(GH51')-8. Plant code of M3(GH52') are M3(GH52')-9, M3(GH52')-11, M3(GH52')-14, M3(GH52')-16, M3(GH52')-20. Control plant variety is Cempo Ireng. The M3 plant population for each strain was 372 plants. The experimental design used was a Randomized Complete Block Design without replication. Data were analyzed descriptively and used a t-test, at significance level of 5%. The variables of observation consisted of flowering period, harvesting period (day), plant height (cm), panicle length (cm) weight of 100 seeds per clump (gram), seed weight per plant, and number of productive tillers (tiller).

3 Results and Discussion

3.1 Phenotype of black rice

The study showed a difference in the flowering and harvesting period between M3 black rice in comparison to the control plants. Some of the individual plants of M3 have early flowering up to 7-9 days and harvesting up to 6-7 days (Table 1). The harvesting period of M3 rice showed that flowering characteristics of the black rice met the requirement of early-harvesting rice. Early harvesting rice has a flowering period ranging from 105-124 days [9]. Flowering is one of the characteristics that is affected by gamma-ray exposure. Gamma-ray exposure can cause changes in enzymes, proteins, and DNA related to flowering periods [10].

The results show that Gamma-ray irradiation causes different effects on the plant height. There is a difference in plant height in each of M3 as a result of gamma irradiation. M3(GH8') plant has the shortest plant with a minimum height of 100 cm with an average of 129,9 cm. Meanwhile, M3(GH51') plant has the lowest plant height of 117 cm with an average of 136,07 cm. Compared to the control plant with the lowest plant height of 115 cm with an average of 133 cm. The reduction in plant height M3(GH8') is more visible compared to M3(GH51') (Table 1). Gamma-ray irradiation can induce genetic mutation in plants, resulting in genetic variations. This genetic variation can be reflected in plants' physiology, one of which is plant height [11].

Table 1. Flowering time, harvesting time, plant height, and number of productive tillers of black rice.

Genotype	Flowering time (day) Min-Max	Harvesting time (day) Mean	Plant height (cm)		Number of productive tillers (tiller)	
			Min-Max	Mean	Min-M	Mean
Cempo Ireng	71-74	119	115-150	133.63 ± 8.07	9-29	16.50 ± 5.34
M3(GH8')	63-71	112	100-148	129.90 ± 9.48	7-37	17.50 ± 4.05
M3(GH51')	65-74	113	117-146	136.07± 6.89	7-27	16.93 ± 4.37
M3(GH52')	65-71	113	113-149	133.17 ± 8.48	7-26	15.77 ± 4.59

Note : Numbers followed by a sign (*) are significantly different from the control as the results of T-test (α=0.05).

The number of productive tillers of M3 black rice, namely M3(GH8') has the highest maximum value of productive tillers compared to the control plant which is 37 tillers. The average black rice productive tillers showed no significant difference. However, the highest average count of productive tillers is M3(GH8') with 17,5 tillers (Table 1). The results of productive tillers illustrate the number of rice tillers that produce panicles. The panicles that cannot form panicles during the vegetative stage are called unproductive tillers [12].

Table 2. Panicle length and number of seeds per panicle.

Genotype	Panicle length (cm)		Number of seed per panicle (seed)	
	Min-Max	Mean	Min-Max	Mean
Cempo Ireng	23-29	26.61 ± 1.33	87-236	155.60 ± 33.79
M3(GH8')	23-28	25.50 ± 1.57*	72-193	131 ± 30.05
M3(GH51')	22-29	25.60 1 ±.55	67-193	134.23 ± 31.93
M3(GH52')	23-28	25.76 ± 1.26	71-191	129.30 ± 30.34

Note : Numbers followed by a sign (*) are significantly different from the control as the results of T-test (α=0.05).

The results on average panicle length showed a significant difference between M3 black rice compared to the control plant. Decreasing in panicle length of M3 black rice is due to the gamma-ray irradiation effect. The initial stages of gamma-ray irradiation affected more on quantitative plant characteristics, which lead to morphological abnormalities like plant height and panicle length [13]. Gamma-ray irradiation causes plants to repair themselves due to chromosome damage leading to inhibition of plant growth [14]. Decreasing of panicle length resulting in fewer grains formed on the plants. Irradiation causes a reduction in the number of filled grains due to alteration in cell structure related to carbohydrate translocation, indicating a decrease in carbohydrate translocation from roots to panicles [15].

The weight of 100 grains per panicle showed a significant difference between M3 black rice and the control plant. However, this result indicates a decrease in M3 grain shape compared to control plants. The weight of 100 grains represents grain shape, lower weight of 100 grains suggests a reduction in rice length. Slender grains have a heavier weight compared to round-shaped grains [16]. The grain weight per plant depends on the number of rice grains formed on productive tiller, heavier grains means more rice grains formed. The average weight results show there is an increase and decrease in M3 black rice yield compared to control plants.

Table 3. Weight of 100 seeds and weight seed per plant.

Genotype	Weight of 100 seeds (gram)		Weight seed per plant (gram)	
	Min-Max	Mean	Min-Max	Mean
Cempo Ireng	2.08-2.63	2.38 ± 0.12	14.32-68.42	39.33 ± 15.95
M3(GH8')	2.04-2.6	2.30 ± 0.11	13.02-69.76	34.47 ± 14.23
M3(GH51')	2.10-2.43	2.27 ± 0.09*	5.75-68.88	35.53 ± 13.46
M3(GH52')	1.96-2.59	2.28 ± 0.14	11.52-79.63	39.82 ± 14.74

Note : Numbers followed by a sign (*) are significantly different from the control as the results of T-test (α=0.05).

3.2 Brown planthopper infestation

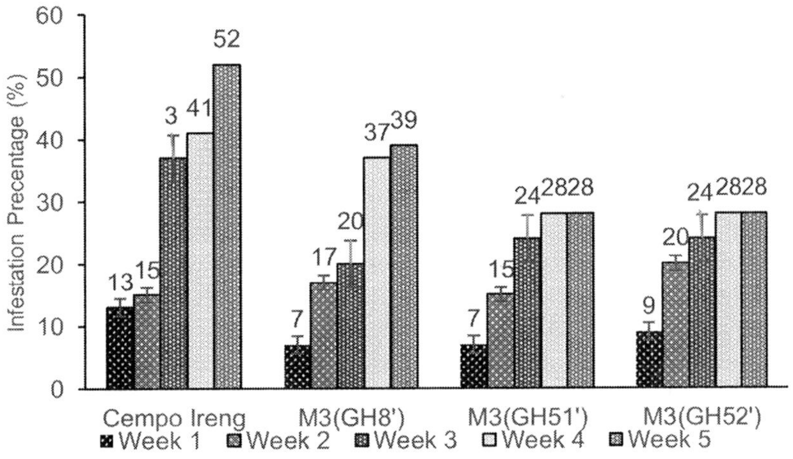

Fig. 1. Percentage of brown planthopper infestation on black rice.

The study showed that the percentage of plants infested by brown planthoppers increased each week. The final week of observation shows that the control plant has a 52% damage percentage, 39% on M3(GH8'), and 28% on M3(GH51') and M3(GH52'). Lower in pest infestation percentage of M3 black due to genetic changes by gamma-ray irradiation. Gamma-ray irradiation can produce generations of plants more resistant to pest attacks compared to parental plants due to genetic improvement from gamma-ray irradiation treatment [17]. Gamma-ray rice mutant improves pest resistance because gamma irradiation can help plants produce antimicrobial compounds, in rice plants by increase of phenolic acids [18].

Brown planthoppers attack rice by piercing and sucking sap from the rice stem causing direct damage and facilitating rapid transmission of plant diseases [19]. Prevention in brown planthopper feeding by increasing the hardness and thickness of the cell walls. Plant epidermis is considered to have an important role against insect attack, and determined brown planthopper feeding responses [20]. Thick cell wall structure makes it more challenging for brown planthoppers to penetrate, effectively hindering their ability to feed from rice [21]. Gamma-ray irradiation alters compounds in rice related to cell wall thickness, such as lignin [22]. Lignin is a phenolic polymer found in the cell walls of vascular plants and mainly thickened secondary walls [23]. High lignin content in rice plants enhances the thickness of the cell wall and increases the size of vascular tissue to make plant stems stronger and more resistant to stem-boring pests [24].

3.3 Individual selection

The results showed eight individual M3 black rice with shorter plant height and early flowering that were selected for M4 generation. Selected individuals of M3 black rice met the criteria of early harvest, short plant height, high productive tillers, and high weight per panicle (Table 4).

Table 4. Selected plant.

No	Accession number	Plant height (cm)	Harvesting period (days)	Number of productive tillers (tiller)	Panicle length (cm)	Weight of seed per plant (gram)
1	M3(GH8')-16	115	112	26	27.5	56.13
2	M3(GH8')-17	100	112	27	25	54.59
3	M3(GH8')-19	125	112	25	28	62.5
4	M3(GH51')-7	120	113	22	25	46.19
5	M3(GH52')-1	124	113	24	27	53.97
6	M3(GH52')-9	130	113	17	26.5	51.33
7	M3(GH52')-13	119	113	22	26.5	50.08
8	M3(GH52')-15	130	113	13	25.5	39.08

4 Conclusion

This study showed changes in the growth components, yield components, and yield of M3 compared to a control plant. Individual selection resulted in 8 individual plants that met the criteria of early harvest, short plant height, high productive tillers, and high weight per plant, namely M3(GH8')-16, M3(GH8')-17, M3(GH8')-19, M3(GH51')-7, M3(GH52')-1, M3(GH52')-9, M3(GH52')-13, M3(GH52')-15.

The authors would like to thanks UNS Black Rice Research Team for providing financial support and assistance through the research and article preparation.

References

1. S. M. Jain, Rom. Biotechnol. Lett. **15**, 88 (2010)
2. A. Shahzad, S. Ullah, A. A. Dar, M. F. Sardar, T. Mehmood, M. A. Tufail, A. Shakoor, and M. Haris, Environ. Sci. Pollut. Res. **28**, 14211 (2021)
3. C. Sen Ma, W. Zhang, Y. Peng, F. Zhao, X. Q. Chang, K. Xing, L. Zhu, G. Ma, H. P. Yang, and V. H. W. Rudolf, Nat. Commun. **12**, 1 (2021)
4. F. Sarsu, Aust. J. Crop Sci. (2021)
5. V. C. Ito and L. G. Lacerda, Food Chem. **301**, 125304 (2019)
6. Y. Dwiningsih and J. Alkahtani, J. Biomed. Res. Environ. Sci. **4**, 426 (2023)
7. S. Vasudevan, A. Dhanarajan, A. Raina, Y. Kasim, G. Balasubramani, S. Gurunathan, and B. Venkatachalam, Plant Sci. Today **10**, 127 (2023)
8. W. Xu, D. Wang, P. Matina, B. Lin, J. Liu, J. Huang, and X. Li, J. Radiat. Res. Appl. Sci. **16**, 100622 (2023)
9. Y. A. Kencana, E. D. Mustikarini, and T. Lestari, J. AGRO **8**, 48 (2022)
10. R. S. Hartanti, T. A. N. Putri, F. Zulfa, Sutarno, and Suranto, IOP Conf. Ser. Mater. Sci. Eng. **193**, (2017)
11. I. Ishak, Biodiversitas **24**, 3300 (2023)

12. T. H. Ayat, J. Pangan **31**, 209 (2022)
13. H. Harmini, S. Sajimin, A. Fanindi, and A. Husni, J. Ilmu Nutr. Dan Teknol. Pakan **18**, 62 (2020)
14. D. Mishra, L. Bhoi, S. K. Tripathy, M. P. Behera, and T. K. Mishra, Int. J. Curr. Microbiol. Appl. Sci. **7**, 3393 (2018)
15. T. Journal and F. Education, **2**, 1 (2014)
16. Sarwanto, S. Samudin, and A. Ete, J. Agrotekbis **6**, 274 (2018)
17. T. G. P. Sari, I. Suliansyah, and N. Akhir, J. Agroteknologi Univ. Andalas **2**, 10 (2018)
18. T. D. Gajbar, M. Kamble, S. Adhikari, N. Konappa, P. Satapute, and S. Jogaiah, Anal. Biochem. **618**, 114121 (2021)
19. A. Raina, S. Khan, P. K. Sahu, and R. Sao, Rice Res. Qual. Improv. Genomics Genet. Eng. Vol. 1 Breed. Tech. Abiotic Stress Toler. **1**, 753 (2020)
20. L. Yan, T. Luo, D. Huang, M. Wei, Z. Ma, C. Liu, Y. Qin, X. Zhou, Y. Lu, R. Li, G. Qin, and Y. Zhang, Int. J. Mol. Sci. **24**, (2023)
21. W. Zhang, L. Yang, M. Li, B. Ma, C. Yan, and J. Chen, Int. J. Mol. Sci. **16**, 28746 (2015)
22. L. Listihani, P. E. P. Ariati, I. G. A. D. Yuniti, and D. G. W. Selangga, Biodiversitas **23**, 4696 (2022)
23. B. Usman, G. Nawaz, N. Zhao, S. Liao, B. Qin, F. Liu, Y. Liu, and R. Li, Int. J. Mol. Sci. **22**, 1 (2021)
24. O. M. Terrett and P. Dupree, Curr. Opin. Biotechnol. **56**, 97 (2019)

The effect of antagonistic microbial and seed bulb-size on fusarium wilt and yield of shallot

Gracela Nanda Puspita Sari[1*], *Arlyna* B. Pustika[2], *Chimayatus* Solichah[1], *Danar* Wicaksono[1], *Setyorini* Widyayanti[2], Sudarmaji[2], and *Kiki* Yolanda[3]

[1]University of Pembangunan Nasional "Veteran" Yogyakarta, Indonesia
[2]Research Center for Food Crops, Research Organization for Agriculture and Food, National Research and Innovation Agency, Cibinong Science Center, Jl. Raya Jakarta-Bogor, KM. 46, Cibinong, Bogor, West Java, 16911, Indonesia
[3]Institute for Agricultural Instrument Standard Implementation Yogyakarta, Jl. Stadion Maguwoharjo No. 22. 55584, Wedomartani, Sleman, Yogyakarta, Indonesia

Abstract. The effort to suppress the development of fusarium wilt disease is by a consortium of antagonistic microbes. This study aimed to determine the effect of microbial consortium and bulb size on the development of fusarium wilt disease and the effect on plant resistance of small bulbs to fusarium wilt. This research was conducted on coastal land, Sri Gading Village, Sanden District, Bantul Regency. The experiment used a split-plot with 2 factors : seed treatment and bulb size with 8 treatments. The factors include K1U1: Consortia microbial and small bulbs, K1U2: Consortia microbial and medium bulbs, K1U3: Consortia microbial and big bulbs, K1U4: Consortia microbial and mixed bulbs, K2U1: Without consortia microbial and small bulbs, K2U2: Without consortia microbial and medium bulbs, K2U3: Without consortia microbial and big bulbs, K2U4: Without consortia microbial and mixed bulbs. Data were analyzed statistically SAS (Statistical Analysis System) ANOVA 5% and further tested with DMRT (Duncan's Multiple Distance Test) at the 5% level of significance. The percentage data which is not normally distributed are transformed to transformation arcsin$\sqrt{x}$. The results showed there was no interaction between the use of microbial consortium and bulb size on the development of fusarium wilt disease and yield of shallot. The giving of consortium antagonists could not increase plant resistance to fusarium wilt and yield. The decline in production is generally related to climate change which has an impact on increasing plant diseases. There is a need for cultivation adaptation in dealing with the effects of climate change.

1 Introduction

Shallot (*Allium ascalanicum* L.) is a horticulture commodity with high economic value and shallot cultivation is carried out in almost all regions of Indonesia. Pests and diseases in shallot farming are often the problem of declining productivity. One of the obstacles that

* Corresponding author: gracelananda@gmail.com

farmers often experience in shallot cultivation is the attack of fusarium wilt disease caused by the pathogenic fungus *Fusarium oxysporum* sp. *Fusarium oxysporum* is one of the soil-borne fungi that easily spread from one area to another [1].

Fusarium wilt disease is characterized by wilting, rotten roots, drooping plants that look like they are going to collapse, and white fungal colonies at the base of bulbs. Microorganisms such as pathogenic microorganisms that cause disease and also insect pests that cause damage to agricultural crops, plantations and forests, have a high sensitivity to global climate change. Pathogens and insect pests have a very strong mechanism to carry out a rapid adaptation process to environmental changes. Similarly, disease-causing pathogens that belong to the rust fungus group will be able to make cytoplasmic and genetic adaptations so that their pathogenicity increases, their spread becomes wider, and their sporulation time is faster [2]. The increase in F. *oxysporum* was also caused by the habit of farmers who continuously planted shallots without crop rotation. The use of non-selective seeds, the use of infected seeds, and low soil organic content can also trigger an increase in F. *oxysporum* attacks [3].

Fusarium wilt disease on shallots is a difficult disease to control in the field. Efforts used to avoid environmental pollution can be using biological agents (biological control agents). The ability of antagonistic microbes to suppress the development of pathogens can be increased by means of a consortium or merger. The use of antagonistic microbial consortia is effective for suppressing disease progression compared to single use, because the use of antagonistic microbial consortia can be synergistic [4]. The use of antagonistic microbial consortia also has advantages that can increase plant growth, suppress the development of pathogens, increase plant resistance, as well increase crop yields [5]. The use of seeds for planting small bulbs is not recommended because they are susceptible to fusarium wilt, so efforts are needed to increase resistance, microbial consortium is used to increase plant resistance to fusarium wilt. The relatively big bulb size has more layers causing the plant's ability to grow stronger. Big bulbs have a wider cross-sectional area of the roots, so they can produce more roots and increase plant growth [6].

Some microbes that are effective as biological control agents of plant pests and diseases that have been reported in various studies include fungi, namely the genus *Trichoderma* sp. and *Gliocladium* sp. While from the bacterial group, among others, from the genus *Bacillus*, *Pseudomonas*, Agrobacterium and so on [7]. Antagonistic microbial consortium *Bacillus* sp., *Pseudomonas fluorescens* Migula, and *Trichoderma* sp. at a dose of 100 ml/plant was able to reduce the incidence of bacterial wilt disease (*Ralstonia solanacearum*) in tomato plants by 8.47% [8]. The use of *Trichoderma harzianum* Rifai. and *Bacillus subtilis* Cohn. able to suppress the growth of *Ralstonia solanacearum* disease in tobacco plants (*Nicotiana tabacum*). Antagonistic microbes from the genus *Trichoderma* and *Gliocladium* are some examples of microbes that have been successfully isolated and evaluated for their effectiveness as biological agents for plant disease control [9]. The use of microbes of the genus *Gliocladium* is quite effective in inhibiting the development of *Fusarium oxysporum* f.sp. capsici disease in chili plants [10]. The purpose of this study was to determine the effect of antagonistic microbial consortium and bulb size on fusarium wilt and yield of shallot.

2 Research method

This research was conducted on the coastal land in the Pasir Makmur Farmers Group, Sri Gading Village, Sanden District, Bantul Regency. The type of soil in the research location is sand with an altitude of ± 17 meters above sea level. This research was conducted in June – October 2021. The field experiment plot was designed in factorial and laid out in a split-plot with 2 factors. The main plot consists of:

K1 : Antagonist Microbial Consortium

K2 : Without Antagonist Microbial Consortia

Subplots consist of:
U1 : Small bulbs (Ø = <1.5 cm or <5 g)
U2 : Medium bulb (Ø = 1.5-1.8 cm or 5-10 g)
U3 : Large bulbs (Ø = >2 cm or >10 g),
U4 : Mixed size bulbs (small : medium : large) (1:1:1) [11]

There were 8 treatments with 4 replications, a total of 32 experimental plots with a study area of 500 m^2 per replicate. In a 40 m^2 sub-plot, there are 8 plots with a size of 1 m x 5 m. The distance between plots is 30 cm. The spacing used was 20 cm x 20 cm with a plant population of 1,000 shallot per treatment. In each sub-plot, 10 plants samples were taken. The shallot used in this study was the shallot bulbs of the Bima Brebes variety, a microbial consortium from the Bogor Soil Research Institute which consisted of 3 bacterial strains (*Bacillus* RRp 27.3, *Pseudomonas* RRp 21.15 and *Bacillus* RRp 42.4) and 1 *Trichoderma* strain.

Soil analysis was carried out after the field experiment. The shallot seeds used are grouped based on bulb size, namely small, medium, big, and mixed bulbs. Planting is done by dipping the prospective growing point into the water for 10 seconds and applying the consortium treatment to the prospective shallot growing point for 10 seconds by smearing it. Testing the percentage of bulbs carrying fusarium by using komada media after harvest. Sampling of bulbs was done randomly after harvest in each treatment. Each treatment was isolated with 4 replications. Each replication used 3 bulbs. So there are a total of 12 bulbs per treatment.

Three bulbs from each replication were cut into small pieces and then put into chlorox for 10 seconds and then immersed in sterile water for 10 seconds. One piece for each bulb was placed on the komada media in a petridish so that there were 3 pieces of shallots per petri dish from 3 different bulbs. Observations were made by calculating the percentage of bulb pieces per petridish that were overgrown by Fusarium. Observations were made 2 days after isolation. The plants were harvested 60 DAP. Harvesting was done using a 1 m x 1 m tile technique to estimate yield. The parameters of observing the intensity of fusarium wilt disease, the incidence of fusarium wilt disease, the percentage of shallot carrying fusarium after harvest (%), the number of *Fusarium* sp colonies. in soil after harvest (CFU/g), and the average yield of shallots (Kw/Ha). The data obtained in this research were analyzed statistically using variance (ANOVA) followed by Duncan's Multiple Range Test to determine a significant difference within the factor's level.

3 Result and discussion

3.1 Intensity of fusarium wilt

The results of the analysis in Table 1. showed that microbial consortium antagonists and bulb size had no significant effect on all periods of intensity of fusarium wilt disease. Shallot in the seed treatments, whether given consortium or without consortium at 10, 20, 30, 40, 50, and 60 DAP showed no significant difference.

Table 1. The average intensity of fusarium wilt disease (%).

Age of Plants (DAP)	Bulb Treatment	Bulb Size				Average
		U_1 (Small)	U_2 (Medium)	U_3 (Big)	U_4 (Mixed)	
		------------------------ % ------------------------				
10	K_1	0	0	0	0	0
	K_2	0	0	0	0	0
	Average	0	0	0	0	(-)
20	K_1	0.5	0	0	0	0.13
	K_2	0	0	0	0	0
	Average	0.25	0	0	0	(-)
30	K_1	5	0	0	0	1.25
	K_2	1	0	0	0	0.25
	Average	3	0	0	0	(-)
40	K_1	8	1.5	3	5	4.38
	K_2	3	1.5	2.5	1.5	2.13
	Average	5.50[a]	1.50[a]	2.75[a]	3.25[a]	(-)
50	K_1	10	2	4	6.5	5.63
	K_2	4.5	1.5	3.5	2	2.88
	Average	7.25[a]	1.75[a]	3.75[a]	4.25[a]	(-)
60	K_1	12	2.5	4.5	7.5	6.63
	K_2	4.5	2	4.5	2.5	3.38
	Average	8.25[a]	2.25[a]	4.50[a]	5.00[a]	(-)

Note : DAP (Days After Planting). The number followed by the same letter per observation time in the same column shows that it is not significantly different according to the 5% DMRT level test.
No interaction.

As a result of climate change, there are more frequent extreme weather events that can infect plants. One of the endemic diseases in lowland regions with extreme climates is fusarium wilt disease. The occurrence of a disease that attacks plants is influenced by three main factors: the pathogen, the environment, and the host plant. These three factors are known as the disease triangle concept [12].

3.2 Incidence of fusarium wilt

The results of the analysis in Table 2. Showed that microbial consortium antagonists and bulb size had no significant effect on all periods of incidence of fusarium wilt disease.

Table 2. The average incidence of fusarium wilt disease (%).

Age of Plants (DAP)	Bulb Treatment	Bulb Size U_1 (Small)	U_2 (Medium)	U_3 (Big)	U_4 (Mixed)	Average
		----------	----------	%	----------	
10	K_1	0	0	0	0	0
	K_2	0	0	0	0	0
	Average	0	0	0	0	(-)
20	K_1	2.5	0	0	0	0.63
	K_2	0	0	0	0	0
	Average	1.25	0	0	0	(-)
30	K_1	12.5	0	0	0	3.13
	K_2	2.5	0	0	0	0.63
	Average	7.5	0	0	0	(-)
40	K_1	17.5	2.5	2.5	2.5	6.25
	K_2	7.5	2.5	5	7.5	5.63
	Average	12.50[a]	2.50[a]	3.75[a]	5.00[a]	(-)
50	K_1	17.5	2.5	5	2.5	6.88
	K_2	7.5	2.5	5	7.5	5.63
	Average	12.50[a]	2.50[a]	5.00[a]	5.00[a]	(-)
60	K_1	17.5	2.5	5	2.5	6.88
	K_2	7.5	2.5	5	7.5	5.63
	Average	12.50[a]	2.50[a]	5.00[a]	5.00[a]	(-)

Note : DAP (Days After Planting). The number followed by the same letter per observation time in the same column shows that it is not significantly different according to the 5% DMRT level test.
No interaction.

Shallot in the seed treatments, whether given consortium or without consortium at 10, 20, 30, 40, 50, and 60 DAP showed no significant difference. The increase in the incidence of Fusarium wilt disease every week is related to the availability of inoculum sources. This is because the consortium is not necessarily able to suppress the development of pathogens, microbial consortia are not able to compete with microbes that have colonized the roots first. The absence of early microbes will make it easier for the bacterial consortium to colonize the rhizosphere [13].

3.3 The percentage of shallot carrying fusarium after harvest

The results of the analysis microbial consortium antagonists and bulb size had no significant effect on the percentage of shallot carrying fusarium after harvest. The average percentage of shallot carrying fusarium after harvest in the treatment of various shallot sizes showed no significant difference in small, medium, big, or mixed shallot. There was no difference in the treatment of seeds, whether given to a consortium or without a consortium.

Table 3. The percentage of shallot carrying fusarium after harvest (%).

Bulb Treatment	Bulb Size				Average
	U_1	U_2	U_3	U_4	
K1	58,33	100	66,67	75,00	75,00^p
K2	41,67	66,67	41,67	50,00	50,00^p
Average	50,00^a	83,33^a	54,17^a	62,50^a	(-)

Note : The number followed by the same letter per observation time in the same column shows that it is not significantly different according to the 5% DMRT level test.
No interaction.

In this research, to see whether the pathogen is virulent or not, it can be seen from the amount of pathogens in the shallot after harvest. Based on the results of the observations, it appears that the shallot still carries *Fusarium oxysporum*. Therefore, the consortium treatment and shallot size do not free the new shallots from Fusarium. The high population of Fusarium in Komada's media can be caused by the fact that the media used to grow Fusarium is a selective medium for Fusarium fungi which is very suitable for Fusarium from the pH and nutrients needed, Fusarium can grow optimally. This is supported by the experimental results of which showed that F. *oxysporum* grew well in media with a pH range of 5-8 [14].

3.4 The amount of *Fusarium* sp. colonies in the soil after harvest

The results of the analysis microbial consortium antagonists and bulb size had no significant effect on the amount of *Fusarium* sp. colonies in the soil after harvest. The average amount of *Fusarium* sp. colonies in the soil after in the treatment of various shallot sizes showed no significant difference in small, medium, big, or mixed shallot. There was no difference in the treatment of seeds, whether given to a consortium or without a consortium. The microbial community had the opportunity to colonize the roots earlier than the growth-promoting bacterial consortium that was given at the age of 11 hst. This resulted in the possibility that the growth-promoting microbial consortium was unable to compete with the microbes that had colonized the roots first. The absence of initial microbes will make it easier for the bacterial consortium to colonize the rhizosphere [13].

Table 4. The amount of *Fusarium* sp. colonies in the soil after harvest (CFU/g).

Bulb Treatment	Bulb Size				Average
	U_1	U_2	U_3	U_4	
K1	10.85	11.100	5.225	5.700	8.218,75^p
K2	1.900	6.650	5.900	7.000	5.362,5^p
Average	6.375^a	8.875^a	5.562,5^a	6.350^a	(-)

Note : The number followed by the same letter per observation time in the same column shows that it is not significantly different according to the 5% DMRT level test.
No interaction.

In this research, to see whether the pathogen is virulent or not, it can be seen from the amount of pathogens in the shallot after harvest. Shallot testing was also carried out by counting Fusarium colonies in the soil and it was found that in the soil after harvest there was no effect of consortium and the size of the shallots because the Fusarium was still there.

3.5 Average yield productivity (Kw/Ha)

Data on the average shallot yield obtained were then analyzed to determine the effect of microbial consortium antagonists and bulb size, which can be seen in Table 5. The average shallot yield in the treatment of various shallot sizes showed no significant difference in small, medium, big, or mixed shallot. There was no difference in the treatment of seeds, whether given to a consortium or without a consortium.

Table 5. Average shallot yield (Kw/Ha).

Bulb Treatment	Bulb Size				Average
	U_1	U_2	U_3	U_4	
K1	1,62	2,57	2,10	1,93	$2{,}05^{q}$
K2	2,04	2,93	2,63	2,24	$2{,}46^{p}$
Average	$1{,}83^{b}$	$2{,}75^{a}$	$2{,}36^{ab}$	$2{,}08^{b}$	(-)

Note : The number followed by the same letter per observation time in the same column shows that it is not significantly different according to the 5% DMRT level test.
No interaction.

Providing seed treatment with a consortium of antagonistic microbes and bulb size did not have an influence on yield shallot. This is influenced by biotic and abiotic environmental factors on the activity of biological agents. This is probably because differences in plant growth responses can be caused by the microbial status of the soil [13]. The results of the research related to *Bacillus subtilis* and *Lysinibacillus* sp. reported many such bacteria as potential bioagents that can have a positive impact on crop protection. Some antagonistic microbe can control pathogens with resistance induction mechanisms, such as *Bacillus subtilis* and *Lysinibacillus* sp. [15][16].

4 Conclusion

The results showed that there is no interaction between the use of the microbial consortium antagonist and bulb size in increasing the growth of shallot plants. Further research needs to be done on how to apply the antagonistic microbial consortium can be done by spraying at more frequent intervals or mixing in the soil so as to get an effective application to suppress the development of fusarium wilt disease against shallot plants directly in the field. In addition, further research can be carried out on the size of the bulbs by using large size bulbs or using small size bulbs but in their use can be treated with antagonistic microbes first.

This research was funded by the Australian Centre for International Agricultural Research (SLAM/2018/145). Field experiment was conducted by a collaboration with students of UPN Pembangunan University (Dzaky Razan Nur Asyam, Akhmad Rudy Fikri Setiyawan). The extension team of Ministry of Agriculture and Dinas Pertanian Bantul team as follow : Evy Pujiastuti, Ahmad Yunan, Fibrianty, Agung Iswadi, Dwi, and Tejokumolo. Laboratorium work was conducted with support from Dini Sundari and Elfrida Indriani Reno Wulan. Appreciation is also delivered to the Farmer Group Pasir Makmur (Sudantoro and Marno) and Manunggal (Subandi and Pardi) Sanden Bantul for the great support during field experimental.

References

1. A. Hadisoeganda, W. Widjaya, *Aplikasi pestisida biorasional agonal 866 untuk mengendalikan hama dan penyakit bawang merah, Jurnal Hortikultura*. **18**, 1 (2008)
2. R. Menéndez, How are insects responding to global warming?. Tijdschrift voor Entomologie, **150**, 2 (2007)
3. Azzam, *Pengendalian penyakit layu F. oxysporum pada tanaman bawang merah*, (2017)
4. R.L. Berendsen, G. Vismans, K. Yu, Y. Song, RD Jonge, WP, Burgman, M Burmolle, J. Herschend, PAHM Bakker, GMJ Pieterse, Disease-induced assemblage of a plant beneficial bacterial consortium. *Journal The ISME,* **12** (2018)
5. U. Aiman, T.B. Sriwijaya, *Pemberian macam konsorsium bakteri hasil isolasi tumbuhan pantai pada kangkung (Ipomoea reptans Poirs.), Jurnal Agrosains*, **5**, 1 (2017)
6. Y. Sufiyati, S. Imran, Fikrinda, *Pengaruh Ukuran Fisik dan Jumlah Umbi Per Lubang Terhadap Pertumbuhan dan Hasil Bawang Merah (Allium ascalonicum L.). J. Floratek*, **1**, 2 (2006)

7. I.K. Suada, *Mikroba Potensial Dalam Pengendalian Biologi Patogen Tumbuhan.,Pelawa Sari*, (2006)
8. A. Manan, E. Mugiastuti, L. Soesanto, *Kemampuan Campuran Bacillus sp., Pseudomonas fluorescens, dan Trichoderma sp. untuk Mengendalikan Penyakit Layu Bakteri pada Tanaman Tomat, Jurnal Fitopatologi Indonesia*, **14**, 2 (2018)
9. A.E. Prihatiningrum, A.K. Jalaluddin, *Uji Kemampuan Trichoderma Harzianum Dan Bacillus Subtilis Sebagai Agen Pengendali Layu Bakteri (Ralstonia solanacearum) Pada Tanaman Tembakau (Nicotiana tabacum). Nabatia,* **8**, 1 (2020)
10. A. Afriani, M. Heviyanti, F.S. Harahap, *Efektivitas Gliocladium virens untuk mengendalikan penyakit Fusarium oxysporum F. sp. capsici pada tanaman cabai, Jurnal Pertanian Tropik*, **6**, 3 (2019)
11. N. Sumarni, A. Hidayat, *Panduan Teknis Budidaya Bawang Merah. Balai Penelitian Tanaman Sayuran. Pusat Penelitian dan Pengembangan Pertanian. Badan Penelitian dan Pengembangan Pertanian*, (2005)
12. KS. Islam, MHA. Miah, SU. Ahmed, Effect of mulch and different levels of N and K on the growth and yield of onion. Progress Agric., **21**, 1-2 (2010)
13. E. Hidayati, Sarkono, Faturrahman, *Respon Pertumbuhan Tanaman Jagung Dan Kedelai Terhadap Pemberian Konsorsium Bakteri Rhizosfer Asal Lahan Kering Lombok Utara. Mataram. Journal of Biological Sciences* **9**, 2 (2022)
14. Achmad, E.P. Sari, *Pengaruh media terhadap pertumbuhan cendawan Fusarium oxysporum. Buletin RISTRI,* **1**, 4 (2009)
15. C. Leiwakabessy, M.S. Sinaga, K.H. Mutaqien, T. Trikoesoemaningtyas, G. Giyanto, The Endophytic Bacteria, Salicylic Acid, and their Combination as Inducers of Rice Resistance Against *Xanthomonas oryzae* pv. *oryzae*. *Agrivita*, **40**, 1 (2018)
16. C.M. Pieterse, C. Zamioudis, R.L. Berendsen, D.M. Weller, S.C. Van Wees, P.A. Bakker, Induced systemic resistance by beneficial microbes. *Annual review of phytopathology*, **52**, (2014)

Soybean growth and yield on corn cob compost application

Maria Theresia Sri Budiastuti[1*], *Desy* Setyaningrum[2,3], *Djoko* Purnomo[1], and *Fauzan* Wahidurromdloni[4]

[1]Department of Agrotechnology, Faculty of Agriculture, Universitas Sebelas Maret, Surakarta, Indonesia
[2]Department of Agribusiness, Vocational School, Universitas Sebelas Maret, Surakarta, Indonesia
[3]Doctoral Program of Agricultural Science, Faculty of Agriculture, Universitas Sebelas Maret, Surakarta, Indonesia
[4]Master Program of Agronomy, Faculty of Agriculture, Universitas Sebelas Maret, Surakarta, Indonesia

Abstract. The continuous use of chemical fertilizers contributes to gas emissions that cause global warming, climate change and low nutrient use efficiency. Compost is an effort to increase nutrient supply to improve soil quality and nutrient efficiency in food crops. The aim of the study was to examine the role of corn cob compost on soybean growth and yield. The study used a complete randomized block design with one factor, namely the dose of corn cob compost with six levels, namely chemical fertilizer 0.225 ton/ha, corn cob compost dose 2; 3; 4;5; 6 tonnes/ha. The dose of compost affects the root biomass. Corn cob compost 5 tons/ha was able to increase soybean yields by 2.28 tons and was not significantly different from chemical fertilizers. Corn cob compost 5 tons/ha produced the highest weight of 100 seeds, namely 41.33 g. Soybean yields were positively correlated with root biomass, number of leaves and crown biomass. Compost from corn cobs is an organic nutrient that can promote soybean growth and yield.

1 Introduction

Climate change has become an urgent global issue and affects various sectors of life, including the agricultural industry, which has a vital role in meeting world food needs [1]. Unstable weather patterns, rising global average temperatures, changes in rain patterns, and natural disasters are increasingly occurring due to climate change. Climate change has caused significant disruption to agricultural productivity worldwide [2,3]. Population growth and changes in consumption patterns cause demand for agricultural production to increase, which will create a gap between production and consumption. One of the dominant approaches used to meet this demand is the use of inorganic fertilizer because it can provide high results instantly [4,5]. The use of inorganic fertilizers can increase plant growth and yield significantly, but the use of these fertilizers also causes several environmental problems. The

* Corresponding author: mariatheresia@staff.uns.ac.id

use of inorganic fertilizers tends to cause a decrease in soil quality in the long term [6,7]. This fertilizer tends to disrupt the balance of microorganisms in the soil, damage the soil structure, and reduce the availability of micronutrients [8]. In addition, using inorganic fertilizers also contributes to releasing greenhouse gasses such as nitrogen oxide (N2O), which accelerates global climate change [9].

Meeting food needs can be done in two ways, namely agricultural intensification and extensification. Agricultural extensification has obstacles, namely the availability of land, which is decreasing due to urbanization, changes in land use, and other environmental factors [10]. Reducing optimal land conditions encourages increased productivity and quality of existing land to meet food needs. One solution that can be taken to overcome this problem is through agricultural intensification, which focuses on improving soil quality [11]. Organic fertilizer can be an alternative because it can increase nutrients and water-holding capacity, improve soil structure, and support soil microbial activity [12,13]. Organic fertilizer includes various types: compost, animal manure, and green fertilizer. This fertilizer is obtained from natural materials such as plant residues, organic waste, and animal materials that are decomposed naturally [14].

The utilization of agricultural waste as raw material for organic fertilizer has not been widely used. One of the agricultural wastes that can be utilized is corn cobs[15]. Corn cobs are not widely used, so they have the potential to be used as organic fertilizer because they are quite abundant. Corn cobs also contain essential nutrients such as nitrogen, phosphorus, and potassium, as well as organic materials that can help increase plant growth and yield [16]. It has been proven from previous research that corn cobs processed into organic fertilizer can support the growth and yield of corn plants [17]. Corn cob fertilizer can also be applied to soybeans. Soybean is a legume plant that has the potential to interact with microbes in the soil. Legume plants are generally able to form a symbiotic relationship with Rhizobium and mycorrhiza bacteria [18,19]. Rhizobium bacteria help in nitrogen fixation, while mycorrhizae help in nutrient absorption and improve overall soil quality. This interaction forms a mutualistic symbiosis, which can increase the availability of nutrients in the soil, thereby supporting plant growth and yield [20].

The use of organic fertilizer has been proven to increase the productivity of several types of plants, such as soybeans, corn, and rice. Previous research states that the application of organic fertilizer can significantly increase soybean yields [21]. Similar analysis was also carried out on corn and rice plants, which provided significant increases in yield [22,23]. The combination of applying organic fertilizers and planting soybeans has been shown to improve soil nutrition and health in the long term [24]. Soil that has high fertility can support plant growth so that it can provide maximum results. High plant productivity is expected to be able to meet increasing global food needs. The novelty of this research is using organic fertilizer from corn cobs to support the growth and yield of soybeans. This research aims to examine the use of organic corn cob fertilizer on soybean growth and yield.

2 Materials and methods

This research was carried out in Bakaran Hamlet, Sukosari Village, Jumantono District, Karanganyar Regency, Central Java, with analysis coordinates LS (07°38'07.04") E (110°57'00.7"), altitude 181 meters above sea level. The materials used in this research were Dena 1 soybean seeds, corn cob compost, cow dung fertilizer, and inorganic fertilizer with Urea, SP36, and KCl. The study used a complete randomized block design with one factor: corn compost dose. Corn compost dosage with six levels, namely inorganic fertilizer 0.225 tons/ha (Urea: 50 kg/ha, SP36: 75 kg/ha, KCl: 100 kg/ha), corn compost dosage 2;3;4;5 and 6 tons/ Ha. Each experimental unit was repeated three times so that there were 18 experimental units. Soybeans are planted at a distance of 20 x 30 cm. The variables observed were plant height, number of leaves, root weight, crown weight, number of pods, fresh weight of pods, one hundred seeds, and soybean yield per hectare. These variables were observed in soybean plants ten days after planting. The data obtained were analyzed using analysis of variance (ANOVA) at a level of 5%. If it had a significant effect, it was continued with the Duncan Multiple Range Test (DMRT) at a level of 5%.

3 Results and discussion

3.1 Soybean growth

The results showed that fertilizer treatment did not affect the height, number of leaves, and crown weight of soybeans (Table 1). Treatment with increasing doses of corn cob compost increased the height of soybean plants. A dose of 2 tons/ha of corn cob compost could not increase plant height compared to inorganic fertilizer treatment. However, with a dose of 5 tons/ha, it showed higher plant height compared to the inorganic fertilizer treatment. Apart from that, the 4-ton/ha treatment also showed many leaves. These results indicate that a 4 tons/ha dose has supported soybean growth. Corn compost can support soybean growth because it contains available nitrogen, potassium, and phosphate elements, which meet the standards as organic fertilizer according to No.70/Permentan/SR.140/10/2011. Corn compost contains 1.44% total nitrogen, 1.43% phosphate, and 2.17% potassium [17]. These nutrients are essential nutrients that play a role in the growth and quality of crop yields because they affect plant physiological processes [25]. A 3 tons/ha dose of corn cob compost has encouraged soybean canopy growth. Shoot biomass at a dose of 3 tonnes/ha was higher than the inorganic fertilizer treatment and increased by 6.5%. The increase in canopy biomass was higher with higher treatment doses. Corn cob compost contains high organic matter that supports the growth of soil microorganisms [26]. The activity of soil microorganisms helps provide good nutrition for plant growth [27,28]. Based on Table 3, plant height and number of leaves are positively correlated with the canopy biomass of soybean plants. The higher and more leaf growth shows higher canopy biomass. The higher growth of the leaves will increase plants' photosynthetic capacity so that biomass production is high [29]. However, nitrogen absorption is supported in photosynthesis because nitrogen is an essential component of chlorophyll and a component of the enzymes involved in photosynthesis [30]. Plant height positively correlates with fresh pod weight, number of pods, and crown weight (Table 3). Plants that grow taller mean they get more nutrients to support crown growth and pod formation [31]. Taller soybeans usually have more nodes from which to flower and produce pods.

Corn compost treatment affects root biomass (Table 1). Corn compost at a dose of 4 tons/ha has shown higher root biomass than inorganic fertilizers. However, the two and 3-ton/ha corn compost treatments showed the lowest root biomass. These results indicate that a dose of 4 tons/ha has encouraged the growth of soybean plant roots. Soybean root growth is supported by the high availability of organic matter in corn compost. Compost can promote the growth of beneficial soil microorganisms in helping to break down organic matter and increasing the availability of nutrients for plant roots [32,33]. Corn cobs compost contains 85.71% organic matter and 62.21% high organic carbon [17]. Organic matter plays a role in improving soil structure and increasing the activity of soil microorganisms [29]. Organic matter can increase soil porosity and water-holding capacity so that roots can easily penetrate the soil, thereby increasing the roots' absorption of water and nutrients [34]. Absorption of root nutrients is correlated with root biomass and density [35]. Higher root biomass can transport nutrients in more significant quantities, thereby supporting the nutritional needs of plants—higher root biomass results in higher root density and greater surface area available for nutrient uptake.

Table 1. Soybean growth.

Treatment	Crown Weight	Number of Leaves	Plant Height	Root Weight
Inorganic fertilizer 0.225 tons/ha	58.22	28.67	28.83	0.72 b
Corn compost dosage 2 tons/ Ha	43.31	22.17	20.50	0.40 a
Corn compost dosage 3 tons/ Ha	62.30	34.67	26.87	0.65 a
Corn compost dosage 4 tons/ Ha	64.80	40.33	34.07	0.75 ab
Corn compost dosage 5 tons/ Ha	80.87	41.33	38.83	0.85 b
Corn compost dosage 6 tons/ Ha	68.04	33.17	33.90	0.98 b

Note: numbers followed by different letters in one column, indicate a significant difference in DMRT ($\alpha = 0.05$).

3.2 Soybean yield

Corn cob compost did not affect soybean yield (Table 2). The number of pods and the yield of soybeans per hectare were high in the treatment of inorganic fertilizers. Corn compost at a dose of 2 to 4 tons/ha has yet to be able to increase the number of pods, fresh weight of pods, weight of one hundred seeds, and soybean yield per hectare. This is due to the slow-release nature of organic fertilizers [36]. However, corn compost treatment at a dose of 5 tons/ha produced a higher number of pods, fresh pod weight, one hundred seeds, and soybean yield per hectare compared to the inorganic fertilizer treatment. Doses of 5 tons/ha have increased by 7% of pods compared to inorganic fertilizer treatment. This pod increase was supported by the high content of phosphor and potassium in corn cob compost [17]. Phosphorus is a component of adenosine triphosphate used in nutrient absorption, cell division, and the synthesis of organic compounds. A high phosphate content will support efficient energy transfer, plant metabolic activity, and plant growth and yield [37]. Plants with intensive development and short cycles, such as soybeans, require higher amounts of phosphate in solution and adding P, which is adsorbed more quickly than annual plants [38]. Based on Table 3, the number of pods positively correlates with root biomass. Roots have an essential role in absorbing nutrients from the soil. Soybeans with more and more roots have a wider absorption field. This causes soybeans to get more water and nutrients for growth and yield [39]. These results are supported by high organic matter and organic carbon in corn compost. Organic matter and organic carbon can improve soil physics, chemistry, and biology. Organic matter can improve soil structure to increase the activity of microorganisms and the availability of nutrients for plants [40].

The weight of one hundred seeds reflects the quality of the soybean seeds. The application of several different doses of fertilizer gives the same results on the weight of one hundred seeds. Application of 2 tons/ha of corn cob fertilizer produced the highest weight of one hundred seeds and increased the weight by 44.16% compared to the application of inorganic fertilizer. Organic fertilizers contain high levels of phosphate and potassium to support seed formation [41]. In addition, organic fertilizers also have a neutral pH so that they can improve soil conditions. The availability of phosphate and potassium at a neutral pH is higher to meet soybeans' needs [42,43]. One hundred seed weight positively correlates with the number of leaves and shoot weight (Table 3). This is because the increasing number of leaves causes the photosynthesis process to be more optimal so that it can produce more photosynthesis.

Many leaves can provide good photosynthetic quality so that they can be used to produce quality seeds.

The yield of soybeans per hectare in treating 5 tons/ha of corn cob compost was 2.28 tons/ha. This yield is low compared to the potential yield of the Dena I variety, which is 2.9 tons/ha. However, 5 tons/ha of corn cob compost gave the highest soybean yield and increased by 10.14% compared to inorganic fertilizers. These results indicate that corn compost and inorganic fertilizers can increase soybean yields. Corn cob compost has a high nitrogen, phosphate, and potassium content, so it can support the process of forming soybean pods [44]. Organic fertilizers that are slow-releasing guarantee the availability of nutrients until harvest because they are not easily washed away by water [36]. In addition, compost can improve soil quality, add nutrients to the soil, increase the activity of microorganisms, and increase its water-holding capacity [12,13]. This can support soybean growth so that soybeans can grow optimally and provide the best results.

Table 2. Soybean yield.

Treatment	Fresh Weight of Pods	Number of Pods	Soybean Yield per Hectare	Weight of One Hundred Seeds
Inorganic fertilizer 0.225 tons/ha	12.73	23.67	2.07	28.67
Corn compost dosage 2 tons/ Ha	9.93	15.67	1.22	22.17
Corn compost dosage 3 tons/ Ha	11.96	16.67	1.51	34.67
Corn compost dosage 4 tons/ Ha	11.38	18.33	1.81	40.33
Corn compost dosage 5 tons/ Ha	16.39	25.33	2.28	41.33
Corn compost dosage 6 tons/ Ha	15.91	21.67	2.00	33.17

Table 3. Correlation of soybean growth and yield.

	Soybean yield per hectare	Weight of 100 seeds	Fresh weight of pods	Number of pods	Root weight	Header weight	Number of leaves	Plant height
Plant height	-0.105	0.259	0.529*	0.518*	0.406	0.759**	0.259	1
Number of leaves	-0.303	1.000**	0.156	0.104	-0.061	0.826**	1	
Header weight	-0.261	0.826**	0.421	0.378	0.210	1		
Root weight	0.262	-0.061	0.576*	0.494*	1			
Number of pods	0.420	0.104	0.847**	1				
Fresh weight of pods	0.366	0156	1					
Weight of one hundred seeds	-0.303	1						
Soybean yield per hectare	1							

Note: ** = significantly very different
* = significantly different

4 Conclusion

The dose of compost affects the root biomass. Corn compost at a dose of 4 tons/ha has shown higher root biomass than inorganic fertilizers. The dose of corn compost did not affect soybean yield. Corn cob compost 5 tons/ha was able to increase soybean yields by 2.28 tons and was not significantly different from chemical fertilizers. Corn cob compost 5 tons/ha produced the highest weight of 100 seeds, namely 41.33 g. Soybean yields were positively correlated with root biomass, number of leaves and crown biomass. Compost from corn cobs as an organic nutrient that can promote soybean growth and yield.

References

1. S. K. Balasundram, R. R. Shamshiri, S. Sridhara, and N. Rizan, Sustainability **15**, 5325 (2023)
2. E. Moyo, L. G. Nhari, P. Moyo, G. Murewanhema, and T. Dzinamarira, Eco-Environment & Health **2**, 74 (2023)
3. A. Tun Oo, D. Boughton, and N. Aung, Climate **11**, 124 (2023)
4. M. Arif, S. Ali, M. Ilyas, M. Riaz, K. Akhtar, K. Ali, M. Adnan, S. Fahad, I. Khan, S. Shah, and H. Wang, Soil Use Manag **37**, 104 (2021)
5. L. J. Wan, Y. Tian, M. He, Y. Q. Zheng, Q. Lyu, R. J. Xie, Y. Y. Ma, L. Deng, and S. L. Yi, Agriculture (Switzerland) **11**, (2021)
6. L. Chi, S. Han, M. Huan, Y. Li, and J. Liu, Int J Environ Res Public Health **19**, (2022)
7. C. Li, O. O. Aluko, G. Yuan, J. Li, and H. Liu, Sci Rep **12**, (2022)
8. L. C. Dincă, P. Grenni, C. Onet, and A. Onet, Applied Sciences (Switzerland) **12**, (2022)
9. H. Gong, J. Li, Z. Liu, Y. Zhang, R. Hou, and Z. Ouyang, Land (Basel) **11**, (2022)
10. N. Isnaeni, D. Arista, A. Dhienar Alifia, H. Mubarok, I. Made, S. Dwi Arta, D. Novira Rizva, and A. I. Wicaksono, *Availability and Potential for Expansion of Agricultural Land in Indonesia* (2023)
11. P. M. Kopittke, N. W. Menzies, P. Wang, B. A. McKenna, and E. Lombi, Environ Int **132**, (2019)
12. S. Tian, B. Zhu, R. Yin, M. Wang, Y. Jiang, C. Zhang, D. Li, X. Chen, P. Kardol, and M. Liu, Soil Biol Biochem **165**, (2022)
13. W. Zhou, Q. Ma, L. Wu, R. Hu, D. L. Jones, D. R. Chadwick, Y. Jiang, Y. Wu, X. Xia, L. Yang, and Y. Chen, Agric Ecosyst Environ **326**, (2022)
14. T. B. Singh, A. Ali, M. Prasad, A. Yadav, P. Shrivastav, D. Goyal, and P. K. Dantu, in *Contaminants in Agriculture: Sources, Impacts and Management* (Springer International Publishing, 2020), pp. 61–77
15. C. sheng Li and R. rong Cai, Science of the Total Environment **838**, (2022)
16. S. Syamsia, A. Idhan, and K. Kasifah, Unri Conference Series: Community Engagement **1**, 362 (2019)
17. M. T. S. Budiastuti, D. Purnomo, B. Pujiasmanto, and D. Setyaningrum, Agriculture (Switzerland) **13**, (2023)
18. R. Ashwin, D. J. Bagyaraj, and B. Mohan Raju, Plant Stress **4**, (2022)
19. N. O. Igiehon, O. O. Babalola, X. Cheseto, and B. Torto, Microbiol Res **242**, (2021)
20. Y. Qin, Y. Yan, L. Cheng, Y. Lu, J. Chen, F. Liu, and J. Tan, J Soil Sci Plant Nutr **23**, 2723 (2023)
21. T. B. Hammed, E. O. Oloruntoba, and G. R. E. E. Ana, International Journal of Recycling of Organic Waste in Agriculture **8**, 81 (2019)
22. C. Gao, A. M. El-Sawah, D. F. Ismail Ali, Y. A. Hamoud, H. Shaghaleh, and M. S. Sheteiwy, Agronomy **10**, (2020)
23. J. Liu, A. Shu, W. Song, W. Shi, M. Li, W. Zhang, Z. Li, G. Liu, F. Yuan, S. Zhang, Z. Liu, and Z. Gao, Geoderma **404**, (2021)

24. M. B. Fontana, L. E. Novelli, M. A. Sterren, W. G. Uhrich, S. M. Benintende, and P. A. Barbagelata, Geoderma **385**, (2021)
25. I. S. Setubal, A. S. de Andrade Júnior, S. P. da Silva, A. C. Rodrigues, A. Bonifácio, E. H. F. M. da Silva, P. F. de M. J. Vieira, R. de S. Miranda, N. Cafaro La Menza, and H. A. de Souza, Plants **12**, (2023)
26. Y. Yue, X. Gong, Y. Zheng, P. Tian, Y. Jiang, H. Zhang, and H. Qi, Agronomy **13**, (2023)
27. P. Rawat, S. Das, D. Shankhdhar, and S. C. Shankhdhar, J Soil Sci Plant Nutr **21**, 49 (2021)
28. N. C. M. Pereira, F. S. Galindo, R. P. D. Gazola, E. Dupas, P. A. L. Rosa, E. S. Mortinho, and M. C. M. T. Filho, Front Environ Sci **8**, (2020)
29. J. Han, Y. Dong, and M. Zhang, Applied Soil Ecology **165**, (2021)
30. B. Qiang, W. Zhou, X. Zhong, C. Fu, L. Cao, Y. Zhang, and X. Jin, Res Sq 1 (2022)
31. O. O. Aluko, C. Li, Q. Wang, and H. Liu, Int J Mol Sci **22**, (2021)
32. Y. Wang, S. Peng, Q. Hua, C. Qiu, P. Wu, X. Liu, and X. Lin, Front Microbiol **12**, (2021)
33. P. Palanivell, O. H. Ahmed, O. Latifah, and N. M. A. Majid, Applied Sciences (Switzerland) **10**, (2020)
34. G. Murtaza, Z. Ahmed, M. Usman, W. Tariq, Z. Ullah, M. Shareef, H. Iqbal, M. Waqas, A. Tariq, Y. Wu, Z. Zhang, and A. Ditta, J Plant Nutr **44**, 1677 (2021)
35. R. Xiong, S. Liu, M. J. Considine, K. H. M. Siddique, H. M. Lam, and Y. Chen, Physiol Plant **172**, 405 (2021)
36. H. Shaji, V. Chandran, and L. Mathew, in *Controlled Release Fertilizers for Sustainable Agriculture* (Elsevier, 2021), pp. 231–245
37. N. Bechtaoui, M. K. Rabiu, A. Raklami, K. Oufdou, M. Hafidi, and M. Jemo, Front Plant Sci **12**, (2021)
38. A. C. M. Lino, S. Buzetti, M. C. M. T. Filho, F. S. Galindo, P. R. Maestrelo, and M. A. De Carvalho Rodrigues, Semina:Ciencias Agrarias **39**, 99 (2018)
39. K. Liu, T. Li, Y. Chen, J. Huang, Y. Qiu, S. Li, H. Wang, A. Zhu, X. Zhuo, F. Yu, H. Zhang, J. Gu, L. Liu, and J. Yang, Field Crops Res **258**, (2020)
40. L. Shang, L. Wan, X. Zhou, S. Li, and X. Li, PLoS One **15**, (2020)
41. B. A. Khan, A. Hussain, A. Elahi, M. Adnan, M. M. Amin, M. Danish Toor, A. Aziz, M. K. Sohail, A. Wahab, and R. Ahmad, International Journal of Applied Research **6**, 1 (2020)
42. K. lou Liu, T. fu Han, J. Huang, S. Asad, D. ming Li, X. chu Yu, Q. hai Huang, H. cai Ye, H. wen Hu, Z. hua Hu, and H. min Zhang, Soil Tillage Res **197**, (2020)
43. C. J. Penn and J. J. Camberato, Agriculture (Switzerland) **9**, (2019)
44. Y. Fan, Z. Wang, D. Liao, M. A. Raza, B. Wang, J. Zhang, J. Chen, L. Feng, X. Wu, C. Liu, W. Yang, and F. Yang, Sci Rep **10**, (2020)

Enhancing soil quality with coal fly ash amendment: a waste recycling approach for climate change mitigation in Indonesia - a review

Jubaedah [1, 3*], Iskandar [2], *Dwi Tejo P.* Baskoro [2], *Budi* Mulyanto [2] and *Neneng L.* Nurida [3]
[1] Graduate School, Soil Science Program, IPB University, Jalan Meranti, Kampus Darmaga, Bogor.
[2] Department of Soil Science and Land Resources, IPB University, Jalan Meranti, Kampus Darmaga, Bogor.
[3] National Research and Innovation Agency, Jalan Jakarta-Bogor No. 46, Cibinong Science Center, Bogor.

Abstract. Coal fly ash (CFA), a byproduct of coal combustion (CCP) from electric/thermal power plants, was classified as hazardous and toxic waste under Government Regulation Number 104/2014 in Indonesia. However, a significant shift in this regulatory viewpoint developed with the introduction of Government Regulation Number 22/2021. This regulation fundamentally altered the treatment and classification of CFA. Managing CFA as a waste has been both extensive and costly. In various countries where CFA is classified as non-hazardous and non-toxic waste, its utilization spans various sectors. Agriculture is a notable sector benefiting from CFA, as it is applied to soil amendment. As a soil amendment, CFA enhances nutrient availability and soil fertility, promotes plant development, and positively alters soil physical properties. In addition, CFA application to soil improves soil carbon storage, contributing to climate change mitigation and adaptation. This paper aims to review Indonesia's state of CFA utilization before the enactment of Government Regulation Number 22/2021, in order to present insights into the potential for expanded CFA utilization in soil amendment following the new regulation (Government Regulation Number 22/2021).

1 Introduction

Fly ash bottom ash (FABA) is categorized as hazardous-toxic waste under Government Regulation No. 85/1999 on Hazardous-Toxic Waste Management. This regulation mandates a comprehensive life-cycle approach for managing hazardous-toxic waste, from generation to final disposal, with strict handling protocols. Specialized permits are required at each phase of FABA waste management, including storage, collection, transportation, treatment, utilization, and landfill operations [1]. However, classifying FABA as hazardous-toxic waste contradicts Indonesia's Coal Development and Utilization roadmap [2]. In contrast, some countries label FABA as non-hazardous, non-toxic waste, enabling its full-scale utilization. Notably, Indonesia's major coal export destinations—China, India, and South Korea—do not

* Corresponding author: jubaedah@brin.go.id

categorize FABA as hazardous-toxic waste. China and India classify it as solid waste, while South Korea defines FABA as a specific byproduct [3,4].

In 2021, the Government of Indonesia officially delisted FABA from the hazardous-toxic waste category, in accordance with Government Regulation No. 22/2021 on the Implementation of Environmental Protection and Management. This policy revision was rendered on the basis that FABA satisfies the criteria for classification as nonhazardous-nontoxic waste. FABA exhibits neither flammable nor explosive characteristics, demonstrates a lack of reactivity, and is non-corrosive in nature. Empirical assessments, including the Toxicity Characteristic Leaching Procedure (TCLP), evaluations of heavy metal content, and Lethal Dose 50 (LD 50) tests, further confirm that FABA complies quality standards for non-hazardous, non-toxic waste [5]. Despite its classification as nonhazardous-nontoxic waste, producers of such waste are still comply with the established standards and technical criteria as stipulated in the approved environmental documentation, issued by the government (central or local). The removal of FABA from the hazardous-toxic waste classification is expected to increase FABA utilization.

Nevertheless, this manuscript will primarily focus on exploring the potential application of coal fly ash (CFA) as a soil amendment to enhance soil quality, particularly after the revocation of its hazardous-toxic waste status after the Government Regulation application. The implementation of the regulation will expand opportunities for CFA reuse/recycling, not only in the construction sector. Additionally, the paper will elucidate CFA's role in climate change mitigation strategies and enhancing carbon stock in soil. This projection is consistent with Yao et al. [6] that reuse/recycle CFA as an alternative to landfill disposal has the potential to yield substantial economic and environmental advantages.

2 Elemental properties of CFA and its effect on soil chemical properties

Numerous scientific investigations have confirmed the effectiveness of CFA as a promising soil amendment, enhancing soil nutrient availability, thereby ameliorating soil fertility and promoting plant development [7]. Significant amounts of Si, Al, Fe, Ca, K, S, Na, and Mo, along with trace elements, are present in CFA. Moreover, Yu et al. [8] reported that CFA incorporates both macro and micro nutrients including P, Mg, B, Co, Cu, Mn, Ni, and Zn which are essential for plant growth and development. CFA with elevated Ca concentration serves as an effective liming agent [9], an important role as soil amendment. However, contrasting observations were presented by Jackson et al. [10], who indicated that the macro-nutrient concentration in CFA is lower, requiring its combination with additional organic materials like manure or sewage sludge for soil amendment applications. While the P content in CFA is modestly elevated relative to its concentration in native soil, it predominantly exists in a phase that is not readily bioavailable. This is because it is largely bound to aluminosilicate minerals, making it less soluble and therefore less available to plants [11–13]. Conversely, Ca, Mg, K, and Na concentrations are significant and bioavailable [11,14,15].

The American Society for Testing Materials (ASTM) classifies CFA into two classes: Class F contains more than 70% combined silica (SiO_2), alumina (Al_2O_3), and iron oxide (Fe_2O_3), while Class C comprises these elements in a range of 50 to 70% [16]. Class C CFA are distinguished by their higher CaO content when compared to Class F CFA. Generally, Class C CFA contains > 15% CaO, while Class F CFA typically contains < 15% CaO [6]. Class C CFA with higher calcium content is generated through lignite or sub-bituminous (soft coal) combustion, while class F CFA is produced through combustion of anthracite and bituminous (hard coal). Class F CFA exhibits pozzolanic characteristics, indicating that it contains Si compounds that will react with $Ca(OH)_2$ at ambient temperatures to create a new compounds with cementitious properties. On the contrary, class C CFA (higher content of

CaO) exhibits inherent self-cementing property. When it reacts with water, class C CFA will harden and increase its strength over time [16].

CFA effectively raises soil pH, largely due to its inherent alkalinity. In addition, replacing lime with CFA has the potential to reduce CO_2 emissions generated during lime production, thereby serving as dual approach for climate change adaptation and mitigation. Class C CFA has been empirically demonstrated to elevate soil pH and mitigate soil acidity [17]. Manoharan et al. [18] reported an elevation of approximately two pH units in topsoil (0-10 cm depth) with an application dose of 108 t ha^{-1} of CFA. The gradual increase of soil pH after CFA application could be mediated by the consumption of H^+ during the weathering of alumino-silicate minerals and the dissolution of calcium from CFA [19,20]. Elevations in soil pH also influence the solubility and bioavailability of soil nutrients [18,21]. Comprehensive compositional analysis of Indonesian-sourced CFA has been conducted to catalog its constituent elements, and CFA that was used for soil amendment in Indonesia and other countries (Table 1). In Table 1, the soil amendments under consideration exhibit high/alkaline pH levels, despite presenting a relatively low concentration of Ca.

Table 1. Elemental CFA content as for identification and as for soil amendment.

Constituents and Characteristics	Unit	CFA [a]		CFA [b]	CFA [c]	CFA [d]
		East Java	West Java	North Halmahera	Bangka	South Africa
SiO_2	%	41.3	51.3	nm	85.86	55.1
Al_2O_3	%	29.5	34.6	nm	5.40	28.6
Fe_2O_3	%	11.55	5.11	2.96	10.30	5.60
CaO	%	9.13	4.48	2.56	3.57	2.75
MgO	%	2.46	1.81	0.53	2.38	0.75
K_2O	%	1.14	0.48	nm	0.21	0.66
Na_2O	%	1.73	0.69	nm	1.09	0.068
MnO_2	%	0.038	0.20	nm	nm	nm
SO_3	%	0.82	nt	nm	nm	nm
P_2O_5	%	0.24	nt	nm	0.10	0.44
TiO_2	%	1.25	0.13	nm	nm	1.66
CuO	ppm	nm	nm	11.1	nm	nm
MnO	ppm	nm	nm	1263	nm	470
CdO	ppm	nm	nm	0.27	nm	nm
PbO	ppm	nm	nm	7.60	nm	nm
As_2O_3	ppm	nm	nm	<0.0001	nm	nm
Cr_2O_3	%	nm	nm	nm	nm	0.050
NiO	%	nm	nm	nm	nm	0.012
pH		nm	nm	nm	9.91	10
Electric Conductivity (EC)	dSm^{-1}	nm	nm	nm	nm	0.24
Calcium Carbonate Equivalent (CCE)	%	nm	nm	nm	nm	10

Note: a: CFA identification [21]
b,c,d: CFA for soil amendment [22–24]
nt: not detected
nm: not measured

3 CFA effects on soil physical property

The incorporation of CFA into soil has positive effects on soil physical properties, particularly in too sandy or clayey soils. Adriano and Webber [25] noted that silt-sized particles in CFA make it a viable substitute for topsoil, enhancing water retention in surface mine lands. This modification of soil fractions leads to enhanced aggregation and infiltration [9]. Textural improvements also impact soil aeration. Furthermore, Chang et al. [26] discovered increased hydraulic conductivity with CFA application, reducing soil surface crusting. Hydraulic conductivity decreases when more than 20% CFA (by volume) is added to calcareous soils or more than 10% to acidic soils. This decline results from the pozzolanic reaction of CFA, binding soil particles when wet, reducing water flow. Interestingly, the pozzolanic effect is more pronounced in acidic soils, although the mechanism is not fully understood.

Table 2. CFA effects on selected soil quality physical parameters [12].

Soil quality parameters	Value for measurement in agricultural soil	Soil amended with		Crops
		Decomposed ash	Fresh ash	
Bulk density	1.3 g cm^{-3}	Decrease than 1.3 g cm^{-3}	Decrease than 1.3 g cm^{-3}	Soybean, barley
Aeration	High	Increase	Increase	Grass
Water retention (whc)	High	Increase	Increase	Corn
Plant available water	High	Zero to minimal effect	Zero to minimal effect	Horticultural crops (carrot, radish, beet, lettuce)
Hydraulic conductivity	High	Rise by minimal dose	Rise by minimal dose	Barley
Modulus of rupture	High	Decrease	Decrease	Barley

Table 2 confirmed CFA application has impact on the physical properties of soil. Fail and Wochok [27] reported that the augmented growth of soybeans treated with CFA were directly associated with alterations in soil texture and an enhancement in water-holding capacity. Such modifications in soil texture subsequently influence other physical parameters such as water-holding capacity, bulk density, and soil porosity. In the context of the Barley cropping system, Chang et al. [26] stipulated that an enhancement in water-holding capacity was exclusively observed when CFA was applied at concentrations exceeding 25%. This specific dosage was also found to lower the soil's bulk density. It is noteworthy to mention that an increase in water-holding capacity does not unequivocally correlate with an elevation in plant-available water.

High CFA application rates (up to 560 and 1120 t ha^{-1}) significantly improve water-holding capacity but should be used cautiously due to potential negative impacts on other soil properties (Table 3) [25]. One adverse consequence includes an escalation in the soil electrical conductivity (EC) [15], attributed to an excessive influx of soluble salts. Table 3 shows that applying CFA to sandy soils can lead to heightened levels of EC. This increase in EC can subsequently reduce the soil's water-holding capacity due to elevated osmotic pressure. Additionally, CFA application lowers the modulus of rupture, affecting soil particle cohesiveness, potentially impacting soil structure and stability [12,26].

Table 3. CFA doses and its impact on sandy soil [15].

Treatments combination	CFA dose	Soil texture	Results of observations
CFA+soil	0, 10, 20, 30, 40% (ww^{-1})	Clayey, sandy clay loam, sandy, sandy loam	Incorporating CFA into soils with varying textures alters both the physical and physico-chemical conditions of the soil. The changes can subsequently impact crop production. Hydraulic conductivity is decreased by 25%, while soil water retention is enhanced.
CFA+soil	0, 280, 560, 1120 t ha^{-1}	-	Enhanced soil physical characteristics (capacity to hold water, water available for plant uptake, ability to retain moisture).
CFA+soil	0, 1, 2.5, 5, 10 and 15 t ha^{-1}	Sandy, sandy loam	Enhanced soil physical characteristics and increased rice growth and yield at 10 tha^{-1} CFA dose.
CFA+soil	0, 3, 6, 12 and 30% (ww^{-1}) dry weight basis	Sandy, sandy loam	The increased electrical conductivity might limit the soil's capacity to hold water due to higher osmotic pressure.

4 Reuse/recycle CFA as waste management strategy and to mitigate climate change

Reuse/Recycle CFA as waste management is imperative, particularly within the Indonesian framework where CFA is no longer classified hazardous-toxic waste. Given its substantial production volumes at both national and global scales, CFA serves as an economically valuable resource. Moreover, CFA application as a soil amendment is correlated with carbon sequestration capacity in soil [28]. Specifically, the incorporation of CFA into soil leads to improvements in soil properties, subsequently augmenting the soil ability to sequester carbon. This mechanistic pathway elucidates that CFA, acting as a soil amendment, elevates soil productivity, which in turn boosts plant yield. This leads to an increase in soil carbon stocks, facilitated by an optimized environment and enhanced root growth.

Above ground, the improved plant growth results in higher biomass production, which, when returned to the soil, can maintain or even increase current carbon stocks. The concentration of C stocks depends on variables like biomass turnover and decomposition rates. Ukwattage et al. [28] posits that the introduction of organic carbon with extended residence time (recalcitrant) into the soil is requisite for effective climate change mitigation and adaptation. According to Amonette et al. [29], the presence Fe oxide in alkaline CFA has the capability to stabilize the tyrosinase enzyme, thereby augmenting the humification process. Furthermore, alkaline nature and Fe oxide fractions within CFA are contributory factors in enhancing humification.

Palumbo et al. [30] delineates that one way for stabilizing soil organic C involves a chemical stabilization process, particularly when labile organic materials are complexes with clay minerals via polyvalent cation bridges. Subsequent soil aggregation offer additional physical protection to organic C, when organic C was covered and adsorbed by soil aggregate. Moreover, Yunusa et al. [31] identifies the role of unburnt carbon and cenosphere particles in carbon adsorption. The two components serve as the primary adsorbents in CFA. Dissolved organic matter (DOM) is anticipated to be effectively adsorbed by cenospheric particle of CFA. Yunusa et al. [31] also notes that the application of alkaline CFA mitigated carbon loss by up to 36%, while acidic CFA reduced total C loss threefold.

5 Conclusions

The enactment of Government Regulation No. 22/2021 represents an innovative policy intervention by the Indonesian government. The removal of coal fly ash bottom ash from the hazardous-toxic waste classification is anticipated to increase the rate of coal fly ash bottom ash utilization especially as soil amendment. CFA is a promising soil amendment, enhancing soil nutrient availability, thereby ameliorating soil fertility and promoting plant development, and also impacted positively on soil physical properties. In addition, Reuse/recycle CFA as an alternative to landfill disposal offers significant economic and environmental benefits, with its environmental consequences are thoroughly documented. Furthermore, utilizing biomass ash (including CFA) as a soil amendment and augmenting soil carbon reserves are effective approaches for mitigating climate change.

The Author (J) thanks to the Indonesian Agency for Agricultural Research and Development (IAARD), the Ministry of Agriculture, and the National Research and Innovation Agency (BRIN). This manuscript is part of author's doctoral program at IPB University. The program was funded by IAARD from 2019 to 2022 and is currently being continued by BRIN (2023).

References

1. T. Prasetiawan, Info Singk. Kaji. Singk. Terhadap Isu Aktual Dan Strateg. **XII**, (2021)
2. Kementerian Energi dan Sumber Daya Mineral, *Road Map Pengembangan Dan Pemanfaatan Batubara* (Direktorat Jenderal Mineral dan Batu bara, Jakarta, 2021)
3. A. K. Mandal and O. P. Sinha, J. Inst. Eng. (2014)
4. H. Cho, S. woo Ji, H. young Shin, and H. Jo, Sustain. **11**, (2019)
5. K. L. H. dan Kehutanan, (n.d.)
6. Z. T. Yao, X. S. Ji, P. K. Sarker, J. H. Tang, L. Q. Ge, M. S. Xia, and Y. Q. Xi, Earth-Science Rev. **141**, 105 (2015)
7. R. Kaur and D. Goyal, Part. Sci. Technol. **33**, 76 (2015)
8. C. L. Yu, Q. Deng, S. Jian, J. Li, E. K. Dzantor, and D. Hui, Environ. Pollut. **250**, 137 (2019)
9. J. Skousen, J. E. Yang, J. S. Lee, and P. Ziemkiewicz, Geosystem Eng. **16**, 249 (2013)
10. B. P. Jackson, W. P. Miller, A. W. Schumann, and M. E. Sumner, J. Environ. Qual. **28**, 639 (1999)
11. S. M. Shaheen, P. S. Hooda, and C. D. Tsadilas, J. Environ. Manage. **145**, 249 (2014)
12. D. C. Adriano, A. L. Page, A. A. Elseewi, A. C. Chang, and I. Straughan, J. Environ. Qual. **9**, 333 (1980)
13. A. L. Page, A. A. Elseewi, and I. R. Straughan, Residue Rev. **VOL. 71**, 83 (1979)
14. P. Kishor, A. K. Gosh, and D. Kumar, Asian J. Agric. Res. (2010)
15. V. C. Pandey and N. Singh, Agric. Ecosyst. Environ. **136**, 16 (2010)

16. A. Bhatt, S. Priyadarshini, A. Acharath Mohanakrishnan, A. Abri, M. Sattler, and S. Techapaphawit, Case Stud. Constr. Mater. **11**, e00263 (2019)
17. I. A. M. Yunusa, P. Loganathan, S. P. Nissanka, V. Manoharan, M. D. Burchett, C. G. Skilbeck, and D. Eamus, Crit. Rev. Environ. Sci. Technol. **42**, 559 (2012)
18. V. Manoharan, I. A. M. Yunusa, P. Loganathan, R. Lawrie, C. G. Skilbeck, M. D. Burchett, B. R. Murray, and D. Eamus, Fuel **89**, 3498 (2010)
19. V. Manoharan, I. a M. Yunusa, and R. Lawrie, 1 (2007)
20. S. Seoane and M. C. Leiro, Water **30**, 1420 (2001)
21. R. Damayanti, J. Teknol. Miner. Dan Batubara **14**, 213 (2018)
22. A. Zuhriansah, I. Mansur, and S. Budi R., J. Teknol. Miner. Dan Batubara **16**, 93 (2020)
23. N. Faoziah, Iskandar, and G. Djajakirana, J. Ilmu Tanah Dan Lingkung. **24**, 1 (2022)
24. J. R. Harper and I. Mbakwe, South African J. Plant Soil **37**, 101 (2020)
25. D. C. Adriano and J. T. Weber, J. Environ. Qual. **30**, 596 (2001)
26. A. C. Chang, L. J. Lund, A. L. Page, and J. E. Warneke, J. Environ. Qual. **6**, 267 (1977)
27. J. L. J. Fail and Z. S. Wochok, Plant Soil **48**, 473 (1977)
28. N. L. Ukwattage, P. G. Ranjith, and M. Bouazza, Fuel **109**, 400 (2013)
29. J. E. Amonette, J. Kim, C. K. Russell, A. V Palumbo, and W. L. Daniels, Int. Ash Util. Symp. Cent. Appl. Energy Res. Univ. Kentucky Paper #47 (2003)
30. A. V Palumbo, J. F. Mccarthy, J. E. Amonette, L. S. Fisher, and S. D. Wullschleger, **8**, 425 (2004)
31. I. A. M. Yunusa, G. Blair, A. Zerihun, S. Yang, S. C. Wilson, and I. M. Young, Clim. Change **131**, 559 (2015)

Effect of jajar legowo planting system on microclimate factors, growth, and rice yield

Imawati Maghfiroh[1*], *Supriyadi* Supriyadi[1], and *Endang Setia* Muliawati[1]

[1]Study Program of Agronomy, Faculty of Agriculture, Sebelas Maret University, Indonesia

Abstract. Climate change, especially global warming, affects plant physiology and crop yields. Jajar Legowo (Jarwo) planting system, with the existing planting alleys that some farmers have applied, needs to be sufficiently identified for changes in microclimate factors, growth, and rice yield. This study aims to identify the microclimate factors, growth, and rice yield in this planting system. The research was designed through field experiments in rice fields with the Jarwo 2:1, 3:1, and 4:1 planting systems. The research variables included microclimate factors (temperature, humidity, and light intensity), rice growth, and yield. The results showed that the sunlight intercept in each Jarwo planting system was not different. The highest average temperature was recorded in the Jarwo 2:1 system and the lowest in the Jarwo 4:1. The highest average humidity was recorded in the Jarwo 4:1 system and the lowest in the Jarwo 2:1 system. The study's results noted no significant difference in the variables of plant height, total number of tillers, number of productive tillers, and grain weight per panicle. The observation of grain weight per clump Jarwo 3:1 is significantly different from Jarwo 4:1. The existence of alleys in the Jarwo planting system does not significantly affect plant growth and yield.

1 Introduction

Climate change impacts various aspects, one of which is agriculture. Climate change, especially global warming, causes changes in weather, temperature, and humidity, which will affect plant physiology [1–3]. Rice, as the staple food of Indonesian society, is widely cultivated with various cultivation technologies, including the Jajar Legowo (Jarwo) planting system [4,5]. There are several Jajar Legowo planting systems, namely 2:1, 3:1, 4:1, and 5:1. Research on this system includes observing the microclimate [6], as well as the effect of the microclimate on rice tillers and plant height where the number of tillers in the Jajar Legowo system is higher than that in the System of Rice Intensification (SRI) [7]. In each of these systems, there is the potential for differences in the formed microclimate. Microclimate is a local climate condition that directly influences certain physical characteristics of the environment [4].

Several studies related to applying Jajar Legowo have been carried out. In previous studies, the microclimate parameters measured in the Jajar Legowo system were temperature, relative humidity, and sunlight intensity [6]. This study observed various Jajar Legowo

* Corresponding author: imawati_m@student.uns.ac.id

planting systems, namely 2:1, 3:1, 4:1, 5:1, and 6:1, compared to the conventional system as a control. This study concluded that several Jajar Legowo systems did not affect the microclimate in temperature, relative humidity, or sunlight intensity. This research aims to identify the effect of Jajar Legowo 2:1, 3:1, and 4:1 planting systems on microclimate factors, growth, and rice yields.

2 Methodology

This research was conducted in Nglangun Hamlet, Karanglo Village, Polanharjo District, Klaten Regency, Central Java, from October to December 2022. Geographically, the research area is at 7°36' 51.9" South Latitude, 110° 39' 32.4" East Latitude ± 200 masl. The research was conducted on a site of 1800 m^2 by planting local Klaten Menthik Wangi rice. The materials used in this study were Menthik Wangi rice seeds, meters, and cameras. The research was carried out by field experiments, with the Jajar Legowo treatment 2:1, 3:1, and 4:1. Each Jajar Legowo system uses a spacing of 25 cm x 12.5 cm with a Legowo width of 50 cm. The study was conducted on a single plot with an area of 600 m^2 for each plot with random placement of the plots. The total land area for the three Jajar Legowo systems is 1800 m^2.

Microclimate observations used a lux meter, minimum-maximum thermometer, and thermohygrometer. Observations of air temperature and humidity were observed between 07.00 and 09.30 (Western Indonesia Time, WIB) using a thermohygrometer. Observations of air temperature and humidity under the canopy of each Jajar Legowo plot were carried out three times at different points. Light intensity was observed using a lux meter to measure the light intensity above and below the plant canopy. The light intercept on each plot is then calculated.

Plant growth observations were made once a week in each plot. Observations of temperature, humidity, and sunlight intensity, as well as observations of plant height, total tillers, number of productive tillers, grain weight per panicle, and grain weight per clump for each plot, were carried out in clumps on the edge of the Jajar Legowo alley and the middle clumps between the Jajar Legowo alley, 30 clumps each. Plant height was observed in 90 edge clumps and 60 middle clumps from each plot weekly. Observational data were analyzed descriptively with the T-test and displayed with a histogram analysis of each research variable.

3 Results and Discussions

3.1 Microclimatic observation

The lowest air temperature was obtained on the 3rd week of observation at 29°C, while the highest was on the 11th week at 37.5°C. The lowest air humidity in the 5th week was 53.67%, and the highest in the 7th week was 90%. Both observations can be seen in Figure 1. Microclimatic observations showed the highest average temperature in Jajar Legowo 2:1 and caused the average humidity in the Jajar Legowo planting system to be the lowest compared to other planting systems. Meanwhile, the lowest average temperature was found in the Jajar Legowo 4:1 planting system with the highest average humidity. This is due to the large number of Legowo alleys in the 2:1 planting system, so much sunlight enters the plant canopy. This condition is in accordance with previous research that the wider the Legowo distance will cause an increase in air temperature, lower humidity, and increase the intensity of sunlight [8].

 https://doi.org/10.1051/e3sconf/202346701009

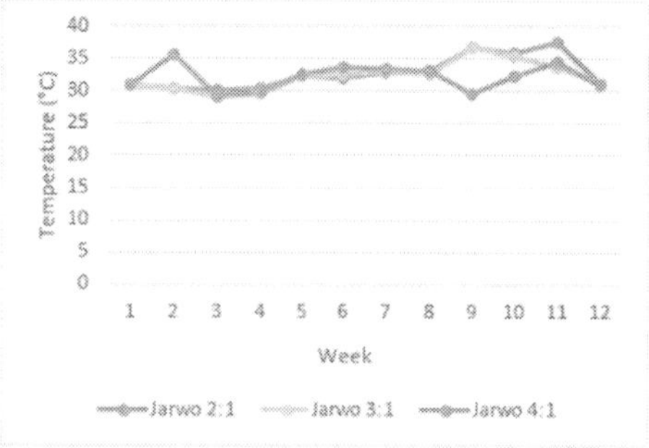

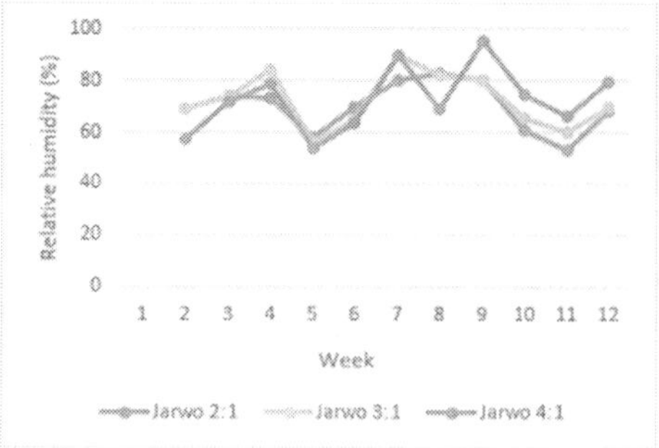

Fig 1. Temperature and humidity observation

Light intercept is the percentage of sunlight that reaches the plant [9]. Interception of light during the study period ranged from 70% to 95%. Observations of light interception showed that at the 9th week of observation or the age of the plant around ten weeks after planting (WAP), there was an increase in light interception but tended to decrease closer to harvest. This is in accordance with previous research that the amount of radiation intercepted by plants increases with increasing age of the plant and decreases again before harvest [10].

3.2 Plant growth and yield

Based on the results of the T-test, plant height between Jajar Legowo was not significantly different, but Jajar Legowo 4:1 showed the highest average, i.e. 124.65 cm (see Figure 2). This is in accordance with research, which states that the plant height of the local variety Irkasuma Merah showed no significant difference in the treatment of the Jajar Legowo 2:1 and Jajar Legowo 4:1 planting systems [11]. Different results were obtained by research where the planting system significantly affected plant height, where the Jajar Legowo 2:1 treatment had taller plants than the Jajar Legowo 4:1 and tiles (20x20cm) [12]. The conditions that occurred in this study were thought to be caused by the metabolic rate in plants, which significantly determines plant growth during the vegetative phase. In this process, apart from being influenced by the cultivation techniques used, it is also influenced by the varieties of rice planted and the environmental conditions of the plantings.

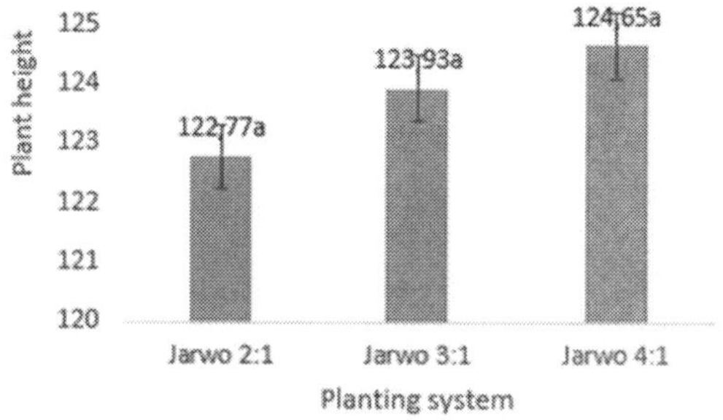

Fig 2. Plant height observation

Plant growth in Jajar Legowo 4:1 was the highest compared to other Jajar Legowo systems. This is in accordance with two previous studies, which stated that plant height in applying Jajar Legowo 4:1 was higher than Jajar Legowo 2:1 and 3:1 [13,14]. This growth was not only influenced by edge plants but also due to plant population density. In addition, the growth in Jajar Legowo 4:1 was higher, allegedly due to the condition of the plot closest to the water source, so plant growth was more fertile than other Jajar Legowo.

Based on the T-test between Jajar Legowo, the total number of tillers did not differ significantly. As seen in Figure 3, the highest number of tillers was in Jajar Legowo 4:1 and the least in Jajar Legowo 2:1. This is different from two previous studies, which stated that higher plants use more assimilate for the formation of stems and leaves than for the formation of tillers [13,15]. In addition to the influence of population density, the formation of tillers is also influenced by water availability so that even though the plant growth is high, the number of tillers is also higher.

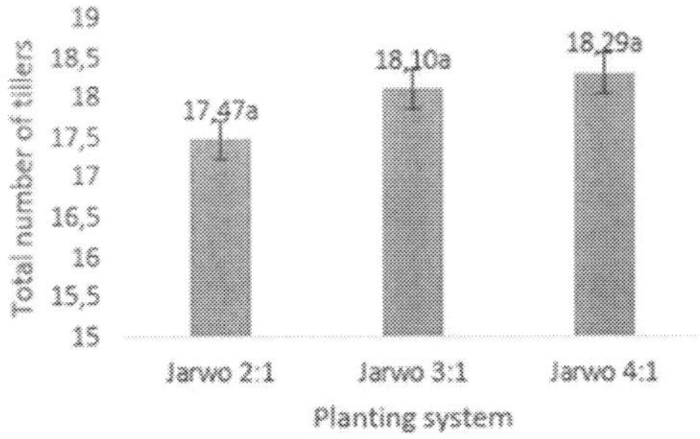

Fig 3. Total number of tillers observation

Based on observations, the average number of productive tillers in edge clumps was 16.23, while the average total tiller was 16.62. This data shows that the percentage of productive offspring to the total number of offspring was 97.65%. Meanwhile, observations in the middle clump showed an average of 16.4 productive tillers and an average of 17.13 total tillers. From this data, the percentage of productive tillers in the total number of tillers in the middle clump was 95.74%. The number of productive tillers in each Jajar Legowo planting system in the edge and middle clumps based on the results of the T-test was not significantly different, and the comparison between Jajar Legowo was also not significantly different.

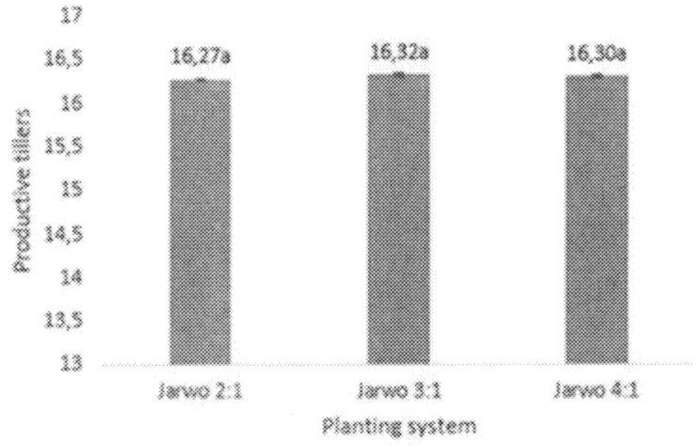

Fig 4. Productive tillers observation

The average number of productive tillers from each Jajar Legowo planting system was relatively the same, with the highest number of tillers in Jajar Legowo 3:1 (see Figure 4). Productive tillers are influenced not only by the capture of solar radiation by the plant canopy but also by environmental conditions, one of which is the availability of water and nutrients. The formation of productive tillers does not only depend on the number of tillers formed but can also be influenced by environmental factors [16]. Other studies have also confirmed that rice plants have the potential to form productive tillers based on the number of tillers, but this is not always the case because tiller formation is also influenced by the environment [17]. Tall plants use more assimilates for developing stems and leaves than for developing tillers [15]. Apart from the influence of population density caused by the use of Jajar Legowo 4:1, the formation of tillers is also influenced by water availability.

 https://doi.org/10.1051/e3sconf/202346701009

The grain weight per panicle was observed just before harvest by taking observational samples from each Jajar Legowo plot of 30 clumps. The weight of grain per panicle in several planting systems of Jajar Legowo based on the T-test was not significantly different. Still, Jajar Legowo 4:1 tended to be the highest, namely 4.39 grams, compared to other Jajar Legowo.

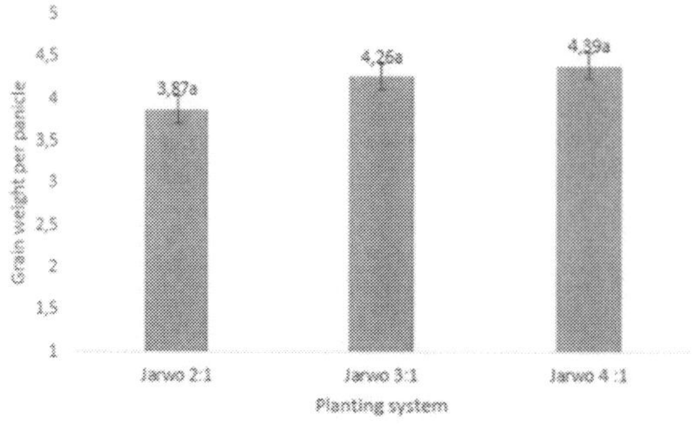

Fig 5. Grain weight per panicle observation

The grain weight per panicle is influenced by the amount of grain per panicle. The highest weight was obtained at Jajar Legowo 4:1, and the lowest was at Jajar Legowo 2:1 (see Figure 5). These results are in accordance with research that Jajar Legowo 2:1 has the lowest number of grains per panicle [14]. Different results were obtained in other studies comparing Jajar Legowo 2:1 and 4:1, that the number of grains per panicle Jajar Legowo 2:1 was more than Jajar Legowo 4:1 [13]. This difference in yield is thought to be influenced not only by environmental conditions but also by the type of variety planted. Although different results were obtained, the Jajar Legowo planting systems were generally not significantly different.

Based on the results of the T-test, there was no significant difference in grain weight per clump between the edge clumps and the middle clumps. As seen in Figure 6, there was a significant difference in grain weight per clump between Jajar Legowo 3:1 and Jajar Legowo 4:1 based on the results of the T-test. The highest grain weight per clump was in Jajar Legowo 4:1 of 59.39 grams, and the lowest was in Jajar Legowo 3:1 of 49.23 grams.

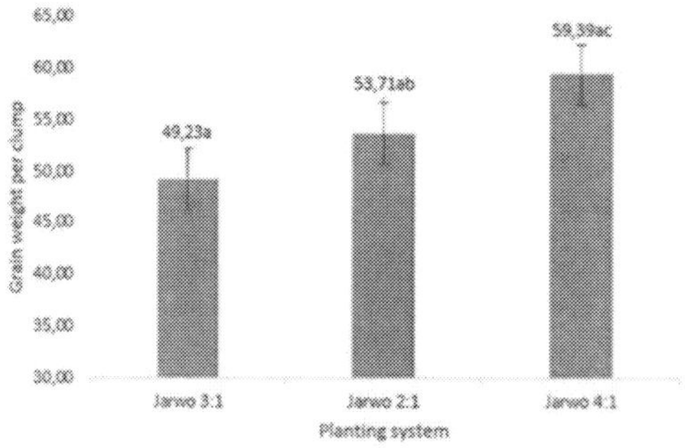

Fig 6. Grain weight per clump observation

These results are in accordance with research, which states that Jajar Legowo 4:1 gives the highest grain yield compared to Jajar Legowo 2:1, 6:1, and 8:1 [18]. Other studies obtained different results; namely, the highest grain yield was obtained at Jajar Legowo 2:1 [13]. This difference in yield is thought to be due to many factors affecting crop productivity. Although the Jajar Legowo system provides the potential for more optimal utilization of sunlight and weed control due to the presence of alleys, environmental factors such as water availability significantly affect rice growth and yield. Similar results were also found in the sorghum study, where a better microclimate resulted in higher yields [19].

4 Conclusion

Applying the Jajar Legowo planting system 2:1, 3:1, and 4:1 did not give a significant difference in the variables of plant height, total number of tillers, and number of productive tillers. However, in the parameter of grain weight per clump, there was a significant difference between the Jajar Legowo 3:1 and 4:1 planting systems. The Jajar Legowo 4:1 planting system gives the highest plant growth and grain yield.

References

1. Y. Dou, R. Huang, L. R. Mansaray, and J. Huang, Int. J. Remote Sens. **41**, 471 (2020)
2. R. G. Pérez-Hernández, M. J. Cach-Pérez, R. Aparicio-Fabre, H. van der Wal, and U. Rodríguez-Robles, Bot. Sci. **99**, 132 (2021)
3. M. Sikma, H. Ikawa, B. G. Heusinkveld, M. Yoshimoto, T. Hasegawa, L. T. Groot Haar, N. P. R. Anten, H. Nakamura, J. Vilà-Guerau de Arellano, H. Sakai, T. Tokida, Y. Usui, and J. B. Evers, J. Geophys. Res. Biogeosciences **125**, 1 (2020)
4. A. Pratiwi, E. R. Imannudin, W. Kusumaningtyas, and S. P. Sugianto, Pros. Semin. Nas. "Membangun Kemandirian Korporasi Petani Indones. Menuju Kedaulatan Pangan Berkelanjutan" **5**, (2018)
5. J. Bobihoe, *Sistem Tanam Padi Jajar Legowo* (Balai Pengkajian Teknologi Pertanian (BPTP) Jambi, 2013)
6. I. W. Adiguna, I. W. Tika, and Sumiyati, J. Beta (Biosistem Dan Tek. Pertanian) **8**, 18 (2020)
7. T. S. Santosa, B. D. A. Nugroho, and R. Tirtalistyani, in *Proc. 2nd Int. Conf. Smart Innov. Agric.* (Atlantis Press, 2021), pp. 342–349
8. I. P. Hendra, Sumiyati, and I. W. Tika, J. Biosist. Dan Tek. Pertan. **2**, 1 (2014)
9. R. Wentasari and A. R. Gusta, J. Penelit. Pertan. Terap. **18**, 199 (2019)
10. Syukri, R. Ridha, and Banjar, in *Semin. Nas. Fak. Pertan. Univ. Samudra Ke-VI* (2021), pp. 105–110
11. Azwir, J. Akta Agrosia **11**, 102 (2008)
12. Yunizar and A. Jamil, in *Pros. Balai Besar Penelit. Tanam. Padi (BB Padi)* (2012), pp. 999–1008
13. A. Kasim and J. Hendri, in *Pros. Semin. Nas. Membangun Pertan. Mod. Dan Inov. Berkelanjutan Dalam Rangka Mendukung MEA* (2016), pp. 127–131
14. Atman and Misran, in *Pros. Semin. Nas. Hari Pangan Sedunia Ke-33* (2015), pp. 237–245
15. Asfaruddin, A. Makmur, D. Sopandie, and H. Aswidinnoor, Evaluasi Ketenggangan Galur-Galur Padi Gogo Terhadap Keracunan Aluminium Dan Efisiensinya Dalam Penggunaan Kalium, Institut Pertanian Bogor, 1997
16. M. Maisura, J. Jamidi, and A. Husna, J. Agrium **17**, 33 (2020)
17. H. Sitinjak and Idwar, J. Online Mhs. Fak. Pertan. **2**, (2015)
18. Misran, J. Penelit. Pertan. Terap. **14**, 106 (2014)
19. S. Thapa, B. A. Stewart, Q. Xue, and Y. Chen, PLoS One **12**, 1 (2017)

Identification of microclimate in jajar legowo planting system and its effect on brown plant hopper and *Lycosa* sp.

Fajar Eka Saputra[1]*, Supriyadi[2], and *Endang Setia Muliawati*[2]

[1]Master Program of Agronomy, Faculty of Agriculture, 57126 Sebelas Maret University, Indonesia
[2]Department of Agrotechnology, Faculty of Agriculture, 57126 Sebelas Maret University, Indonesia

Abstract. Rice was an important food crop that one of the cultivation was the jajar legowo planting system. However, the effect of the planting system on brown planthopper and predatory *Lycosa* has not been well studied, related to microclimate changes. This study aims to identify microclimate factors in the jajar legowo and their effects on brown leafhopper populations and *Lycosa*. This research was designed through field experiments in rice fields with the treatment of legowo planting system (2: 1) and conventional system as a check. The research variables consisted of temperature, humidity, light intensity, brown leafhopper population, and *Lycosa* population. The results of this study showed the highest humidity was at the end of the planting period. The temperature and light intensity recorded in jajar legowo planting systems tends higher. The population of *Lycosa* in the Jajar legowo planting system was highest in the 2 week after transplanting (WAT), while in the conventional planting, the system was found in the 1 WAT The highest brown plant hopper population was recorded at 3 WAT, while in jajar legowo at 2 WAT. The results of this study showed no significant effect of microclimate factors on the *Lycosa* population and brown leafhopper populations. The benefit of this research is knowing the effect of microclimate on *lycosa* and WBC populations in Jajar Legowo and Conventional rice planting systems.

1 Introduction

Rice was one of the local food crops that were most in demand by the people of Indonesia. Many things affect rice production such as limited agricultural land, soil fertility levels, weather, cropping systems and pests. This plant has a shallow root system of about 40-50 cm so it is suitable for cultivation on peatland.

Jajar legowo was a planting system that regulates row spacing in one clump of plants. The system that has a nickname (SI JARWO) applied to rice has several patterns and was considered capable of increasing production. In the organic farming system, chemical fertilizers or pesticides ware not used for management from planting to post-harvest[1] Many factors influence farmers in carrying out agricultural management, including education,

* Corresponding author: Fajar900es@student.uns.ac.id

economy, environment and media.[2] The legowo system 2: 1 and 3: 1 have the highest grain production per hectare.[3]

Brown leafhoppers ware one of the most dangerous pests for rice. Although not every growing season these pests appear. The method in collecting insects uses sweep net, pit fall trap and light trap[4].Brown leafhoppers begin to be found in Inpari 2 and TN 1 varieties about 9-10 days after infestation (HIS) some pests of plants breed very dependent on temperature, so it will develop rapidly if the temperature is suitable.[5]

Lycosa pseudonnata was a type of spider that was a natural enemy of annoying insects on rice plants. The use of natural enemies was our step in realizing environmentally friendly agriculture. There are many negative impacts in the use of pesticides with active ingredients and difficult to degrade, such as reduced predatory insects and damage to ecosystems.[6] In the application of IPM, the main key is to convince farmers that the use of pesticides was carried out if natural control fails or is unable to control existing pests.[7]Predators that look for prey on the ground are more effective in suppressing cereal tick populations.[8]

Microclimate was a branch of science that studies the climate in the area around plants. Because climate was an influential factor in rice production. Some pests and diseases will thrive well in humid conditions. In some cases, drying the rice field area will affect production. Diversity of plant species and management of natural enemy habitats will reduce pest populations.[9] With drying, the existing climate will change. Carrying out habitat change must be carried out on an ongoing basis in order to increase emphasis on natural enemies.[10]

[11]. in their research Tillage Intensity Influences Insect-Pest And Predator Dynamics Of Wheat Crop Grown Under Different Conservation Agriculture Practices In Rice-Wheat Cropping System Of Indo-Gangetic Plain. Says The dynamics of insect pests and their predators are driven by soil habitat-related changes (direct) as well as plant growth-related effects (indirect).The novelty in this research is that researchers try to discuss the influence of the Jajar Legowo planting system on rice microclimate. This study aims to identify microclimate factors in the jajar legowo and their effects on brown leafhopper populations and *Lycosa* sp. As well as its relationship with the population of brown planthoppers and the spider *Lycosa* sp. The benefit of this research is knowing the effect of microclimate on *lycosa* and WBC populations in Jajar Legowo and Conventional rice planting systems.

2 Materials and methods

2.1 Study sites

This research was conducted in Nglangun, Karanglo Village, Polanharjo District, Klaten RegencyProv. Central Java. This research will be carried out on March 21 - June 22 2022. During the rice planting season in the middle of the rainy season. In this study the author used observation parameters, *Lycosa* population, WBC population, temperature, humidity and light.

2.2 Microclimate

Climate is an important factor in the breeding and spread of living things. The development of pests and predators is affected by changes in habitat and plants.[11] Meanwhile, microclimate influences plant development in a planting area, examples are temperature, humidity, light intensity, and rainfall.

2.3 Planting system

Jajar legowo is a planting system that regulates row spacing in one plant cluster. The system which has the nickname (SI JARWO) which is applied to rice has several patterns and is considered capable of increasing production. This will be compared with the conventional planting system (Tegel). In organic farming systems, chemical fertilizers or pesticides are not used for management from planting to post-harvest.[1]

2.4 Brown leafhoppers

Brown planthopper is one of the main pests that attack rice. Therefore, various methods have been used to deal with these pests, Such as planting resistant varieties, trapping, using natural enemies, changing planting systems and irrigation arrangements. Brown planthopper populations are lower in organic rice systems compared to semi-organic and conventional systems.[12]

2.5 Lycosa pseudonnata

Spiders are important predators in the paddy field ecosystem and are controls that use little capital.[13] *Lycosa* does not make nests in the form of nets to catch prey. But he hunts and pounces on his prey by jumping, so he is classified as a wild spider (wolf spider). The number of predators on land that does not show the use of pesticides shows a larger population.[14]

2.6 Data analysis

Data identification of microclimate in jajar legowo planting system and its effect on brown plant hopper and *Lycosa* sp. were ested for normality and transformed as needed. And presented as and arithmetic standard deviation in histogram using excel 2013 (window 10). Simple regression and simple correlation were estimated for the temperature, humidity, light intensity, brown leafhopper population, and *Lycosa* population similarly method by using SPSS software (10.0J). For statistical analysis. Data brown leafhopper population, and *Lycosa* population presented in graphics using exel 2013 (window 10).

3 Result and discussion

3.1 Brown leafhoppers

From the data in fig.1 we can see that the highest brown planthopper population occurred in the second week of the planting season. It consists of brown planthoppers which are classified as nymphs which are spread evenly almost throughout the planting area. That can happen because in this system there are narrow gaps and wide gaps. The closer the distance between the legowo, the humidity will also increase.[7] Whereas in the conventional system the distance is the same, making it possible for the brown planthopper to prefer the narrow gaps found in the row legowo system which is more humid.

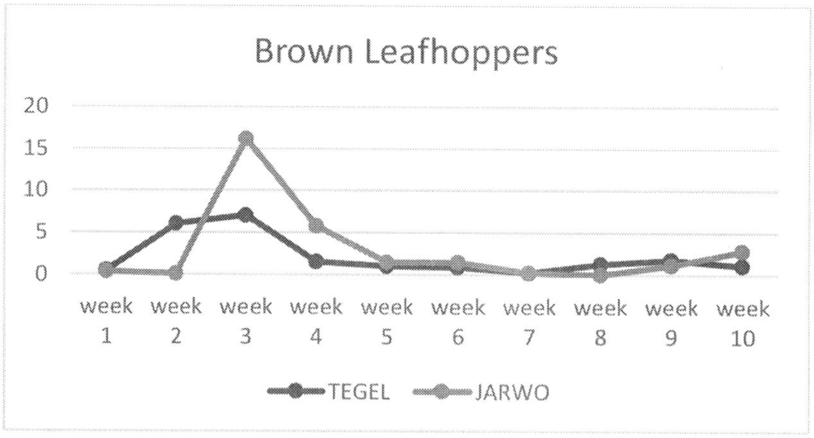

Figure 1. Population Brown leafhoppers.

Based on simple linear regression analysis and the correlation between WBC population and humidity in Jarwo, it appears that the correlation is positif and not significant. (y = 66.011 + 0.660x). Sianipar M.s et al in their research said that the influence of climate has a low correlation with the WBC population.[8] The results of simple linear regression can be seen in Figure. 2.

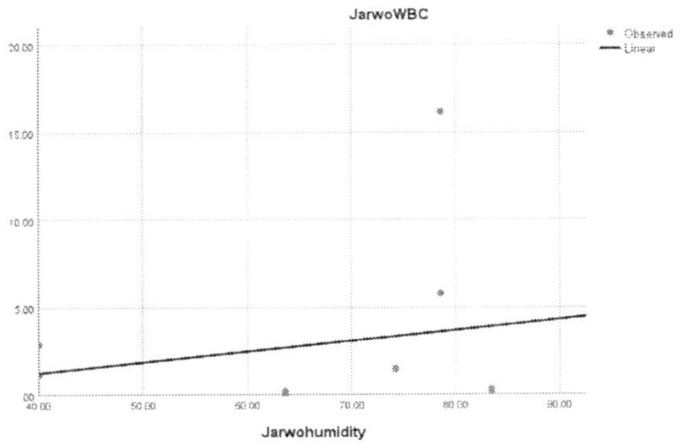

Figure 2. WBC and Microclimate in Jarwo.

Based on simple linear regression analysis and the correlation between WBC population and humidity in conventional it appears that the correlation is positive and not significant. (y = 65.158 + 1.934x). The results of simple linear regression can be seen in Figure. 3.

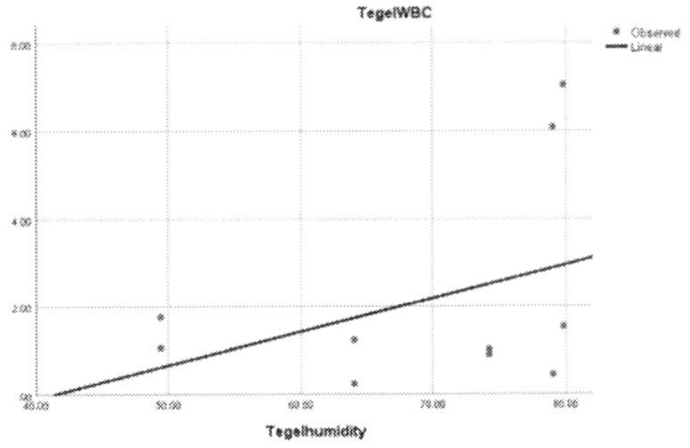

Figure 3. WBC and Microclimate in conventional.

3.2 Lycosa pseudonnata

In the data in figur.4 we can see that the *Lycosa* population increased in the second week due to an increase in the brown planthopper population in the same week. This can happen because *Lycosa* hunts prey more often than making traps. So that where many prey eat *Lycosa* will approach him. Lawalata jj et al. In his research, he said that the presence and level of predation by natural enemies in a particular place really depends on the presence and type of pest or prey of these natural enemies.[15]

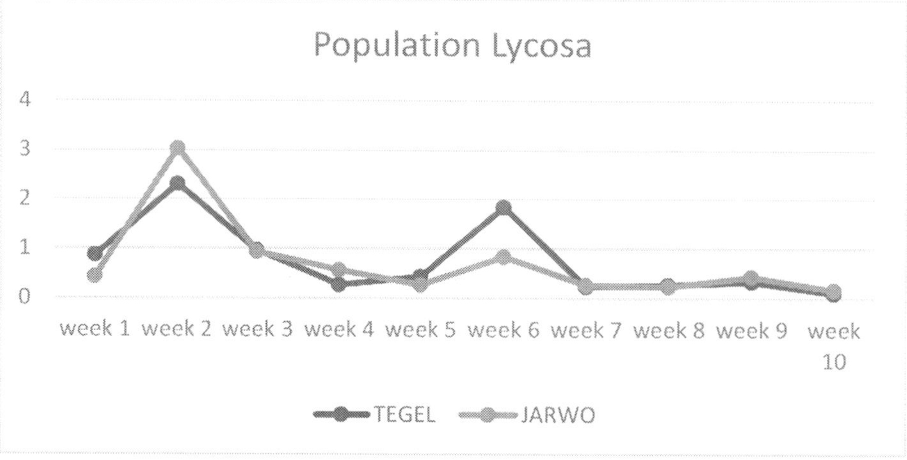

Figure 4. Population Lycosa pseudonnata.

Based on simple linear regression analysis and the correlation between *Lycosa* population and humidity in Jarwo, it appears that the correlation is positif and not significant. (y = 61.723 + 8.747x). [16]In his research, he said that the microclimate that has a significant effect on arthropods is temperature and wind speed.[16] The results of simple linear regression can be seen in Figure 5.

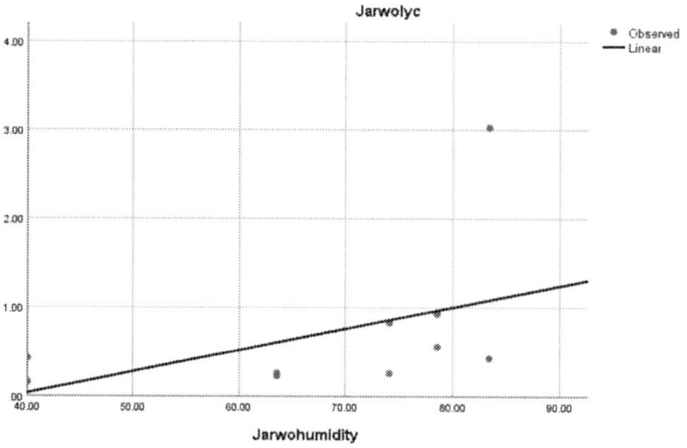

Figure 5. LYC and microclimate in JARWO.

Based on simple linear regression analysis and the correlation between *Lycosa* population and humidity in conventional it appears that the correlation is positive and not significant. (y = 62.797 + 8.575x). The results of simple linear regression can be seen in Figure 6.

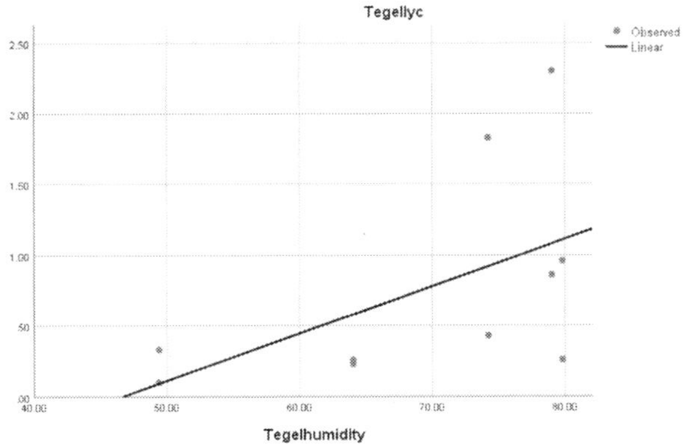

Figure 6. LYC and microclimate in conventional.

4 Conclucion

The conclusion of this research is that the influence of temperature, humidity and light intensity, has no significant effect on the brown planthopper population and the *Lycosa* population. However, the population of hunting predators such as *Lycosa* will increase if there are many food sources such as brown planthoppers.

Thank you much to Prof. Dr. Ir. Supriyadi M.S and Dr. Ir. Endang Setia Muliawati, M.Si. For her guidance in this research. And much thank you to the author's parents because with their prayers, the author was able to finish this paper.

References

1. N. Usyati, N. Kurniawati, A. Ruskandar, and O. Rumasa, "Populasi Hama dan Musuh Alami pada Tiga Cara Budidaya Padi Sawah di Sukamandi," *Agrikultura*, vol. 29, no. 1, p. 35, 2018, doi: 10.24198/agrikultura.v29i1.16924.
2. N. Kesmayanti, "Analisis peran azolla pinnata pada pertumbuhan vegetatif beberapa varietas padi (Oryza sativa L .)," *Pros. Semin. Nas. Penelit. Dan Pengabdi. 2021*, pp. 560–566, 2021.
3. dan I. R. Wahyuddin Abbas, Muh. Riadi, "Respon Tiga Varietas Padi (Oryza sativa L.) Pada Berbagai Sistem Tanam Legowo," *J. Penelit. dan Pengemb. Agrokompleks*, vol. 1, no. 2, pp. 45–55, 2015.
4. Baehaki S.E., Nugraha Budi Eka Irianto, and S. W. Widodo, "Rekayasa Ekologi dalam Perspektif Pengelolaan Tanaman Padi Terpadu Ecological Engeneering on Integrated Crop Management Perspective," *Iptek Tanam. Pangan*, vol. 11, no. 1, pp. 19–34, 2016.
5. P. Anand *et al.*, "Climate Change: Impact on Crop Pests," *Res. Today*, vol. 2, no. 5, pp. 327–329, 2014, [Online]. Available: https://www.weforum.
6. Sumarmiyati, H. Fitri, and Sundari, "Keragaman serangga pada pertanaman padi sawah di Kabupaten Kutai Kartanegara , Kalimantan Timur," *J. Pros Sem Nas Masy Biodlv Indor*, vol. 5, no. 2, pp. 217–221, 2019, doi: 10.13057/psnmbi/m050213.
7. P. Qian *et al.*, "Diversified Bund Vegetation Coupled with Flowering Plants Enhances Predator Population and Early-Season Pest Control," *Environ. Entomol.*, vol. 50, no. 4, pp. 842–851, 2021, doi: 10.1093/ee/nvab027.
8. N. Rosida and D. Arina Fahmi, "Uji Kemampuan Makan Beberapa Predator Terhadap Wereng Hijau (Nephotettix virescens Distant)," *Pros. Semin. Nas. InovasiTeknologi Pertan. Spesifik Lokasi*, no. November 2013, pp. 701–704, 2020, [Online]. Available: https://www.researchgate.net/publication/341930757
9. D. G. Bottrell and K. G. Schoenly, "Integrated pest management for resource-limited

farmers: Challenges for achieving ecological, social and economic sustainability," *J. Agric. Sci.*, vol. 156, no. 3, pp. 408–426, 2018, doi: 10.1017/S0021859618000473.
10. Y. Hadi, M. Soesilohadi, Wagiman, & Rahayuningsih, "Pertanian Organik Suatu Alternatif Pengelolaan Ekosistem Sawah Yang Sehat, Alami Dan Ramah Lingkungan," *Anat. Fisiol.*, vol. XXII, no. 1, pp. 72–77, 2014.
11. P. Jasrotia *et al.*, "Tillage intensity influences insect-pest and predator dynamics of wheat crop grown under different conservation agriculture practices in rice-wheat cropping system of Indo-gangetic plain," *Agronomy*, vol. 11, no. 6, 2021, doi: 10.3390/agronomy11061087.
12. S. L. Clement, "Pea Aphid Outbreaks and Virus Epidemics on Peas in the US Pacific Northwest: Histories, Mysteries, and Challenges," *Plant Heal. Prog.*, vol. 7, no. 1, 2006, doi: 10.1094/php-2006-1018-01-rv.
13. M. Yadav, R. Prasad, P. Kumari, M. Madhu, and A. Kumari, "Potential and Prospects of Natural Enemies in Rice Ecosystem in Jharkhand," no. 7, pp. 3389–3396, 2018, doi: 10.13140/RG.2.2.22058.57287.
14. S. C. A. umA deVi G. Anitha G, ChitrA ShAnker, ShAShibhuShAn V, "Diversityofspidersasinfluencedbycultivationtechniquesinrice." 2019. doi: DoI No.: 10.5958/0974-8172.2019.00173.1.
15. J. J. Lawalata, K. Anam, S. Santo, and T. Aquinas, "Median Volume 12 Nomor 1 Bulan Februari 2020 Pengamatan Jenis Predator Hama Tanaman Padi Di Kampung Karya Bumi Distrik Waibu Kabupaten Jayapura Median Volume 12 Nomor 1 Bulan Februari 2020," vol. 12, pp. 13–20, 2020.
16. dan H. T. Amania Amin*, Ibrohim, "Studi Keanekaragaman Arthropoda Pada Lahan Pertanian TUMPANGSARI UNTUK INVENTARISASI PREDATOR PENGENDALIAN HAYATI DI KECAMATAN BUMIAJI KOTA BATU," *Pertan. Trop. ISSN*, vol. 3, no. 2, pp. 139–149, 2016.

Applications of cow manure and bagasse biochar on the growth and flavonoids of white turmeric (*Curcuma zedoaria*)

Andriyana Setyawati[1*], Muji Rahayu[1], Endang Setia Muliawati[1], and Febriyanti[2]

[1]Department of Agrotechnology, Faculty of Agriculture, Universitas Sebelas Maret, Surakarta, Indonesia
[2]Undergraduate Department of Agrotechnology, Faculty of Agriculture, Universitas Sebelas Maret, Surakarta, Indonesia

Abstract. Cow manure and bagasse biochar are expected to increase yield of white turmeric (*Curcuma zedoaria*) and reduce the use of chemical that can trigger global warming. This study aimed to obtain the response of adding cow manure and bagasse biochar on the growth and flavonoids in white turmeric. This research was held in July 2022 – June 2023, located at faculty agriculture experimental field and CV Che-Mix Pratama. The method used completely randomized design with two factors and 3 replications. The first factor was cow manure dose 0, 100, 200, and 300 g perpolybag and the second was bagasse biochar 0, 80, 160, and 240 g perpolybag. Variables consisted number of leaves, number of tillers, diameter, stem dry weight, rhizomes wet weight, rhizomes dry weight, and flavonoids percentage. The data were analyzed using analysis of covariance with a 5% level test and continued with Duncan's Multiple Range Test at 5% level. Dose 300 g of cow manure was the best treatment to generate the best number of leaves, number of tillers, plant diameter, stem dry weight, rhizomes wet weight and dry weight. The highest flavonoids at 0.06% was found in 100 g of cow manure and 240 g of bagasse biochar.

1 Introduction

White turmeric (*Curcuma zedoaria*) is one of the spices or medicinal plants that is useful for warming the body, maintaining immunity, and as an alternative medicine for tumor diseases [1]. The white turmeric rhizome contains secondary metabolic compounds, one of which is flavonoids which are antioxidant compounds, acting as inhibitors of tyrosine kinase activity, an enzyme that plays an important role in the development of cancer cells [2]. The use of chemical fertilizers in white turmeric cultivation can trigger global warming because their manufacture emits carbon dioxide (CO_2) gas [3]. The potential for eutrophication increases as a result of excessive use of chemical fertilizers especially with high phosphate content. Reducing the use of excess chemicals can be achieved with cow manure and bagasse biochar as a supplier of soil organic matter especially for white turmeric cultivation. Cow manure is

* Corresponding author: andriyanasetyawati@staff.uns.ac.id

able to increase the activity of microorganisms in the soil, and improve soil structure by increasing the number and stability of soil aggregates so as to facilitate root development while supporting sustainable agriculture. Furthermore, biochar is organic waste that can improve soil fertility and is able to retain water and nutrients so that they are more available to plants. Fertilization with organic waste will be in line with aim to preserve the ecosystem [4]. This study aims to determine whether there is a real effect in the use of cow manure and bagasse biochar simultaneously on the growth and yield of white turmeric.

2 Methodology

The research was conducted from July 2022 to June 2023 at faculty agriculture experimental field and CV Che-Mix Pratama. The tools used include hygrometer, campbell stokes, destilator, oven, spectrophotometer, optilab, and plant photosynthesis meter. The materials used were white turmeric rhizomes, cow manure, bagasse biochar, water, polybags, NaOH, HCl, distilled water and aquabidest. The method used was a completely randomized design with two factors and 3 replications. The first factor was cow manure dose 0, 100, 200, and 300 g perpolybag and the second factor was bagasse biochar 0, 80, 160, and 240 g perpolybag. The variables consisted number of leaves, number of tillers, diameter, stem dry weight, rhizomes wet weight, rhizomes dry weight, and percentage of flavonoids. The data were analyzed using analysis of covariance with a 5% level test and if there was a significant difference it is continued with Duncan's Multiple Range Test at 5% level. The research method is experimental which includes preparation, seeding, planting, stress treatment, harvesting, and laboratory testing of flavonoid content.

3 Results and discussion

3.1 Growth variables

Plant growth variables affect assimilate production which in turn has implications for yield formation. Dose of cow manure had a significant effect on the number of leaves, diameter, and number of tillers (table 1).

Table 1. Plant growth with fertilizer and biochar treatments.

Treatment	Number of leaves (strand)	Diameter (mm)	Number of tillers (sproud)
Cow manure dose (g)			
0	6.00 a	12.44 a	1.23 a
100	7.11 ab	15.58 ab	1.65 a
200	8.19 bc	15.29 ab	1.86 ab
300	9.35 c	17.05 b	3.11 b
Biochar dose (g)			
0	7.81 a	15.00 a	1.70 a
100	7.55 a	15.21 a	1.73 a
200	7.64 a	15.02 a	1.86 a
300	7.82 a	15.51 a	1.89 a
Interaction	-	-	-

Notes: Numbers followed by the same letter in one column indicate not significantly different at DMRT 95% confidence level, (-): There is no interaction.

Based on the results of the analysis of covariance showed that the treatment of cow manure had a significant effect on the number of leaves. The number of leaves with the control treatment produced 6.00 leaves, significantly different from the cow manure 300 g perpolybag treatment of 9.35 leaves (an increase of 55.83% from the control treatment). [5], the Nitrogen (N) element in fertilizer as much as that absorbed by plants is used for the formation of protein as a constituent of plant organs, especially in plant tissues that are actively dividing (meristem) both in leaves, branches, roots, and stems.

The plant diameter of the control treatment produced an average of 12.44 mm, not significantly different from cow manure 100 g perpolybag which produced an average of 15.58 mm or cow manure 200 g perpolybag which produced an average of 15.29 mm. However, the control treatment is significantly different from the cow manure 300 g perpolybag treatment which produces an average of 17,05 mm (an increase of 37.06% from the control treatment). [6], pseudo-stems of plants from the Zingiberaceae family play a role in the formation of assimilates which in turn affect the yield in the form of rhizomes.

The number of tillers in the cow manure 300 g perpolybag treatment resulted in an average of 3.11 tillers (an increase of 152.84% from the control treatment), significantly different from the cow manure 100 g perpolybag treatment with an average of 1.65 tillers and the control treatment with an average of 1.23 tiller. The Nitrogen content in the fertilizer contributes to the process of tuber formation through tillers [7]. More N content in the composition of planting media supplied by fertilizers will stimulate the growth of tillers so that the harvest will be obtained with a greater number of rhizomes because the tillering factor affects the number of rhizomes [8].

3.2 Yield variables

Dose of cow manure had a significant effect on stem dry weight, rhizomes wet weight, and rhizomes dry weight (table 2). These results are then used for testing rhizomes for flavonoid content as a secondary metabolite compound.

Table 2. Crop yield with fertilizer and biochar treatments.

Treatment	Stem dry weight (g)	Rhizome wet weight (g)	Rhizome dry weight (g)
Fertilizer dose (g)			
0	30.01 a	111.63 a	33.42 a
100	54.11 b	193.21 b	55.34 a
200	55.02 b	194.04 b	77.69 ab
300	87.88 c	388.76 b	123.32 b
Biochar dose (g)			
0	53.08 a	210.02 a	77.08 a
100	54.13 a	211.91 a	81.66 a
200	55.61 a	229.54 a	73.35 a
300	56.23 a	228.12 a	72.16 a
Interaction	-	-	-

Notes: Numbers followed by the same letter in one column indicate not significantly different at DMRT 95% confidence level, (-): There is no interaction.

Based on the results of the analysis of covariance, it shows that the treatment of cow manure has a significant effect on the dry weight of the stalk. Cow manure 300 g perpolybag treatment with a weight of 87.88 g (increased 192.83% from the control treatment) was significantly different from the control with a weight of 30.01 g and cow manure 100 g perpolybag with a weight of 54.11 g. The weight of stalks as plant biomass is influenced by the manipulation of nutrient-rich planting media, one of which is supplied by fertilizers. Plant dry weight reflects the accumulation of organic compounds that plants successfully synthesize from organic and inorganic compounds, especially water and carbon dioxide. [9]

macronutrient content in fertilizers is an important component that affects soil quality to support plant growth.

Plant growth with cow manure 300 g perpolybag produces a wet weight of 388.76 g (an increase of 248.26% from the control treatment) was significantly different from cow manure 200 g perpolybag with a yield of 194.04 g, cow manure 100 g perpolybag with a yield of 193.21 g, and the control treatment with a wet weight of 111.63 g. The application of manure 300 g perpolybag in addition to providing better nutrient availability can also improve soil chemical, physical, and biological properties. [10], efforts to apply fertilizer to the soil can increase rhizome yield and growth parameters.

Cow manure 300 g perpolybag treatment that produced rhizomes dry weight of 123.32 g (an increase of 269% from the control treatment) was significantly different from cow manure 100 g perpolybag of 55.34 g and the control treatment with a weight of 33.42 g. [11] nutrients especially Nitrogen (N), Phosphate (P), and Potassium (K) in cow manure play a role in general plant growth and function in the production of assimilates. [12], high N and K elements in fertilizer can result in the dry weight of a plant and higher yields.

3.3 Flavonoid content of white turmeric rhizome

The test results showed different flavonoid content responses in each treatment. Nutrient and water stress is one of the causes of activating secondary metabolite compounds as a self-defense effort against unwanted conditions. The test results are shown in Figure 1.

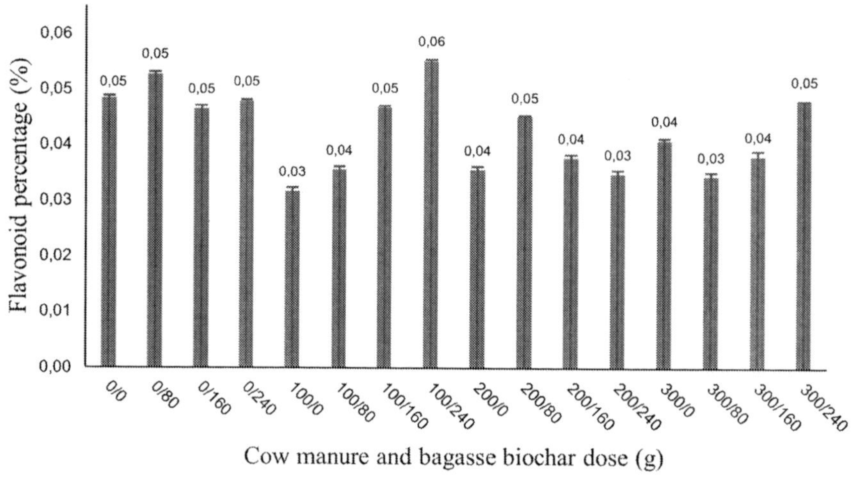

Fig.1. Flavonoid content of white turmeric rhizomes.

The highest result of flavonoid content was in the P1B3 (100 g perpolybag fertilizer and 240 g perpolybag biochar) at 0.06%. [13], in stressful conditions when the availability of water and nutrients supplied by cow manure is very little, it causes a plant response to unwanted conditions by activating secondary metabolite compounds as a self-defense effort. [14], when plants interact with pathogens, pests or biotic and abiotic stresses, plants will activate various defense mechanisms, including the induction of secondary metabolite biosynthesis.

4 Conclusion

The use of cow manure and bagasse biochar together did not increase the growth and yield variables of white turmeric. The dose of cow manure 300 g perpolybag can increase the number of leaves, diameter, number of tillers, stem dry weight, rhizomes wet weight, and rhizomes dry weight of white turmeric. The highest flavonoid content was in P1B3 (100 g perpolybag cow manure and 240 g perpolybag bagasse biochar).

This research was funded by Universitas Sebelas Maret through the Fundamental Scheme Research Program 2023.

References

1. D. SUBOSITI and S. WAHYONO, Biodiversitas J. Biol. Divers. **20**, (2019)
2. T. K. Lee, D. Lee, S. R. Lee, Y.-J. Ko, K. Sung Kang, S. J. Chung, and K. H. Kim, Bioorg. Chem. **87**, 117 (2019)
3. D. Kumar, M. Singh, S. Kumar, R. K. Meena, R. Kumar, M. R. Yadav, M. Kushwaha, G. Makarana, S. Bhattacharjee, S. Kashyap, B. Biswal, Rajesh, and P. Kumar, Agronomy **13**, 981 (2023)
4. T. Rodríguez-Espinosa, I. Papamichael, I. Voukkali, A. P. Gimeno, M. B. A. Candel, J. Navarro-Pedreño, A. A. Zorpas, and I. G. Lucas, Sci. Total Environ. **876**, 162666 (2023)
5. S. Jena, A. Ray, A. Sahoo, S. Sahoo, B. Dash, B. Kar, and S. Nayak, 3 Biotech **10**, 17 (2020)
6. U. Zeeshan, M. Q. Barkat, and H. Khalid, Matrix Sci. Pharma 2, 15 (2018)
7. A. Herawati, J. Syamsiyah, M. Mujiyo, M. Rochmadtulloh, A. A. Susila, and M. R. Romadhon, SAINS TANAH - J. Soil Sci. Agroclimatol. **18**, 73 (2021)
8. C. Pornaro, S. Macolino, and M. D. Richardson, Acta Agric. Scand. Sect. B — Soil Plant Sci. 69, 657 (2019)
9. R. Yuniarti, S. Nadia, A. Alamanda, M. Zubir, R. A. Syahputra, and M. Nizam, J. Phys. Conf. Ser. **1462**, 012026 (2020)
10. M. J. Guji, H. T. Yetayew, and E. D. Kidanu, Agric. Food Secur. **8**, 5 (2019)
11. A. Bashir, M. Rizwan, M. Zia ur Rehman, M. Zubair, M. Riaz, M. F. Qayyum, H. F. Alharby, A. A. Bamagoos, and S. Ali, Chemosphere **246**, 125809 (2020)
12. A. M. Peña-Gutiérrez, J. Pérez-Flores, N. del Rivero-Bautista, and A. O. L. Santos, J. Exp. Agric. Int. **30**, 1 (2019)
13. H. M. I. C. Herath, T. D. C. M. K. Wijayasiriwardene, and G. A. S. Premakumara, Sri Lankan J. Biol. **3**, 24 (2018)
14. Y. M. Muflihah, G. Gollavelli, and Y.-C. Ling, Antioxidants **10**, 1530 (2021)

Soil properties of agricultural sloping land in Banjarnegara Regency, Indonesia

Rahmah Dewi Yustika[1*], Maswar[1], *Ai* Dariah[1], *Neneng Laela* Nurida[1], *Jelly Amalia* Santri[2], *Ladiyani Retno* Widowati[2], and *Wiwik* Hartatik[1]

[1]BRIN, Research Organization for Agriculture and Food, Jl. Raya Jakarta-Bogor, Cibinong 16915, Indonesia
[2]BPSI Tanah dan Pupuk, Jl. Tentara Pelajar 12, Bogor 16124, Indonesia

Abstract. Intensive farming in highland sloping land can lead to more frequent soil erosion, thus exacerbating soil deterioration. Good agricultural practices on sloping land can prevent soil degradation. This study aimed to identify soil properties within transects consisting of the shoulder slope (forest) and upper and lower slopes (agricultural field) in steep slope ranges between 25 – 40%. The study area was located in Sumberejo village, Banjarnegara Regency, Indonesia, with young coffee tree intercropping carrot land use previously planted with horticulture crops. The result showed that the concentration of potential P and K, pH, N, exchangeable cation (Ca, Mg, K), and CEC within the transect in sloping land resulted in higher concentration in the upper and lower slope than shoulder slope because agriculture area got influence from fertilizer application. Higher soil organic carbon content and dehydrogenase enzyme also lower bulk density value in shoulder slope because the forest had undisturbed condition. This study informs that agricultural activities influence soil characteristics. Implementing proper land management on sloping lands can reduce soil degradation and mitigate the effects of climate change.

1 Introduction

The hilly and mountainous area has steep slopes vulnerable to soil erosion, threatening environmental sustainability. It is neglectable that hilly and mountainous areas are often cultivated as farming systems of horticulture crops [1]. The pressure of farming systems to support higher productive systems and improve farmers' income caused the spreading of cultivation up to the top of mountains.

Banjarnegara regency in Central Java of Indonesia has become the center of horticulture crops as its fertility can support farming systems. Population growth and limited available land implicate encroachments of high slopes for agricultural activities supporting people's livelihood. The most important crop in Banajarnegara Regency is potato, which has the

* Corresponding author: rahm063@brin.go.id

highest production in Central Java Province in 2020 [2] and supports Central Java as the number 3 biggest potato production in Indonesia in 2021 [3].

The intensive farming system of horticulture crops on steep slopes of highlands will lead to soil degradation that can lead to physical degradation (sealing, compaction, and erosion), chemical degradation (loss of organic matter, contamination), and biological degradation (decrease in biodiversity) [4]. The highland with high precipitation is suitable for vegetable farming activities and is vulnerable to soil erosion due to the slope of > 25% [5]. Further, the steep slope can trigger landslides [6]. A study by Haryati [7] mentioned that soil characteristics in upland vegetable farming areas had high soil organic carbon (SOC) content, low bulk density (BD), high soil permeability, and high porosity. Climate change will worsen farming systems on steep slopes because the high precipitation intensity will result in soil erosion and surface runoff [8]. Besides, climate change caused high runoff and N loss [9].

The government should start educating farmers to change planting systems from horticulture to perennial trees in steeply sloping areas to improve environmental sustainability and soil health. This study aimed to know soil properties within a transect from shoulder to lower slopes. The information from this study can describe soil properties in agricultural land previously used for horticulture crops and with the recent condition of young coffee plantation intercropping with carrots.

2 Methods

2.1 Study area

The study area was located in Sumberejo Village, Batur District, Banjarnegara Regency (Figure 1). Soil sampling was conducted in December 2021 along a transect of shoulder, upper and lower slopes. The land use on the shoulder was forest (-7°11'50", 109°49'58"), while the land use on the upper (7°11'53", 109°49'57") and lower slopes (7°11'57", 109°49'57") was agricultural fields. Soil samples were collected from three points at each location. The chemical and biological properties of the soil were identified through composite sampling at three points. The soil type in the study area was Typic Hapludand with parent material consisting of andesite and basalt.

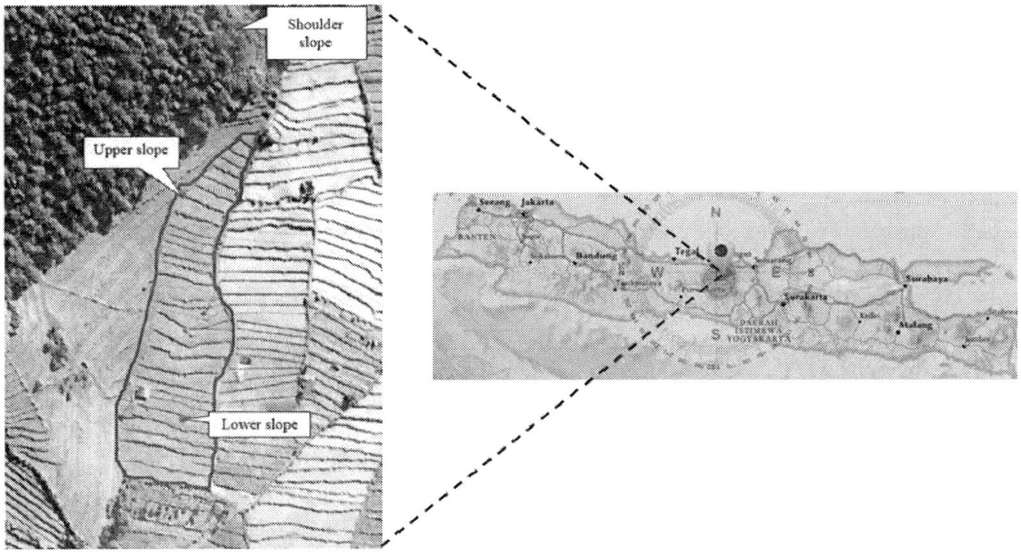

Figure 1. Study area.

The agricultural field area was 0.74 ha recently in 2021 and was planted with coffee intercropping with carrots, but previously with horticultural crops, especially potatoes. The fertilizer dose for the coffee plantation was 100 g Phonska (N 15%, P_2O_5 10%, K 12%, and S 10%)/tree, while the carrot was 300 kg Phonska/ha/planting season. When horticultural crops were still grown in the agricultural area, the fertilization rate of organic fertilizer for potato cultivation was 600 kg Phonska/ha/planting season, in addition to applying manure and dolomite. In October 2021, replacing horticultural crops with coffee began. The length from the forest to the lower boundary of the agricultural field is 252 m. The shoulder slope (forest) where the sample was collected was >40%. In addition, the slope in upper and lower slopes (agricultural area) ranged from 25 to 40%. Terrace systems were applied to minimize soil erosion in agricultural areas. The altitudes on the shoulder, upper, and lower slopes were 1803, 1773, and 1716 m above sea level.

2.2 Soil analysis

The parameters of soil physical properties collected were bulk density (core method) in 0 – 15 cm and 15 – 30 cm depths. These parameters were analyzed based on the West European Method [10]. Soil chemical properties were collected in 0 – 15 cm depth with composite. The soil chemical characteristics consist of pH (ratio 1:5 of soil: water suspension), organic C (Walkey and Black methods), total N (Kjeldal methods), exchangeable cation (Ca, Mg, K), cation exchange capacity (CEC), potential P and K, and texture were measured according to Soil Survey Staff [11]. Collecting soil biology properties consisted of analyses of dehydrogenase enzyme [12].

3 Results and discussion

3.1 Soil chemical properties

Soil pH (Table 1) indicated increasing trends in both depths from the shoulder slope (5.4 – 5.5) to the upper (6.6) and lower slopes (6.8). Shoulder slope with land-use forest has soil in its original condition, but agriculture fields usually use input such as dolomite; therefore, pH in the agriculture field is higher than in the forest. The lower slope possibly got accumulated dolomite from the upper slope. Therefore, the pH on the lower slope was higher than on the upper slope. The study in potato farming with legume intercrops revealed that the pH of the eroded soil is higher than that of the original soil [13].

Soil texture in the shoulder and upper slopes in both depths were silt loam and loam, respectively. In lower slopes, the soil texture in the first depth was silt loam, and in the second depth was loam. The SOC had a decreasing trend from shoulder to lower slopes. SOC reduced after forest conversion. The study in recent clearing forest presented high SOC in A horizon >5% [14]. SOC of Andisol in Indonesia varied from 1.24% – 22.46% [15]. N concentration ranged from 0.33 – 0.36% in the study area. N concentration was higher in the upper and lower slopes compared to the shoulder slope because of the application of urea fertilizer in the agricultural field.

Table 1. Soil chemical properties.

Position	Depth (cm)	Texture			pH	C%	N%	Potential P (mg/100 g)	Potential K (mg/100 g)	Ca	Mg	K	CEC
		Sand	Silt	Clay						cmol$_c$/kg			
Shoulder slope	0 – 15	25	57	18	5.4	5.57	0.33	250	119	8.91	1.97	2.15	27.55
	15 – 30	18	66	16	5.5	5.93	0.34	259	111	8.63	1.93	1.76	27.45
Upper slope	0 – 15	37	44	19	6.6	5.45	0.36	1,375	153	24.98	2.04	2.66	32.01
	15 – 30	40	46	14	6.6	5.17	0.35	1,704	159	20.60	1.97	2.79	26.54
Lower slope	0 – 15	34	52	14	6.8	5.32	0.36	2,025	137	25.91	2.12	2.44	31.45
	15 – 30	35	49	16	6.8	5.27	0.35	2,052	134	26.46	2.39	2.41	31.99

Potential P and K showed higher values in upper and lower slopes than shoulder slopes (Table 1) due to fertilizer application affecting nutrient availability in soil. The agricultural fields that get inorganic fertilizers for farming activities will have higher nutrients compared to the natural conditions of the forests.

Exchangeable cations Ca, Mg, K, and CEC showed increasing trends from shoulder to lower slopes (Table 1). The value of exchangeable Ca in the first depth from shoulder, upper, and lower slopes in the first depth were 8.91, 24.98, and 25.91 $cmol_c/kg$, respectively. Trends of exchangeable Mg in the first depth were 1.97 $cmol_c/kg$ in the shoulder slope, 2.04 $cmol_c/kg$ in the upper slope, and 2.12 $cmol_c/kg$ in the lower slope. The concentration of K in the first depth from shoulder, upper, and lower slopes in the first depth were 2.15, 2.66, and 2.44 $cmol_c/kg$, respectively. The CEC values at first layer depth in shoulder, upper, and lower slopes were 27.55, 32.01, and 31.35 $cmol_c/kg$, respectively. Agricultural fields in upper and lower slopes had higher concentrations than forests (shoulder slope) because higher fertilizer input is usually added in farming practices. The deposition of nutrients exchangeable Ca and Mg had the highest concentration in the lower slopes because runoff brought nutrients up to down; therefore, the concentration of potential exchangeable Ca and Mg increases along lower positions. Soil erosion transport nutrient in topsoil, thus, could trigger nutrient loss [16].

3.2 Soil physical properties

BD in the surface layer (0 – 15 cm) showed the lowest value in shoulder slope 0.35±0.05 g/cm^3, and as gradually down, BD values increased in upper and lower slopes that were 0.63±0.03 and 0.69±0.04 g/cm^3, respectively (Table 2). Lower BD in forest areas indicates better physical soil conditions because forest condition has minimal disturbance. Besides, the trend showed that intensively cultivated land causes the soil to become denser and have a higher BD. A study in volcanic highland Toba in Indonesia shows ranges of 0.31 – 0.86 g/cm^3 in horizon A of soil with andic properties [14]. Based on Keys to Soil Taxonomy [17], andic properties had BD 0.9 g/cm^3 or less if measured at a pressure of 33 kPa water retention.

Table 2. Soil BD.

Position	Depth (cm)	BD (g/cm^3)
Shoulder slope	0 – 15	0.35±0.05
	15 – 30	0.42±0.16
Upper slope	0 – 15	0.63±0.03
	15 – 30	0.57±0.10
Lower slope	0 – 15	0.69±0.04
	15 – 30	0.64±0.07

3.3 Soil biology properties

Identification of soil biology properties was conducted in 0 – 15 cm depth. The dehydrogenase enzyme value in the shoulder slope was the highest of the 38.04 μg TPF/g sample.hour (Table 3). Because of the abundance of organic matter and root exudates, the high dehydrogenase activity in forest soils can stimulate microbial activity and the production of dehydrogenase enzymes. This enzyme activity is crucial in maintaining soil health and ecosystem functioning. Dehydrogenase in forests is higher than in degraded soil [18].

Table 3. Soil biology properties.

Position	Depth (cm)	Dehydrogenase (μg TPF/g sample.hour)
Shoulder	0 – 15	38.04
Upper slope	0 – 15	35.29
Lower slope	0 – 15	36.55

4 Conclusion

Soil properties of SOC, BD, and dehydrogenase had better conditions in the shoulder slope than on the upper and lower slopes because the forest had undisturbed conditions. Soil chemical characteristics of potential P and K, N, exchangeable cations, and CEC had larger concentrations in upper and lower slopes compared to shoulder slopes because of enrichment from fertilizer application in the agriculture area. Application of input for farming activities, dolomite increases the pH value. Land management strategy in sloping land could minimize runoff that could provoke soil erosion, transporting nutrients in the soil. Land management to mitigate land degradation in sloping areas is essential for improving soil health, increasing carbon stocks, and resilience to climate change through the application of soil conservation techniques (bench terraces) and planting perennial crops such as coffee that are part of the agroforestry system.

This study was funded by the RPIK Project 2021 by the Ministry of Agriculture. We thank farmers and village heads Bapak Ibrahim and Bapak Mahyono for their valuable support on fieldwork.

References

1. P. Grogan, F. Lalnunmawia, and S. K. Tripathi, Agrofor. Syst., **84**, 2 (2012)
2. Statistics of Jawa Tengah Province, *Harvested Area and Production of Potato 2018-2020* (2020)
3. Statistics Indonesia, *Production of Vegetables 2021*, (2021)
4. C. S. S. Ferreira, S. Seifollahi-Aghmiuni, G. Destouni, N. Ghajarnia, and Z. Kalantari, Sci. Total Environ., **805** (2022)
5. U. Haryati, J. Sumberd. Lahan, (2016)
6. B. Harjadi, S. Abdiyani, I. G. S. Sukartono, E. Hesthiati, P. H. Zaki, and M. H. Ismail, Sains Tanah, **19**, 2 (2022)
7. U. Haryati, J. Sumberd. Lahan, **8**, 2 (2014)
8. J. R. Rodysill, J. M. Russell, M. Vuille, S. Dee, B. Lunghino, and S. Bijaksana, Quat. Sci. Rev., **225**, (2019)
9. L. Øygarden *et al.*, Agric. Ecosyst. Environ., **198**, (2014)
10. The West European Working Group on Soil Structure, *West European Method for Soil Structure Determination* (Ghent: The State Faculty of Agricultural Sciences, 1967)
11. Soil Survey Staff., *Kellogg Soil Survey Laboratory Methods Manual. Soil Survey Investigations Report No. 42, Version 5.0.* (Department of Agriculture, Natural Resources Conservation Service., 2014)
12. Soil Science Society of America, *Methods of Soil Analysis, Part 2: Microbiological and Biochemical Properties* (Madison, WI, 1994)
13. S. O. Nyawade, C. K. K. Gachene, N. N. Karanja, H. I. Gitari, E. Schulte-Geldermann, and M. L. Parker, *Geoderma Reg.*, **17**, (2019)
14. B. H. Prasetyo, N. Suharta, and D. E. Yatno, J. Tanah Dan Iklim, **29**, (2009)
15. Sukarman and A. Dariah, *Tanah Andosol di Indonesia*. Bogor, Indonesia: Balai Besar Penelitian dan Pengembangan Sumberdaya Lahan Pertanian, (2014)
16. Z.-A. Su, J.-H. Zhang, and X.-J. Nie, Pedosphere, **20**, 6 (2010)

17. Soil Survey Staff, *Keys to Soil Taxonomy, 13th edition*. USDA Natural Resources Conservation Service, (2022)
18. S. Kumar, S. Chaudhuri, and S. K. Maiti, Middle East J. Sci. Res., **13**, 7 (2013)

The impact of climate change in the spread of *Pythium myriotylum* in ginger: a critical review

Rani Yosilia[1*], *Auliana* Afandi[2], *Lindiana*[3], *Sischa* Febriani Yamesa Away[4], *Marlina* Kamelia[1], *Annisyia* Zarina Putri[5], and *Romy* Dwipa Yamesa Away[6]

[1]Biology Department, Raden Intan State Islamic University, Jln. H. Endro Suratmin, Sukarame, Bandar Lampung, Lampung, Indonesia
[2]Research Center for Horticultural and Estate Crops, National Research and Innovation Agency, Jln. Raya Jakarta-Bogor, Cibinong, Bogor, Indonesia
[3]Indonesian Institute for Testing Instrument Standard of Spices, Medicinal and Aromatic Plants, Ministry of Agriculture, Jln. Tentara Pelajar No. 3, Bogor, West Java, Indonesia
[4]Agribusiness Department, Pasir Pengaraian University, Jl.Tuanku Tambusai, Kumu Desa Rambah, Rokan Hulu, Riau, Indonesia
[5]Sustainable Agriculture, Food, and Energy (SAFE) Innovation Center, Indonesia
[6]Chemistry Department, State University of Padang, Jl Hamka no 16 Air Tawar Barat, Padang, West Sumatra, Indonesia

Abstract. Climate change is a growing concern due to its potential impact on biotic stressors, particularly fungal infections, such as *Pythium myriotylum*, which causes significant damage to ginger plants. Among the several yield-reducing fungal diseases targeting ginger, *Pythium myriotylum* is known to cause significant damage through rhizome rot and damping off. Scientific studies have explored the survival and spread of *P. myriotylum* in gingers and other crops that are thought to be caused by climate change. Oospores, the primary inoculum source, are crucial for the *P. myriotylum*'s life cycle and disease progression. To mitigate the effects of *P. myriotylum* in ginger, adaptation techniques such as crop rotation, pesticide use, and breeding of resistant varieties can be developed. However, understanding the current spread of *P. myriotylum* and its limitations is crucial for future research. This review discusses the current status of *P. myriotylum*'s importance in ginger, the effects of climatic variables on its life cycle, survival, and spread, adaptation strategies, and the effects of climate change on it. It also identifies significant knowledge gaps for future research.

1 Introduction

Pythium myriotylum, a fungal pathogen, poses a recurring threat to the ginger industry over time. This pathogen is known to cause ginger soft rot, a destructive disease that leads to significant yield losses. The occurrence of *Pythium myriotylum* in ginger cultivation has been observed to vary over time. In a study conducted in Australia in 2009 [1], *Pythium myriotylum* was identified as the causal agent responsible for a severe outbreak of soft rot in ginger crops. Then *P. myriotylum* appeared in China in recent research as a major cause of

* Corresponding author: raniyosilia@radenintan.ac.id

Pythium soft rot of ginger [2]. The presence of *Pythium myriotylum* in ginger cultivation is influenced by various factors such as temperature, soil moisture, and agricultural practices. Temperature plays a crucial role in the prevalence of *Pythium myriotylum* in ginger crops. It has been observed that *Pythium myriotylum* tends to be more prevalent in warmer climates, with temperatures between 26 to 30°C providing ideal conditions for its growth [3]. Additionally, humidity plays a role in the spread and severity of *P. myriotylum* in ginger crops.

High humidity in the form of continuous rainfall or saturated soil provides a conducive environment for the spores of *P. myriotylum* to move through flowing water and infect the ginger plants. Once the ginger fields become infested with the spores of *P. myriotylum*, the pathogen can persist in the soil and pose a long-term threat to ginger cultivation. Moreover, the persistence of infected rhizomes can also contribute to the reemergence of the pathogen in subsequent plantings of ginger. *P. myriotylum* has been found to infect various parts of the ginger plant, including the roots, collar, and succulent parts of the rhizome [4]. The spread of *P. myryotylum* can be serious thread for ginger plantation as it can result in significant yield losses, ranging from 5 to 30%, and in some cases, complete crop failure [5]. Previous study provide data about biology and epidemiology of *P. myryotylum* while this study discusses the relationship between the biology and epidemiology of *P. myryotylum* and climate change.

2 Climate change and *Pythium myriothilum*

Climate change has the potential to impact the proliferation of the pathogen *Pythium myriothilum*. Infectious diseases, including those caused by pathogens like *Pythium*, are known to be sensitive to temperature, rainfall, and humidity, all of which can be influenced by climate change [6]. Climate warming can increase the development and survival rates of pathogens, as well as disease transmission and host susceptibility [6]. Studies have shown that rising mean water temperatures associated with global change phenomena can promote *Pythium* infection of plants like *Phragmites australis* during flooding events [7]. Similarly, future climate changes, including extended hot summers, may increase the risk of serious disease problems caused by high-temperature species of *Pythium* [8].

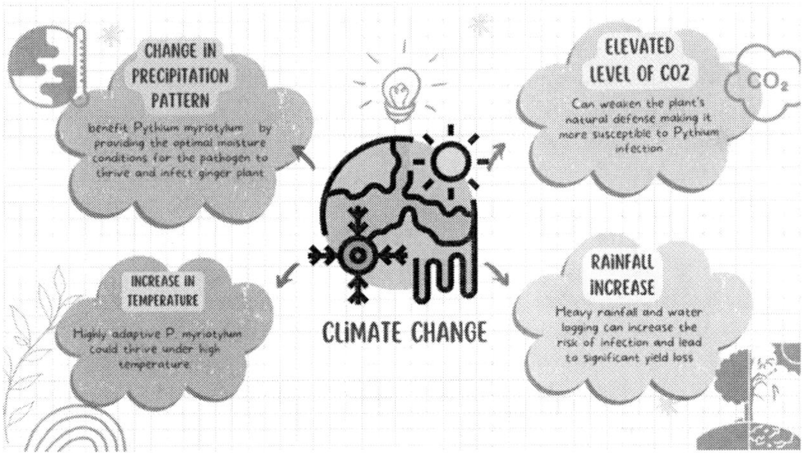

Figure 1. The impact of climate change in *P. myriotylum*

Pythium disease is more severe in locations with excessive rainfall or thick clay soils with poor drainage [9]. *Pythium* oospores found in soil or infected seed rhizomes would be the principal source of inoculums [10]. If the infection is seed-borne, the rhizome rot infection appears early. Zoospores produced by sporangia are chemotactically attracted to host roots, where they encyst and create germ tubes that infect the host roots. Lesions occur approximately 72 hours around the entry spot under optimum circumstances. The zoospores generated by the sporangia on the host may be transferred passively or actively to healthy plants, where they may induce fresh infection, resulting in disease transmission [11].

Additionally, increased humidity prolongs the survival of *Pythium myriotylum* spores in the soil. In relation to the control of plant diseases caused by fungi on ginger with seed treatment, the impact of climate change cannot be ignored. As the prevalence of high-temperature species of *Pythium* increases due to climate change, the control of fungal diseases on ginger becomes even more critical [12]. Therefore, implementing effective seed treatment measures becomes imperative to combat the spread and severity of fungal diseases on ginger. Given the impacts of climate change on the frequency of suitable infection conditions and the proliferation of new pathogens, it is crucial to prioritize effective seed treatment measures for the control of plant diseases caused by fungi on ginger [13].

Changes in precipitation patterns can benefit *Pythium myriotylum* growth in greenhouse float systems by providing the optimal moisture conditions for the pathogen to thrive and infect ginger plant [14]. This is particularly concerning as heavy rainfall and water logging can increase the risk of infection and lead to significant yield losses in ginger crops. On the other hand, rising temperatures associated with climate change can have a negative effect on the survival and growth of beneficial microorganisms in the soil that help control the population of *Pythium myriotylum* [15]. This reduction in beneficial microorganisms can further contribute to the spread of *Pythium myriotylum* and its impact on ginger crops. Furthermore, studies have shown that the spread of *Pythium myriotylum* in ginger is strongly influenced by soil physic-chemical properties [16]. Specifically, changes in soil pH and organic matter content due to climate change can create a more favorable environment for the growth and survival of *Pythium myriotylum* [17].

Elevated levels of carbon dioxide can weaken the plant's natural defenses, making it more susceptible to infection by *P. myriotylum* and leading to more severe disease outbreaks in ginger crops. When there is an increase in atmospheric carbon dioxide concentration, it disrupts plant physiology and biochemical aspect that compromises their ability to resist attacks from patogen [18] making them more susceptible to *P. myriotylum* infection.

3 Epidemiological implications

Pythium group F, which is likely identical with *P. dissotocum*, has been found in tomato soilless culture [19]. This pathogen was discovered to be widespread in tomato soilless culture, and another *Pythium* species, such as *P. ultimum*, *P. sylvaticum*, and *P. intermedium*, have also been found in similar environments [19]. This raises concerns about the epidemiological implications of *Pythium myriothilum* in agricultural settings, particularly in soilless growing methods like hydroponics. The prevalence and persistence of *P. myriothilum* in agricultural soils, particularly in soilless growing methods like hydroponics, have significant implications for crop health and productivity.

Pythium root rot, produced by a variety of *Pythium* species, is a major issue with hydroponically grown crops. Cucumber, tomato, sweet pepper, spinach, lettuce, nasturtium, arugula, rose, and chrysanthemum are all susceptible to this disease [20]. *Pythium aphanidermatum*, *Pythium dissotocum*, *Pythium* group F members, and *Pythium ultimum* var are the primary causes of Pythium root rot in hydroponics [20]. The disease can lead to reduced crop yields and economic losses in commercial greenhouses.

The cultivation of ginger involves both rainfed and irrigated methods. According to Stirling *et al.* [1], ginger thrives in a humid climate with an air temperature ranging between 28-30°C, which also creates favorable conditions for the development of rhizome rot disease. The largest occurrence of plant infection was observed when the relative humidity reached 90% and the air temperature ranged from 27-28°C. Zoospore formation is facilitated by the presence of abundant soil water and reduced soil temperature, as these conditions promote the dispersal of zoospores across long distances through the utilization of soil water [27].

Additionally, the combination of high soil temperature and high rainfall has been found to expedite the occurrence of rhizome rot in ginger. The presence of a warm and humid atmosphere increases the susceptibility of the plant to infection during the sprouting stage due to the delicate and moisture-rich nature of its tissues. The dissemination of soil-borne diseases can be attributed to the occurrence of moderately intense and evenly distributed precipitation throughout the agricultural cultivation period, which spans from June to October [21].

In terms of host specificity, research has demonstrated that various *Pythium* isolates can differ in terms of pathogenicity and mortality in prospective host species [22]. While *Pythium* species are not confined to a single host species, they do demonstrate an intermediate level of specialization, validating the Janzen-Connell assumption of host specificity [22]. The presence of vulnerable plant tissue frequently constrains the spread of plant disease epidemics. By reducing the number of seeds planted or incorporating resistant plants in between, the extent of disease can be minimized through the limitation of plant-to-plant transmission.

Historically, *Pythium myriotylum* was primarily documented in tropical and subtropical regions of the world [23]. However, recent reports and studies have indicated that the geographic range of this pathogen is expanding. This expansion is characterized by its detection in regions previously considered unsuitable for its survival and proliferation. Recent report showed that *P. myriotylum* were detected in Shandong, Henan, and Anhu Province of China [2] which has average temperature in summer at 27°C. Furthermore, the occurrence of *P. myriothilum* in crops such as hemp [32, 33] and green bean [34] highlights its ability to infect a diverse range of plants and potentially cause widespread damage to agricultural systems. The accurate identification and understanding of *Pythium* species in agricultural environments are crucial for assessing their epidemiological implications.

4 Adaptation and management strategies

Ginger plants have evolved defense mechanisms against the invasion of *Pythium myriotylum*. One key defense mechanism is the formation of jasmonate, a signaling chemical involved in plant defense against pathogens [25]. Jasmonate is required for plant defense against *Pythium* species, according to research done on Arabidopsis, a model plant species [25]. Jasmonic acid (JA) and ethylene, which have emerged as important regulators for beneficial soil-borne microorganisms, can induce an enhanced defensive capacity in above-ground plant parts, providing protection against a wide range of microbial pathogens and even insect herbivores [26].

Plants have been demonstrated to be protected from *Pythium* infections when methyl jasmonate, a synthetic version of jasmonate, is applied exogenously. In a research done by [25], methyl jasmonate considerably reduced disease incidence in mutant plants, putting it close to the level reported in wild-type controls. Jasmonate's protective impact was mediated through the development of plant defense mechanisms rather than by direct antifungal action, and jasmonate-responsive defense genes were activated in wild-type plants but not in jasmonate-deficient mutants upon *Pythium* challenge [25].

One critical factor in mitigating the danger of infection caused by *Pythium* spp. is the selection of seedlings that are in good health and free from any signs of disease [27]. Various techniques for seed management, including germ enrichment, seed disinfestation, and germ sterilizing, are employed to ensure the production of high-quality seedlings. All of the aforementioned techniques demonstrate potential in the therapeutic management of infections while ensuring the safety of embryos and preserving the viability of seeds [28].

5 Early detection and monitoring

Development of rapid detection system for *P. myriotylum* are a major step for understanding and preventing disease of ginger caused by this pathogen. Some conventional method for detecting *P. myriotylum* are based on baiting and isolation of *P. myriotylum* on selective media [8]. The conventional method is inexpensive but its time consuming. Moreover, it is difficult to recover the *P. myriotylum* using conventional methods before planting season due to the dormant oospore.

Fukuta *et al.* [29] developed a loop-mediated isothermal amplification (LAMP) technique for *P. myriotylum* identification. The LAMP test is a quick and easy approach for detecting pathogens in field samples in about 30 minutes. The primer set derived from *P. myriotylum*'s internal transcribed spacer (ITS) region shown good sensitivity and specificity in identifying the pathogen. The LAMP technique was as sensitive as PCR, with a detection limit of 100 fg of genomic DNA. Furthermore, the LAMP approach detected *P. myriotylum* in hydroponic solution samples, with findings that matched those of the traditional plating method.

Another research about the rapid detection of *P. myriotylum* based on PCR reaction were initially worked by Wang et al [30]. Their study was developed specific PCR primer from ITS1 variable region of *P. myriotylum* that capable to differentiate *P. myriotylum* and other cause of ginger soft rot. Presently the loop-mediated isothermal amplification assay for the detection of *P. myriotylum* were developed [29]. The primer set designed from ITS region of *P. myriotylum* can detect the pathogen in field sample with a fast and convenient method. Analysis of the annealing curve of the LAMP reaction products increases the reliability of the LAMP diagnosis. This study shows that the diagnostic method using the LAMP assay is useful for monitoring *P. myriotylum* in the field.

Early detection of *Pythium* keratitis, a condition caused by *P. myriotylum* in the eye, is critical for rapid treatment. Vishwakarma *et al.* [31] presented Pythium keratitis' clinical profile, laboratory diagnosis, therapy, and histology. For a better prognosis, they underlined the need of early identification and treatment with antibacterial medications such as linezolid and azithromycin. The pathogenesis and epidemiology of *Pythium* root rot in hydroponic crops were explored by Sutton *et al.* [20]. They emphasized the use of visual inspection, antibody-based dipsticks, immunoassay-based diagnostic kits, and DNA microarrays in tracking *Pythium* infections in hydroponic systems.

6 Conclusion

In conclusion, it is imperative to recognize the significant impact of climate change on *Pythium myriotylum's* geographical distribution and the heightened severity of associated diseases. The climate change can change the precipitation pattern, increase the temperature and rainfall as well wich benefit the *P. myriotylum* . The elevated level of CO_2 due to human activity could suppress the ginger defensive mechanism which make it more susceptible for *P. myriotylum* infection. A comprehensive understanding of the interplay between climatic factors and pathogen dynamics is essential for devising efficient strategies for disease control. Subsequent research endeavors should prioritize the implementation of monitoring systems, mathematical modeling techniques, and predictive tools to effectively anticipate and mitigate against the ever-evolving threat posed by *P. myriotylum* in response to a changing climate.

References

1. G. R. Stirling, U. Turaganivalu, A. M. Stirling, M. F. Lomavatu, and M. Smith, Australasian Plant Pathology **38**, 453 (2009)
2. P. Daly, Y. Chen, Q. Zhang, H. Zhu, J. Li, J. Zhang, S. Deng, L. Wang, D. Zhou, Z. Tang, and L. Wei, Plant Disease **106**, 510 (2022)

3. Y. Liu, L. Wu, X. Wu, H. Li, Q. Liao, X. Zhang, Z. Sun, and W. Li, Scientifica **2017**, 1 (2017)
4. P. Thomas and A. C. Sekhar, Frontiers in Microbiology **7**, (2016)
5. G. Gamal, M. R. Samak, and M. A. Shahba, Atmosphere **12**, 1589 (2021)
6. C. D. Harvell, C. E. Mitchell, J. R. Ward, S. Altizer, A. Dobson, R. S. Ostfeld, and M. D. Samuel, Science **296**, 2158 (2002)
7. J. Nechwatal, A. M. Wielgoss, and K. Mendgen, Hydrobiologia **613**, 109 (2008)
8. H. Watanabe, K. Kageyama, Y. Taguchi, H. Horinouchi, and M. Hyakumachi, Journal of General Plant Pathology **74**, 417 (2008)
9. M. Gupta and J. Tennyson, in *IntechOpen EBooks* (2020)
10. T. J. Burr and M. E. Stanghellini, Phytopathology **63**, 1499 (1973)
11. C. Walker and P. Van West, Fungal Biology Reviews **21**, 10 (2007)
12. C. Benavent-Celma, A. Puértolas, D. McLaggan, P. Van West, and S. Woodward, Journal of Fungi **7**, 479 (2021)
13. S. Sharma, N. P. Dohroo, V. Shanmugam, S. Phurailatpam, N. Thakur, and A. N. Yadav, International Journal of Current Microbiology and Applied Sciences **6**, 3580 (2017)
14. B. A. Fortnum, J. W. Rideout, S. B. Martin, and D. T. Gooden, Plant Disease **84**, 289 (2000)
15. C. S. Schmidt, F. Agostini, C. Leifert, K. Killham, and C. E. Mullins, Phytopathology **94**, 351 (2004)
16. B. R. Sharma, S. Dutta, S. Roy, A. Debnath, and M. Roy, Plant Pathology Journal (Suwon) **26**, 198 (2010)
17. Y. Serrano, J. M. Melero-Vara, M. L. Guirado, E. Segundo, and J. Gómez, Plant Pathology **57**, 776 (2008)
18. A. K. Romanova, Russian Journal of Plant Physiology **52**, 112 (2005)
19. G. L. Floch, J. T. Tambong, J. Vallance, Y. Tirilly, A. Lévesque, and P. Rey, FEMS Microbiology Ecology **61**, 317 (2007)
20. J. C. Sutton, C. R. Sopher, T. N. Owen-Going, W. Liu, B. Grodzinski, J. C. Hall, and R. L. Benchimol, Summa Phytopathologica **32**, 307 (2006)
21. G. N. Dake and S. Edison, Indian Phytopathology **42**, 116 (1989)
22. C. K. Augspurger and H. T. Wilkinson, Biotropica **39**, 702 (2007)
23. A. J. Van Der Plaats-Niterink, Studies in Mycology (1981)
24. P. Daly, S. Chen, T. Xue, J. Li, T. M. M. Sheikh, Q. Zhang, X. Wang, J. Zhang, D. A. Fitzpatrick, J. McGowan, S. Xiu-Juan, S. Deng, M. Jiu, D. Zhou, I. S. Druzhinina, and L. Wei, Frontiers in Microbiology **12**, (2021)
25. P. Vijayan, J. Shockey, L. Ca, C. Rj, and J. Browse, Proceedings of the National Academy of Sciences of the United States of America **95**, 7209 (1998)
26. S. Van Der Ent, S. C. M. Van Wees, and C. M. J. Pieterse, Phytochemistry **70**, 1581 (2009)
27. N. P. Dohroo, S. Kansal, and N. Ahluwalia, Indian Phytopathology **68**, 93 (2015)
28. M. Rai, A. P. Ingle, P. Paralikar, N. Anasane, R. M. Gade, and P. Ingle, Applied Microbiology and Biotechnology **102**, 6827 (2018)
29. S. Fukuta, R. Takahashi, S. Kuroyanagi, Y. Ishiguro, N. Miyake, H. Nagai, H. Suzuki, T. Tsuji, F. Hashizume, H. Watanabe, and K. Kageyama, Letters in Applied Microbiology **59**, 49 (2014)
30. P. H. Wang and C.-Y. Chang, Letters in Applied Microbiology **36**, 157 (2003)
31. P. Vishwakarma, A. Mohanty, A. Kaur, S. Das, S. R. Priyadarshini, S. Mitra, R. Mittal, and S. K. Sahu, Indian Journal of Ophthalmology **69**, 1544 (2021)
32. C. S. McGehee, P. V. Apicella, R. E. Raudales, G. A. Berkowitz, Y. Ma, S. N. Durocher, and J. D. Lubell, Plant Disease **103**, 3288 (2019)
33. J. Hu, Plant Disease **105**, 2736 (2021)

34. Y. Serrano, J. M. Melero-Vara, M. L. Guirado, E. Segundo, and J. Gómez, Plant Pathology **57**, 776 (2008)

The use of Silica fertilizer to increase the effectiveness of N,P,K fertilizer on rice plant growth

Jauhari Syamsiyah[1]*, *Ganjar* Herdiansyah[1], *Sri* Hartati[1] , *Lidya Zaela* Wijaya[2]

[1]Department of Soil Science, Faculty of Agriculture, Sebelas Maret University, Surakarta, Indonesia.
[2]Master Program of Soil Science, Faculty of Agriculture, Sebelas Maret University, Surakarta, Indonesia.

Abstract. One of the impacts of climate change is drought, which will disrupt plant growth. Silica can increase plant growth even in drought-stress conditions, but few rice farmers use Si fertilizer. This study aims to determine the effect of silicate combined with NPK fertilizer on the growth of rice plants. The study was conducted experimentally using a Randomized Complete Block Design (RCBD) with 9 treatments and 3 replications, namely 7 treatments of NPK fertilizer and Si fertilizer combinations and 2 treatments in the form of control and NPK fertilizers (Urea 300 kg ha-1 , SP 20 180 kg ha-1, and KCL 100 kg ha-1). The results showed that the use of Si (silica fertilizer) combined with NPK fertilizer increased the growth of rice plants, as indicated by the growth of plant height 4.30% higher in the application of ¾ NPK Fertilizer + 1 Silica Fertilizer dose and the number of tillers 1.55% more in the application of ¾ NPK fertilizer + ¼ silica fertilizer dose compared to the standard NPK treatment at the maximum vegetative period.

1 Introduction

Increasing population growth demands that the fulfillment of food supplies must also increase. Rice is a staple food for the Indonesian population, whose products must continue to be improved to meet increasing needs. However, reality shows that rice production fluctuates due to unfavorable environmental conditions. Floods and drought events as a result of climate change greatly affect rice production [1]. At present, the efforts made by the government to overcome this problem are innovations in cultivation through plant technology engineering and nutrition.

Plant nutrition is a very important factor in improving the quality and maintaining crop yields' stability. One of the important nutrients that farmers in Indonesia have forgotten is the nutrient Silica (Si). The positive effect of the nutrient Silica on rice plants has been widely reported in various developed countries such as the United States, Japan, China, South Korea, Taiwan, and India [2]. Silica is one of the nutrients needed by Gramineae family plants, such as rice, corn, sugar cane, and these elements accumulate inthe surface of leaves, stems, and

* Corresponding author: ninukts@staff.uns.ac.id

grain (rice). However, the role of Silica as a beneficial element, which is needed by rice species, has not received serious attention. This can be seen in Indonesian farmers not used to applying Si fertilization to their rice plants. However, considering that the productivity of intensified rice plants has experienced leveling off over a long period, Si fertilization is one of the efforts to overcome this phenomenon. A common symptom of Si deficiency in rice plants is that plants are susceptible to pests and disease. Another symptom is that the plant stems are not strong, so the plant easily collapses [3]. However, rice plants that collapse, apart from being caused by low Si content, are also caused by small stem base diameter, biomass weight, water management systems, and plant management [4]. [4] Research shows that applying Si to rice plants can reduce blast disease attacks.

One of the impacts of climate change is drought, which will disrupt plant growth. Silica is known to increase plant growth even in drought stress conditions, but not many rice farmers use Si fertilizer. Si status in the soil is also related to the level of soil weathering and soil moisture. Soils with advanced weathering and high rainfall generally experience Si leaching, so their availability status is low. Research [5] states that the critical point for Si content in paddy soil is 300 mg SiO_2 h^{-1}. Some researchers say that Si is considered low if the Si content is <600 ppm and deficient if <300 mg SiO_2 ha^{-1} [6]. [7] even suggested a lower critical level, namely 86 mg SiO_2 kg^{-1}. Research results [8] show that around 22.5% of intensified rice fields in West Java have low Si status, most of which are in Karawang and Subang. It was further reported that 76% of 92 paddy fields in West Sumatra also had Si content <300 mg SiO_2 kg^{-1}. This study aims to determine the effect of silicate combined with NPK fertilizer on the growth of rice plants.

2 Methods

This research was an experiment with a Randomized Completely Block Design experimental design of 9 treatments (Table 1) with 3 replications, namely A = Control, B = NPK Fertilizer (Urea 300 kg ha-1, SP 20 180 kg ha-1, and KCL 100 kg ha-1), C = ¼ NPK + 5 L ha-1 Si, D=½ NPK + 5 L ha-1 Si, E=¾ NPK + 5 L ha-1 Si, F=1 NPK + 5 L ha-1 Si, G=¾ NPK + 1.25 L ha-1 Si, H=¾ NPK + 2.5 L ha-1 Si, I=¾ NPK + 3.75 L ha-1 SThe number of plots is 27 plots, with a plot size of 2,5 m x 2,5 m (Figure 1). The plant height number of tillers was observed every two weeks until the plants entered the maximum vegetative phase. This study used rice plants of the Inpari 32 variety as plant indicators.

Table 1. Treatment of Si and NPK fertilizers.

Code	Treatments	Silica fertilizer	NPK fertilizer		
		L ha^{-1}	Kg ha^{-1}		
			Urea	SP-20	KCl
A	Control	0	0	0	0
B	NPK fertilizer	0	300	180	100
C	¼ NPK + 1 Silica fertilizers	5	75	45	25
D	½ NPK + 1 Silica fertilizers	5	150	90	50
E	¾ NPK +1 Silica fertilizers	5	225	135	75
F	1 NPK + 1 Silica fertilizers	5	300	180	100
G	¾ NPK + ¼ Silica fertilizers	1,25	225	135	75
H	¾ NPK + ½ Silica fertilizers	2,5	225	135	75
I	¾ NPK + ¾ Silica fertilizers	3,75	225	135	75

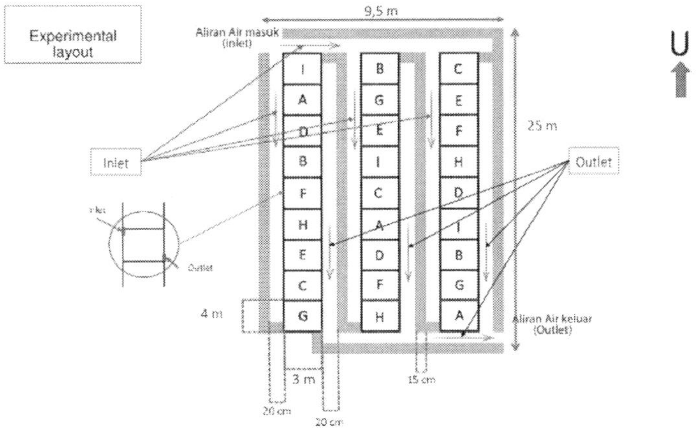

Figure 1. Experimental layout.

The research process began with thorough preparation, which involved observing the research location, designing the research layout, and gathering the necessary tools and materials. Following this, soil tillage was carried out by plowing the land until it was level, creating 27 treatment plots separated by bunds with water inlets. Each plot measured 3 m x 4 m. The seeding phase involved sowing 10 kg of seeds on a 3 m x 5 m land area previously treated with goat manure compost. After 2 weeks and the emergence of two leaves, the seeds were transplanted to the field. The planting process spaced rice seeds 20 x 20 cm apart, with 2-3 plants in each hole, resulting in 300 rice plants per plot. Fertilizer application included the spreading of silica fertilizer 2 weeks after planting, followed by the application of NPK fertilizer after 14 days. Observational data were tested with ANOVA (F test) to determine differences in response to rice plant growth from each treatment. Further tests were conducted with Duncan's Multiple Range Test (DMRT) at a 5% level.

3 Results and Discussion

3.1 Plant Height

The growth of rice plants is influenced by several factors, one of which is fertilization (Table 2). Proper and balanced fertilization can increase plant growth and yield. Element N is a nutrient plants need in large quantities compared to other retreats because N plays a role in photosynthetic activity, affecting growth and yield.

Providing doses of NPK and Silicate fertilizer did not significantly affect plant height at 14 DAP; however, a significant effect on rice plant height was seen in plants aged 28 DAP, 42 DAP, and 56 DAP. Table 2 shows that plant height in the ¾ NPK Fertilizer + 1 Silica Fertilizer treatment significantly differed from other treatments when the plants were 28 DAP, 42 DAP, and 56 DAP. This is because the fertilizer given can contribute nitrogen to growing rice plants. In accordance with the statement [9], available nitrogen affects plant height, number of tillers, leaf width, and number of grains. In addition, according to [10] the addition of silica fertilizer can increase the stability of the root membrane so as to prevent root damage due to water stress in rice plants which can cause reduced nutrient uptake by plants and affect plant growth. The supply of Si also helps the leaves become more upright under the influence of high nitrogen fertilization conditions, so that it can increase the photosynthesis rate up to 10 percent higher than rice plants that are not given silica [11].

Si can also increase the activity of antioxidant enzymes, which in turn can reverse oxidative stress and reduce the damaging effects of the element Boron on plants which can cause a decrease in photosynthetic activity, the amount of chlorophyll [12], and the activity of enzymes that activate CO_2 assimilation [13]. Increased photosynthetic efficiency due to Si fertilization makes the absorption of nutrients by plants in the vegetative period more optimal and is able to increase growth determined by the parameter of plant height [14]. A number of studies have also reported that silica application can also induce plant resistance where when silica is absorbed by rice plants in large quantities, plant resistance to diseases and fungi will also increase [15].

Table 2. Rice plant height at 14, 28, 42 and 56 days after planting (DAP).

Treatments	Plant height			
	14 DAP	28 DAP	42 DAP	56 DAP
Control	47,27 a	63,4 a	83,2 a	92,73 a
NPK fertilizer	51,4 a	66 b	88,2 b	102,2 ab
¼ NPK + 1 Silica fertilizers	49,33 a	68,87 b	89,87 b	103,93 ab
½ NPK + 1 Silica fertilizers	48,73 a	71,33 b	93,2 b	102,87 ab
¾ NPK +1 Silica fertilizers	52,6 a	69,73 b	92,47 b	106,6 b
1 NPK + 1 Silica fertilizers	49,67 a	69,67 b	91,4 b	106,33 b
¾ NPK + ¼ Silica fertilizers	48,33 a	67,53 b	86,8 b	99,67 ab
¾ NPK + ½ Silica fertilizers	52,4 a	67,33 b	89,8 b	102,87 ab
¾ NPK + ¾ Silica fertilizers	49,73 a	65,4 ab	92,53 b	105,67 b

3.2 Number of tillers

The effect of the combination treatment of NPK fertilizer and Silica fertilizer on the number of tillers of rice plants at plant ages 14, 28, and 42, up to 56 DAP (Table 3). Observation results showed that NPK fertilizer and Silica fertilizer significantly affected the number of rice tillers at 14 and 28 DAP. It can be seen in Table 3 that the ¾ NPK fertilizer + ¼ silica fertilizer treatment is significantly different from the other treatments.

Table 3. Number of Tillers at 14, 28, 42 and 56 days after planting (DAP).

Treatments	Number of tillers			
	14 DAP	28 DAP	42 DAP	56 DAP
Control	17,87 a	31 a	38,43 a	37 a
NPK fertilizer	20,73 ab	34,27 b	40,67 a	42,53 a
¼ NPK + 1 Silica fertilizers	22,33 b	33,6 b	38,7 a	39,2 a
½ NPK + 1 Silica fertilizers	22,13 c	36,53 b	40,8 a	43 a
¾ NPK +1 Silica fertilizers	23,33 bc	33,4 b	40,8 a	39,93 a
1 NPK + 1 Silica fertilizers	22,53 b	34,4 b	40 a	40,33 a
¾ NPK + ¼ Silica fertilizers	23,4 b	37,73 b	40,93 a	43,2 a
¾ NPK + ½ Silica fertilizers	24 c	35,07 b	38,5 a	39,47 a
¾ NPK + ¾ Silica fertilizers	21,87 b	34,47 b	40,77 a	40,67 a

Fertilization treatment with a dose of ¾ NPK fertilizer + ¼ Silica fertilizer produced the highest average number of tillers at 14, 28, 42, and 59 HST compared to other treatments (Table 3). According to [16], the soil's availability of nitrogen and phosphorus nutrients greatly influences the number of tillers. If there are enough nitrogen nutrients in the soil, the plant can produce lots of tillers, although not all of the shoots that grow will produce panicles. The addition of Si fertilizer also plays a role in increasing the number of tillers because of its role in activating the formation of tillers which is determined by cell division activities. According to [17], the number of tillers increases with the increase in P uptake due to the application of Si fertilizer where phosphorus elements are needed by plants in the process of cell division and become energy in all plant metabolic processes.

According to [18], tillers begin to form from the age of 10 days and reach a maximum of 50-60 days after planting. After getting the maximum, the number of tillers will decrease. In Table 3, it can be seen that the number of tillers decreased at the plant age of 56 HST. The correlation analysis results showed no relationship between plant height and number of tillers with a value of $p = 0.401$ ($p = > 0.05$).

4 Conclusion

The results showed that using Si (silica fertilizer) combined with NPK fertilizer increased the growth of rice plants, as indicated by higher plant growth compared to standard NPK treatment. Silica fertilizer is effective when combined with standard fertilizer to increase rice plant growth and mitigate Climate Change.

We would like to thank the Rector of Sebelas Maret University, the Dean of the Faculty of Agriculture and the Head of the Soil Science Department for providing the opportunity and facilities for this research activity.

Reference

1. E. Sumastuti and N. S. Pradono, J. Econ. Educ. **5**, 31 (2016)
2. S. D. Candra, Ngatimun, and J. Suharsono, Probolinggo Penerbit LPPM UPM 50 (2020)
3. J. F. Ma and N. Yamaji, Trends Plant Sci. **11**, (2006)
4. I. G. M. Subiksa, J. Tanah Dan Iklim **42**, (2020)
5. H. Sumida, Bull. Tohoku. Agric. Exp. Stn. **85**, (1992)
6. S.-M. YANG, L. XIE, S.-L. ZHENG, J. LI, and J.-C. YUAN, ACTA Agron. Sin. **35**, (2009)
7. W. J. Zhang, L. M. Wu, Y. F. Ding, F. Weng, X. R. Wu, G. H. Li, Z. H. Liu, S. Tang, C. Q. Ding, and S. H. Wang, J. Integr. Agric. **15**, (2016)
8. Husnain, Aflizar, Darmawan, and M. T, Proc. 5th Int. Conf. Silicon Agric. (2011)
9. Z. Zahra'a, E. M. Harahap, and Sarifuddin, J. Pertan. Trop. **5**, (2018)
10. A. Yuniarti, I. H. Hermawan, R. Sudirja, and D. S. Sara, Soilrens **19**, 10 (2022)
11. A. Amrullah, D. Sopandie, S. Sugianta, and A. Junaedi, J. Pangan **23**, 17 (2014)
12. C. Liu, W. Lu, Q. Ma, and C. Ma, J. Plant Nutr. **40**, 2458 (2017)
13. C. Kaya, M. Ashraf, A. A. Al-Huqail, M. A. Alqahtani, and P. Ahmad, Chemosphere **257**, 127241 (2020)
14. N. Fitriyah and M. A. Prayogo, J. Buana Sains **21**, 2527 (2021)
15. W. Islam, M. Tayyab, F. Khalil, Z. Hua, Z. Huang, and H. Y. H. Chen, Pestic. Biochem. Physiol. **168**, 104641 (2020)
16. T. H. Nguyen and H. Shindo, Agric. Sci. **02**, (2011)
17. Zulputra, Wawan, and Nelvia, J. Agroteknologi **4**, 1 (2014)
18. T. Bustomi, B. N. Fitriatin, and T. Simarmata, Soilrens **18**, 11 (2021)

Heat shock protein 90 alpha family class B member 1 (HSP90AB1) gene polymorphism and its effect on milk production traits in friesian holstein cattle

Muhammad Mumtaz Adzdzakiy[1], *Septiana* Widayanti[2], *Rebecca* Vanessa[2], *Galih* Pambuko[3], *Ari* Susilowati[2], *Sigit* Prastowo[3]*, and Sutarno[2]

[1]Department of Biology, Faculty of Mathematics and Natural Sciences, Universitas Sebelas Maret, Surakarta, Indonesia
[2]Graduate School of Bioscience, Faculty of Mathematics and Natural Sciences, Universitas Sebelas Maret, Surakarta, Indonesia
[3]Faculty of Animal Science, Universitas Sebelas Maret, Surakarta, Indonesia

Abstract. Increasing temperature caused by climate change is main contributor of heat stress in dairy cattle. Heat stress has a major impact on the milk production efficiency in dairy cattle. The HSP90AB1 gene is involved in overcoming heat stress response in cattle. The purpose of this study is to find single-nucleotide polymorphism (SNP) of HSP90AB1 gene and its relationship to milk production traits in Friesian Holstein (FH) cattle. In this study, 50 FH blood samples were used. Allele Specific Polymerase Chain Reaction (AS-PCR) method was used to successfully identify three genotypes: CC, CT, and TT. The CT genotype is the most common in the sample population. When compared to the C-allele, the T-allele is more common. Hardy-Weinberg (HW) analysis using Chi-Square method revealed that the population was not equilibrium ($P<0.05$). Association study between genotypes and milk production trait was not significant ($P>0.05$), but the CC genotype had a trend of higher mean milk yield in the first and second lactation. It was determined that the HSP90AB1 gene could be used as molecular marker for FH cattle heat stress response, in order to increase milk production capacity.

1. Introduction

The climate change causes the rising global temperatures. The rising of earth temperature is directly and indirectly affecting the production of livestock, as indicated by heat stress. Heat stress can cause major problems with the health, reproductive and milk production performance of dairy cows [1]. That problems lead to profitability loses. In the normal range of temperature, it is possible for bovines to keep their body temperatures down with a balance between metabolic warmth production and heat's flow from its surface into the environment by means of thermoneutral conditions. Further, when the temperature is above the comfort

* Corresponding author: prastowo@staff.uns.ac.id

zone the cattle become hyperthermia. hyperthermia is caused by high ambient temperature above a thermal comfort zone, as well as extremely humilities and strong solar radiation that can affect the effectiveness of animal heat exchange only or in combination [2].

The mammalians have the mechanism to control the heat exposure from the environment. In the cellular system, the adaptation response to high temperature is the activation and aggregation of the protein set, namely Heat Shock Protein (HSP). Some families classify HSP based on the weight and molecule mass, such as HSP60, HSP70, and HSP90. Further, the 90 kilodaltons heat shock proteins (HSP90) have prominent function in Molecular chaperones that are constitutively expressed as a result of heat or stress induction [3]. two isoforms called HSP90α or HSP90AA1 (inducible) and HSP90β or HSP90AB1(constitutive) [4]. The Heat Shock Protein 90 Alpha Family Class B Member 1 (HSP90AB1) gene plays a role in the physiological response of the body to heat shock. HSP90AB1 located in Bos taurus autosome (BTA) 23 and consist of 12 exons. The polymorphism in this gene has association in heat tolerant. One of the HSP90AB1 SNP that associate with the tolerant is SNP g.4338T>C with reference ID rs109251249. The g.4338T>C has significant effect on heat tolerant and milk production [5–7].

Increasing global temperature especially in Indonesia lately can affect dairy cattle production. The annual temperatures ranged from 26.8°C to 26.9°C between 2000 and 2024, with a projected increase to 27°C in 2024 [8]. The ideal temperature for dairy cattle is 13°C to 25°C [9]. Furthermore, the ideal Temperature Humidity Index (THI) for the dairy cattle is 68 to 74, above that the cattle can experience stress and decreasing the milk productivity [10]. Therefore, this study aimed to determine the genotype variant and milk production association at g.4338T>C in the HSP90AB1 gene in the Friesian Holstein population. The findings from this research can potentially serve as a molecular marker for temperature-tolerant dairy cattle selection programs.

2. Material and methods

2. 1. Blood collection and DNA extraction

Fifty blood samples of FH cows were obtained from the Dairy Cattle Breeding Center in Central Java Province, Indonesia. The blood was collected by venipuncture from the coccygeal vein and stored with EDTA for blood preservation. All the blood collection procedure was supervised by a veterinary and also comply with the Indonesian Law number 19, 2009 regarding animal husbandry and health. Furthermore, the DNA Extraction was performed by Promega Wizard® Genomic DNA Purification Kit, Promega. Blood Collection and DNA extraction were used following the previous study, according to [11].

2. 2. HSP90AB1 gene genotyping

HSP90AB1 genotyping was done by the Allele Specific-Polymerase Chain Reaction (AS-PCR) method. Each allele is amplified with specific primer sets (**Table 1**). The T allele was amplified with the F-T and R primer set, while the F-C and R primer set was used to amplify the C allele. Then, the amplification products were electrophorized and visualized by agarose 2%. The HSP90AB1 gene genotyping were used following the previous research, according to [11].

Table 1. Primer list for amplification HSP90AB1 gene fragment and genotyping.

Primers (5'-3')		Tm (°C)	PCR Cycle	Product Size (bp)	Reference
F-T	CTGGAGTCACACTGAGGAA<u>T</u>	58.9	35	561	[7]
F-C	CTGGAGTCACACTGAGGAA<u>C</u>	57.9			
R	TGTTGGAGATCGTCACCTG				

2. 3. Data Analysis

2. 3. 1. Milk production data

The milk production data were acquired from the first lactation record. The 305 2X Mature Equivalent (ME) factor was used for milk production data standardization. The Kg/lactation was used for unit measurement.

2. 3. 2. Genotype-Allele frequency and Hardy-Weinberg Equilibrium (HWE) analysis

The genotype-allele frequency and HWE analysis by Chi-Square (χ^2) was calculated by R with Genetics Package [12,13].

2. 3. 3. Association milk production trait with HSP90AB1 genotype

The association of milk production with the HSP90AB1 genotype were analyzed with ANOVA. Duncan's Multiple Range Test (DMRT) was used for further testing. P<0.05 was set for statistical differences. The Analysis was carried out by R with Agricolae Package [13,14].

3. Result and discussion

The AS-PCR amplified HSP90AB1 specifically with a 561 bp amplification product size. The genotyping discovered three variation of genotypes (TT, CC, and CT) and two alleles (C and T) in the FH population (**Fig 1**). The TT genotype just amplified T-allele primer set and the amplification product presented in well 1^T and no amplification in well 1^C. Otherwise, only the C-allele primer set amplified 2^C successfully, and 2^T had no band present in the CC genotype. Further, C and T primer sets were amplified successfully in 3^C and 3^T wells in the CG genotype. The homozygous genotype only amplifies one of the allele-specific primer sets, while in the heterozygote genotype, all the allele-specific primer sets are amplified well. The previous study on the local Indonesian Friesian Holstein and Sahiwal cattle confirmed this result, but the study on Thai Crossbreds Frisian Holstein TT genotype was not found, while in White Lumphun (Thai indigenous cattle) only found the TT genotype [5,7,11].

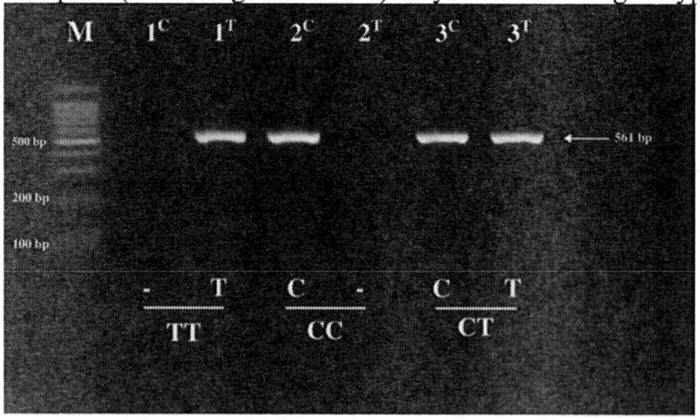

Fig 1. HSP90AB1 Genotyping by AS-PCR Result. Note: (M): Marker 100 bp, (1^C-3^C): C-allele samples, (1^T-3^T): T-allele samples, (TT, CC, CT): the HSP90AB1 genotype.

The CC genotype had the lowest genotype frequency (0.06), while the CT genotype had the highest (0.66). Further, the T allele (0.61) was higher than the C allele (0.39). The HWE analysis revealed the FH population was not in equilibrium (P-value = 0.0072) with a chi-square value of 7.49 (**Table 2**). The highest in CT genotype and the lowest TT genotype result was found in local Indonesian Friesian Holstein and Sahiwal cattle; in another study on the Frieshwal cattle the TT genotype was the lowest genotype [7,11]. The inequilibrium state in this study caused by the selection activity and non-random mating with elite bulls in FH population [15].

Table 2. HSP90AB1 HWE analysis in FH population.

n	Genotype Frequency			Allele Frequency		HWE Analysis	
	CC	CT	TT	C	T	χ^2	P-Value
50	0.06	0.66	0.28	0.39	0.61	7.49	0.0072*

*Inequilibrium = (P<0.05)

The association analysis of milk production with the HSP90AB1 genotype in the FH population revealed the CC genotype had the highest milk production with 4126.26±1600.23 kg. The TT genotype had the lowest milk production with 2950.50±1143.37 kg. There was no association between milk production and HSP90AB1 genotype (P-value > 0.05). Although there was no association, the CC genotype performs better in milk production than the CT and TT genotypes.

Table 3. Association of milk production with HSP90AB1 genotypes in FH population.

Trait	Genotype			P-Value
	CC	CT	TT	
Milk Production (Kg)	4126.26±1600.23	3687.49±1459.16	2950.50±1143.37	0.19

Significant = (P<0.05)

The previous studies, the genotype variation in HSP90AB1 g.4338T>C had better heat tolerance in the TT genotype, following CT genotype and CC genotype as the lowest heat tolerant among others genotype in Frieswal and Sahiwal cattle [7]. Here in after, T allele in g.4338T>C intensified the heat tolerant in Thai crossbred Friesian Holstein, White Lumphun and White Mountain cattle [5]. The TT genotype in HSP90AB1 g.4338T>C was associated with better milk production performance and the CC genotype was the lowest milk production in Sahiwal and Frieswal cattle [7]. On other study had different result, the CT genotype had the higher milk production than CC genotype in Sahiwal cattle [6]. In this association study, there was an inverse trend in milk production. The CC genotype had the highest milk production, followed by the CT genotype, and the TT genotype had the lowest milk production. The various breeds of cattle cause the differences in genotype associated with milk production in HSP90AB1.

The milk production belongs to quantitative trait that influence of the many gene interaction. Furthermore, environmental condition and climate have major contribution to the milk production performance of the cattle. The high temperature climate will perish the milk production in dairy cattle[1]. In this study, the population of Friesian Holsteins came from a subtropical climate country, then the cattle were transported to Indonesia during pregnancy and give birth in Indonesia. In this situation, cattle are being affected the stress by changes in the environment from that of a subtropical to an equatorial climate. As a consequence, the milk production which is not as genetically potent has been disrupted and inconsistent, this has been explained in previous study [16, 17].

4. Conclusion

The rising environmental temperatures due to climate change can disrupt dairy cattle milk production. Implementing a selection program for temperature tolerance adaptation in dairy cattle is essential to maintain milk production. The HSP90AB1 variation in g.4338T>C in Friesian Holstein population had three genotypes and two alleles: CC, CT, TT genotype and C, T allele. The CT genotype had the highest genotype frequency, and the T allele had the highest allele frequency. The Friesian Holstein population was not equilibrium. The milk production was no association with the HSP90AB1 genotypes. Nonetheless, the CC genotype had better milk production among the other genotypes. The HSP90AB1 gene is the potential candidate genetic marker for heat tolerance dairy cattle. The further study is needed in different dairy cattle breeds and more of quantitative traits.

References

1. J. Liu, L. Li, X. Chen, Y. Lu, and D. Wang, Asian-Australas J Anim Sci **32**, 1332 (2019)
2. S. Foroushani and T. Amon, Int J Biometeorol **66**, 1811 (2022)
3. S. Maiti and D. Picard, Biomolecules **12**, (2022)
4. L. Sailo, I. D. Gupta, A. Verma, R. Das, M. V. Chaudhari, and S. Singh, Indian J Anim Res **50**, 856 (2016)
5. R. Charoensook, K. Gatphayak, A. R. Sharifi, C. Chaisongkram, B. Brenig, and C. Knorr, Trop Anim Health Prod **44**, 921 (2012)
6. S. Lalrengpuii, D. G. I, V. Archana, D. Ramendra, and V. C. M, African Journal of Biochemistry Research **9**, 99 (2015)
7. B. Sajjanar, R. Deb, U. Singh, S. Kumar, M. Brahmane, A. Nirmale, S. K. Bal, and P. S. Minhas, Anim Biotechnol **26**, 45 (2015)
8. (n.d.)
9. D. H. Lim, T. Il Kim, S. M. Park, K. S. Ki, and Y. Kim, J Anim Sci Technol **63**, 872 (2021)
10. P. Herbut, S. Angrecka, and J. Walczak, Int J Biometeorol **62**, 2089 (2018)
11. S. Prastowo, M. Mumtaz Adzdzakiy, R. Vanessa, G. Pambuko, Purwadi, A. Susilowati, and Sutarno, in *E3S Web of Conferences* (2021)
12. G. Warnes, G. Gorjanc, F. Leisch, and M. Man, (2021)
13. R Core Team, (2023)
14. F. de Mendiburu and M. Yaseen, (2020)
15. R. S. Waples and F. Allendorf, Journal of Heredity **106**, 1 (2015)
16. S. Prastowo, T. Nugroho, N. Mahfudhoh, F. Y. Putra, Subiakti, A. Ratriyanto, A. Susilowati, Sutarno, and N. Widyas, IOP Conf Ser Mater Sci Eng **633**, (2019)
17. N. Widyas, N. Mahfudhoh, Subiakti, and S. Prastowo, AIP Conf Proc **2202**, 1 (2019)

'Pot cultivation' as a form of soil and water conservation technique for adaptation to soil condition and climate change in karst area

Arif Anshori[1 *], *Tri Endar* Suswatiningsih[2], Mujiyo[3], *Sigid* Handoko[4], and *Helena Lina* Susilawati[4]

[1]Research Center for Food Crops, Research Organization for Agriculture and Food, National Research and Innovation Agency, Bogor, Indonesia
[2]Department of Agribusiness, Faculty of Agriculture, Institut Pertanian Stiper, Yogyakarta, Indonesia
[3]Department of Soil Science, Faculty of Agriculture, Universitas Sebelas Maret, Surakarta, Indonesia
[4]Research Center for Horticultural and Estate Crops, Research Organization for Agriculture and Food, National Research and Innovation Agency, Bogor, Indonesia

Abstract. Shallow soil is found in Gunungsewu karst area of Gunungkidul Indonesia, often referred as 'soilly rock'. Its the main problem are limited of growing medium and soil loss due to erosion. 'Pot cultivation' (PC) is term to describe technique of cultivating crop on small plot of soil between rock. Study aimed to determine 1) the practice of soil and water conservation (SWC) techniques of PC as a form of adaptation to soil condition, 2) the potency of PC for adaptation and mitigation of climate change in karst area. Research was conducted in Gunungsewu karst area, from January to August 2020. Study method was observation and interview. Results showed that practice of PC was carried out as form of adaptation to shallow soil in karst area. Soil is managed to fill the rocks, so sufficient as plant growth medium. Rock functions as a 'pot'. Rock 'pots' function like terraces in SWC. PC is effective way in increase of growing medium, preventing soil from erosion, preserving soil organic carbon, and as an action of adaptation and mitigation co-benefit. The results of this study are expected to be useful for development of karst area which have shallow soil.

1 Introduction

Karst area is vulnerable to environmental changes [1], such as climate change [2], erosion [3] and uncontrolled exploitation [4], cause degradation. The slopes of karst hill are generally steep, can be 60 to 70 degrees [5], accelerate the process of soil erosion and degradation [6], so forming the critical land [7]. Soil erosion and degradation in karst area interfere with nutrient cycling, food and fuel supply, water purification, soil retention, climate and flood regulation, carbon sequestration, and habitat for organisms. The negative consequences are disruption of hydrology, nutrient, carbon and biodiversity cycle [8]. Critical land with high

* Corresponding author: arifanshori@yahoo.com

slopes and shallow soil solum are found in the Gunungsewu Gunungkidul karst area. Without appropriate soil and water conservation (SWC) practices, critical land will expand.

SWC in karst area adapts to land use, vegetation cover, slope construction, and the application of clean energy [9]. SWC practices depend on technological capabilities, which can increase productivity, reduce land degradation, rehabilitate land, as well as economic and environmental benefits [10]. SWC benefit add up over time [11]. Social and economic factors influence the level in adoption of SWC technology [12]. SWC practices are adaptive to environmental conditions [13] and lead to climate change adaptation and mitigation [14]. SWC is expected to increase food security [15], with low greenhouse gas emissions in management techniques [16] and applications that prioritize an ecological approach [17].

In the karst area of Gunungsewu Gunungkidul, there are various forms of SWC practices according to environmental condition, especially soil condition. Its one form is PC, which developed in shallow soil on karst hill. This study aimed to determine 1) the practice of SWC technique of PC as a form of adaptation to soil condition, 2) the potency of PC for adaptation and mitigation of climate change in karst area.

2 Material and method

The Gunungsewu karst area is more than 1,300 km², with more than 10,000 hills (in 1968, Balazs estimated there were 40,000 hills), with a density of about 30 km^{-2}, with very varied morphology. The Gunungsewu landscape is described as kegel karst, with steep slopes on karst hills [5], which can accelerate soil erosion, making it vulnerable to formation of critical land. SWC practices are necessary to preserve soil, such as PC on shallow soil.

The research was carried out in the karst area of Gunungsewu, Gunungkidul Indonesia, at the PC location, from January to August 2020. The research locations were Semanu, Tepus, Saptosari and Girisubo. Observation and interview were carried out in this study. Observation was determining the PC condition, soil, conservation technique, water and supporting agricultural activities. Interviews were conducted with 6 key farmers and 4 agricultural extension officers. Interviews were conducted to explore 1) the conservation technique of the PC, 2) the knowledge of farmer about conservation and climate change, 3) the socio-economic condition of farmers, and 4) opportunities for the preservation and development of PC. Data from observation and interview are narrated always referring in ecological aspect.

3 Result and discussion

3.1. Practice of 'pot cultivation' as soil and water conservation

Karst of Gunungsewu Gunungkidul has valleys bounded by moderate to steep slopes. Hills are found between valleys, cone-shaped or long flat, serrated and serrated without basin. The valley is in the form of deep soil. Shallow soil is found in karst hill [5]. Management of the soil, water and crop are carried out by farmer as a form of adaptation to the environmental condition [13]. One of them is PC in shallow soil on karst. PC is a term for crop grown in a small volume of soil between rocks, often referred as 'soilly rock'. Soil is managed to fill the rock, so sufficient for growing medium. Rocks function like a pot. The depth and width of the 'pot' varies, influenced by the rock condition. Soil in 'rock pot' can be in situ or native, or from soil material collected so that it is enough as a growing medium. The 'pot' visualization in karst area can be seen in Fig. 1a, 1f and 1g.

Monthly rainfall in the Gunungsewu karst area shows a seasonal pattern. High rainfall in November to March, low rainfall May to September, April and October transition [18]. Crop commodities are grown according to rainfall patterns, as [13]. Based on rainfall, PC can be used to grow annual and seasonal crops. The dominant annual is teak tree (Fig. 1c, 1d and 1e). The seasonal crops cultivated are corn (Fig. 1f), groundnut (Fig. 1g) or cassava (Fig. 1g, 1h and 1i). Seasonal crops are planted a maximum of twice in a year.

Fig. 1. 'Pot cultivation' in karst hill (a, b), with teak tree (c, d, e), with corn (f), with multiple cropping of groundnut and cassava (g), with cassava (h, i).

Soil formation in karst areas is slow with low vegetation cover, making it susceptible to land degradation, including by erosion. Erosion causes the loss of soil layer and the exposure of rocks [9]. Conservation practices are needed to maintain and improve soil condition [19]. PC is a form of soil and water conservation action in the Gunungsewu karst area. Rock 'pots' function like terrace, namely reducing water run off, increasing water infiltration, and preventing soil erosion and landslide. Crops in 'pots' reduce the damage by rainwater.

3.2. 'Pot cultivation' for adaptation and mitigation of climate change

Management of the soil, water and crop according to conservation principle will increase crop productivity, maintain soil and environmental health [20]. Management in PC adjusts the water availability, from no crop to seasonal or annual crop. Benefits are soil retention, reduced erosion, provision of food, feed and fuel, improved the infiltration, nutrient, hydrological cycle, biomass cycle, organism habitat and carbon absorption. Ecologically, there are difference in nutrient and biomass cycle in seasonal and annual crop. The cycle in seasonal crop is open. Cycle in annual crop is more closed (Fig. 2). Crop productivity is difficult to measure, due to the rock 'pot' and crop density in irregular. PC management does not contradiction with climate change adaptation principles such as in [21], so PC has the potential to be a climate change adaptation technology in karst area. Carbon absorption as the mitigation co-benefit.

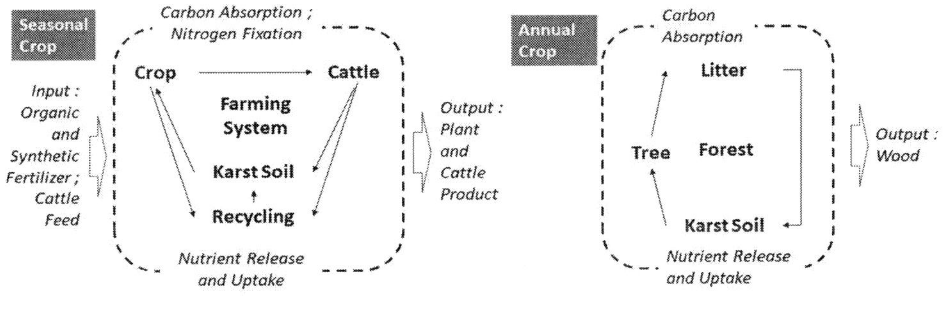

Fig. 2. Flow of biomass and nutrient in 'pot cultivation' with seasonal and annual crops.

PC of seasonal crop has developed into a farming system, with input and output. The flow of nutrient and biomass are open, input in the form of organic and synthetic fertilizers, as well as animal feed input. Crop roots provide organic input in the form of root exudates [22]. Output is in the form of crop and cattle products. PC with annual crop, developing into community forest, with closed of the nutrient and biomass cycle, without input. Output is in the form of wood, harvested between 10 to 15 years.

PC reduces soil erosion, in according to [23] reduces soil carbon loss by erosion. In PC with annual crop, soil cultivation will be eliminated, which is synonymous with reduced energy and fuel consumption, reduced working time and equipment depreciation. All of these things, lead to reduced emission in the operation and manufacture of agricultural equipment. Indirectly, PC reduces greenhouse gas emissions. PC also contribute to carbon absorption, so the accumulation of litter in the soil will increase soil fertility and health. As [13], PC has co-benefit as mitigation for climate change in karst area.

PC in karst areas has the potential to be developed as an adaptive and mitigative agriculture to climate change. The implementation of PC must be in accordance with the socioeconomic of farmer and fulfilled with principle of sustainability. PC quality can be improved through improving inputs, such as application of rice husk charcoal and organic fertilizer as in [26, 27, 28], to increase water absorption and improve nutrients, as an alternative technology, especially for seasonal crops. Another alternative technology, management of irrigation water [29, 30, 31] and cropping pattern [32, 33, 34], with the potential for irrigation water in karst area from spring or underground river [35, 36].

3.3. Socioeconomic of 'pot cultivation'

Perception of erosion and erosion hazard greatly determines the accelerate in adoption of SWC technology [12]. Farmers are aware about the erosion and soil fertility problems in karst hill. Farmers' understanding of erosion is related to soil loss, soil becoming shallow, decreasing soil fertility, forming furrow and ditched, stunting crop growth, falling production, soil deposit in downstream, and murky of river water. Erosion can be controlled by SWC. Farmers understand the shift of season, flood and drought, but no about climate change. PC is practiced by farmers to overcome the negative effect of erosion and increase of soil fertility, but they don't understand it related to climate change mitigation.

The population is increasing, so the population pressure on land is getting higher. Farmer use land to produce biomass, food, feed and energy. According to [35] farmer at the research location are classified as peasant, with narrow land area. Farmer adapt to the environmental condition to survival [36]. Farmers do not see only from the aspect of crop production, but also about the health and sustainability in karst area. Farmer are indirectly familiar with environmental economic. [37] states that, economic value can be directly in the form of consumption products, but can also be in the form of indirect value as non-consumption. Non-consumption value in the form of long-term benefit of SWC.

3.4. Sustainability of 'pot cultivation'

PC is a form of land resource management for agriculture, furthermore it is the karst are management. According to [38] the system survives with loyality, synergistic process, lots of activities, places that profitable, and policies with a sense of belonging. Five components are involved, namely loyality, synergy, lots of activities, profitable, and sense of belonging. Related these components, sustainability of PC is shown in Fig. 3.

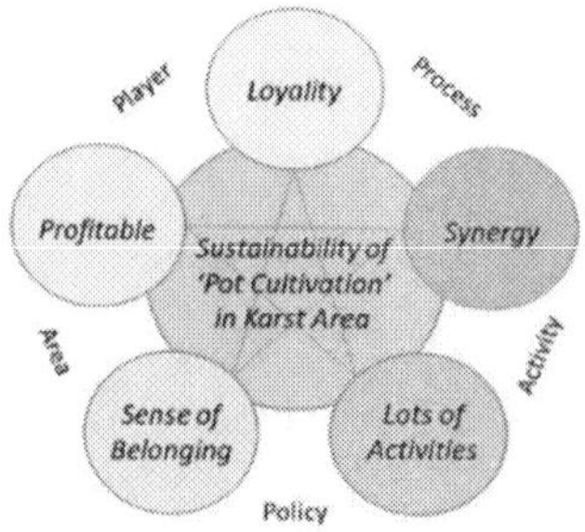

Fig. 3. Construction and relationship of component for sustainable 'pot cultivation'.

Loyality means faithfulness, growing without coercion in PC. Synergy is needed to combine various activities to achieve results. Lots of activity means there is a lot of activity related to the PC. The sense of belonging includes maintaining, advancing, repairing, maintaining as one's own or together. PC provides short term benefits as a product of consumption, and long term as conservation benefit. The sustainability of PC is largely determined by these components, according to [13] while still emphasizing farmer as the main factor. It is hoped that the research result will be useful for the development and sustainability of karst area that have shallow soil.

4 Conclusion

The practice of SWC PC is carried out in karst area on 'soilly rock' or shallow soil. Soil is managed to fill the rock, increasing the soil volume, so sufficient as medium for plant growth. The rock functions as a 'pot'. Rock 'pots' function like terraces, reducing surface runoff and increasing infiltration, as well as preventing soil erosion and landslides. PC management is in accordance with the principles of climate change adaptation, so it has potential as a climate change adaptation. PC reduces soil carbon loss and greenhouse gas emissions, and contribute to carbon absorption, so co-benefit as mitigation for climate change. The sustainability of PC is determined by loyality, synergy, lots of activity, profitable, and a sense of belonging, while still emphasizing farmer as the main factor. The development and sustainability of karst area with shallow soil can refer to the result of this research.

References

1. M. T Quine, D Guo, S M Green, C Tu, I Hartley, X Zhang, J Dungait, X Wen, Z Song, H Liu, H Buss, T Barrows, R Evershed, P Johnes, J, Acta Geochim. **36**, (2017)
2. E Haryono, T N Adji, *Geomorfologi Dan Hidrologi Karst* (UGM Yogyakarta, 2017)
3. H Čustović, M Misilo, M Marković, Soil Sci. Plant Nutr. **60**, 100 (2014)
4. Guntarto, Bul. Geol. Tata Lingkung. **13**, 101 (2003)
5. E Haryono, M Day, J. Cave Karst Stud. **66**, (2004)
6. Q Dai, Z Liu, H Shao, Z Yang, Solid Earth **6**, 985 (2015)
7. A Anshori, Pros Semin. LPPM UMP Purwokerto, Buku 1 197 (2015)
8. J P Field, D D Breshears, D J Law, J C Villegas, L L Hoffman, P D Brooks, J Chorover, G A B Gafford, R E Gallery, M E Litvak, R A Lybrand, J C McIntosh, T Meixner, G Y Niu, S A Papuga, J D Pelletier, C R Rasmussen, P A Troch, Vadose Zo. J (2015)
9. Y Sun, Y Zhang, S Li, N Zhang, J Wang, E3S Web Conf. **145**, (2020)
10. Y Abebe, A Bekele, Sci. Technol. Arts Res. J **3**, 198 (2014)
11. Y Ademe, T Shafi, T Kebede, A Mullatu, J. Dev Agric Econ **9**, 37 (2017)
12. A J Tenge, J D Graaff, J P Hella, L. Degrad. Dev. **15**, 99 (2004)
13. A Anshori, T E Suswatiningsih, Mujiyo, H L Susilawati, IOP Conf. Ser. Earth Environ. Sci. **1165**, (2023)
14. J A Delgado, R Li, Int. Soil Water Conserv. Res **4**, 148 (2016)
15. FAO, *Climate Smart Agriculture: Sourcebook* (FAO Book, 2013)
16. A Anshori, B H Sunarminto, E Haryono, A Pramono, Mujiyo, IOP Conf. Ser. Earth Environ. Sci. **724**, (2021)
17. D M Kadyampakeni, Agron. J. **106**, 100 (2014)
18. A Cahyadi, E Haryono, T N Adji, M Widyastuti, I A Riyanto, D T N Muhammad, N F Tastian, Indones. J. For. Res. **8**, 23 (2021)
19. A M Misebo, J. Agric. Sci. **10**, 227 (2018)
20. T K Das, Y S Saharawat, R Bhattacharyya, S Sudhishri, K K Bandyopadhyaya, A R Sharma, M L Jat, F. Crop. Res. **215**, 222 (2018)
21. P Karing, A Kallis, H Tooming, Clim. Res. **12**, 175 (1999)
22. J Ni, D H Luo, J Xia, Z H Zhang, G Hu, Solid Earth **6**, 1209 (2015)

23. T B Sapkota, M L Jat, J P Arya, R K Jat, A Khatri-Chhetri, J. Integr. Agric. **14**, 1524 (2015)
24. A Anshori, A Iswadi, Sunarya, D Riyanto, AgriHealth J. Agri-Food, Nutr. Public Heal. **2**, 1 (2021)
25. A Anshori, T E Suswatiningsih, N A Viandari, Mujiyo, H Purwaningsih, S W Budiarti, A B Pustika, H Cahyaningrum, IOP Conf. Ser. Earth Environ. Sci **1168**, (2023)
26. N A Viandari, A Anshori, IOP Conf. Ser. Earth Environ. Sci **672**, (2021)
27. A Anshori, T E Suswatiningsih, AgriHealth J. Agri-Food, Nutr. Public Heal. **3**, 124 (2022)
28. A Anshori, E Srihartanto, Fibrianty, T E Suswatiningsih, S W Budiarti, D Riyanto, H Cahyaningrum, Suradal, Caraka Tani J. Sustain. Agric. **38**, 1 (2023)
29. A Anshori, T E Suswatiningsih, N A Viandari, Rajiman, Mujiyo, IOP Conf. Ser. Earth Environ. Sci **824**, (2021)
30. E Srihartanto, A Anshori, AGRIC **34**, 115 (2022)
31. D Riyanto, Yustisia, A Anshori, E Srihartanto, E3S Web Conf **232**, (2021)
32. D Riyanto, T Cahyono, A Anshori, Yustisia, B K Udiarto, IOP Conf. Ser. Earth Environ. Sci **1172**, (2023)
33. T N Adji, M. A Mujib, E Haryono, H Fatchurohman, A N Kholis, Kuwait J. Sci. **in Press**, (2023)
34. A Cahyadi, E Haryono, T N Adji, M Widyastuti, I A Riyanto, D T N Muhammad, A P K Adji, AIP Conf. Proc **2683**, (2023)
35. B M Campbell, P Bradley, S E Carter, Agric. Hum. Values **14**, 159 (1997)
36. E K Zimmerman, J C Tyndall, L A Schulte, G L D Larsen, Water Resour. Res. **55**, 3796 (2019)
37. D Pearce, D Morgan, *The Economic Value of Biodiversity* (Earthscan, 2009)
38. I Aliyah, *Pasar Tradisional : Kebertahanan Pasar Dalam Konstelasi Kota* (2020)

Growth of mint (*Mentha spicata* L.) on various biochar and liquid organic fertilizers

Muji Rahayu[1*], *Amalia Tetrani* Sakya[1], *Andriyana* Setyawati[1], and *Bayutama* Rahmawan[1]

[1]Department of Agrotechnology, Faculty of Agriculture, Universitas Sebelas Maret, Surakarta, Indonesia

Abstract. Climate change has an impact on plant growth, including mint. Mint grows optimally in environments with low temperatures and high humidity. Erratic climate change has the potential to reduce mint growth. Biochar is a soil enhancer that can improve soil fertility and soil water availability. Liquid Organic Fertilizer (LOF) is able to provide the nutrients needed by plants so that it has the potential to increase mint growth. This research was conducted in August 2022 - June 2023 in the greenhouse of the Agriculture's Faculty, Sebelas Maret University, Surakarta. The study used a factorial Randomized Complete Group Design (RCBD), which consisted of two factors, namely biochar and fish waste LOF. The types of biochar consisted of no biochar treatment, palm fiber biochar, and rice husk biochar. Fish waste LOF consisted of no LOF treatment, 5 ml/L water, 10 ml/L water, and P3 = 15 ml/L water. The results showed that the application of palm fiber biochar and 5 ml/L fish waste liquid fertilizer can increase plant fresh weight and plant dry weight. Giving palm fiber biochar can increase plant fresh weight, while the addition of rice husk biochar can increase leaf area and plant dry weight. This research has implications to knowing the interaction of biochar type and LOF concentration, getting the appropriate biochar type to increase mint growth, and getting the appropriate LOF concentration to increase mint growth. concentration that is suitable for increasing mint growth.

1 Introduction

Mint (*Mentha spicata* L.), is native to Europe and southern Asia, stretching from France in the west to southern China in the east. [1] stated that mint is one of the essential oil-producing plants. Essential oil of *Mentha spicata* L. or Spearmint is widely used as a raw material in the food, beverage and pharmaceutical preparation industries. The main components of spearmint are carvone, limonen, cineol, dihydrocarveol, myrcene, and menthol with a level of 0.5%. The components contained in spearmint have properties as herbal medicines.

Mint have benefits that have been utilized by humans in various industrial fields so that their existence is starting to be sought after so it is necessary to carry out cultivation, care, and efforts to increase yields. One important effort in maintenance is to fertilize plants using fertilizers from organic materials. Organic fertilizers can improve the physical and chemical properties of soil so that it is better in providing nutrients. Organic fertilizers can come from

* Corresponding author: mujirahayu@staff.uns.ac.id

plant and animal residues, such as compost, goat urine fertilizer, and can also come from livestock and fish waste.

Fish waste fertilizer is a by-product of fish. The resulting fish waste can be in the form of leftovers or fish that are usually discarded by fishermen due to limitations in how fish waste is managed. This fish waste can be used in agriculture as a fertilizer that can be used after going through a decomposition process so that it can be used as a nutrient provider. The use of fish waste liquid organic fertilizer has the potential to increase plant growth because it contains nutrients that plants need. [2] stated that fish waste fertilizer provides nutrients, namely N (Nitrogen), P (Phosphorus) and K (Potassium) which are the constituent components of organic fertilizers.

The nutrient content contained in liquid organic fertilizer waste will be optimized by the use of biochar. Biochar used in this research comes from palm waste in the form of palm fiber and rice waste in the form of rice husks. [3] said that biochar as a soil improver is useful in increasing growth and nutrient uptake in plants. [4] in his research mentioned that the use of biochar as a soil conditioner can increase crop yields, nutrient absorption efficiency by plants, improve acidic soil, and increase soil organic matter content.

Previous research by [5] showed that giving rice husk biochar was able to increase plant height, leaf length, and leaf width of the mint. Previous research by [6] by giving LOF paitan can increase the number of leaves, number of branches, fresh weight of plants, and dry weight of mint plants (*Mentha arvensis* L.). The novelty of this research compared to previous research is the addition of factors in the form of biochar, so that it can examine the interaction of biochar and LOF fish waste on mint (*Mentha spicata* L). The use of LOF fish waste and palm fiber biochar for mint is also still small, even though LOF fish waste has the potential to provide nutrients needed by plants, the use of palm fiber biochar also has the potential to become a soil improver so as to reduce waste from palm production.

The addition of biochar and fish waste liquid organic fertilizer is expected to increase and optimize the growth of mint plants. Research with these treatments aims to obtain an effective type of biochar and the right concentration of fish waste liquid organic fertilizer. Another goal is to determine the interaction between the effective type of biochar and the concentration of fish waste liquid organic fertilizer on the growth of mint plants.

2 Materials and methods

This research was conducted from July 2022 to January 2023 at the Agriculture Faculty's Greenhouse, Sebelas Maret University, Surakarta, Central Java. The design used was a factorial Randomized Complete Group Design (RCBD), which consisted of two factors, namely biochar and fish waste liquid organic fertilizer. The type of biochar consists of no biochar treatment, palm fiber biochar, rice husk biochar. Fish waste liquid organic fertilizer consists of no LOF treatment, 5 ml/L water, 10 ml/L water, P3 = 15 ml/L water. There were 12 treatment combinations that were repeated 3 times so that there were 36 treatment units. Each experimental unit was repeated 4 times so there were a total of 144 polybags. The parameters observed in this study were plant height, number of branches, number of leaves, leaf area, plant fresh weight, and plant dry weight. Observation data were analyzed using Analysis of Variance (ANOVA) in accordance with the research design. Single treatment that has a real to very real effect, then continued with the DMRT test at the 5% level. Gowth observation is done once a week, from 1 to 12 week after planting (WAP).

3 Results and Discussion

3.1. Plant Height

Plant height is an indicator of plant growth that is easily observed and measured. The results of the analysis of variance showed that there was no interaction between the provision of biochar and fish waste liquid fertilizer on plant height. Table 1 shows that the provision of rice husk biochar treatment is not significantly different from without the provision of biochar and palm fiber biochar. Rice husk biochar gave the highest results on plant height. Each type of biochar treatment is able to increase plant height but not significantly different from each other. This is because biochar has not worked optimally so that it has not been able to retain nutrients in the soil. According to [7], the increase in plant height is not much different due to environmental factors in the form of biological and soil properties that are not yet optimal because the application of biochar does not directly provide nutrients without any process.

Table 1. Effect of biochar type on mint (*Mentha spicata* L.) growth at 12 WAP.

Treatment	Plant Height (cm)	Number of branches (pieces)	Number of leaves (strands)
Biochar Type			
No Biochar	44,31 a	60,25 a	445,28 a
Palm Fiber	44,49 a	55,25 a	468,88 a
Rice Husk	44,53 a	60,42 a	522,69 a
LOF Concentration (ml/L)			
0	43,52 a	62,56 a	479,19 a
5	45,92 a	51,67 a	463,67 a
10	44,24 a	67,78 a	507,35 a
15	44,09 a	52,56 a	465,59 a

Description: Numbers followed by the same letter in each treatment show no significant difference based on DMRT at 5%.

Table 1 shows that the concentration of fish waste liquid organic fertilizer has no significant effect on each treatment. Fish waste liquid organic fertilizer with a concentration of 5 ml/L gave the highest results on plant height. Fish waste liquid organic fertilizer has N content in the soil which helps increase plant height. According to [8], N element is useful for forming plant parts during the vegetative phase such as plant height.

3.2. Number of branches

Branches are plant organs that play an important role because they determine the position of the leaves. The results of the analysis of variance showed that the provision of biochar and fish waste liquid fertilizer and the interaction of the two did not significantly affect the number of branches. Table 1 shows that the provision of rice husk biochar treatment is not significantly different from without the provision of biochar and palm fiber biochar. Rice husk biochar gave the highest results on the number of branches. This is thought to be due to rice husk biochar which is able to help provide N nutrients to plants. This is in accordance with the statement [9] that rice husk biochar can help increase soil N uptake, provide N nutrients available to plants and can increase crop production. Table 1 shows that the concentration of fish waste liquid organic fertilizer has no significant effect on each treatment. Giving fish waste liquid organic fertilizer with a concentration of 10 ml/L gives the highest results on the number of branches. This happens because fish waste liquid organic fertilizer is able to provide nutrients needed by plants. [10] stated that liquid organic fertilizer from fish waste contains N (0.311%), P (0.154%), and K (0.036%). [11] added that the availability of nutrients and in sufficient quantities can launch plant metabolic activities so that the process of cell division, cell elongation and also tissue formation has increased which also increases plant growth in this case is the number of branches.

3.3. Number of leaves

Leaves are an important part of plants and function as a place for photosynthesis. The number of leaves will contribute to the amount of light that the plant receives. The results of the analysis of variance showed that the provision of biochar and fish waste liquid fertilizer and the interaction of the two did not significantly affect the number of leaves. Table 1 shows the effect of biochar type and LOF concentration on the number of leaves. Giving biochar rice husk gives the highest results on the number of leaves which is 522.69. The increase in the number of leaves is related to the increase in plant height. Giving rice husk biochar in this study also gave the highest results in plant height parameters followed by palm fiber biochar. According to [12], the number of leaves is related to plant height, because the higher the plant, the more leaves are formed. The treatment of fish waste LOF with a concentration of 10 ml/L gave the highest results on the number of leaves, namely 507.35. Fish waste LOF contains N and P elements that play a role in increasing the number of leaves. According to [12], Leaf formation is related to the content of nutrients such as nitrogen and phosphorus available to plants.

3.4. Leaf area

Leaf area will influence plant growth. The wider the leaves of a plant, the greater the amount of light it can receive and this can increase the rate of photosynthesis and photosynthate. The results of the analysis of variance showed that the provision of biochar types had a significant effect on the leaf area of mint (*Mentha spicata* L.).

Table 2. Effect of biochar type on mint (*Mentha spicata* L.) growth at 12 WAP.

Treatment	Leaf Area (cm^2)	Plant Fresh Weight (g)	Plant Dry Weight (g)
Biochar Type			
No Biochar	937,36 b	102,49 a	19,77 b
Palm Fiber	1018,64 ab	106,98 a	21,92 a
Rice Husk	1187,98 a	109,69 a	22,46 a
LOF Concentration (ml/L)			
0	1121,59 a	109,15 ab	22,38 ab
5	1163,81 a	117,62 a	23,43 a
10	903,88 a	104,08 ab	20,51 bc
15	1002,69 a	94,71 b	19,21 c

Description: Numbers followed by the same letter in each treatment show no significant difference based on DMRT at 5%.

Table 2 shows that the provision of rice husk biochar gives the best results and is significantly different from without biochar but not significantly different from palm fiber biochar on leaf area. This is explained by [13] that the provision of rice husk biochar is able to retain water, nutrients, especially N, P, K and other nutrients. According [14], K nutrients in plants make the leaves of plants not easily fall so that it affects the number of leaves which in turn affects the leaf area. The element K is needed by plants for the formation of leaf strength, leaf thickness, and leaf enlargement. Plants that get enough potassium can produce wider leaves.

3.5. Plant fresh weight (g)

The results of the analysis of variance show there is an interaction between the type of biochar and the concentration of LOF on the fresh weight of plants at 12 WAP. The results of further tests on the effect of biochar type and LOF concentration on plant fresh weight are presented in Table 2, while the results of further tests on the interaction of biochar type and LOF concentration on plant fresh weight at 12 WAP are presented in Table 2.

Table 2 show that biochar type has no significant effect on plant fresh weight at 12 WAP, while fish waste LOF treatment has a significant effect at 12 WAP. Organic matter in palm fiber biochar is able to make plants absorb water optimally in the soil thus increasing the water absorbed by plants. According to [15], Fresh weight of plants is influenced by the water content absorbed by plants. The more water in the soil that is absorbed, the higher the fresh weight of the plant. The increase in plant fresh weight is also caused by organic matter which affects photosynthate so that the fresh weight of plants increases. The use of appropriate fertilizer concentrations also affects plant fresh weight. According to [16], the use of appropriate fertilizers such as the right type, dose, concentration, time and method can increase crop yields both in quantity and quality. [17] stated that the more nutrients absorbed by plants, the more organic matter is used for photosynthesis. [18] states that optimal photosynthesis stimulates the growth of plant organs.

3.6. Plant dry weight

Plant dry weight describes the plant's ability to accumulate photosynthates produced in the process of photosynthesis. The results of the analysis of variance showed that there was an interaction between the provision of biochar and fish waste liquid fertilizer on plant dry weight. Based on the results of the analysis of variance in Table 2 shows that the provision of rice husk biochar gives the best results and is significantly different from without biochar but not significantly different from palm fiber biochar on plant dry weight. Table 2 shows that the application of fish waste liquid organic fertilizer with a concentration of 5 ml/L gives the best results and is significantly different from the concentration of 10 ml/L and 15 ml/L but is not significantly different from the concentration of 0 ml/L on plant dry weight. Based on the DMRT results in table 3, it shows that there is an interaction between the type of biochar and the concentration of fish waste liquid organic fertilizer. This is thought to be because biochar is able to increase plant uptake of nutrients provided by fish waste LOF. Palm fiber biochar combined with fish waste LOF increases soil fertility which will have an impact on plant dry weight. According to [19] the interaction between biochar and LOF has the potential to improve the physical, chemical and biological properties of the soil so that nutrients are more available in the soil. According to [20], biochar is able to increase water holding capacity and provide nutrients in improving nutrient uptake by plants thus increasing soil fertility. [21] states that optimal plant growth can be achieved if plants are water-sufficient.

4 Conclusion

Based on the research that has been done, the following conclusions can be drawn. The addition of palm fiber biochar and fish waste LOF concentration of 5 ml/L was able to increase the wet weight of plants and dry weight of mint plants. The addition of palm fiber biochar was able to increase the fresh weight of plants and the chlorophyll content of mint. The addition of rice husk biochar is the best type of biochar to increase leaf area and dry weight of mint plants. Fish waste LOF concentration of 5 ml/L was able to increase plant fresh weight and plant dry weight.

References

1. P. Y. Pratiwi, A. Mardiyaningsih, and E. Widarti, J. Ris. Kefarmasian

 https://doi.org/10.1051/e3sconf/202346701017

Indones. **1**, 148 (2019)

2. N. Hapsari and T. Welasih, Envirotek J. Ilm. Tek. Lingkung. **3**, (2013)
3. B. Satriawan and E. Handayanto, **1**, 69 (2014)
4. M. A. Baquy, M. A. Al Mamun, S. Mia, M. Alam, and M. S. H. Khan, **19**, 145 (2022)
5. W. Jiang, L. Xu, Y. Liu, W. Su, J. Yan, and D. Xu, Agronomy **12**, (2022)
6. P. Pangestu and Y. Tyasmoro, **7**, 1115 (2019)
7. V. H. M. Karamina, Hidayati, Bambang Siswanto, J. Ilm. Hijau Cendekia Vol. **10**, 379 (2022)
8. Darmawan, M. Yusuf, and I. Syahruddin, J. Agroplantae, **4**, 13 (2015)
9. W. Lestari, A. Aryunis, and A. Akmal, J. Agroecotania Publ. Nas. Ilmu Budid. Pertan. **5**, 13 (2022)
10. W. Lepongbulan, V. M. A. Tiwow, and A. W. M. Diah, J. Akad. Kim. **6**, 92 (2017)
11. N. Nasrulloh, T. Mutiarawati, and W. Sutari, Kultivasi **15**, 26 (2016)
12. D. Haryadi, H. Yetti, and S. Yoseva, Jom Faperta **2**, 33 (2015)
13. Hadili, Astina, and M. Safwan, J. Sains Pertan. Equator **10**, 1 (2014)
14. I. S. Rahmawan, A. Z. Arifin, and Sulistyawati, **3**, 17 (2019)
15. Muharam, J. Agrotek Indones. **2**, 44 (2017)
16. G. Jayantie, A. Yunus, B. Pujiasmanto, and Y. Widiyastuti, Agrotech Res J. **1**, 1 (2017)
17. Y. H. Listin Fitrianah, Siti Fatimah, Agrovigor J. Agroteknologi **5**, 34 (2012)
18. Yekti Puji Lestari, N. Helilusiatiningsih, and Eva Pebriana, J. Soilscape Agric. **1**, 32 (2022)
19. A. U. Rahmah and S. Anwar, **8**, 106 (2013)
20. M. Haris, M. Ali, W. Ahmad, M. Tariq, R. Alrowais, and M. Bilal, J. Build. Eng. **57**, 104850 (2022)
21. F. R. Ratmadanti and M. M. Maryani, J. Trop. Biodivers. Biotechnol. **2**, 1 (2017)

Growth and yields of some peanut varieties with application of palm fiber biochar

Samanhudi[1] *, *Muji* Rahayu[1], *Amalia Tetrani* Sakya[1], *Gani Cahyo* Handoyo[1], and *Bramantyo Bagus* Priambodo[1]

[1]Department of Agrotechnology, Faculty of Agriculture, Universitas Sebelas Maret, Surakarta, Indonesia

Abstract. Climate change negatively affects the growth of crops, including groundnuts. Groundnuts are susceptible to short-term weather changes, as well as climate change. Climate change can reduce groundnuts production because it affects soil fertility. Biochar is a kind of soil enhancer that can increase soil fertility. This study aims to determine the appropriate dose of palm fiber biochar for the growth and yield of various varieties of groundnuts. This research was conducted in July - December 2022 at the Experimental Field of the Faculty of Agriculture UNS Jumantono, Karanganyar. The study used a factorial Complete Randomized Block Design consisting of 2 factors, namely the type of peanut variety (Takar 2; Hypoma 1) and the dose of palm fiber biochar (0 t.ha-1; 7.5 t.ha-1; 15 t.ha-1; 22.5 t.ha-1). The variables observed were number of branches, number of flowers, number of pods, number of seeds, seed weight per plant, and harvest index. Data analysis used analysis of variance and was further tested using 5% Duncan's Multiple Range Test The results showed that the addition of palm fiber biochar at a dose of 7.5 t.ha-1 could increase the number of peanut flowers. Treatment of the Takar 2 variety had a higher number of flowers, number of pods, number of seeds, seed weight, and yield index than Hypoma 1.

1 Introduction

Climate change has a negative impact on soil conditions and texture. Climate change will decrease land productivity due to environmental imbalance, so land degradation cannot be avoided [1]. Land degradation also affects the physical condition of the soil, which results in changes in the properties and conditions of the soil. Apart from that, climate change can also cause a decrease in the quality, fertility, and carrying capacity of land so that the productivity of agricultural products also decreases. Peanut plants are one of the food crops that are very vulnerable to the impacts of climate change, where they can experience a decrease in production due to climate change [2]. One type of soil that is often used for peanut cultivation is alfisol soil. Alfisol soil is a type of soil that has low nutrient and organic material content. Apart from that, this soil has a clay texture, which causes a lack of oxygen supply and inhibits root growth [3]. Climate change can worsen the condition of

* Corresponding author: samanhudi@staff.uns.ac.id

alfisol soil, so that soil improvement efforts are needed, which can be carried out using soil-improving materials, one of which is biochar [4].

Biochar is a carbon-rich product defined as a solid material obtained through a pyrolysis process at temperatures ranging from 250ºC to >900ºC and oxygen availability below the limit [5]. Biochar has benefits as a soil amendment. Applying biochar to the soil can potentially increase soil C levels, water retention, and nutrients in the soil. Another advantage of using biochar is that the carbon in biochar is stable so that it can be stored for thousands of years in the soil [6]. Biochar plays a role in improving poor soil conditions such as leaching, acidity, and contamination. Besides that, biochar can reduce carbon emissions in the atmosphere directly and become renewable energy [7].

Biochar can be made from various agricultural waste, including palm fiber. The palm fiber used comes from palm tree trunks, which have been made into palm fibers. Palm fiber is one of the ingredients that can be used as biochar because it has a high cellulose content. The presence of cellulose in palm fiber biochar makes palm fiber absorb and store more water in the soil. In addition, the form of fibrous palm fiber biochar will increase the oxygen supply for plants.

Using palm fiber biochar as a soil conditioner for peanut cultivation can increase the growth and yield of peanuts. However, the correct dose of biochar fertilization for the growth of peanuts has yet to be found. This study aims to determine and understand the effect of administering various doses of palm fiber biochar on the growth and yield of various peanut varieties to obtain the optimal dose of biochar for peanut cultivation.

2 Materials and method

This research was conducted in July-December 2022 at the Experimental Field of the Faculty of Agriculture, Sebelas Maret University, Jumantono, Karanganyar. The method used was a factorial randomized complete block design (RCBD) consisting of 2 factors, namely the type of groundnut variety (Takar 2; Hypoma 1) and the dose of palm fiber biochar (0; 7.5; 15; 22.5 tons/ha). The experiment was repeated three times, obtained 24 units; in each unit, there were three plants, a total of 72 plants. The variables observed were the number of branches, number of flowers, number of pods, number of seeds, seed weight per plant, and harvest index. Data analysis used analysis of variance and was further tested using 5% Duncan's Multiple Range Test.

3 Results and discussions

3.1 Number of branches

Branches is a plant organ that functions as a place for leaves to grow and an intermediary or place to channel water and nutrients from the stem to the plant leaves. Table 1 shows that the treatment of a variety of types did not significantly affect the number of branches in plants aged 12 WAP. Giving various doses of palm fiber biochar also did not significantly affect the number of branches.

The Takar 2 variety had slightly more branches, namely 56.42 branches, compared to the Hypoma 1 variety, which only had 53.06 branches (Table 1). Providing biochar at various doses did not affect the number of branches throughout the plant's life. The genetic factors of each variety influence the number of plant branches. The number of branches can lead to more intense competition for light sources between leaves, resulting in lower net photosynthetic rates, smaller leaf areas, and shorter stem lengths per branch as leaves compete for light and resources used to produce biomass. To eliminate the negative effects of such competition as much as possible, plants can grow additional leaves as well as enlarge their total leaf area to make better use of limited light sources [8].

 https://doi.org/10.1051/e3sconf/202346701018

Table 1. Effect of giving various doses of palm fiber biochar on peanut growth at 12 WAP

Treatment	Number of branches (branches)	Number of flowers (flowers)
Variety		
Takar 2	56.42±1.19 a	4.14±0.70 a
Hypoma 1	53.06±1.19 a	2.83±0.23 b
Biochar (t.ha^{-1})		
0	57.67±2.07 a	2.33±0.58 b
7.5	56.83±1.48 a	3.56±0.28 a
15.0	53.39±0.95 a	3.44±0.21 a
22.5	51.06±2.60 a	3.28±0.09 a

Description : Numbers followed by the same letter in each treatment show no significant difference based on DMRT at 5%.

Branches on peanut plants are one of the places where gynophores grow, where gynophores are plant organs that will develop into peanut pods. Therefore, indirectly, a high or low number of branches will affect the yield of peanuts. The number of branches will affect the yield components, namely the number of plant pods, the yield of plant pods, the yield of plant seeds, and the weight of 100 seeds [9]. The number of branches positively correlates with pods and the number of filled pods per plant. The more the number of branches, the more the number of leaves, so the process of photosynthesis increases. Increased photosynthesis will also increase photosynthate, a material for filling pod seeds [10].

3.2 Number of flowers

Flowers are one of the plant organs that play a role in the formation of peanut pods and seeds. Table 1 shows that the treatment of a variety of types and the administration of various doses of palm fiber biochar did not have a significant effect on the number of flowers in plants aged 12 WAP. The Takar 2 variety significantly affected the number of flowers at 12 WAP, with an average of 4.14 flowers (Table 2). Varietal treatment is very influential, especially in forming peanut plant flowers. Each plant variety has a different genotypic response to the environment in which the variety grows. High-quality varieties generally have advantages that can be seen in high yields, response to fertilization, and resistance to pests and plant diseases [11].

Table 2 also shows that the application of palm fiber biochar significantly affects the number of flowers at 12 WAP compared to no biochar. Palm fiber biochar with a dose of 7.5 t.ha^{-1} produced the highest number of flowers, with an average of 3.56. Providing the correct dose of biochar has been proven to increase the number of flowers in peanuts. Biochar has excellent potential to increase the efficiency of using plant fertilizers by expanding the availability of nutrients in the soil [12]. Providing biochar into the soil can directly retain plant macronutrients such as N, P, K, and Ca [7]. Flower formation is influenced by the element phosphorus, where the element phosphorus can stimulate flowering. The presence of the element P can accelerate the flowering and ripening of cultivated plants' seeds and increase yield production [13]. Biochar is a stable carbon material that can last a long time in the soil. The use of biochar is good as a soil amendment for managing soil. In the long term, biochar can maintain the stability of soil organic C [14]. At 12 WAP, applying biochar can increase the number of flowers but cannot increase the number of branches because peanuts will shed some leaves and branches when they enter the harvest period [15]. Peanuts focus more on the growth of generative organs such as flowers and pods.

3.3 Number of pods

Peanut pods are one of the yield parameters to determine the generative growth process of plants and the production results from peanut cultivation. Table 2 shows that the treatment of varieties had a significant effect on the number of pods per plant. Giving palm fiber biochar had no significant effect on the number of pods.

Table 2. Effect of giving various doses of palm fiber biochar on peanut yield at 12 WAP

Treatment	Number of pods (pods)	Number of seeds (seeds)	Fresh weight of seeds (g)	Harvest index (g)
Variety				
Takar 2	24.08±2.02 a	43.00±3.49 a	27.25±1.33 a	0.88±0.12 a
Hypoma 1	18.36±2.02 b	33.14±3.49 b	23.48±1.33 b	0.54±0.12 b
Biochar (t.ha^{-1})				
0	18.89±1.65 a	34.39±2.60 a	22.95±1.71 a	0.71±0.00 a
7.5	22.61±0.98 a	40.83±1.95 a	26.38±0.72 a	0.73±0.01 a
15.0	22.61±0.98 a	40.33±1.60 a	27.39±1.43 a	0.73±0.01 a
22.5	20.78±0.31 a	36.72±0.95 a	24.73±0.45 a	0.67±0.03 a

Description : Numbers followed by the same letter in each treatment show no significant difference based on DMRT at 5%.

The Takar 2 variety treatment significantly affected the number of pods per plant, namely 24.08. The Takar 2 variety is one of the superior varieties with the advantage of being resistant to various diseases and having high productivity. The Takar 2 variety has the most number of branches but is low in plant height parameters, where the number of branches is also a factor affecting the number of peanut pods [16]. The dose of palm fiber biochar did not significantly affect the number of pods per peanut plant. One of the factors that significantly influence the formation of peanut pods is the availability of nutrients in the soil. Peanut plants are different from other legume plants in that peanuts produce flowers in the air and pods in the ground. The pods in the soil will absorb nutrients directly from the soil for their growth and development [17].

3.4 Number of seeds

Peanut seeds are the main production product in peanut cultivation and can be called peanuts. After fertilization, a gynophore will be formed, which acts as an organ that carries the peanut seeds into the soil. Table 2 shows that the variety type treatment had a significant effect on the number of seeds per plant. Giving palm fiber biochar had no significant effect on the number of seeds.

Based on Table 2, the treatment of the Takar 2 variety significantly affected the number of seeds per plant, namely 43 seeds. In comparison, biochar application did not significantly affect the number of seeds per plant. One factor that influences yield productivity, especially the number of peanut seeds, is the genotype of the cultivated variety. Using superior peanut varieties can increase the productivity of peanuts [18]. As its name implies, the Takar 2 variety has the advantage of being resistant to leaf rust disease and has high yield potential. Several factors affect the formation of peanut seeds, including the number of flowers and gynophores that are formed, environmental conditions, and what is quite important is the process of photosynthesis, which provides food and energy for the development of peanut seeds.

3.5 Fresh weight of seeds

The weight of peanut seeds can indicate the oil and protein content and various nutrients that are beneficial to health, especially the fatty acid composition. Tabel 2 show that the treatment of varieties had a significant effect on the fresh weight of the seeds. Giving palm fiber biochar had no significant effect on fresh weight of seeds.

Based on Table 2, variety treatment significantly affected the fresh weight of peanut seeds, whereas the Takar 2 variety had a higher fresh seed weight, namely 27.25 g (Table 2). Varieties play an important role in determining yield components because achieving high productivity is largely determined by the yield potential of the superior varieties planted [19]. One of the superior varieties of peanuts is the Takar 2 variety. The superiority of the Takar 2 variety is that it is resistant to leaf rust and wilt disease and has good pod quality. Treatment of palm fiber biochar doses did not significantly affect the fresh weight of peanut seeds. However, the treatment of palm fiber biochar with an amount of 15 $t.ha^{-1}$ had the highest average seed fresh weight of 27.39 g. The application of biochar will improve soil's physical and chemical properties, which include soil friability, soil water binding capacity, increased cation exchange capacity, increased soil organic matter content, and the ability to retain nutrients against leaching so they can be available for plants [20].

3.6 Harvest index

The harvest index is one of the observation variables that can show the productivity of peanut plants. The harvest index value indicates the level of efficiency of the plant in using the results of photosynthesis. Table 2 shows that the treatment of varieties had a significant impact on the harvest index, but not by administering various doses of palm fiber biochar.

Based on Table 2, the Takar 2 variety treatment significantly affected the harvest index, namely 0.88 g. The application of palm fiber biochar did not significantly affect the harvest index. These data show that the Takar 2 variety is more efficient in utilizing photosynthesis results, as seen from the high harvest index. The harvest index value indicates the level of efficiency of the plant in using the results of photosynthesis. The higher the harvest index value, the higher the seed yield produced. The differences in different harvest indices are caused by differences in varieties and the ability of each variety to adapt to the environment [21]. The genetic superiority of each variety is different, so each variety has a different level of productivity, depending on the nature of the plant variety itself [14]. The genetic factors of each variety greatly influence the results. Differences in genetic structure cause variations in the growth and yield of each variety studied [22].

4 Conclusion

Climate change has a negative effect on soil conditions and texture, which causes land degradation and affects crop productivity. The provision of palm fiber biochar acts as a soil enhancer that can improve soil problems. The use of superior varieties is also a method used to increase productivity. Based on the research results, providing 7.5 t.ha-1 palm fiber biochar can increase the number of peanut flowers. The Takar 2 variety treatment has a higher number of flowers, number of pods, number of seeds, seed weight and peanut harvest index than Hypoma 1.

References

1. D. N. Utami, J. Alami **3**, 122 (2019)
2. M. D. M. Kadiyala, S. Nedumaran, J. Padmanabhan, M. K. Gumma, S. Gummadi, S. R. Srigiri, R. Robertson, and A. Whitbread, Sci. Total Environ. **776**, 145996 (2021)
3. Kaharuddin, Dahlan, A. Farhanah, and A. Pona, J. Agrisistem **16**, 114 (2020)
4. I. N. Safitri, T. C. Setiawati, and C. Bowo, Techno J. Penelit. **7**, 116 (2018)
5. G. Bonanomi, F. Ippolito, G. Cesarano, B. Nanni, N. Lombardi, A. Rita, A.

Saracino, and F. Scala, Front. Plant Sci. **8**, (2017)
6. F. Fadhlina, J. Jamidi, and U. Usnawiyah, J. Agrium **14**, 26 (2017)
7. M. A. Al Baquy, M. A. Al Mamun, S. Mia, M. M. Alam, M. S. H. Khan, and S. M. Rahman, Sains Tanah **19**, 145 (2022)
8. H. Huan, X. Xu, G. Liu, J. He, Y. Yang, and K. Huang, Acta Ecol. Sin. **34**, 823 (2014)
9. A. A. Rahmianna, A. Wijanarko, J. Purnomo, and Y. Baliadi, Biodiversitas **21**, 5747 (2020)
10. N. Noertjahyani, C. Akbar, A. Komariah, and H. Mulyana, J. Agro **7**, 102 (2020)
11. A. A. Y. Astari, A. A. N. M. Wirajaya, and L. Kartini, Gema Agro **24**, 29 (2019)
12. Y. DING, Y. LIU, S. LIU, X. HUANG, Z. LI, X. TAN, G. ZENG, and L. ZHOU, Pedosphere **27**, 645 (2017)
13. A. Y. Setiawan, W. Eko, and T. Islami, **6**, 1171 (2018)
14. B. Suroso and A. J. Sodik, Agritrop J. Ilmu-Ilmu Pertan. **14**, 124 (2016)
15. Syahrani, Magrobis J. **14**, 46 (2014)
16. D. E. Puspita, Serambi Saintia **VI**, 26 (2018)
17. S. R. Kadirimangalam, G. Sawargaonkar, and P. Choudhari, J. Agric. Food Res. **9**, 100320 (2022)
18. N. Safira, S. Sumadi, and D. S. Sobarna, J. Agroteknologi **11**, 55 (2017)
19. D. N. Pandiangan and A. Rasyad, Jom Faperta **4**, 1 (2017)
20. A. K. Berek, P. O. Tabati, U. U. Keraf, E. Bere, R. Taekab, and A. Wora, Savana Cendana **2**, 56 (2017)
21. A. Karuniawan, R. Aulia, H. MAulana, D. Ustari, and N. Rostini, J. Agro **7**, 24 (2020)
22. A. Suriadi, F. Zulhaedar, and M. Nazam, Ilmu Pertan. (Agricultural Sci. **6**, 99 (2021)

Optimization of lowland productivity through fertilization as a food buffer against the impact of climate change

NP. Sri Ratmini[1]*, *IGM.* Subiksa[1], *Agus* Suprihatin[1], *IW.* Suastika[1] and *Damasus* Riyanto[2]

[1]Research Center for Food Crops, Research Organization for Agriculture and Food, National Research and Innovation Agency, Jl. Raya Bogor-Jakarta, Cibinong, Bogor 16911, West Java, Indonesia
[2]Institute of Agriculture (Intan) Yogyakarta – Magelang Street Km. 5.6 P.O Box 1059 Yogyakarta Special Region, Indonesia.

Abstract. Lowland is an agroecosystem that can be cultivated for agricultural production activities. Its wide potential and planting time are generally carried out in the dry season resulting in lowland areas being the main national food buffer during El-Nino. Its management, which is not optimal results in low production. Increased productivity of lowland land can be done by applying the optimum fertilizer. This study aims to determine the effect of NPK fertilizer on the growth and productivity of rice in lowland fields in South Sumatra. The study used a completely randomized block design with four treatments and six replications. The indicator plant uses rice variety Inpari 42. The results showed that the highest production was obtained in the combined treatment of 250 kg NPK ha^{-1} + 150 kg urea ha^{-1} + 1,000 kg dolomite ha^{-1} with a yield of 5.81 t ha^{-1}. These results indicate that NPK and urea fertilization combined with dolomite can increase the productivity of lowland rice fields as a buffer zone for rice production against the impacts of climate change.

1 Introduction

Agricultural land revitalization is one of the fundamental challenges that must be faced since it is important to meet national food needs. The Indonesian population increase quite high at a rate of 1.6% per year [1] and projected to reach 350 million people in 2050 [2]. It is due to increasing of food consumption per capita that have an impact on the increasing of national food supply. This condition could result in an increase of national food scarcity [3].

The lowland development as agricultural land is one of the efforts that can be carried out in increasing rice production to support national food security, especially on the impact of climate change. The specific advantage of lowland is as a reserve for rice production in the event of crop failure in other agricultural land, especially during El-Nino [4]. The lowland is characterized by deep puddles soil with high rainfall that varies between 50-200 cm [5] [6] at least 3 months a year. Currently, the cropping pattern management of the lowland has not been maximized resulting in low productivity namely around 3 t ha^{-1} in average [7]. This

* Corresponding author: nps_ratmini@yahoo.com

productivity is still far from the provincial rice yield average in around 5 t ha^{-1} [8] and planted with rice only once a year (CI 100) [9]. The use of technological innovations in lowland can increase rice yield, such as high-yielding varieties can increase the productivity of lowland rice in the ranges from 6-8 t ha^{-1} [10]

In addition to the use of varieties, fertilization is another effort that can be carried out to increase the productivity of lowland with low fertility status [11]. Generally, the organic-C content ranges from high to very high, the total N is classified as medium-very high, meanwhile available P, CEC and base saturation are classified as low-medium [12]. The lowland nutrients status is mainly influenced by the nutrients content of river water from the upstream area through the water runoff. The study results conducted by Syahri and Utami [13] show that the balance fertilization can increase paddy yields to 8.2 t ha^{-1} at the first planting season. Productivity in the second planting season is generally very low (< 3 t ha^{-1}), so there are few farmers planting rice in this season. Therefore, efforts are needed to increase rice production in the second planting season using balanced fertilizer and the addition of soil ameliorant in lowland rice fields. It is hoped that increasing rice productivity in the second planting season will support national food security, especially in the face of climate change.

This research aims to determine the effect of a combination of treatments between NPK and urea fertilizer and dolomite on the growth and yield of lowland rice in South Sumatra.

2 Materials and methods

The study was conducted on farmer's field at Gelebag Dalam Village, Rambutan, Banyuasin District, South Sumatra (-3,08929S; 104,85329E) (Figure 1), the annual average temperature range was 23°C-39°C. This experiment was carried out from November 2021 to February 2022. It was conducted in the second planting season (CI 200). The research location is classification as shallow non tidal swamps.

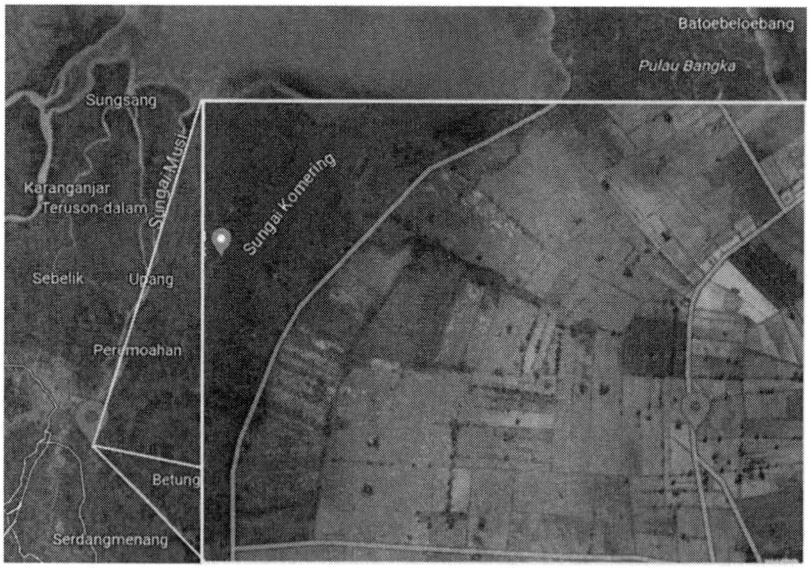

Fig. 1. Research site map at Gelebag Dalam Village, Rambutan, Banyu Asin District, South Sumatra.

Field experiments were conducted on farmers' land with a plot size of 640 m^2 and arranged using a randomized complete block design with four treatments and four replications. The treatment is shown in Table 1. The fertilization rate was determined based on the results of soil analysis using the Swamp Soil Test Kit (*PUTR*) prior to the study started. The plant indicator was paddy (*Oryza sativa*) Inpari 42 variety with a planting distance of 25 cm x 25 cm. The study used the NPK compound fertilizer (15:15:15) as basal fertilizer for all treatments. The Dolomite as ameliorant was given after the first tillage. The NPK and KCl fertilizers were applied at 7 days after planting (DAP), while the urea was applied twice at dose 50% of total for each, at the age of 7 and 30 DAP respectively.

The plant growth parameters observed were plant height and number of tillers at 30 DAP and harvest, productive tillers and yield components include panicle length, percentage of hollow grain and productivity. The statistical analysis used SPSS 22.0 including Analysis of Variance (ANOVA) and further tests with Duncan's Multiple Range Test (DMRT) at 5% level.

Table 1. List of treatments in this study.

Treatments	Remaks (kg ha^{-1})
P0 (Farmers practice)	150 Urea + 0.03 Ecofarming
P1 (Recommendation on PUTR)	250 NPK + 150 Urea + 1,000 Dolomite
P2 (75% Recommendation on PUTR)	150 NPK + 200 Urea + 1,000 Dolomite
P3 (125 % Recommendation on PUTR)	350 NPK + 100 Urea + 1,000 Dolomite

3 Results and discussion

3.1 Soil properties

Based on the initial soil sample analysis results using PUTR, it showed that the pH value of the soil is acidic, the score for N is classified as low and P and K values are moderate (Table 2). The soil sample taken after harvest showed that the average soil pH value in the treatment of P1, P2 and P3 was 5.41. This indicates an increase in pH due to the lime application. The lime application can suppress the solubility of aluminum and hydrogen to increase the soil pH. The liming process causes aluminum ions (Al^{3+}) to be replaced in the soil lattice by Ca and Mg to form Ca $(HCO_3)_2$ and Mg $(HCO_3)_2$ resulting in an increased soil pH [14]. The increasing of soil pH will affect changes in base saturation, Soil Cation Exchange Capacity, mineralization of organic matter, N transformation, nitrification, and denitrification, which further affects the production of N_2O in the soil [15] as well as the activity of microorganisms in the soil. Improvements will affect the availability of several nutrients, thereby increasing plant growth and production.

Table 2. Initial soil analysis of study site at Gelebag Dalam, Rambutan Banyuasin District, South Sumatra.

Soil chemical properties	Category
pH	Acid
N	Low
P	Moderate
K	Moderate

3.2 The effect of treatment on plant growth

The effect of fertilization on plant height, both at 30 days after planting (DAP) and prior to harvest, is presented in Fig. 2. The fertilization could increase the plant height both at the age of 30 DAP and harvest. The highest plant height at the age of 30 DAP and harvest were obtained in the P1 treatment, which was 97.60 cm and 109.56 cm respectively, while the lowest farmer's treatment was 81.20 cm. The number of productive tillers and tillers was also affected by NPK fertilization (Fig. 3). The highest number of one-month-old tillers and productive tillers were obtained in the P1 treatment, namely 24.70 stems and 20.13 stems respectively.

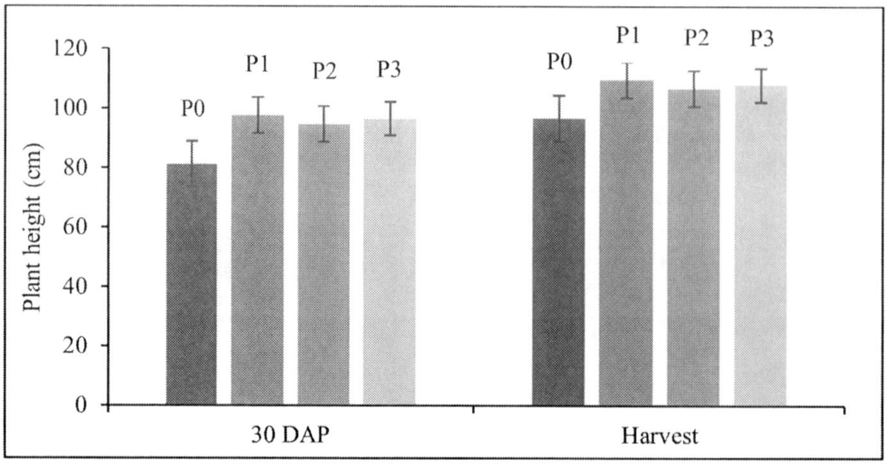

Fig. 2. The average of of plant height at 30 DAP and prior to harvest at the fertilization treatment for low land paddy plant.

Plant height parameter is one of the selection criteria in paddy plants, but the high plant height does not guarantee the level of production. Plant height has a great influence on the relationship between panicle length and yield. Well growing plants can absorb nutrients in large quantities. The availability of nutrients in the soil affects the increase in plant photosynthesis activity to increase the growth and components of plant yield [16].

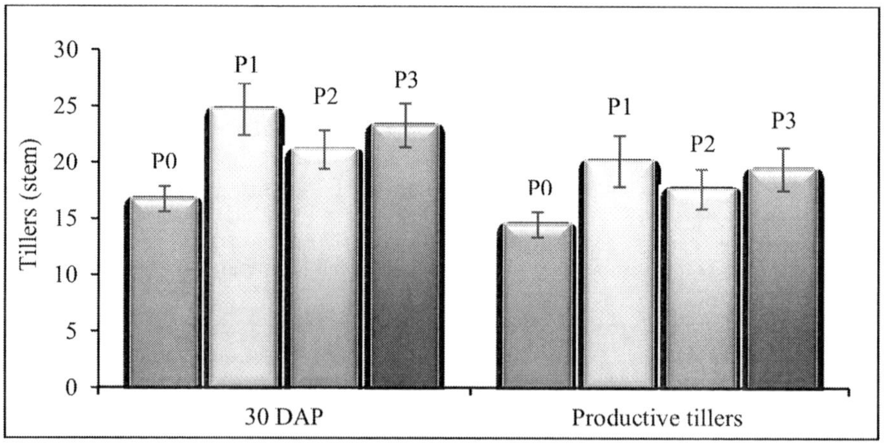

Fig. 3. The average number of tillers at 30 days after planting and the number of productive tillers at the fertilization treatment for lowland paddy plant.

3.3 The effect of treatment on yield and productivity components

The effect of fertilization on rice yield and yield components is presented in Table 3. The highest productivity obtained in the P1 treatment of 5.81 t ha^{-1} which was significantly different from the P0 (control) treatment, but not significantly different from the P2 and P3 treatments. The application of NPK fertilizer is expected to supply the availability of P and K nutrients. According to Wang et al. [17], the production can be increased by applying the balanced fertilizers. The P is very important because some of the main processes of plants include photosynthesis, root development, plant yield, and quality [18], [19], [20]. Other research results show that P fertilization can increase paddy yield, the higher the P is given, the more the production achieves [21]. The yield components of panicle length and percentage of hollow grain showed no noticeable effect on the control, but the fertilization tended to have a positive effect on the observed yield component.

Table 3. The effect of treatment on production components and paddy productivity.

Treatment	Panicle length (cm)	Hollow grain (%)	Productivity (t ha^{-1})
P0 (control)	20,22 a	16,28 a	4,82 a
P1	20,87 a	22,12 a	5,81 c
P2	20,68 a	17,97 a	5,34 b
P3	20,74 a	20,62 a	5,44 bc

Notes: P0=150 kg Urea ha^{-1} + 0.03 Ecofarming kg ha^{-1} ; P1=250 kg NPK ha^{-1} + 150 kg urea ha^{-1} + 1,000 kg dolomite ha^{-1} ; P2=150 kg NPK ha^{-1} + 200 kg urea ha^{-1} + 1,000 kg dolomite ha^{-1} ; P3=350 kg NPK ha^{-1} + 100 kg urea ha^{-1} +1,000 kg dolomite ha^{-1}. Means followed by different letters are significantly different (DMRT, $\alpha = 0.05$).

4 Conclusions

Lowland productivity can be increased by applying balanced and optimal fertilization. The application of NPK compound fertilizer combined with dolomite in accordance with the soil fertility status increased soil pH value and the plant growth, as well as rice yield up to 5.81 t ha^{-1}. These results showed that fertilizing NPK coupled with dolomite application is an opportunity to increase productivity of lowland as buffer zone in rice production against climate change impact.

References

1. G. Irianto, *Proceedings of the National Seminar on Integrated lowland Land Management* (Banjarbaru, Swamp Land Research Institute, (2006)
2. R. Saparita, J. Mathematic, Science and Technology **7**, 24 (2006)
3. S. Rusdiana and A. Maesya, Agriekonomika J **6**, 12 (2017)
4. A. A. Sulaiman, Y. Sulaeman, and B. Minasny, Resources **8**, (2019)
5. M. Noor and Y. Sulaeman *Pemanfaatan dan pengelolaan lahan Rawa* (Gadjah Mada University Press, 2021)
6. Haryono, M. Noor, Syahbuddin, M. Syarwani *Swampland research and development indonesia agency for agricultural research and development* (Jakarta: IAARD Press, 2013)
7. Suparwoto. dan Waluyo J. Pemb. Manusia **5**, 39 (2011)
8. S. S. Province and I. Figures, P*rovinsi dalam angka* (2022)
9. Sudana, J. Agricultural Policy Analysis **3**, 141 (2005)
10. NPS. Ratmini and Herwenita, in *IOP Conf Ser Earth Environ Sci* (IOP Publishing Ltd, 2021)
11. Djamhari, J. Hidrosfir Indonesia **5**, 1, (2009)
12. Waluyo, S. Djamhari, J. Sains dan Teknologi Indonesia **13**, 204 (2011)
13. Syahri, RU. Somantri, J. Assessment and Development of Agricultural Technology **17**, 211 (2014).
14. R. Paradelo, I. Virto, and C. Chenu, Agric Ecosyst Environ **202**, 98 (2015)
15. M. Shaaban, Q. an Peng, R. Hu, Y. Wu, S. Lin, and J. Zhao, Environmental Science and Pollution Research **22**, 19961 (2015)
16. S. Yoshida *Fundamental of Rice Crop Science*. (IRRI, Philippines, 269, 1981)
17. L. Wang, U. Ashraf, C. Chang, M. Abrar, and X. Cheng, J Soil Sci Plant Nutr **20**, 557 (2020)

18. D. P. Schachtman, R. J. Reid, and S. M. Ayling, *Update on Phosphorus Uptake Phosphorus Uptake by Plants: From Soil to Cell* (n.d.)
19. G. Henrique Demari, I. Ricardo Carvalho, M. Nardino, V. Jardel Simone Morgan Dellagostin, T. Corazza da Rosa, D. Nicolau Follmann, M. Andrade Monteiro, C. José Basso, T. Pedó, and T. Zanatta Aumonde, *International Journal of Current Research* **8**, 36629 (2016)
20. M. S. Imanudin, M. E. Armanto, and Bakri, Irrigation and Drainage **68**, 488 (2019)
21. P. Massawe and J. Mrema, Asian Journal of Advances in Agricultural Research **3**, 1 (2017)

Growth and yield of soybean varieties with biochar application

Amalia Tetrani Sakya[1*], *Muji* Rahayu[1], Samanhudi[1], *Gani Cahyo* Handoyo[1], and *Muhammad Dio Nugroho Anindra* Putra[1]

[1]Department of Agrotechnology, Faculty of Agriculture, Universitas Sebelas Maret, Indonesia

Abstract. Climate change cause a temperature increase, decreasing soil fertility and causing a reduced microbial and nutrient availability, and increased salinization, so it cannot support the growth of soybean plants. Soybeans are vulnerable to climate change, which can lead to lower yields. Biochar is an organic material that can improve soil's physical, chemical, and biological fertility. This study aims to obtain soybean varieties and biochar types that increase growth and yield in low soil fertility cultivation. This research was conducted in August 2022 – June 2023 at the Agricultural Faculty Experimental Field of Universitas Sebelas Maret. The research was designed in randomized complete block design factorial with three replications. The first factor was soybean varieties, and the second factor was types of biochar. Data analyzed using analysis of variance (ANOVA), followed by DMRT 5% if there was a significant difference. The results showed that Demas 1 variety significantly affects growth and yield with biochar addition compared to Dena 1 and Gepak Kuning. The addition of rice husk biochar can increase the growth and yield of 3 soybean varieties, compared to the addition of sugar palm fiber biochar and without the addition of biochar.

1 Introduction

Climate change can have a negative impact on soil properties and processes, which could threaten global food security since global demands for food and fiber for an increasing population are met by the soil. Climate change has potential negative impacts on soil health such as soil's organic matter, texture, hydrology, salinity, and fertility [1]. Soybean plants belong to the category of crops that are susceptible to the effects of climate change, as they experience reduced yields under its influence [2]. Low soil fertility resulted from climate change that caused by reduced microbial activity, reduced nutrient availability, and increased salinization inside the soil could also negatively affect soybean growth and yield, especially in alfisol soils that are already poor in nutrients and organic matter [3]. Alfisol soil also has a clay-like texture that has a low porosity and slow drainage, making it prone to expansion and shrinking that could happen from temperature changes. This condition will be exacerbated due to the impact of climate change.

* Corresponding author: amaliatetrani@staff.uns.ac.id

The challenges posed by climate change-induced soil fertility issues in alfisol soils can be addressed by incorporating soil-amending substances to enhance their quality. One of the soil-amending substance that can be used is biochar [4]. Biochar is an organic material synthesized through a pyrolysis process that comes from agricultural waste that could be easily obtained. The utilization of biochar as agricultural waste is a significant endeavor in promoting sustainable agriculture [5]. Biochar significantly affects soil fertility in degraded soil by fixing the chemical, biological, physical characteristics of the soil, and bind plant macronutrients [6]. The results showed that biochar increased soybean yields by increasing soil nutrient levels and soybean nutrient uptake. Applying biochar to plantings can provide a positive response to the stability of soil aggregates, soil CEC, soil organic C content, water and nutrient retention due to an increase in soil carbon [7]. Applying biochar such as rice husk improves soil penetration resistance because the formation of stable soil aggregates causes an increase in soil porosity and reduces soil density [8].

The use of biochar in crop cultivation is not to replace the role of fertilizer, but as a complement to fertilizer. Applying biochar can provide nutrients for plants, thereby increasing plant growth and yields. The information presented highlights the need for research on diverse biochar variations. This study aims to obtain soybean varieties and biochar types that increase growth and yield in low soil fertility cultivation

2 Materials and method

This research was conducted in August 2022 - June 2023 at the Agricultural Faculty Experimental Field of Universitas Sebelas Maret, Jumantono, Karanganyar. The study used a randomized complete block design with three replications. The treatment factors were soybean varieties, namely Dena 1, Demas 1, Gepak Kuning and biochar types, namely control (0 tons/ha), rice husk (15 tons/ha), and sugar palm fiber (15 tons/ha). Data analyzed using analysis of variance (ANOVA), followed by a DMRT level of 5% if the data was significant. Variables observed were plant height, number of leaves, stem diameter, flowering age, dry weight of stover, number of pods per plant, number of seeds per pod, weight of 100 seeds, number of root nodules, and harvest index.

3 Results and discussions

3.1 Soybean growth components

Plant growth which includes the vegetative and generative phases plays an important role in the life cycle of plant. Vegetative growth involves an increase in the volume, number, shape and size of vegetative organs such as leaves, stems, and diameter. Generative growth involves the growth of the generative organs starting from the formation of the initial flower.

Analysis of variance showed that the use of varieties significantly affected stem diameter, flowering age, and dry weight of stover. In contrast, the use of biochar had a significant effect on stem diameter (Table 1). The Dena 1 variety treatment showed the highest plant height growth at 10 WAP of 38.78 cm but not significant if compared to the Demas 1 and Gepak Kuning varieties; meanwhile, rice husk biochar gave the highest results on plant height of 40.06 cm but not significant if compared to other types. According to [9], genetic factors include plant innate characteristics such as tolerance and adaptation, and environmental factors such as climate, physical, biological conditions, and nutrient availability in the soil. Applying rice husk biochar can provide nitrogen nutrients plants need and encourage plant growth mechanisms [10].

Table 1. The effect of biochar application on plant height, stem diameter, and number of leaves of several soybean varieties at the age of 10 week after planting.

Treatment	Plant Height (cm)	Stem Diameter (mm)	Number of Leaves
Varieties			
Dena 1	38.78a	8.04b	82.67b
Demas 1	38.31a	9.69a	98.67ab
Gepak Kuning	36.10a	7.81b	108.00a
Biochar			
Control	36.37a	7.91b	89.04a
Sugar Palm Fiber	36.76a	8.44b	96.00a
Rice Husk	40.06a	9.19a	104.30a

Numbers followed by the same notation in the same column group show no significant difference at the DMRT level of 5%.

Demas 1 variety and rice husk biochar gave the highest stem diameter on 10 WAP of 9.19 mm compared to other treatment combinations. The Demas 1 variety had a significantly different stem diameter of 9.69 mm compared to the Dena 1 variety of 8.04 mm and Gepak Kuning of 7.81 mm (Table 1). The application of rice husk biochar significantly affect stem diameter of 9.19 mm compared to the other types of biochar (Table 1). Stem diameter rises with increased water and nutrient absorption, as stated by [11]. According to [12], nitrogen and hosphorus nutrients can help in the vegetative growth of plants, especially in plant stems. [13] stated that adding rice husk biochar increases soil nitrogen content, promoting plant growth. Optimal plant growth occurs with the right biochar quantity, demonstrating its crucial role.

The Gepak Kuning variety had the highest number of leaves with 108.00 leaves, but not significantly different from the Demas 1 variety, with 98.97 leaves. The type of biochar did not significantly affect the number of leaves, but the type of rice husk gave results on the number of leaves of 104.30 leaves, not significant if compared to the treatment of the type of sugar palm fiber biochar and without biochar (Table 1). Genetics influence leaf formation. [14] stated it as a factor shaping adaptability and organ development. Leaf number depends on photosynthate production influenced by nitrogen, impacting assimilated yield [15]. Table 2 below shows the effect of biochar application on flowering time and stover dry weight of several soybean varieties.

Table 2. The effect of biochar application on flowering time and stover dry weight of several soybean varieties.

Treatment	Flowering Age (WAP)	Stover Dry Weight(g)
Varieties		
Dena 1	6.07b	20.41b
Demas 1	7.33a	39.95a
Gepak Kuning	6.26b	17.35b
Biochar		
Control	6.52a	20.59a
Sugar Palm Fiber	6.70a	29.00a
Rice Husk	6.44a	28.11a

Numbers followed by the same notation in the same column group show no significant difference at the DMRT level of 5%.

The Demas 1 variety had the longest flowering age compared to other varieties, around 7.33 WAP, while the application of biochar did not have a significant effect on the flowering period of the three soybean varieties (Table 2). Less efficient phosphorus absorption alters flower appearance [16].Genetics and environment influence flowering age through vegetative growth and development [17].

The Demas 1 variety had a significantly different dry weight of stover of 39.95 g compared to Dena 1 of 20.41 g and Gepak Kuning of 17.35 g. The application of biochar type did not have a significant effect on the dry weight of the stover (Table 2). Genetic and environmental factors, unique to each plant, mutually influence stover dry weight due to varying photosynthetic abilities and conditions [18]. Soybean varieties adapt differently to the same environment based on their genetics, altering physical and biological traits for optimal growth [19].

3.2 Soybean yield components

Soybean yield components is one of the indicator to determine soybean productivity. Soybean yield components observed were number of pods per plant, number of seeds per pod, 100 seed weight, number of root nodules, and harvest index. Table 3 below shows the effect of biochar application on number of pods per plant, number of seeds per pod, and 100 seed weight of several soybean varieties.

Table 3. The effect of biochar application on number of pods per plant, number of seeds per pod, and 100 seed weight of several soybean varieties.

Treatment	Number of Pods per Plant	Number of Seeds per Pod	100 Seed Weight (g)
Varieties			
Dena 1	68.63b	1.94a	4.23a
Demas 1	122.41a	1.21c	2.06b
Gepak Kuning	96.89ab	1.60b	2.53b
Biochar			
Control	77.44 b	1.74a	3.08a
Sugar Palm Fiber	93.11ab	1.52b	2.61a
Rice Husk	117.37a	1.50b	3.13a

Numbers followed by the same notation in the same column group show no significant difference at the DMRT level of 5%.

The results of the analysis of variance showed that the use of varieties had a significant effect on the number of pods per plant, the number of seeds per pod, the weight of 100 seeds, and the yield index, while the type of biochar had a significant effect on the number of pods per plant and the number of pods per seed (Table 3). The Demas 1 variety produced the highest number of pods per plant with 122.41 pods, but was not significantly different from the Gepak Kuning variety with 96.89 pods. The application of rice husk biochar gave the highest number of pods per plant of 117.37 pods, significantly different from the treatment without biochar of 77.44 pods but not significantly different from sugar palm fiber of 117.37 pods (Table 3). Number of pod per plant rises due to favorable environmental conditions and enhanced photosynthetic processes, leading to increased assimilation production [20]. [21] stated that nitrogen plays a vital role in the formation of pods in plants.

The Dena 1 variety with rice husk biochar gave the highest yield on the number of seeds per pod compared to other treatment combinations. The Dena 1 variety had a significantly different number of seeds per pod of 1.94 seeds compared to the Gepak Kuning variety of 1.60 seeds and the Demas 1 variety of 1.21 seeds. The treatment without biochar addition had a significantly different effect on the number of seeds per pod of 1.74 seeds compared to rice husk and sugar palm fiber biochar (Table 3). The ability of plants to form seeds in pods is determined predominantly by the adequate nutrient status of plants [22].

The Dena 1 variety had a significantly different weight of 100 seeds of 4.23 g compared to the Gepak Kuning and Demas 1 variety, which were 2.53 g and 2.06 g. The type of biochar did not significantly affect the weight of 100 soybean seeds (Table 3). Seed weight indicates seed size, influenced by each variety's unique photosynthetic ability, impacting overall production [23]. Dena 1 seeds fall into the large category, with a weight of ± 14.3 g, while Demas 1 and Gepak Kuning seeds are in the medium category, weighing ± 13 g and ± 8.25 g, based on variety descriptions [24]. Table 4 below shows the effect of biochar application on number of root nodules and harvest index of several soybean varieties.

Table 4. The effect of biochar application on number of root nodules and harvest index of several soybean varieties.

Treatment	Number of Root Nodules	Harvest Index
Varieties		
Dena 1	11.33b	0.82a
Demas 1	25.26a	0.25c
Gepak Kuning	18.44ab	0.79b
Biochar		
Control	11.33b	0.62a
Sugar Palm Fiber	25.67a	0.57a
Rice Husk	18.04ab	0.68a

Numbers followed by the same notation in the same column group show no significant difference at the DMRT level of 5%.

Treatment of varieties and types of biochar significantly affected the number of root nodules. The Demas 1 variety significantly affected the number of root nodules of 25.26, not significant if compared to Gepak Kuning of 18.44. Sugar palm fiber biochar led to 25.67 nodules, significantly different from the control of 11.33, but not from rice husk biochar of 18.04 (Table 4). Root nodule formation varies due to soil nutrient availability, aeration, root size, and plants nitrogen-fixing ability, leading to differences between varieties [25]. Rhizobia facilitate biological nitrogen fixation in root nodules, converting nitrogen into usable forms [26].

The Dena 1 variety gave a significantly different harvest index of 0.82 compared to the Gepak kuning and Demas 1 varieties, which were 0.82 and 0.25. Biochar treatment did not significantly affect harvest index (Table 4). Genetic differences cause differences in yield index between varieties. [27] stated that the effect of crop production is influenced by genetic factor of the soybean. Each soybean has a different size and seed weight so it can affect the yield. Sugar palm fiber has a larger fiber texture and pores compared to rice husk, so the resulting aeration is also better. According to [28], this allows nitrogen-fixing bacteria to adapt and work better.

4 Conclusion

Climate change can have a negative impact on soil fertility, which could threaten global food security. Low soil fertility resulted from climate change could also negatively affect soybean growth and yield, especially in alfisol soils that are already poor in nutrients and organic matter. Biochar can be used as a sustainable solution to overcome the problem with its soil-fixing abilities. Based on the results of the research, the use of Demas 1 variety significantly affects growth and yield with biochar addition compared to Dena 1 and Gepak Kuning. The addition of rice husk biochar can increase the growth and yield of 3 soybean varieties, compared to the addition of sugar palm fiber biochar and without the addition of biochar.

References

1. S. Mondal, in (Springer Nature, Switzerland, 2021), pp. 551–569
2. Ruminta, Kultiv. **15**, 37 (2016)

3. A. Membretti, S. J. Taylor, and J. Delves, *Sustainable Futures in Southern Africa's Mountains* (Springer International Publishing, Turin, 2023)
4. A. Rohman, *Ensiklopedi Kedelai, Deskripsi, Filosofi, Manfaat, Budidaya, Dan Peluang Bisnisnya* (Penerbit KBM Indonesia, Yogyakarta, n.d.)
5. N. Nurida, Jubaedah, and Dariah, PPTP **3**, 67 (2019)
6. A. Tomczyk, Z. Sokołowska, and P. Boguta, Rev Environ Sci Biotechnol **19**, 191 (2020)
7. S. Bahri, N. Novianto, and S. Sumini, Adimas: J Pengabdian Kepada Masyarakat **4**, 1 (2020)
8. H. Asadi, M. Ghorbani, M. Rezaei-Rashti, S. Abrishamkesh, E. Amirahmadi, C. Chengrong, and M. Gorji, Rice Sci **28**, 325 (2021)
9. Sufardi, *Pengantar Nutrisi Tanaman* (Universitas Syiah Kuala, Aceh, 2020)
10. G. Abel, R. Suntari, and A. Citraresmini, Jurnal Tanah Dan Sumberdaya Lahan **8**, 451 (2021)
11. Kisman, Sumarjan, A. F. Hemon, S. M. Dewi, L. E. Susilowati, and B. W. Gunawan, IOP Conf Ser Earth Environ Sci **1107**, 012031 (2022)
12. F. Xu, C. Chu, and Z. Xu, Sci Rep **10**, 1033 (2020)
13. N. E. Ebido, I. G. Edeh, B. O. Unagwu, A. L. Nnadi, O. V. Ozongwu, S. E. Obalum, and C. A. Igwe, J. Soil Sci. Agroclim. **18**, 177 (2021)
14. W. Liu, B. Ming, R. Xie, G. Liu, K. Wang, Y. Yang, X. Guo, P. Hou, and S. Li, Agriculture **10**, 411 (2020)
15. M. D. Duaja, E. Kartika, and B. Buhaira, IOP Conf Ser Earth Environ Sci **391**, 012015 (2019)
16. Y. Zhang, B. Liu, F. Kong, and L. Chen, Front Plant Sci **14**, (2023)
17. S. Yuan, Y. Wang, J. Wang, C. Zhang, L. Zhang, B. Jiang, T. Wu, L. Chen, X. Xu, Y. Cai, S. Sun, F. Chen, W. Song, C. Wu, W. Hou, L. Yu, and T. Han, Front Plant Sci **13**, (2022)
18. A. Syaputra, Nurhayati, and C. Ichsan, Jurnal Ilmiah Mahasiswa Pertanian **3**, 128 (2018)
19. M. Herisva, T. Hidayat, and Zaitun, Jurnal Ilmiah Mahasiswa Pertanian Unsyiah **1**, 90 (2016)
20. M. Nazaruddin and Irmayanti, Jurnal Agrium **17**, 57 (2020)
21. Suhardianto, A. Karyawati, and S. Sitompul, Jurnal Produksi Tanaman **6**, 1 (2018)
22. D. Pandiangan and A. Rasyad, JOM FAPERTA **4**, 1 (2017)
23. A. Aziez, A. Budiyono, Suprapti, A. Prasetyo, and F. Pranantyo, Indian J. Agr. Res. **56**, 551 (2022)
24. Balitkabi, *Deskripsi Varietas Unggul Kedelai 1918-2016* (Kementerian Pertanian Republik Indonesia, Malang, 2016)
25. N. P. N. Lusiana, A. A. N. G. Suwastika, I. W. D. Atmaja, and A. A. I. Kesumadewi, Agrotrop **11**, 189 (2021)
26. K. Lindström and S. A. Mousavi, Microb. Biotechnol. **13**, 1314 (2020)
27. A. Amin, M. Munauwar, B. Bakar, A. Aziz, and M. Ismail, Agrotop **9**, 307 (2021)
28. A. Manasikana and K. Kusrinah, J. Biol. Appl. Biol. **2**, 28 (2019)

Gamma irradiation effects on seed germination, growth and morphological traits of kemiri sunan (*Reutialis trisperma* (Blanco) Airy Shaw)

Nur Ajijah[1*], *Cici* Tresniawati[1], *Nur Kholilatul* Izza[1] and *Azri Kusuma* Dewi[2]

[1]Research Center for Horticultural and Estate Crops, National Research and Innovation Agency. Cibinong, Bogor, Indonesia
[2]Research Center for Radiation Process Technology, National Research and Innovation Agency. Jakarta, Indonesia

Abstract. Kemiri Sunan is a biodiesel producing tree that has an ecological function as a reforestation plant since it can adapt to the marginal, dry and ex-mining land. It is necessary to increase the genetic diversity of kemiri sunan through mutation in order to support its breeding program. Dried seeds of Karmindo1 and M03 varieties were exposed to gamma sources with doses ranging from 50 to 1000 Gy to evaluate the effect of gamma irradiation on the germination and growth of kemiri sunan seeds. The percentage of germination, as well as physiological parameters of seedlings (shoot length and number of leaves) decreased with increasing irradiation doses. However, the two varieties showed a difference in sensitivity responses to gamma rays. The recommendation doses for the two varieties are 100 – 200 Gy and 50 – 100 Gy for Karmindo1 and M03 respectively. No morphological variation was found in seedlings except for one mutant with early branch formation. The development of this plant is important, not only as an alternative source of renewable energy, but also as one of the crops that can adapt to the threat of global climate change such as drought.

1 Introduction

Kemiri sunan (*Reutealis trisperma* (Blanco) Airy Shaw) is one of the estate crops that has strategic and economic value as a biodiesel producing plant. It also has ecological function as a reforestation or revegetation plant on marginal and post-mining land since it is able to adapt to marginal land. According to [1], one hectare of kemiri sunan plantation with a density of 100-150 trees/ha has the potential to produce 8-9 tons of crude oil or the equivalent of 6-8 tons of biodiesel/ha/year. Kemiri Sunan also has the potential to be developed as a mixed tung oil factory since its oil has an iodine number that varies between 122 – 160. With a fairly high iodine number, kemiri sunan has the potential to be developed as a producer of tung oil by mixing it with oil from its relatives (Euphorbiaceae), namely *Vernicia fordii* and *Vernicia montana*.

* Corresponding author: jijah_ridwan@yahoo.co.id

Tung oil is the name or designation for oil produced from the seeds of *V. fordii* or *V. montana* (formerly known as *Aleurites fordii* and *A. montana*) originating from China. Tung oil is sold in the market in the form of pure oil from *V. fordii* or *V. montana* seeds or in mixtures with other oils or with a catalyst. Tung oil is widely used in the paints, inks, plasticizers, pharmaceuticals and various chemical reagents industries [2]. Currently, the largest tung oil producing country is China which produces about 80,000 tons of oil per year or supplies around 70%-80% of tung oil in the world market [2]. However, the current production of tung oil has not been able to meet the needs of the international market [3]. The trade value of Tung oil in the international market is estimated to increase to US$ 330 billion in 2025, from around US$ 270 billion in 2018. Therefore, it is important to develop research on kemiri sunan as both biodiesel and tung oil producing plant [3].

Kemiri sunan is native to the Philippines (Luzon, Negros, Mindanao) and is the only species of the genus Reutealis. Although it isn't native to Indonesia, this plant is found in West Java, and has been developed on dry land with a dry climate in NTT and ex-tin mining land in Bangka Belitung [1]. Until now, there are four superior varieties of kemiri sunan that have been released as a result of selection from natural populations in West Java. The varieties namely karmindo 1, karmindo 2, kemiri sunan 1 and kemiri sunan 2 with a productivity range of 76.38 – 157.17 kg of dry seeds/ha/year and 45-50% oil yield in the kernel [1]. The productivity and oil yield are still far below expectations, namely <25 tons of dry seeds/ha and oil yield <60% (Herman, Personal communication). Therefore, genetic improvement of kemiri sunan is important to increase its productivity and quality.

As an introduced plant, the genetic diversity of kemiri sunan germplasm is narrowed. Therefore, efforts are needed to increase its genetic diversity through induced mutation or introduction. The availability of adequate genetic variability is required for the success of breeding program. Induced mutagenesis has contributed greatly to breeding programs of crop plants by creating mutant varieties with better and desirable genetic alterations in agronomic important traits [4]. Mutation induction can be carried out *in vivo* or *in vitro* using chemical mutagens or physical mutagens to artificially increase the number of mutations and expand genetic variability that can be exploited in breeding programs [5]. Gamma rays are the most widely used mutagen, more than 80% of mutant varieties were developed from mutants induced by gamma irradiation [4]. It is hoped that induced mutation can induce genetic variability of kemiri sunan and support its breeding program. This study aims to determine the effect of gamma ray irradiation on germination, growth and morphological characters of kemiri sunan.

2 Materials and methods

Four hundred and forty seeds of Karmindo 2 were irradiated using gamma rays at doses of 0, 100, 200, 300, 400, 500, 600, 700, 800, 900, and 1000 Gray (Figure 1). Furthermore, the same number of seeds from M03 accession was irradiated at doses of 0, 50, 100, 150, 200, 250, 300, 350, 400, 450 and 500 Gray. The size of the irradiated seeds ranged from 5-7 grams as many as 40 seeds per radiation dose. Radiation was carried out at the Research Center for Radiation Process Technology, Jakarta.

The irradiated seeds were then soaked in water for one night to break their dormancy. The next day the seeds were planted in plastic box using soil, sand and manure in a ratio of 1:1:1. Observations were made on the percentage of germination, the percentage of leaf forming seedling, hypocotyl length, seedling height, root length, number of leaves and leaf length and width. Seedling morphology was also observed. Determination of LD50 is done using CurveExpert Basic 2.2 software.

https://doi.org/10.1051/e3sconf/202346701021

3 Results and discussions

3.1. Seeds germination and seedling growth

Radiosensitivity assay is required to determine the appropriate radiation dose to induce the highest mutation rate with the most negligible effects on the gene complex [6]. Gamma ray radiation suppressed the germination and growth of Karmindo 2 and M03 seeds, except for gamma ray doses of 100 and 200 Gray for Karmindo 2, tended to encourage seed germination compared to controls (Table 1). Root length, plant height, hypocotyl length, number of leaves and leaf size were also longer, higher and greater at a radiation dose of 100 Gray compared to control and other radiation doses. The percentage of seedling forming leaves was also higher at a radiation dose of 200 gray compared to control and other radiation doses. There was no seed of Karmindo 2 germinated at radiation doses of 500-1000 Gray, while at a dose of 400 Gray the seeds could germinate but could not develop further into leaves (Table 1). Low radiation doses can stimulate the induction of metabolites and biochemical changes involved in plant regeneration, while high radiation doses can induce radio inhibition by affecting growth promoters and eventually tissue destruction [6]. It can also cause loss of regenerative capacity and malformation of plant tissues as well as tissue destruction [6].

Based on the results of Karmindo 2, then radiation was carried out on the seeds of M03 accession with a dose of 0-500 Gray with an interval of 50. The results showed M03 has different responses, all growth and development parameters of M03 tended to be depressed in irradiated seeds compared to control. No seeds germinated at radiation doses of 200-500 Gray (Table 2). At a dose of 150 Gray the seeds were able to germinate but could not develop further to form leaves (Figure 1B). This result is in line with the former results [7] that gamma radiation depressed the germination rate of Kemiri sunan 2 variety seeds, the higher the dose of gamma radiation, the lower the germination rate of the seeds.

Table 1. Gamma irradiation effect on seed germination and seedling growth of Karmindo 2.

Gamma Rays Doses (Gray)	Germination Rate (%)	Seedlings Forming Leaves (%)	Roots Length (cm)	Plant Height (cm)	Hypocotyl Length (cm)	Leaves Number	Leaf Length (cm)	Leaf Width (cm)
0	35.0	27.5	22.4	25.6	20.3	6.3	11.6	10.8
100	47.5	27.5	26.7	29.2	22.5	6.8	11.7	117
200	45.0	35.0	12.7	23.1	17.3	3.6	8.1	9.4
300	7.5	7.5	6.5	17.1	9.6	2	3.9	2.3
400	2.5	0	-	-	-	-	-	-
500	0	0	-	-	-	-	-	-
600	0	0	-	-	-	-	-	-
700	0	0	-	-	-	-	-	-
800	0	0	-	-	-	-	-	-
900	0	0	-	-	-	-	-	-
1000	0	0	-	-	-	-	-	-

https://doi.org/10.1051/e3sconf/202346701021

Table 2. Gamma irradiation effect on seed germination and seedling growth of M03.

Gamma Rays Doses (Gray)	Germination Rate (%)	Seedlings Forming Leaf (%)	Roots length (cm)	Plant Height (cm)	Hypocotyl Length (cm)	Leaves Number	Leaf Length (cm)	Leaf Width (cm)
0	23.33	23.33	15.30	22.17	15.71	5.57	10.80	10.06
50	6.66	6.66	16.65	16.42	12.25	6.00	10.00	9.00
100	10.00	10.00	14.17	13.83	8.83	3.68	8.40	7.67
150	6.66	0.00	7.50	8.65	5.65	0.00	-	-
200	0	0	-	-	-	-	-	-
250	0	0	-	-	-	-	-	-
300	0	0	-	-	-	-	-	-
350	0	0	-	-	-	-	-	-
400	0	0	-	-	-	-	-	-
450	0	0	-	-	-	-	-	-
500	0	0	-	-	-	-	-	-

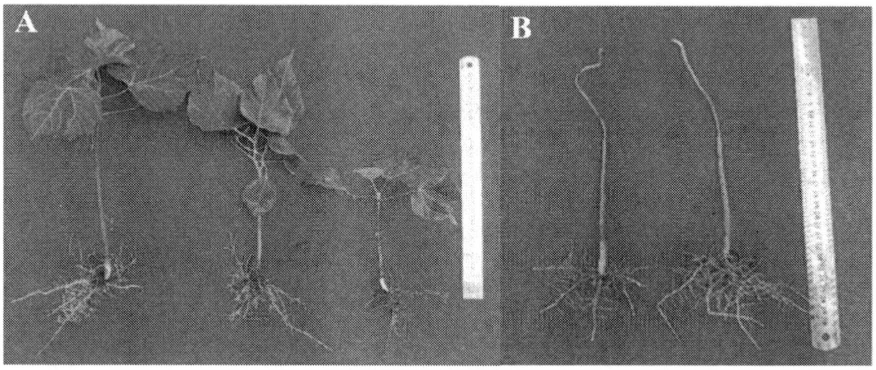

Figure 1. M03 seedlings three months after irradiation: A. From left to right, irradiation doses of 0, 50 and 100 Gray, respectively. B. Seedlings from radiation dose of 150 Gray didn't form leaves.

3.2. LD50 determination

Efficient induced mutagenesis by gamma radiation requires determination of the optimal radiation dose; that is, the dose that reduces 50% of the population (median lethal dose, LD50) [8]. The LD50 of Karmindo 2 and accessions M03 were different, as well as the LD50 of the Kemiri Sunan 2 variety reported by [7]. In this study, the LD50 for karmindo 2 was 293.92 Gray, while for M03 it was 60.56 Gray. Meanwhile, [7] has reported that the LD50 of Kemiri Sunan 2 is 93.86 Gray, approaching the LD50 of M03. According to [8] LD50 is influenced by the type of tissue (seed, callus or meristem), the stage of development and the water content in the tissue. M03 is an accession from Majalengka Regency, West Java, as well as the superior variety Kemiri Sunan 2. While Karmindo 2 is a superior variety selected from germplasm originated from Garut Regency, West Java. The recommended optimum dose could vary significantly from one study to another since the mutagenic effect can be influenced by genetic susceptibility and its physiologic status. For example, the successful dose for groundnut varies from 100-450 Gray, both for X and gamma rays [4].

3.3. Morphological variation

Morphological variation wasn't found in irradiated seedling except for early branching in M03 plant resulted from 50 Gray radiation (Figure 2). Meanwhile, [7] did not report any morphological variations in the Kemiri Sunan 2 resulted from irradiated seed except for variations in growth characters. However, in this study, observations were only made on the M1 generation as well as in [7]. According to [4], screening and selection should be started in the M2 generation, the first generation is not suitable for evaluation when multicellular material is mutagenized because the resulted plants will be genotypically heterogeneous (chimeric). The first non-chimeric generation in a seed-mutagenized and seed-propagated material is the M2. Visual screening is the most effective and efficient method for identifying mutant phenotypes [4].

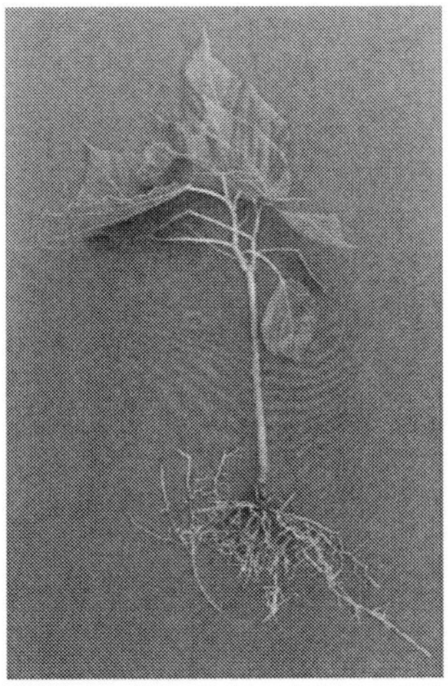

Figure 2. Morphological variation of M03 seedling resulted from 50 Gray radiation showing early branching.

4 Conclusions

Gamma radiation affected germination and growth of kemiri sunan Karmindo 2 and accession of M03. The two varieties have different responses to the gamma radiation doses. LD50 for Karmindo 2 was about 293.92 Gray and it was about 60.56 Gray for M03. Gamma radiation could also induce morphological change in M1 accession of M03.

References

1. D.Pranowo, M.Herman, Syafaruddin. Perspektif **14** 87–102 (2015)
2. Y.H.Chen, J.H. Chen, C.Y.Chang, C.C. Chang. Bioresour. Technol. **101** 9521–6 (2010)
3. Z.Zhan, Y. Chen, J.Shockey, X.Han, Y.Wang.. Molecules **21** 1–11 (2016)
4. M.Zakir. Int. J. Res. Stud. Agric. Sci. **4** 2454–6224 (2018)
5. E.Mullins, J.L.Bresson, T.Dalmay, I.C.Dewhurst, M.M.Epstein, L.G.Firbank, P.Guerche, Hejatko J, F.J. Moreno, H.Naegeli, F.Nogué, J.J.Sánchez Serrano, G.Savoini, E. Veromann, F.Veronesi, J.Casacuberta, P.Lenzi, I.Munoz Guajardo, T.Raffaello, N.Rostoks. EFSA J. **19** (2021)
6. M. De Lourdes, I.Castillo-ju, M. Soto. Plant **11** 1–19 (2022)
7. Parnidi, A.Ridhawati, M.Murianingrum. Prosiding Seminar Peripi komda Jatim pp 249–56 (2017)
8. A.Álvarez-Holguín, C.R.Morales-Nieto, C.H.Avendaño-Arrazate, R.Corrales-Lerma, F.Villarreal-Guerrero, E.Santellano-Estrada, Y.Gómez-Simuta. Rev. Mex. Ciencias Pecu. **10** 227–38 (2019)

Impacts of seawater level and human activities on Yeh Gangga Beach's coastal area at Tabanan Town

Anom Rajendra[1*], *I Wayan* Wiryawan[1], *I Ketut* Mudra[1]

[1]Udayana University, Architecture Department, Bukit Jimbaran, Badung, Bali

Abstract. The coastal environment of Bali has undergone significant changes due to rising seawater levels and human activities associated with tourism. As a result, public beach areas have become narrower, which has marginalized public interest and religious activities, particularly in coastal tourist resorts. To address this issue, this research focuses on Yeh Gangga's coastal area in Tabanan Regency, which is a crucial ritual site for the Tabanan Town Community, a traditional fisherman's territory, and a location of conflicts of interest among parties in the use of the beach. Empirical studies in the field, including observation, institutional data, and theories, were conducted to gather data. From the results of in-depth interviews conducted by a number of residents, they said that erosion has been going on for a long time and it is estimated that 20-50 meters of the coast towards the mainland has become the sea. Hence, the main objective of this research is to provide factual evidence of the decreasing public coastal areas as public spaces in Yeh Gangga Beach and to suggest that the local authorities build an organization body for integrated coastal zone management. The findings of this research will benefit academic purposes, urban planners, and local authorities.

1 Introduction

The impacts of rising sea levels and erosion have already wrought significant damage on the coastal environment, even in highly visited locations such as Bali Island. It is imperative that we take steps to address these issues to safeguard these regions and the communities that depend on them. It has been noted by Ewing that Sea Level Rise (SLR) can lead to erosion and is caused by extreme temperatures, storms, and global warming, which are all significant factors in coastal areas [1]. Recent risk studies and climate change adaptation plans have revealed that the island of Bali has experienced an SLR rate of around 3.5 mm/year to 8.0 mm/year based on historical data, including tidal data in Benoa and altimetry data [2]. This has hurt the development of coastal areas, causing changes in coastlines due to erosion [3].

In addition to natural factors, human activities such as land clearing and mineral exploitation on coastal land can also contribute to changes in the coastline and affect its balance. The coastline is dynamic and can shift according to tidal conditions, making it

* Corresponding author: rajendra@unud.ac.id

crucial for development planning and coastal protection efforts [4]. The Bali coastal area experiences various geographic processes that contribute to coastal damage through erosion and accretion, with the vulnerability of the beach itself being a major factor. The interaction between oceanographic aspects can also cause changes in coastlines [5]. Understanding changes in coastlines and coastal vulnerability is critical for effective management of Bali's coastal areas.

Moreover, the conversion of agricultural land to accommodate the growing population in Bali has led to the development of built environments and tourist accommodations along the coast. This has resulted in the shrinking of coastal areas and negative impacts on the environment and public interests. Unfortunately, land use practices that prioritize short-term gains without conservation efforts have led to the marginalization of public interests in coastal areas and the degradation of the environment. Weak policies and management have exacerbated these issues, causing losses in the agriculture and fisheries sectors. Research results have also clarified that the marginalization of public interests in coastal areas and the degradation of the coastal environment have been commonly caused by weak policies and management [6], bringing implications for losses in the fisheries and agriculture sectors [7].

To address these problems, Yeh Gangga Beach was chosen as a research location due to its strategic location for Tabanan community activities, traditional fishing practices, and the developing tourism industry. Despite severe erosion, the beach now accommodates tourist, religious, and fishing activities, while public space on the coast has decreased. This has led to conflicts of interest among parties. To address these issues, this research aims to determine the extent of erosion and its future impact, as well as the management of coastal areas as public spaces.

2 Literature review

There are several important concepts from the literature that will guide this research, including the definitions of coast and beach, sea level rise and surge storm, and coastal area.

2.1 Coast and beach

The coast is an interface between land and sea ecosystems that holds immense potential for natural resources and environmental services [8], [9]. Law No. 27 of 2007 recognizes coastal areas as transitional zones that are influenced by changes in land and sea. These areas can be defined as the meeting point of land and sea or the downstream areas with clear boundaries on the side that borders the sea. Although the terms "coast" and "beach" are often used interchangeably, they have different meanings.

2.2 Sea Level Rise (SLR) & storm surge

The rise of sea levels due to global warming plays a significant role in the changes happening in coastal areas [1]. It is one of the factors accelerating erosion rates, altering coastlines, and increasing inundation areas along the coast. Marine ecosystems and beaches are also at risk when sea levels and temperatures exceed the limits of marine biota to adapt [10]. Storm surges, especially from tropical storms, can cause serious disasters in coastal areas, particularly in tropical and sub-tropical zones [11]. These surges, caused by wind and pressure from cyclones, frequently hit coastal waters in the Indian Ocean, including Java, Bali, and West Nusa Tenggara. They can lead to flooding, relocation of people, and disruption of tourism and economic activities in the area. A simulation study revealed that the vulnerability to storm surge events is highest along the south coast of Java, Bali, and West Nusa Tenggara , where the highest surge was recorded at 19.0 cm [12].

2.3 The coast/beach erosion

Coastal erosion occurs when currents and waves cause the land to move to another location. If a sandy beach is affected, the sand will be transported parallel to the coast or into the deep sea. Beach erosion occurs when sand is moved due to water and wind factors and also as the result of water flow breaking and transporting sand over the surface [8]. The erosion process consists of three parts: detachment, transportation, and sedimentation [13]. The size of the eroded sand depends on the intensity and type of erosion, as well as the variables that affect the process. Climate, coastal material, topography, and vegetation cover are the four factors that determine erosion, and they are used to calculate erosion size with the Universal Soil Loss Equation (USLE) [14].

3 Research method

This research uses a descriptive quantitative method, in which a selected site was based on several reasons fulfilling the research objective. The data that has been obtained was analyzed systematically to generate coastal erosion process data and its impacts. To get a treatment strategy for coastal erosion. it was carried out through literature studies. The data in this research includes secondary data and primary data. Data secondary are from the existing, such as from related research, previous measurements, and observations. These secondary data were expected to support the research. The secondary data included wind data, tidal data, and research results which have been done before. Meanwhile, primary data was obtained through survey activities in the form of observation such as measurement of coastal morphological data, wind speed and wave, and current coastline data, whilst direct interview activities were carried out in the field.

Yeh Gangga Beach is a picturesque destination located in Sudimara Village, Tabanan District, Tabanan Regency.

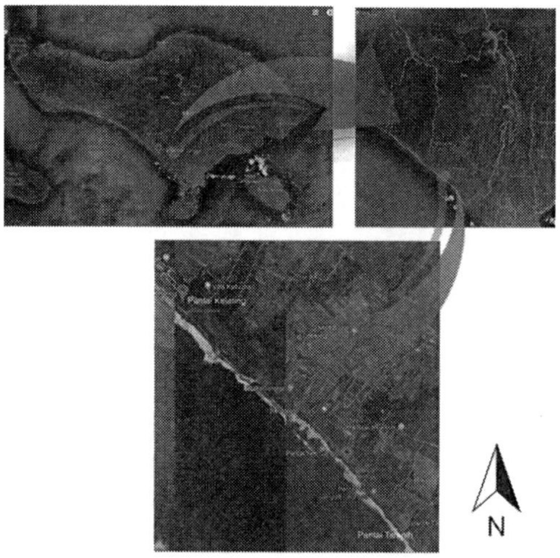

Source: Google Map

Figure 1. Map of Research Location

It is approximately 11 km away from the center of Tabanan City, 9 km from the renowned Tanah Lot Temple, and roughly 31 km from I Gusti Ngurah Rai International Airport (Figure 1). This stunning beach is situated in the southern part of Tabanan Town facing the Indian Ocean, boasting a coastline that spans 2.16 km. Interestingly, only 7.16% of the entire beach length which is a total of 30.18 km. It falls within the Tabanan Regency (Table 1).

4 Results and discussion

4.1 Result

Based on research conducted by PT Parama Kridha Pratama Consultants, erosion has affected Yeh Gangga Beach over a distance of 1.13 km at a rate of 1.58m/year. Unfortunately, only

60% of this erosion was addressed in 2015. This erosion has resulted in severe environmental damage, with the sand loss categorized as heavy at 1.58m/year. The prioritized weight for treatment is between 400 to 499, emphasizing the urgent need for treatment at Yeh Gangga Beach.

Table 1. List of beach section names according to their village in Tabanan Regency.

No	Name of Coast Section	Village	Coastline - Citra SPOT (km)
1	Nyanyi and Tanah Lot	Beraban	3.18
2	Kedungu	Belalang	0.99
3	Pangkung Tibah	Pangkung Tibah	2.18
4	**Yeh Gangga**	**Sudimara**	**2.16**
5	Kelating	Kelating	1.39
6	Tibu Biyu	Tibu Biyu	2.36
7	Beraban, Pasut and Abian Kapas	Beraban	1.19
8	Tegal Mengkeb and Klicung	Tegal Mengkeb	2.07
9	Brembeng	Brembeng	0.15
10	Soka and Srijong	Antap	6.70
11	Lalanglinggah, Mekayu and Balian	Lalang Linggah	4.80
12	Selabih	Selabih	3.01
	Tabanan Regency		30.18

Source: Consultan Engnineering of PT. Parama Kridha Pratama.

Since being designated as a special tourist attraction area (*Kawasan Daerah Tujuan Wisata Khusus/KDTWK*) under Tabanan Regency Regional Regulation No. 11/2012, Yeh Gangga Beach has experienced rapid development in its tourism industry. What was once a location for fishing and rituals has now become a destination for both domestic and foreign tourists. As the number of visitors to the area has increased, so too has the growth in tourism facilities, including hotels, villas, restaurants, and recreational activities. This trend began in 2000 and has reached its peak in the last decade (Figure 2).

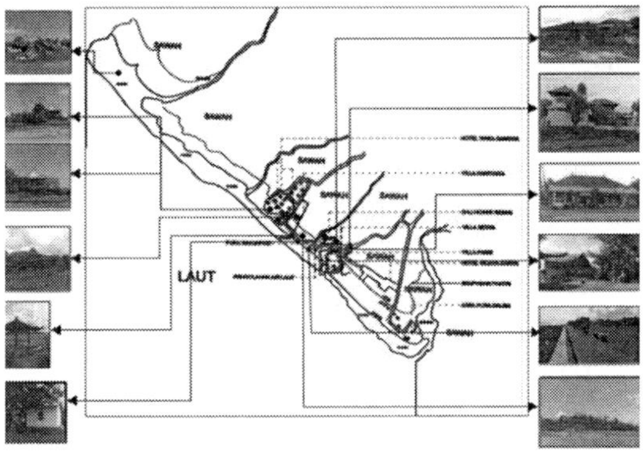

Figure 2. The images of building facilities at Yeh Gangga Beach

4.2 Discussion

4.2.1 The impact of erosion

It was discovered through interviews with residents and the head of Sudimara Village that the Yeh Gangga coastline has retreated by 20-50 meters. Studies using commercial GIS and

CAD software have also revealed that erosion has affected over 50% of the 2.16 km-long Yeh Gangga Beach, equating to 1.15 km in total. The average yearly loss of coastal area is 1.58 meters, indicating that the beach will lose approximately 47.4 meters of coastal area within the next 30 years. This will result in a loss of 48,111 square meters (or 4,811 hectares) of coastal area, using the length of erosion as 1.15 km. If the estimated price range for the coastal area is 7 million rupiah/m2, the total loss would amount to a staggering 336,777 billion rupiah. These losses are not just economic, but also social, cultural, and political, as they result in a loss of open spaces for public use. Furthermore, beaches affected by erosion will become longer and wider, leading to more severe environmental damage. It is certainly a sad outcome that will have far-reaching impacts.

4.2.2 The impact of human activities

Development in the Yeh Gangga Beach Area was not followed by a strict zoning plan by considering and implementing the coastal border in accordance with existing regulations. The weak building permitting process has resulted in the construction of buildings that are not in accordance with existing regulations, especially the implementation of the coastal border. Based on Presidential Decree No.32 / 1990 and other regulations, it is stated that the coastline is 100m measured from the highest tide point towards the mainland. This proves how weak the application of regulations is in granting building permits. Because of the violation of this boundary, the building directly takes up the beach space in front of it. Loss of beach space is not only caused by erosion but also by utilization by individuals/groups at the expense of the public interest. In fact, the beach space that is still in front of it is used for the benefit of the owner of the building behind it for the sole purpose of commercial interests, ignoring the public interest.

4.2.3 The eroded beach handling strategies

The issue of erosion and its impact on coastal areas is a pressing concern that requires immediate action. To effectively resolve this problem, it is important to establish a management body that is representative of all parties involved in coastal use. This body's primary objective is to prevent violations of coastal boundaries and preserve the area. Based on the experience of managing beaches in other parts of Bali, the management body should consist of four elements: indigenous communities, tourism service entrepreneurs, local government, and expert groups or academics to ensure democratic decision-making. Since handling erosion requires a significant amount of funds, cooperation and donations from all parties are necessary. Furthermore, continuous and sustainable conservation measures must be implemented to ensure long-term success. Without taking these steps, proper handling of erosion will be impossible.

5 Conclusion

Erosion taking place in Yeh Gangga Beach will never stop as long as climate change still exists on the earth. Climate change will raise the sea water level and then cover the land to become part of the sea. Without real protection, erosion will become larger and more widespread, resulting in greater loss of land as coastal space. The existence of fishing activities, rituals, and recreational activities for the community will be threatened. Whilst, human activities, especially physical development to support their commercial and tourist activities, which tend to violate coastal boundaries in regulating the distance from buildings to the beach, have become another threat to the loss of coastal space. Coastal space is very crucial in its existence at this time, where its existence in plants will be lost without efforts and efforts to overcome it. To handle this problem, an integrated management body is needed which consists of four elements: local indigenous communities, tourism and other commercial service entrepreneurs, local government, and experts/academics.

The majority of this article derives from historical and empirical studies and observations. We would like to thank to Tabanan Government and the head of Sudimara Village for supporting additional documents. Also special thanks to the Department of Architecture, Faculty of Engineering, Udayana University for their support in encouraging the completion of this article, as well as the Rector of Udayana University for their support.

References

1. L. Ewing, in *Handb. Coast. Ocean Eng.* (World Scientific, 2010), pp. 997–1021
2. H. Bachtiar, J. Sumber Daya Air **8**, 157 (2012)
3. F. M. Fajrin, M. R. Muskananfola, and B. Hendrarto, Manag. Aquat. Resour. J. **5**, 43 (2016)
4. N. Anggraini, S. Marpaung, and M. Hartuti, J. Penginderaan Jauh Dan Pengolah. Data Citra Digit. **14**, (2018)
5. S. Sakka, P. Paharuddin, and E. Rupang, Torani J. Fish. Mar. Sci. **24**, (2014)
6. J. Juhadi, J. Geogr. (2007)
7. J. Sihite, Southeast Asia Policy Res. Work. Pap. 91 (2001)
8. J. R. Clark, Ocean Coast. Manag. **37**, 191 (1997)
9. O. S. Djunaedi, Bandung: Widya Padjajaran (2011)
10. H. Latief and O. Dominic, Minist. Environ. Indones. (2010)
11. I. P. Samskerta, H. Bachtiar, and F. Riandini, Kolok. Pus. Penelit. Dan Pengemb. Sumber Daya Air 1 (2011)
12. N. S. Ningsih, S. Hadi, A. B. Harto, M. D. Utami, and A. P. Rudiawan, ILMU Kelaut. Indones. J. Mar. Sci. **15**, 179 (2010)
13. M. E. R. Alie, (2015)
14. C. Asdak, *Hidrologi Dan Pengelolaan Daerah Aliran Sungai* (UGM PRESS, 2023)

The effectiveness of acid-tolerant antagonists in the control of oil palms root necrotic caused by *Ganoderma* sp. in peat soils

Supriyanto[1,3]*, Purwanto[2], Susilo H Poromarto[2], and Supyani[2]

[1]Department of Agricultural Science, Graduate School, Sebelas Maret University, Surakarta, Indonesia
[2]Faculty of Agriculture, Universitas Sebelas Maret, Jl. Ir. Sutami 36A Kentingan, Surakarta, Central Java, 57126 Indonesia.
[3]Department of Agriculture Cultivation, Faculty of Agriculture, Universitas Tanjungpura, Jl. Prof. Dr. H. Hadari Nawawi, Pontianak, Indonesia.

Abstract. Oil palm is the main plantation commodity in Indonesia. In the last two decades, Indonesian palm oil has always been associated with global warming because most of its plantation areas are located on peatlands and the cultivation method is considered not environmentally friendly. One of the problems of oil palm cultivation on peatlands is stem rot disease caused by the fungi *Ganoderma* sp. This disease is transmitted through root contact. There is no effective way to control this disease. Biological control is an alternative method that is environmentally friendly. This study aims to determine the effectiveness of acid-tolerant antagonists in controlling oil palm root necrotic on peat soils. The study tested 5 isolates of the acid-tolerant antagonists in reducing oil palm root necrosis caused by *Ganoderma* on peat soils in experimental gardens. The results showed that the acid-tolerant antagonists were effective in controlling root necrosis in oil palm seedlings on peat soils. Its effectiveness reached 16.67%. Differences in peat soil pH and peat maturity did not affect the effectiveness of root necrotic control caused by *Ganoderma* sp. These results indicate that using acid-tolerant antagonistic fungi can help reduce Ganoderma sp attacks on oil palms on peat soils.

1 Introduction

Oil palm (*Elaeis guninensis* Jaqc.) is an important contributor to global vegetable oil production. In the last two decades, Indonesian palm oil has always been associated with global warming because most of its plantation areas are located on peatlands and the cultivation method is considered not environmentally friendly. Indonesia, the largest palm oil-producing country in the world, has a large area of oil palm plantations in peatlands, which is 2.046 million ha, equivalent to 14.58% of the total area of Indonesian palm oil, which until 2017 was 14.03 million ha [1, 2, 3]. A serious problem of oil palm plantations in peatlands is the high incidence of basal stem rot (BSR) disease caused by *Ganoderma* sp.

* Corresponding author: supriyantountan2013@gmail.com

This disease is transmitted through root contact and causes palm death [4, 5]. Nowadays, it is stated that BSR disease is a very threatening disease for the sustainability of oil palm plantations on peatlands due to its increasing intensity [6].

Until today, there are no effective methods to control BSR disease in peatlands [7, 8]. Biological control is an alternative control method that is the focus of current development [9]. However, biological control of BSR disease in peatlands has not been widely reported. Tropical peatlands in the lowlands of Kalimantan have an average pH of 3.3 - 4.3 [10]. Allegedly, low soil pH causes the antagonist to be unable to develop or causes its functional ability as an antagonist to not function properly. Based on the results of preliminary research, it is known that in the peatlands of West Kalimantan there are acid-tolerant fungi and bacteria which are also capable of acting as an antagonist to Ganoderma in vitro. Some fungi are known to be able to grow at pH 2, although their functional ability as an antagonist can only be active at pH 3 – 6 [11]. Likewise, bacteria are known to be able to grow at pH 2, even though the growth rate has decreased significantly [12]. However, this ability has not been tested in the field. This research was conducted to determine the effectiveness of the use of indigenous acid-tolerant antagonists in the control of oil palm root necrotic in peat soils.

2 Materials and methods

The research was conducted at the Plant Disease Laboratory and experimental garden at the Universitas Tanjungpura, Pontianak, Indonesia (0° 3'27,875" N, 109°20'54,3 E).

2.1 Experimental design

The experiment was carried out in natural peat soil, sterile peat soil, peat soil that was pH-changed, and mineral soil as a comparison. The peat soils used were hemic peat soil and sapric peat soil. A total of 3 isolates of Ganoderma-antagonistic fungi, namely isolates 13EJ15, B3J19, and E4J8 and 2 isolates of Ganoderma-antagonistic bacteria, namely isolate E4B6 and isolate E2B12 were tested in this study. Each antagonist was obtained from oil palm plantations on West Kalimantan peat land [11,12]. Each antagonist isolate was applied singly and mixed isolates. *Ganoderma* sp. from peatlands, namely *Ganoderma* G301 isolate was used as the test pathogen. As a comparison, positive and negative control treatments were used in this experiment. All treatments were arranged in factorial completely randomized design and replicated 3 times.

2.2 Peat soil preparation and oil palm seedlings

The soil was made friable and weighed in air-drying condition, and as much as 5 kg of peat soils were placed in a 30 x 40 cm polyethylene bag (polybag). Mineral soils were measured by volume to equal with the volume of peat soil in polybag in air-drying conditions due to the difference of their bulk density. Sterile peat soil was obtained by sterilization with the steaming method for 6 hours. To obtain a different peat pH, the peat soil was added with dolomite ($CaMg(CO3)_2$) at a dose of 12.5 g/kg and 25 g/kg. Soil pH was measured at the beginning of medium preparation, 2 weeks after liming by the dolomite, and at the end of the experiment [13]. This research used commercial oil palm seeds of Marihat variety from PPKS Medan, Indonesia. Oil palm sprouts were grown in 10 x 20 cm polybags containing a mixture of peat soil and mineral soil in a 1:1 ratio. Before it was used, the medium was sterilized by steaming for 6 hours. Seedlings were cultivated for two months under 50% shade.

2.3 Preparation of inoculum and inoculation of oil palm seedlings with *Ganoderma*

The Ganoderma inoculum was prepared by growing the *Ganoderma* G301 onto 6 x 6 x 12 cm rubber woodblocks. The inoculum was ready to use after the mycelium has covered all the rubber woodblock surfaces [14]. Ganoderma inoculation of oil palm seedlings carried out by rubber wood blocks that have been overgrown with Ganoderma mycelium were buried in the soil in polybags 1 week before the oil palm seedlings were planted at a depth of 10 cm [15]. The 2-month old oil palm seedling was planted on the medium, placed on a rubber woodblock and the main root was made sure to sit on the rubber wood block, and then backfilled. Oil palm seedling was then cultivated for 24 weeks.

2.4 Preparation and inoculation of antagonists onto peat soil medium

The antagonistic fungi and bacteria were rejuvenated. The antagonistic fungal inoculum was prepared in a corn-water medium (1:1 ratio) for 2 weeks. The bacterial inoculum was prepared in a bacterial suspension with a minimum density of 10^8 cfu/ml. The antagonist inoculum was given into the medium according to the predetermined treatment. Fungal isolates were buried and mixed into the planting medium 3 days after Ganoderma inoculation. Antagonistic bacteria were given in a bacterial suspension by soil drenching a day before planting oil palm seedlings.

2.5 Observation of Root Necrotic Intensity

At the 6th month, the roots are removed by dismantling the growing medium in the pots. The necrotic symptom on the roots was scored (Table 1) [16].

Table 1. Criteria for the scoring of oil palm seedling root necrotic.

Score	Criteria
0	No necrotic symptoms
1	$1 < x < 20\%$ necrotic progress less than 20%
2	$20 \le x < 40\%$ necrotic develops 20 – 40%
3	$40 \le x < 60\%$ necrotic develops 40 – 60%
4	$x > 60\%$ necrotic develops more than 60%

$$NI = \frac{\sum_{i=1} (k \times nk)}{Z \times N} \times 100\% \tag{1}$$

NI = Necrotic intensity
nk = number of plants with a score of k (k=0,1,2,3,4)
k = Score used.
Z = Highest score
N = Number of plants observed

3 Results and discussion

Based on the observation of root necrosis, except in the treatment of B3J19 isolate and the treatment of consortium of bacteria, the intensity of root necrosis was higher than the control treatment. The intensity of necrosis was varied, ranging from 16.67% to 50%. The lowest average of intensity for various treatments of hemic peat soil medium was obtained in the treatment of B3J19 isolate, i.e. 14.59%. The highest average of intensity was obtained in the consortium treatment of fungi and bacteria. Whereas in sapric peat soil, all the acid-tolerant antagonist treatments resulted in a lower intensity of root necrosis than the control treatment except for the B3J19 isolate fungus antagonist treatment. The lowest average intensity on various treatments of sapric peat soil medium was obtained in the E2B12 isolate treatment (25%), and the highest was in the B3J19 isolate treatment (Figure 1).

Observations on the roots after the demolition of the oil palm seedling growth medium also demonstrated that root infection by pathogenic Ganoderma only occurred in secondary roots. These results are different from those reported before where the infection occurred at the primary root [18]. This is presumed to be related to differences in the aggressiveness of the Ganoderma used in this experiment. Ganoderma was known to have different levels of aggressiveness [19, 20, 21].

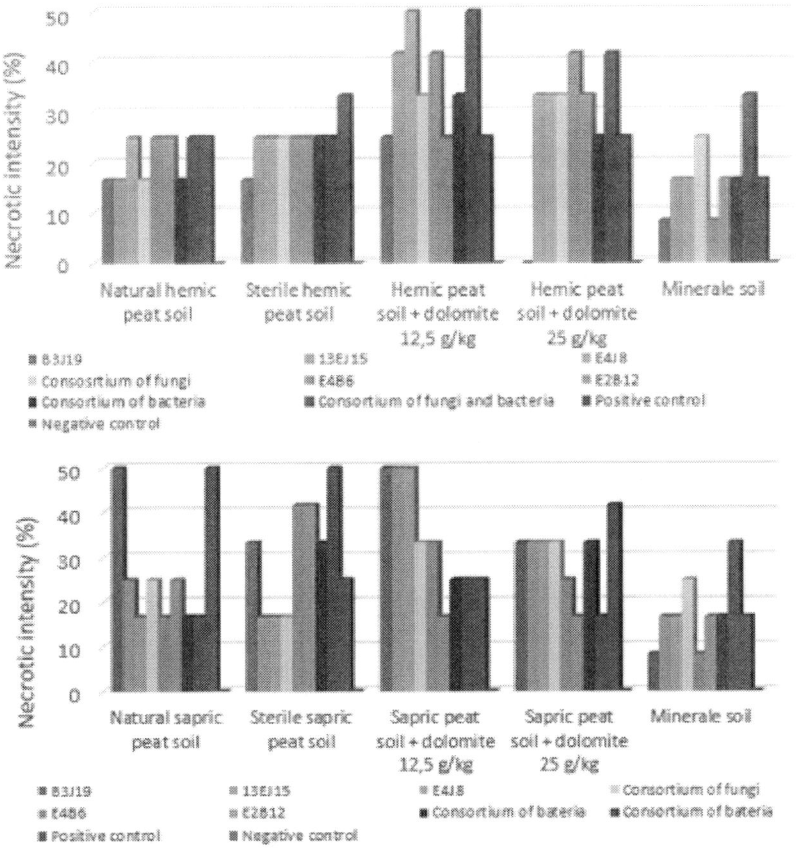

Figure 1. The effect of acid-tolerant antagonists on the intensity of necrosis in oil palm seedling roots in hemic peat (top) and sapric peat (bottom).

3.1 Effect of the peat soils on the effectiveness of acid-tolerant antagonists

Based on observations on root necrosis, acid-tolerant antagonists were only able to reduce the root necrosis symptoms by 16.67%. This result was lower than that of sapric peat soil which 52.08%. Differences in peat pH also did not affect the effectiveness of controlling root necrosis in hemic peat soil, but it did affect the effectiveness in sapric peat soil (Table 2).

Table 2. Effect of peat soils treatment on the average effectiveness of acid-tolerant antagonists.

Growth medium	pH medium	Root necrosis (%)
Mineral soil	4.0	0
Natural hemic peat soil		16.67
Hemic peat soil + dolomite 12.5 g/kg	2.9 – 3.5	0
Hemic peat soil + dolomite 25 g/kg	4.1 – 5.3	0
Sterile hemic peat soil	6.4 – 6.3	28.12
Natural sapric peat soil	2.9 – 3.3	52.08
Sapric peat soil + dolomite 12.5 g/kg		0
Sapric peat soil + dolomite 25 g/kg	2.7 – 3.4	0
Sterile sapric peat soil	5.7 – 5.5	32.0

Ganoderma-antagonistic fungi isolates from peatlands have different characteristics and effectiveness. The B3J19 isolate. in vitro. has an ideal growth range at pH 4-5 [6]. However. in this control effectiveness test. it was only relatively effective on hemic peat soils. The 13EJ15 isolate. in vitro. has an ideal growth range at pH 3-6 and is relatively effective in almost all medium conditions. Meanwhile. E4J8 isolate which had an ideal growth range at pH 3-5 in vitro. was only effective on natural sapric peat. The use of a consortium of isolates also did not affect the effectiveness compared to single isolates. These results indicate that the more tolerant an isolate is to various pH levels of the growing medium. the more effective the isolate is in various peat soil environments. Based on the in vitro test results. most of the antagonistic fungi and bacteria from West Kalimantan peatlands were able to grow on a medium with a pH spectrum of 3-6 [6]. In this experiment. the addition of dolomite to hemic peat soil resulted in a pH ranging from 4.1 to 6.4. This pH range is thought to be tolerated by the antagonists. although the higher the resulting pH causes a decrease in its ability.

Meanwhile. the difference in pH in sapric peat soil which affects the control effectiveness is thought to be related to differences in the chemical characteristics of the peat soil as a result of the addition of dolomite. The decomposition rate of organic matter in sapric peat soil was higher than that of hemic peat. so the addition of the same amount of dolomite could have a relatively different impact. This could be seen at the pH level after the incubation period which tends to be higher in sapric peat soil than in hemic peat soil (Table 2). This is thought to also cause a change in chemical composition that is different from that of hemic peat soil. The addition of dolomite lime donates Ca^{2+} (or Ca^{2+} and Mg^{2+}) alkaline cations into the soil with the result that it could replace the position of H^+ and or Al^{3+} ions contained in the soil absorption complex. A decrease in the concentration of H^+ ions increases OH- ion which causes an increase in soil pH and an increase in soil saturation [22]. However. an increase in pH that was too high causes the decomposition process to be too fast [23]. A pH level that was close to neutral would damage the natural properties of the peat so that it can change the chemical reactions and the availability of nutrients in the soil [24]. Increasing pH in peat has been shown to increase the activity of phenol oxidase enzymes. as well as to increase the abundance of actinobacteria and fungi [25]. Phenolic acid was allelopathic for plants [26]. These phenolic acids have the effect of inhibiting plant root development and nutrient supply in the soil [27. 28]. Generally. the impact is also accumulative [29].

4 Conclusion

The acid-tolerant antagonists tested in this study were effective in controlling Ganoderma in oil palm seedlings on peat soil. The effectiveness in reducing root necrosis is 16.67%. The difference in the pH of the peat soil in this study did not affect the effectiveness. Further studies are needed to determine environmental factors in peat soil other than pH that affect the effectiveness of control by acid-tolerant antagonists.

References

1. J. Miettinen. C. Shi. SC. Liew. Global Ecology and Conservation. **6**. 67–78 (2016)
2. Direktorat Jenderal Perkebunan. *Statistik Perkebunan Indonesia 2016-2018*. Jakarta. Indonesia. 28p (2017) [Indonesian]
3. USDA. Oilseeds: World Market and Trade. Foreign Agricultural Service. Washington D.C.. USA (2018)
4. KP. Chong. J. Dayou. A. Alexander. A. Detection and Control of Ganoderma boninense in Oil Palm Crop. Springer. Switzerland. . 48p (2017)
5. RM. Rakib. CJ. Bong. A. Khairulmazmi. AS. Idris. MB. Jalloh. OH. Ahmed. J. Phytopathol.. 165. 276-282 (2017)
6. Supriyanto. Purwanto. SH. Poromarto. Supyani. IOP Conf. Series: Earth and Environmental Science 423. 012064 (2020) doi:10.1088/1755-1315/423/1/012064.

7. CG. Fee. The Planter. **87**. 1022. 325-339 (2011)
8. H. Priwiratama. AE. Prasetyo. A. Susanto. J. Fitopatol. Indones. 10. 1. (2014) DOI: 10.14692/jfi.10.1.1 [Indonesian]
9. MS. Muniroh. SA. Nusaibah. G. Vadamalai. Y. Siddique..Curr. Plant. Biol. 20. 100116 (2019) DOI: 10.1016/j.cpb.2019.100116.
10. JP. Andriesse. Nature and Management of Tropical Peat Soils. (FAO). Roma. (1988) http://www.fao.org/3/x5872e/x5872e06.htm#4.3.3%20Acidity.
11. Supriyanto. Purwanto. SH. Poromarto. Supyani. Biodiverstas. **21**. 7 (2020)
12. Supriyanto. Purwanto. SH. Poromarto. Supyani. IOP Conf. Series: Earth and Environmental Science. 724. 012013 (2021) doi:10.1088/1755-1315/724/1/012013
13. RMT. Notohadiprawiro. Selidik cepat ciri tanah di lapangan. Ghalia Indonesia. Jakarta. 114p. (1989)
14. MRM. Rakib. CFJ. Bong. A. Khairulmazmi. AS. Idris. Int. J. Agric. Biol.. **16.** 91–699 (2014)
15. AS. Idris. D. Kushairi. Ariffin. M.W. Basri. MPOB Information Series **314** (2006)
16. D. Risanda. Pengembangan Teknik Inokulasi Buatan Ganoderma boninense Pat. Pada Bibit Kelapa Sawit (Elaeis guineensis Jacq.). Skripsi. undergraduate degrees. Institut Pertanian Bogor. Bogor. (2008)
17. T. Arwiyanto. M. Goto. S. Tsuyumu. Y. Takikawa. Annual Phytopathology Society. Japan 60. 421-430 (1994)
18. RW. Rees. J. Flood. Y. Hasan. U. Potterd. RM. Cooper. Plant. Pathol.. **58**. 982–989 (2009)
19. YK. Goh. FW. Ng. SM. Kok. KJ. Goh. Malays. Appl. Biol.. **43**. 9–16 (2014)
20. MRM. Rakib. CFJ. Bong. A. Khairulmazmi. AS. Idris. J. Oil Palm Res.. **27**. 229 – 240 (2015)
21. RRM. Paterson. Crop. Prot.. **26**. 1369 – 1376 (2007)
22. Basuki. GI. Ichriani. Sutikno. Jurnal AGRIPEAT. **9**. 82 – 87 (2008)
23. IGM. Subiksa. W. Hartatik.F. Agus. Pengelolaan Lahan Gambut Secara Berkelanjutan. Balai Penelitian Tanah.. Bogor. Indonesia. 84p (2011)
24. E. Martani. BH. Sunarminto. A. Supriyo. S. Margino. JPTI. **7.** 22-31 (2001)
25. H. Kang. MJ. Kwon. S. Kim. S. Lee. TG. Jones. AC. Johncock. A. Haraguchi. C. Freeman. Nat. Commun. (2018) DOI: 10.1038/s41467-018-06259-1.
26. M. Djazuli. Perspektif. **10**. 44 – 50 (2011)
27. T. Tadano. KA. Ambak. K. Yonebayashi. T. Hara. P. Vijarnsorn. C. Nilnond. S. Kawaguchi. Nutritional Factors Limiting Crop Growth in Tropical Peat Soils. In Soil Constraints on Sustainable Plant Production in the Tropics. Proc. 24th inter. Symp. Tropical Agric. Res. Kyoto. Japan. p.146-156 (1990)
28. W. Hartatik. IGM. Subiksa. A. Dariah. Sifat Kimia dan Fisik Tanah Gambut. Balai Penelitian Tanah. Bogor. Indpnesia. 12 p. (2011)
29. ZH. Li. Q. Wang. X. Ruan. CD. Pan. DA. Jiang. Molecules. 5. 8933-8952 (2010) doi:10.3390/molecules15128933.

The effect of the land cover plant on infiltration rate in Alas Bromo, Karanganyar

Dwi Priyo Ariyanto[1*], *Prasetya Adi* Sayoga[2], Suryono[1], and *Trijono Djoko* Sulistyo[3]

[1]Soil Science Department, Faculty of Agriculture, Universitas Sebelas Maret, Jl. Ir. Sutami 36a Surakarta, Central Jawa, Indonesia
[2]Bachelor Degree of Soil Science Department, Faculty of Agriculture, Universitas Sebelas Maret, Jl. Ir. Sutami 36a Surakarta, Central Jawa, Indonesia
[3]Agrotechnology Department, Faculty of Agriculture, Universitas Sebelas Maret, Jl. Ir. Sutami 36a Surakarta, Central Jawa, Indonesia

Abstract. Forest management is a way to keep the forest well-preserved. Forest management can mitigate the impacts of climate change. It will affect the balance of the environment such as the balance in the hydrological or water systems aspect. Infiltration is one of the water system factors in hydrology. This research aims to determine the value and classification of the infiltration rate. This research was conducted using a descriptive exploratory method to determine the infiltration rate of each land cover plant. This research is in Alas Bromo, Karanganyar, Central Java which has Alfisol soil types with six land uses: seasonal mixed, annual mixed, mahogany, pine, pine-mahogany, and renewal. Different types of land cover and plant characteristics affect the soil condition and the ability to absorb water into the soil. The calculation of the infiltration rate uses the Philip method. According to the classification of the infiltration rate in Alas Bromo, four classes were obtained including: moderately rapid 9.15 cm.h^{-1} (Mahogany), moderate 3.72, and 2.77 cm.h^{-1} (Pine and Pine-Mahogany), moderately slow 0.87 cm.h^{-1} (annual mixed), slow 0.28 and 0.265 cm.h^{-1} (seasonal mixed and renewal). Therefore, the presence of land-cover plants increases the infiltration rate.

1 Introduction

Forest management is an effort to restore forest conditions, and it is a very necessary action. Forest management can mitigate the impacts of climate change, and it will affect environmental balance from the aspect of hydrological balance or water management. Infiltration is an inseparable part of the hydrological cycle. Infiltration is the process of water entering the soil which is measured by how quickly the water can be absorbed into the soil and is usually expressed in cm per hour [1]. Water that enters the soil will fill the pore space and continue to seep until it reaches the underground layer (aquifer) as a groundwater source. The infiltration rate is determined by the size of the infiltration capacity and the rate of water supply.

* Corresponding author: dp_ariyanto@staff.uns.ac.id

When the soil is dry, the infiltration rate is high. As time passes, the infiltration rate will decrease until it is constant, this indicates that the soil has experienced a water saturated condition. [2] stated that in the infiltration process, when water falls onto the relatively dry or unsaturated soil surface, water movement will occur to the lower soil layer which is influenced by the suction/pull of the soil matrix and gravitational forces. The longer the infiltration process takes, the soil water content increases, and when the soil begins to become saturated, the movement of water underground is only influenced by gravity.

The condition of vegetation diversity and topography on land can influence the infiltration rate. This is supported by research by [3], where the infiltration rate in each land use will be different because it is influenced by the physical properties of the soil and vegetation on the land. Different land cover plants will affect the physical properties of the soil, so this can affect the ability of water to enter the soil. This research aims to determine the value and classification of infiltration rates in different land cover plants. Because at this location no research has ever been conducted before, it is hoped that this research will provide information and a database on the condition of water infiltration into the soil in each land cover so that it becomes a consideration for the management of the Gunung Bromo Forest Education or called Alas Bromo.

2 Research methodology

2.1 Time and location

The research location was in Alas Bromo, Sebelas Maret University, which is in Karanganyar District, Karanganyar Regency, Central Java, Indonesia. Laboratory analysis of soil physical properties was carried out at the Soil Physics and Conservation Laboratory, Faculty of Agriculture, Sebelas Maret University, Surakarta. This research was conducted from July to December 2019. Alas Bromo has an area of 126.291 ha with an altitude of 200-337.5 meters above sea level. The soil on Alas Bromo has the Alfisol soil order. This is supported by [5] that the soil of Alas Bromo is Alfisol and according to Data from the Indonesian Central Bureau of Statistics [6], the soil in that area is Mediterranean Reddish Brown which has the equivalent of Alfisol. The topography and slope of the Alas Bromo area vary observations are made in areas that have a slope of 0 – 8% (flat-sloping). This can also be used as a factor in determining the sample point.

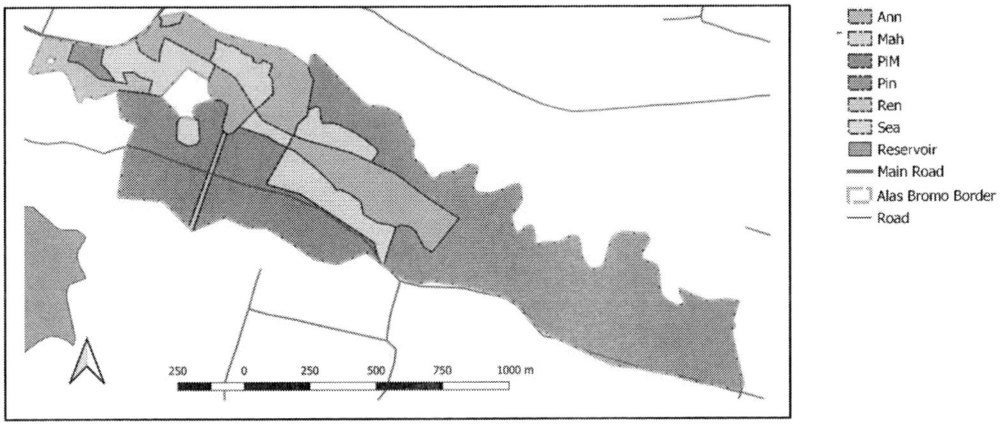

Fig. 1. Map of Alas Bromo Land Cover (Source: Aerial Photo Data, 2020).

Alas Bromo has 6 (six) land covers: Mahogany, Pine-Mahogany, Pine, Mixed annuals (dominated by Kedawung or tree bean (*Parkia timoriana*), flamboyant (*Delonix regia*), Saga or Acacia Coral (*Adenanthera pavonina*), Sonokeling or rosewood (*Dalbergia latifolia)*, and

Kusum (*Schleichera oleosa*) or gum-leac tree), Mixed annuals (dominated by corn, peanuts, and cassava), and Renewal (Figure 1).

2.2 Research Methods

This study used a descriptive exploratory method through field surveys conducted by observing, measuring, and identifying as well as direct testing on Alas Bromo on each land cover. The sample points in this study amounted to 18 based on 6 land cover plants which were replicated three times in each study area. Soil samples for laboratory analysis were taken using purposive sampling at each research observation point. Samples are taken at a depth of 0-20 cm and ensure that they are not on a sloping surface and at a minimum distance of 2 meters from the tree.

Calculation of infiltration rate and capacity using Philip's equation. The Phillips infiltration method considers the sorptivity (s) and conductivity/constant (A) values from field observations which are then shown by an equation (Philip, 1957 in [7]) as follows:

$$I = St^{\frac{1}{2}} + A \tag{1}$$

$$v = \frac{1}{2} St^{\frac{1}{2}} + A \tag{2}$$

Note: I = infiltration capacity (cm^3); v = Infiltration rate ($cm.hour^{-1}$); t = Time; A = Constant that depends on the nature and content of water content; S = Sorptivity of soil

Data analysis was carried out to determine the effect of infiltration rate on each land cover using the F test (ANOVA) and DMRT 95%.

3 Results and discussion

Figure 2 shows that the highest infiltration rate was obtained under Mahogany, followed by Mahogany Pine, Pine, Annual Mixed, Seasonal Mixed, and Renewal. The graph shows the infiltration rate for 60 minutes. This value was obtained based on a formula calculation from the regression results for each land cover. In the Mahogany land cover (49.03 $cm.hour^{-1}$), Pine (12.06 $cm.hour^{-1}$) and Mahogany Pine (19.19 $cm.hour^{-1}$) the initial condition of the infiltration rate is in the fast classification, whereas in Annual Mixed (5.92 $cm.hour^{-1}$), Seasonal Mixed (1.50 $cm.hour^{-1}$) and Renewal (1.19 $cm.hour^{-1}$) have initial conditions that are in the medium class classification. This difference is due to the actual water content of each land cover being different, the Mahogany land cover plant has the highest value and the Renewal land cover plant has the lowest value.

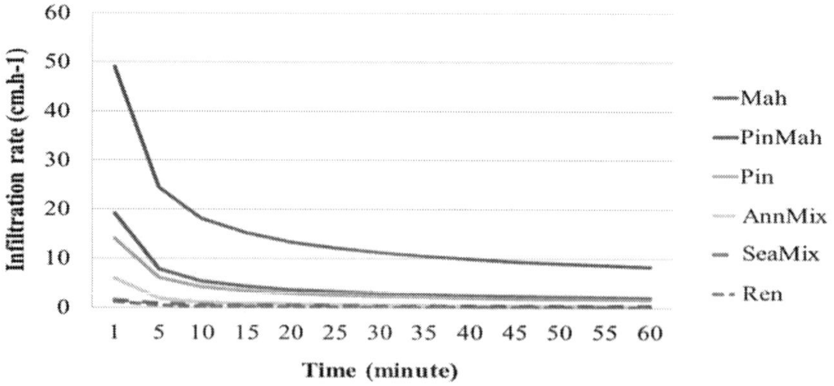

Description: Mah: Mahagony; PinMah: Pine-Mahagony; Pin: Pine; AnnMix: Annual Mixed; SeaMix: Seasonal Mixed; Ren: Renewal

Fig. 2. Results of Infiltration Rate for each Land Cover Plant.

The infiltration rate of Mahogany land cover was the highest, because the vegetation on this land is more than 70 years old, the physical size of the vegetation is very large, the spacing or crowns are also dense so that it affects the physical properties of the soil. According to Elfiati and Delvian [8], forest soil has a high infiltration rate and relatively large macro porosity, accompanied by high soil and root biological activity. The entry of roots into the soil with a certain depth can make the soil aggregates loose, so that it will create gaps for water to enter the soil. One of them is that the water content capacity of this land cover has the highest value. In this land, there are also ground cover plants in the form of shrubs, this makes the land rich in litter which then affects the formation of pore structures in the soil. In addition, the physical properties of the soil in this land use such as low volume weight compared to the others and high porosity values also indicate that these values can support high infiltration rates. This can also be seen in every other land cover crop which shows that the value of the infiltration rate is inversely proportional to volume weight [9] and directly proportional to porosity [10].

Mahogany Pine land use has a character that is not much different from Mahogany or Pine land cover. This land cover has a planting distance between Pine and Pine of 8 m and Mahogany and Mahogany of 4 m, so that at every distance from Pine to Pinus there is a Mahogany tree in the middle. This land has ground cover in the form of bushes and some shrubs. The graph on this land is in the second part after the highest.

Pine land use has the same graph in the middle. It has the character of land cover with a regular spacing of 4 meters, with a plant age of approximately 20 years and has a dense canopy, this can be seen from the lowest intensity of sunlight when compared to the others. Then the mixed land use has quite a lot of and varied vegetation with a dense canopy density. The Mixed land cover graph itself is low.

Plants covering the seasonal mixed have a low infiltration value. The vegetation on the location only has a few plants and only a few trees. With minimal vegetation on the ground on a land, there is a minimal supply of plant remains that can be contributed to the soil. The ability of soil to absorb water can be seen from the presence of the type of vegetation on the surface of a land. Land with denser vegetation will tend to produce more litter than land with little vegetation, this can improve soil pore conditions so that infiltration becomes better. According to Supangat and Putra [11], the presence of many plants can increase soil infiltration capacity due to the role of roots in improving soil physical properties such as structure formation, increasing porosity, and stabilizing soil aggregates.

Renewal land use has the lowest infiltration value compared to other land covers. It has a slow class because in the Renewal land cover there are only pine trees which are still more or less 3 years old with a spacing of 7 meters and there are seasonal crops that have been harvested in the form of cassava, peanuts, and corn. According to the texture analysis obtained, the fraction of sand in this land is the highest compared to the others, this results in when water enters the soil the particles will also be absorbed so that they can prevent water from seeping into the soil. In addition, human intervention for the land to become annual crop land resulted in the characteristics of the soil experiencing compaction. This is also supported by Endarwati et al. [12], who found that the level of infiltration in annual crop fields is higher compared to annual crops.

The condition of Seasonal Mixed and Renewal are open-space with no ground cover plants or large growing vegetation with higher sunlight intensity. The infiltration rate on open land has a low value [13] because water (rain) that falls to the ground surface damaged soil aggregates hence the soil become compact. The topsoil on the surface is washed away by surface runoff, resulting in low infiltration rate values.

After obtaining the infiltration value for each land cover, infiltration rate classification was carried out and the relationship between the influence of land cover plants on the infiltration rate was carried out using statistical analysis using ANOVA. In the ANOVA test, the relationship between land cover crops and infiltration rate showed a very significant value of 0.000, which was then further tested as shown in Table 1.

Table 1. Infiltration Rates in specific Land Cover.

Land Cover	Infiltration rate (cm.hour^{-1})	Classification [14]
Mahagony	9.15 a	Moderately Rapid
Pine-Mahagony	3.72 b	Moderate
Pine	2.77 bc	Moderate
Annual Mixed	0.87 bc	Moderately Slow
Season Mixed	0.28 c	Slow
Re	0.27 c	Slow

Description: Results of DMRT (95% confidence level) obtained between land cover and infiltration rate. Numbers followed by different letters indicate that they are significantly different

Land cover crops have a significant effect on the infiltration rate. Table 1 confirmed the highest infiltration rate value is found in the Mahogany land cover with a value of 9.15 cm.hour^{-1}, which is included in the moderately rapid. Then Mahogany Pine and Pine have a moderately class because they are in the classification value between 1.52 - 5.08 cm.hour^{-1}. Annual Mixed is in the moderately slow with a value of 0.87 cm.hour^{-1}. The Renewal and Seasonal Mixed have a slow classification with an infiltration rate of 0.152 – 0.51 cm.hour^{-1}[14]. Then a correlation test was carried out between land cover plants and the infiltration rate. The correlation value obtained is 0.000, which means it is correlated with a very strong relationship, for the Pearson correlation (-0.86), meaning that the two have a weak relationship or influence. In addition, a negative value means that the high infiltration rate value is not entirely due to land cover plants. Then the infiltration results are correlated with the physical properties of the soil, this aims to determine the relationship of the infiltration rate with the physical supporting properties of the soil.

4 Conclusion

The presence of land cover plants can affect the rate of infiltration. Mahogany land cover has the highest rate with moderately rapid. The land cover plants of Pine-Mahony and Pine have a moderate. Annual Mixed land cover have a moderately slow and the Mixed Seasonal and Renewal land cover have a slow. The land has different infiltration rates due to one of the factors, namely the differences in the characteristics of the land cover plants above it, such as root and leaf factors. Land cover crops correlate with the soil physical properties, especially the actual soil moisture, permeability, and soil bulk density.

Thank you to the Forestry Education and Training Unit of Universitas Sebelas Maret (UPT PPK UNS) as the Authority of the Alas Bromo Forest Education.

References

1. C. Asdak, *Hidrologi dan Pengelolaan Daerah Aliran Sungai* (UGM Press, Yogyakarta, 2020)
2. L. D. Susanawati, B. Rahadi, Y. Tauhid, J. *Sumberd. Alam dan Lingkung.* **5**, 28 (2018)
3. P. Trisnawati, H. Budianto, R. Wirosoedarmo, B. Suharto, *J. Sumberd. Alam dan Lingkung*, **1**, 15 (2014)
4. P. Budianto, R. Wirosoedarmo, B. Suharto, *Sumberd. Alam dan Lingkung*, 1, 5 (2014)

5. M. Nufus, Y. A. B. Pertiwi, A. T. Sakya, IOP Conf. Ser. Earth Environ. Sci. **528**, (2020)
6. BPS, Badan Pus. Stat. Kabupaten Karanganyar, 1 (2019)
7. Mt. S. Budiastuti, *Sains Tanah – J. Ilm. Ilmu Tanah dan Agroklimatologi,* **6**, 15 (2009)
8. D. Elfiati dan Delvian, *J. Hidrolitan*, **1**, 29 (2010)
9. Y. E. Pratiwi, S. Kadir, B. Badaruddin, *J. Sylva Sci.*, **4**, 928 (2021)
10. K. S. Wicaksono, I. Nita, A. N. Putra, W. Widianto, F. H. Rusdianto, P. Ayuningtyas, *J. Tanah dan Sumberd. Lahan,* **9**, 131 (2022)
11. A. B. Supangat, P. B. Putra, *J. Penelit. Hutan dan Konserv. Alam*, **7**, 149 (2010)
12. M. A. Endarwati, K. S. Wicaksono, D. Suprayogo, *J. Tanah dan Sumberd. Lahan*, **4**, 577 (2017)
13. Askoni, S. Sarminah, *ULIN J. Hutan Trop.*, **2**, 6 (2018)
14. Soil Quality Institute, *Soil Quality Test Kit Guide* (United States Department of Agriculture, 1999)

Enhancing zinc uptake and grain yield of Ciherang and Inpari IR Nutri Zinc rice varieties using Zn-solubilizing bacteria

Etty Pratiwi[1*], *Nabila Strata* Ilmi[2], *A. Arivin* Rivaie[1], *Terry Ayu* Adriany[1], and *Antonius Fransisco* Nababan[1]

[1]Research Center for Food Crops, National Research and Innovation Agency, Cibinong, West Java 16911, Indonesia
[2]Agrotechnology Study Program, Faculty of Agriculture, University of Jember, East Java 68121, Indonesia

Abstract. In elevated atmospheric CO_2 conditions, a prominent greenhouse gas contributing to climate change, there is a notable reduction in zinc (Zn) concentration within lowland rice crops. Zn deficiency diminishes rice productivity and lowers Zn content in grains, impacting human health, particularly stunting. This study aims to assess the effects of applying isolated Zn-solubilizing bacteria (ZnSB) strains and $ZnSO_4$ on Zn uptake and yield in two rice varieties: Inpari IR Nutri Zinc (high Zn content) in the first trial and Ciherang (medium Zn content) in the second trial, conducted in an Inceptisol soil. The treatments were organized in a Randomized Block Design with three replications. The first factor involved $ZnSO_4$ fertilizer (Z0 = without $ZnSO_4$; and Z1 = with $ZnSO_4$). The second factor was ZnSB strains (B0 = without ZnSB; B1 = *Klebsiella pneumoniae* Zn2; B2 = *Enterobacter cloacae* CmA12; B3 = *E. bugandensis* CmD2; B4 = *S. marcescens* CmT13; and B5 = *B. thuringiensis* PuT1). The results indicated that applying ZnSB strains without $ZnSO_4$ significantly increased grain yields for both Ciherang and Inpari IR Nutri Zinc varieties (7.67% and 4.12%, respectively) and enhanced Zn uptake (28.05% and 13.58%, respectively). Moreover, combined ZnSB and $ZnSO_4$ application also significantly increased yields for both varieties (12.14% and 25.03%, respectively) and Zn uptake (33.29% and 13.99%, respectively), demonstrating the potential of ZnSB for enhancing Zn uptake and rice yield. These findings suggest promising strategies for addressing Zn deficiency in rice cultivation under elevated CO_2 conditions.

1 Introduction

The rise in average global air temperature or global warming, which causes climate change, has been accompanied by several reports from diverse regions of a substantial drop in food output by 17 to 40%, with food scarcity occurring during severe dry seasons. The

* Corresponding author: ettypratiwi@yahoo.com

concentration of carbon dioxide (CO_2) in the atmosphere, a major greenhouse gas responsible for climate change, is quickly growing and is anticipated to exceed 550 ppm by the end of the century [1]. It has risen by roughly 30% since pre-industrial times, and CO_2 levels might double or treble by the end of the century.

As the CO_2 content in the atmosphere increases daily, it is important to understand the impact of increasing CO_2 on plant growth and development and plant nutritional status. The negative impact of increasing atmospheric CO_2 levels on the quality of cereal grains is widely known, especially in reducing micronutrient levels in rice grains, such as Zn and Fe [2]. It is important to note that rice (*Oryza sativa* L.) is a staple food for more than half of the world's population and contributes more than 70% of total food energy in several developing countries, including Indonesia [3]. Zn deficiency results in stunted child growth or stunting [4].

Zinc deficiency in paddy fields was identified in soil that was given NPK fertilizer intensively, causing the formation of compounds such as $Zn\text{-}PO_4$, $ZnCO_3$, and $Zn(OH)_2$, or due to poor drainage resulting in the formation of insoluble ZnS compounds. Zn deficiency also occurs in rice fields with a pH > 7.0 or in waterlogged areas with high Fe, Ca, Mg, Cu, Mn, and P levels. Most of the paddy fields in Indonesia already contain high P nutrients (2.90 million ha) out of 7.54 million ha of paddy fields [5]. Under these conditions, Zn is fixed with P, making it unavailable to plants.

Several previous studies have investigated the interaction between using of $ZnSO_4$ fertilizer and plant growth-promoting rhizobacterial (PGPR) and the interaction between using of various rice varieties and Zn-solubilizing bacteria. [6] isolated native bacteria increased grain yield and Zn absorption by 17.0% and 52.5%, respectively. Most of the PGPR strains identified so far belong to the genera *Pseudomonas*, *Ochrobactrum*, *Bacillus*, *Azospirillum*, *Azotobacter*, *Rhizobium*, *Stenotrophomonas*, *Serratia*, and *Enterobacteria* [7]. Meanwhile, [8] found that a Zn-solubilizing bacterial strain, namely *Bacillus megaterium* KY687496, could increase Zn solubility in soil. Therefore, inoculating this bacterial strain into paddy soil is a potential strategy to reduce the need for Zn fertilizer use.

Biofortification is one approach to increase the Zn content in rice grains. Inpari IR Nutri Zinc is rice with the highest Zn content in Indonesia, containing around 29-34 ppm Zn [9]. The strategy to increase biofortification in Indonesia through the use of biofertilizers with Zn-solubilizing bacteria active ingredients in rice varieties, such as Ciherang, which has a moderate Zn content (around 24.70 ppm), seems to be an option for farmers to produce grain with higher Zn content. In addition, this approach is considered a new concept in plant biofortification because biofertilizers have various abilities to increase growth and production and Zn uptake. According to [6], adding Zn-solubilizing bacteria to rice plants increased dry weight, productive tillers, and panicle height compared to the treatment without inoculation. Bacterial inoculation significantly increased the Zn concentration in rice grains by 41.6% compared to the control treatment and $ZnSO_4$ fertilization.

Based on the previous studies, to obtain the right adaptation technology to the impacts of climate change, this study was designed to isolate native Zn-solubilizing bacterial strains (ZnSB) from the rhizospheres of reeds and bamboo. This strain has been tested in vitro to determine the potential for solubilizing Zn in paddy soil. The study results showed that these Zn-solubilizing bacteria were found to produce various organic acids and phytohormones indole acetic acid (IAA) [10].

So far, information on to which extent Zn-solubilizing bacterial (ZnSB) strains isolated from Imperata and Bamboo bushes (which represents the novelty of the initial results of this research) could enhance the Zn uptake and yield of Inpari IR Nutri Zinc and Ciherang is still very limited. Therefore, this study aimed to examine the impact of ZnSB strains and $ZnSO_4$ fertilizer application on the yield and Zn uptake of rice varieties, specifically Inpari IR Nutri

Zinc (with high Zn content) and Ciherang (with medium Zn content), especially in Inceptisols.

2 Material and methods

2.1 Bacterial culture

The Zn-solubilizing bacteria used in this study were non-pathogenic bacterial strains previously isolated from the rhizospheres of reeds and bamboo. All strains were selected based on their growth promotion capability and ability to solubilize insoluble Zn in a modified Pikovskaya agar medium (amended with 0.1% ZnO). Based on partial 16S rRNA sequencing, these bacterial strains were identified as *Klebsiella pneumoniae* Zn2, *Enterobacter cloacae* CmA12, *Enterobacter bugandensis* CmD2, *Serratia marcescens* CmT13, and *Bacillus thuringiensis* PuT1. All the bacterial strains were checked previously for their plant growth-promoting rhizobacteria properties, such as N_2-fixing, Zn, P, K-solubilization, production of organic acids and phytohormones IAA [10]. These activities demonstrated the potential of these bacteria to be used as biofertilizers.

2.2 Site description and experimental design

Two experiments were installed, one using the rice variety of Inpari IR Nutri Zinc (high Zn content), the other using the rice variety of Ciherang (medium Zn content) in the next glasshouse at the Indonesia Soil Research Institute (ISRI), Bogor (6°34'32"S - 106°45'15"E) in August 2021. The bulk sample of soil collected from Sukaresmi Village, Megamendung District, Bogor Regency (6°42'12"S - 106°54'27"E), taken from an area of 3 x 3 m^2 with a depth of 20 cm and cleared of branches and roots. The soil is classified as Inceptisols [11], having selected chemical and physical properties that can be seen in Table 1.

Based on analysis using the Kit PUTS, the N, P, and K content were classified as moderate, high, and low, respectively. The recommended dosages of fertilizers are 250 kg/ha of Urea, 100 kg/ha of SP-36, and 100 kg/ha of KCl. The total Zn content of soil used in the experimental pot was 119 ppm, classified as moderate. Whereas the available Zn concentration was only 1.1 ppm, classified as low, indicating that the Zn concentration in the soil solution was insufficient for optimal growth of rice plants. The low available Zn concentration in the soil of this study could be related to the high P availability [**12**] or the formation of insoluble franklinite ($ZnFe_2O_4$) in acidic soils [13].

Table 1. The selected chemical and physical properties of the rice paddy soil of Megamendung District, Bogor Regency.

	Silt (%)	**Dust (%)**	**Clay (%)**	**pH H_2O**	**C-org (%) (Walkley & Black)**	**N (%) (Kjeldahl)**	**P_2O_5 (mg/ 100g) (HCl 25%)**	**K_2O (mg 100/g) (HCl 25%)**	**Zn (ppm) (Total HNO_3)**	**Zn (ppm) (Morgan Wolf)**
Value	9	49	42	5.35	2.17	0.26	142	24	119	1.1
Criteria		Clay		slightly acid	moderate	moderate	high	low	moderate	low

Before the soil was put into the plastic pots, the soil was air-dried and passed through a 5 mm sieve. Each pot contained 5 kg of soil. All pot soils were given basal fertilizers, calculated based on the Paddy Soil Test Kit (PUTS) recommendations. In addition, compost was added at the rate of 5 g/pot or equivalent to 2 t/ha. The pot was watered and flooded for seven days to get the equilibrium. The experiments tested the effects of the isolated ZnSB strains applied with $ZnSO_4$ fertilizer. The treatments were arranged factorially in a Randomized Block Design with three replications. The first factor was fertilizer of $ZnSO_4$ (Z0 = without $ZnSO_4$;

and Z1 = with $ZnSO_4$). The second factor was ZnSB strains (B0 = without ZnSB; B1 = *Klebsiella pneumoniae* Zn2; B2 = *Enterobacter cloacae* CmA12; B3 = *E. bugandensis* CmD2; B4 = *S. marcescens* CmT13; and B5 = *B. thuringiensis* PuT1).

The fertilizer of $ZnSO_4$ was applied seven days before planting at the rate of 10 kg $ZnSO_4$/ha (as recommended). Nitrogen, phosphorus, and potassium were applied as a basal application at 57.5 mg N, 18 mg P_2O_5, and 30 mg K_2O per kg soil.

2.3 Preparation of bacterial inocula

Bacterial inocula for plant inoculation were prepared by cultivating Zn-solubilizing bacteria in Nutrient Broth by placing the inoculated broths on a rotary shaker at 100 rpm and 30°C for 24 to 48 h. Then, finally, the concentration of the bacterial suspension was adjusted to 10^7 CFU/ml.

2.4 Application of bacterial on rice seeds

The rice seeds were surface sterilized using 2% NaOCl for 15 minutes, followed by rinsing with sterilized water. Subsequently, the seeds were inoculated by soaking them for 24 h in different bacterial inocula maintained at 10^7 CFU/ml. Five seedlings were transplanted per pot and grown until the grain yield. Two top dressings of Urea were applied at 21 and 35 days after rice transplantation.

2.5 Harvesting of plants and Zn analysis

All the pot plants were harvested by cutting the shoots approximately 1 cm above the soil surface. After initial air drying, plant samples were placed in paper bags, dried at 60ºC in an electric oven for 48 h, and weighed to determine the dry matter and grain yield. The analysis of Zn uptake was performed using atomic absorption spectrophotometry in the Soil Chemistry Laboratory, ISRI, Bogor.

2.6 Data processing and statistical analysis

The observations included grain weight and grain Zn uptake. An analysis of variance (ANOVA) for a factorially randomized block design was performed. The Duncan's Multiple Range Test (DMRT) at $P < 0.05$, unless otherwise stated, was used to separate the means when the ANOVA results indicated that there were significant treatment effects [14].

3 Results and discussion

3.1 Grain yield

The results in Table 2 showed that the inoculation of Zn-solubilizing bacteria on Ciherang rice seeds significantly increased grain yield compared to control. Likewise, Inpari IR Nutri Zinc rice seed inoculation significantly increased grain yield. Whereas the application of $ZnSO_4$ fertilizer did not significantly affect grain yields of both rice varieties. Meanwhile, the interaction between the inoculation of ZnSB and the application of $ZnSO_4$ fertilizer significantly affected grain yields of Ciherang and Inpari IR Nutri Zinc varieties.

Table 2. The effect of the application of $ZnSO_4$ fertilizer and the inoculation of Zn-solubilizing bacteria on the grain yield of Ciherang and Inpari IR Nutri Zinc varieties in an Inceptisol rice field.

Zn-solubilizing bacteria	Grain yield (g)			
	Ciherang		Inpari IR Nutri Zinc	
	Z0	Z1	Z0	Z1
B0	57.9 ± 5.87 b	61.8 ± 4.10 b	57.1 ± 1.98 d	66.2 ± 8.70 bcd
B1	69.6 ± 1.06 a	65.4 ± 2.97 a	73.9 ± 2.19 ab	71.5 ± 0.64 abc

B2	64.6 ± 2.40 b	65.3 ± 6.79 a	79.7 ± 3.32 a	65.6 ± 10.61 cd
B3	67.5 ± 3.96 a	65.05 ± 2.9 a	73.1 ± 9.05 abc	72.4 ± 0.49 abc
B4	66.2 ± 2.05 a	65.0 ± 5.09 a	71.7 ± 0.28 abc	70.2 ± 1.98 abc
B5	64.9 ± 6.51 a	61.2 ± 2.97 b	72.6 ± 4.24 abc	77.4 ± 4.45 ab

Note: Number within the same column followed by the same lowercase letters are not significantly different by the DMRT at $P < 0.05$ (Z0 = 0 kg $ZnSO_4$/ha, Z1 = 5 kg $ZnSO_4$/ha; B0 = no bacteria; B1 = *K. pneumoniae* Zn2, B2 = *E. cloacae* CmA12, B3 = *E. bugandensis* CmD2, B4 = *S. marcescens* CmT13, B5 = *B. thuringiensis* PuT1)

For the Ciherang variety (medium Zn content) at without the addition of $ZnSO_4$ but with inoculation by B1 (*K. pneumoniae* Zn2), B3 (*E. bugandensis* CmD2), and B4 (*S. marcescens* CmT13), higher grain yields were observed compared to the control by 22.2%, 16.6%, and 14.3%, respectively. Meanwhile, at the addition of 10 kg $ZnSO_4$/ha, the inoculation of B1 (*K. pneumoniae* Zn2), B2 (*E. cloacae* CmA12), and B3 (*E. bugandensis* CmD2) resulted in higher grain yields compared to the control by 5.8%, 5.7%, and 5.2%, respectively.

For Inpari IR Nutri Zinc variety (high Zn content) without the addition of $ZnSO_4$, the inoculation of B2 (*E. cloacae* CmA12), B1 (*K. pneumoniae* Zn2), and B3 (*E. bugandensis* CmD2) gave higher grain yields compared to the control by 39.6%, 29.4%, and 28.0%, respectively. Meanwhile, with the addition of 10 kg $ZnSO_4$/ha, the inoculation of B5 (*B. thuringiensis* PuT1), B3 (*E. bugandensis* CmD2), and B1 (*K. pneumoniae* Zn2) gave the higher grain yields compared to the control by 16.9%, 9.4%, and 8%, respectively. The result of the present study follows the findings of other studies. [6] conducted glasshouse research to investigate the effects of isolated indigenous bacteria and Zn fertilizer on the growth promotion and Zn absorption of Zn responsive (NDR 359) and Zn non-responsive (PD 16) rice cultivars. They found that these three isolates boosted grain yield by 17.0%.

The effects of the inoculation of ZnSB and the interaction between ZnSB and $ZnSO_4$ were significant on the grain yields. However, adding $ZnSO_4$ fertilizer 10 kg/ha did not result in any difference in grain yield when compared to the control. These results indicate that applying Zn-solubilizing bacterial strains on rice field soils may reduce or eliminate the need for $ZnSO_4$ fertilizer, as there was no grain yield response to $ZnSO_4$ fertilizer. In line with this result, [8] reported that *B. megaterium* KY687496 was a potential candidate for enhanced Zn solubilization in soil, which would allow for lower inorganic Zn application rates after isolating an array of bacteria from soils of virgin forest sites and evaluating them for Zn solubilization.

The insignificant effect of applying $ZnSO_4$ fertilizer on the grain yield could be related to the total Zn content of the soil used in this study, which is classified as moderate. However, the concentration of plant-available Zn is low. Therefore, the removal of the readily available form of Zn by the crops in the soil will be replenished from exchangeable Zn or less soluble Zn pools (fractions). It explains why there was no grain yield response to the $ZnSO_4$ fertilizer. Following this result, a study conducted by [15] in Subang showed that $ZnSO_4$ had no significant impact on the yield of 24 rice genotypes despite its influence on the concentration of Zn in rice grains. Furthermore, [16] revealed that Zn fertilizer did not cause a significant increase in grain yield on soils with sufficient to high Zn availability.

3.2 Zn uptake

Inoculation of Zn-solubilizing bacteria on Ciherang rice seeds significantly increased Zn uptake by the rice crops. Similarly, Zn uptake by the Inpari IR Nutri Zinc variety was increased because of the inoculation of the rice seeds compared to the control treatment (no inoculation of ZnSB). Generally, due to the inoculation treatments, the increase in Zn uptake by the Ciherang variety was greater than that by the Inpari IR Nutri Zinc variety. However, the total uptake of Zn by the Inpari IR Nutri Zinc variety was relatively greater than that by the Ciherang variety (Table 3).

Table 3. The effect of the application of $ZnSO_4$ fertilizer and the inoculation of Zn-solubilizing bacteria on Zn uptake by Ciherang and Inpari IR Nutri Zinc varieties in an Inceptisol rice field.

Zn-solubilizing bacteria	Zn uptake (g/pot)			
	Ciherang		Inpari IR Nutri Zinc	
	Z0	Z1	Z0	Z1
B0	B 3.63 f	A 4.30 f	B 5.30 f	A 6.02 e
B1	B 4.54 c	A 4.89 e	B 5.39 e	A 6.54 d
B2	B 4.77 a	A 5.84 a	A 5.90 a	B 5.84 f
B3	B 4.69 b	A 5.55 d	B 5.72 c	A 6.86 b
B4	B 4.33 d	A 5.62 c	B 5.69 d	A 6.63 c
B5	B 4.23 e	A 5.66 b	B 5.73 b	A 7.29 a

Note: Numbers within the same column followed by the same lowercase letters or numbers within the same row followed by the same uppercase letters are not significantly different by the DMRT at $P < 0.05$ (Z0 = 0 kg $ZnSO_4$/ha, Z1 = 5 kg $ZnSO_4$/ha; B0 = no bacteria; B1 = *K. pneumoniae* Zn2, B2 = *E. cloacae* CmA12, B3 = *E. bugandensis* CmD2, B4 = *S. marcescens* CmT13, B5 = *B. thuringiensis* PuT1)

Regarding the difference in the magnitude of Zn uptake increase between the Ciherang and Inpari IR Nutri Zinc varieties, [17] noted that rice genotypes exhibit varying responses to zinc application in zinc-deficient soils, emphasizing the role of genetic factors in this response.

Furthermore, the application of $ZnSO_4$ fertilizer significantly affected the Zn uptake of both varieties. The interaction between the inoculation of ZnSB and the application of $ZnSO_4$ fertilizer also significantly affected Zn uptake by Ciherang and Inpari IR Nutri Zinc varieties. For the Ciherang variety (medium Zn content) at the addition of 0 kg $ZnSO_4$/ha, the inoculation of B2 (*E. cloacae* CmA12), B3 (*E. bugandensis* CmD2), and B1 (*K. pneumoniae* Zn2) gave higher Zn uptake in grain over control treatment by 31.4%, 29.2%, and 25%, respectively. Meanwhile, with the addition of 10 kg $ZnSO_4$/ha, the inoculation of B2 *(E. cloacae* CmA12), B5 (*B. thuringiensis* PuT1), and B4 (*S. marcescens* CmT13) also gave higher Zn uptake in grains when compared to the control by 35.8%, 31.6%, and 30.7% respectively. In other words, the effect of interaction between the inoculation of *E. cloacae* CmA12 and the application of either without or with 10 kg $ZnSO_4$/ha gave the highest Zn uptake in rice plant of Ciherang variety, suggesting that *E. cloacae* CmA12 strain had a greater potential to solubilize Zn than other strains.

For Inpari IR Nutri Zinc variety (high Zn content), at addition of 0 kg $ZnSO_4$/ha, the inoculation of B2 (*E. cloacae* CmA12), B5 (*B. thuringiensis* PuT1), and B3 (*E. bugandensis* CmD2) had higher Zn uptake in grains over control by 11.3%, 8.1%, and 7.9%, respectively. Meanwhile, at the addition of 10 kg $ZnSO_4$/ha, the inoculation of B5 (*B. thuringiensis* PuT1), B3 (*E. bugandensis* CmD2), and B4 (*S. marcescens* CmT13) had higher Zn uptake in grains over control by 21.1%, 13.9%, 10.1%, respectively.

The results of the present study demonstrated that the Zn-solubilizing bacteria strains employed could enhance the Zn uptake by the two rice varieties. This capability is most likely related to the ability of these strains to produce various organic acids, such as acetic, citric, malic, lactic, and oxalic acids [10]. Moreover, they could also produce phytohormone IAA, which influences the number and length of lateral roots, leading to improved nutrient absorption in the soil [10]. This result follows the findings of [6], who found that the isolated indigenous bacteria were able to enhance total Zn absorption (52.5%) and grain yield (17.0%). Furthermore, [18] reported that the Zn-solubilizing bacteria strains, identified as *Bacillus cereus* by 16S rRNA gene analysis, enhanced Zn uptake in grains and increased super basmati rice variety yield by 18–47%.

3.3 Changes in Zn uptake

The data regarding changes in Zn uptake in rice grains due to different treatments, including Zn-solubilizing bacterial strains (ZnSB), $ZnSO_4$, ZnSB + $ZnSO_4$, and the control, generally followed this order of decrease: ZnSB + $ZnSO_4$ > $ZnSO_4$ > ZnSB > control for both Ciherang and Inpari IR Nutri Zinc varieties (see Figure 1). The increase in Zn uptake in the Ciherang variety was greater than in the Inpari IR Nutri Zinc variety. However, the total Zn uptake by the Inpari IR Nutri Zinc variety was relatively higher than that by the Ciherang variety. In line with this result, [17] reported variations in the response of rice genotypes to Zn application in Zn-deficient soils, highlighting the influence of genetic factors in this response.

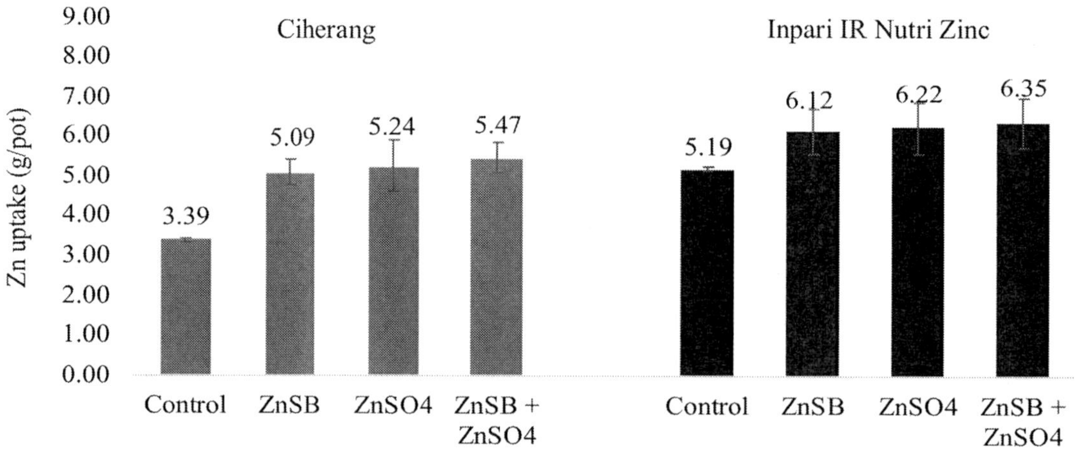

Fig 1. Comparison of the effect of Zn-solubilizing bacterial strains and $ZnSO_4$, and their interaction on the changes in Zn uptake by Ciherang and Inpari IR Nutri Zinc varieties

Figure 1 showed that the Zn uptake in rice grains at the three treatments (ZnSB, $ZnSO_4$, and ZnSB + $ZnSO_4$) was relatively similar. Studies suggested that applying Zn fertilizers to soils is a general strategy to cope with Zn deficiency and enhance grain Zn concentration [19]. However, this approach is not always optimal from an economic perspective [20]. It is one reason why people are now starting to switch to using more biofertilizers than it was.

4 Conclusions

The results of the present study have demonstrated in a glasshouse that the Zn-solubilizing bacteria strains employed could enhance the grain yield and Zn uptake in rice grains of Ciherang and Inpari IR Nutri Zinc varieties, indicating their Zn solubilization potential in rice Zn deficient soils. In turn, they may reduce the need for $ZnSO_4$ fertilizer. However, further research under field conditions is required to see the possibility of reducing the cost incurred on Zn fertilizers in rice cultivation and their effectiveness in augmenting the Zn nutrition of rice.

The increase in Zn uptake in the Ciherang rice grains was greater than in the Inpari IR Nutri Zinc variety. However, the total uptake of Zn by the Inpari IR Nutri Zinc variety was relatively greater than that by the Ciherang variety. Concerning this result, further research is also required to study the possible compatibility of a strain with a particular rice variety.

The authors sincerely thank Dr. Untung Susanto for his generous contribution to the Ciherang and Inpari IR NutriZinc rice seeds. His willingness to provide these seeds was indispensable in facilitating this research.

References

1. A. Lamichaney, A. Maity, Int. J. Biometeorol. **65**, 805-812 (2021)
2. B. Radha, N.C. Sunitha, R.P. Sah, T.P. MA, G.K. Krishna, D.K. Umesh, S. Thomas, C. Anilkumar, S. Upadhyay, A. Kumar, B.C. Marndi, K.H. Siddique, Front Plant Sci.**13**, 996514 (2023)
3. Food and Agricultural Organization of the United Nations (FAO), 'World agriculture: towards 2030/2050. Interim Report'. (2006)
4. J.E. Arsenault, E.A. Yakes, M.B. Hossain, M.M. Islam, T. Ahmed, C. Hotz, B. Lewis, A.S. Rahman, K.M. Jamil, K.H. Brown, J. Nutr. **140**, 1683–1690 (2010)
5. A. Kasno. Pros. Konser Karya Ilmiah Nasional (in Bahasa). 26-38. ISSN 2460-5506 (2017)
6. S.K. Vaid, B. Kumar, A. Sharma, A.K. Shukla, P.C. Srivastava, J. Soil Sci. Plant Nut. **14**, 889–910 (2014)
7. R. Dinesh, V. Srinivasan, S. Hamza, C. Sarathambal, S.J. Anke Gowda, A.N. Ganeshamurthy, S.B. Gupta, V. Aparna Nair, K.P. Subila, A. Lijina, V.C. Divya, Geoderma. **321**, 173-186 (2018)
8. Y. Ma, M. Rajkumar, Y. Luo, H. Freitas. J. Hazard, Mater. **195**, 230–237 (2011)
9. T. Sitaresmi, A. Hairmansis, Y. Widyastuti, Rachmawati, U. Susanto, B.P. Wibowo, M.L. Widiastuti, I.A. Rumanti, W.B. Suwarno, Y. Nugraha, J. Agric. Food Res. **12**, 100602 (2023)
10. N.S. Ilma. Jember University, Thesis. 129 p. (2021)
11. Soil Survey Staff, 12th ed., USDA-Natural Resources Conservation Service, Washington DC. (2014)
12. K.S. Sajwan, W.L. Lindsay, Soil Sci. Soc. Amer. J. **50**, 1264-1269 (1986)
13. B.C. Bostick, C.M. Hansel, M.J. La Force, S. Fendorf, Environ. Sci. Technol. **35**, 3823–3829 (2001)
14. R.G.D. Steel, J.H. Torrie, D.A. Dickey, The McGraw-Hill Companies, Inc., 672 p (1997)
15. U. Susanto, C. Gunarsih, W.R. Rohaeni, IOP Conf. Series: Earth. Env. Sci. **715**, 1-6 (2021)
16. J.X. Guo, X.M. Feng, X.Y. Hu, G.L. Tian, N. Ling, J.H. Wang, Q.R. Shen, S.W. Guo, J. Agric. Sci. **154**, 584-597 (2016)
17. E. Hoffland, C.Z. Wei, M. Wissuwa, Plant Soil **283**, 155–162 (2006)
18. M. Shakeel, A. Rais, M.N. Hassan, F.Y. Hafeez, Front. Microbiol. **6**:1286 (2015)
19. S. Hussain, M.A. Maqsood, Z. Rengel, T. Aziz, Plant. Soil **36**, 279-290 (2012)
20. R.D. Graham, Z. Rengel, Kluwer Academic Publishers, The Netherlands. pp 107–118 (1993)

Paper biosensors utilize silver nanoparticles for onsite pesticide residue detection

Dhony Hermanto[1*], *Nurul* Ismillayli[1], *Saprini* Hamdiani[1], *Siti Raudhatul* Kamali[2], *Rahadi* Wirawan[3], and *Handa* Muliasari[4]

[1]Department of Chemistry, Faculty of Mathematics and Natural Sciences, University of Mataram, Jl. Majapahit 62 Mataram, West Nusa Tenggara 83125, Indonesia
[2]Department of Environmental Science, Faculty of Mathematics and Natural Sciences, University of Mataram, Jl. Majapahit 62 Mataram, West Nusa Tenggara 83125, Indonesia
[3]Department of Physics, Faculty of Mathematics and Natural Sciences, University of Mataram, Jl. Majapahit 62 Mataram, West Nusa Tenggara 83125, Indonesia
[4]Department of Pharmacy, Faculty of Health, University of Mataram, Jl. Majapahit 62 Mataram, West Nusa Tenggara 83125, Indonesia

Abstract. The improper use of pesticides and excessive doses in the long term contribute to climate change and even threaten human health, organisms, and the balance of the ecosystem. A pesticide detection device is needed to monitor its levels to minimize risks to human health and the environment. A paper biosensor was developed in this study to detect organophosphates by utilizing the enzyme acetylcholine esterase (AChE) and silver nanoparticles (AgNP). The Whatman filter paper was used as a visual OP detection zone. AgNP, as an indicator, is adsorbed and gives a brownish-yellow color to the paper, while AChE is immobilized into the film and layered on the paper. The addition of acetylthiocholine chloride as a substrate to the film released thiocholine products which could replace the capping AgNPs causing the AgNPs to aggregate and the paper color to pale. The presence of OP in the sample will inhibit AChE activity so that paper fading is reduced. The biosensor response is quantized as an RGB value, which is determined using an application on a smartphone. The resulting biosensor has excellent performance with a linear range of 0.05-2.00 mg/L, a detection limit of 0.04 mg/L, and a CV of 0.48%. Biosensor measurements on vegetable samples showed conformity with the GCMS results as the standard method. Therefore, this biosensor is suitable for the on-site detection of pesticides offering easy, fast, and inexpensive analysis.

1 Introduction

Pesticides have been used for quite a long time, especially in agriculture, for reducing harmful pest organisms. Using pesticides according to the rules provides benefits, increasing crops and producing a more efficient and economical impact on agriculture. Unfortunately, the improper use of excessive pesticide doses also has negative effects, especially continuous

* Corresponding author: dhony.hermanto@unram.ac.id

application over a long period. The negative effects of pesticides do not only occur in the application area. They can spread due to their mobility through evaporation, leaching, and runoff, contaminating the air, water, and soil [1]. The application of pesticides also increases greenhouse gas emissions. In contrast, increasing temperatures due to climate change encourage the use of more pesticides due to an increase in pests and the rate of pesticide degradation [2]. Pesticides reduce soil biodiversity, cause poisoning of organisms and even chronic diseases such as cancer, and damage the nervous, respiratory, and heart systems. Pesticides must be used carefully and wisely to minimize their impact. Regulations are made to ensure human health, organisms, and environmental quality. It is known that the maximum residue limit for organophosphate pesticides is 0.01 mg/kg/day [3]. Therefore, controlling pesticide levels below the maximum residue limit is very important. For this reason, developing an analytical tool to detect pesticides is needed.

High-Performance Liquid Chromatography (HPLC) or Gas Chromatography (GC) is commonly used to determine the presence of excess pesticides in vegetables [4]. However, the weaknesses of the HPLC and GC methods are the high cost of reagents and the extraction and purification treatment in the laboratory, which requires solvents and a longer analysis time, which can lead to a risk of errors. Therefore, developing a pesticide residue detector that is cheap, fast, and easy to apply is necessary. Paper biosensors are a platform capable of detecting analytes (pesticide residues) that require fewer chemicals, do not require large volumes of samples, and have short analysis times [5,6].

This biosensor works on an enzyme inhibition system by analytes in the presence of an indicator that induces a color change [7]. This system is not new, but with a smaller design, easier to operate, and less expensive, paper-based biosensors have great potential for routine analysis. Although paper-based biosensors are inherently affordable and can be delivered to end users, researchers are still researching to achieve good performance to produce analytical tools capable of rapid and accurate quantitation to enable rapid risk management decision-making. Modifying paper biosensors using nanoparticles is known to improve their performance, obtained from the unique properties of nanomaterials [8]. This research aimed to develop a biosensor as an alternative method for quantizing pesticides with the principle of inhibition of the enzyme acetylcholinesterase on paper microzone plates using Whatman filter paper. Acetylthiocholine substrate is added to produce products from the hydrolysis reaction, namely acetic acid and thiocholine, which interact with silver nanoparticles (AgNP) to indicate color change [9,10]. Hence, monitoring and controlling pesticide residue levels to stay below thresholds is easier to achieve.

2 Experimental

2.1 Chemicals

The paper biosensor in this study was fabricated on paper microzone plates using Whatman grade 41-cellulose filter paper paint no. 1441-125. The color change indicator on the paper biosensor used biogenic AgNP synthesized by Hermanto et al. [11]. It electrochemically used green tea leaf extract in a spherical form of AgNP with an average size of ~17 nm. Other materials used in this study included acetylthiocholine chloride (ATChCl) (Sigma Aldrich) substrate, acetylcholinesterase (AChE) (EC 3.1.1.7, VI-S type, 200 units/mL) (Sigma Aldrich), tris HCL buffer (Merck), and the pesticide profenofos (sigma Pestanal®). The real samples used were vegetables obtained from local markets in Mataram, Indonesia.

2.2 Fabrication process and operation paper biosensor

Paper biosensors are designed using modified filter paper as analytical tools. Whatman 41 filter paper was chosen as the biosensor medium and biogenic AgNPs for chemical color reactions, as seen in the biosensor design (Figure 1). First, with a piece of filter paper cut to the size of 1 x 1 cm^2, 10 μL of 10 μg/mL colloidal biogenic AgNP in 1% PVA solution is absorbed into the filter paper medium, where PVA serves to bind biogenic AgNPs to the filter paper fibers. Absorption was carried out by soaking overnight. The color-adsorbed filter paper from biogenic AgNPs was allowed to dry at a chiller temperature of around 10°C for 24 h. A 10% PVA solution was prepared with Tris HCl buffer solution pH 7.0 and 10 μL of AChE solution (prepared with 200.0 units/mL AChE in 40 μL of Tris-HCl buffer solution pH 7.0). The dried filter paper coating was then coated with a PVA solution containing AChE by immersion for 10 s, and the results were dried at 10°C for 1 h. This coating process was repeated 3 times. This paper-based biosensor uses filter paper as a medium for chemical color reactions and coats it with a particular material, as previously described. Then, the filter paper was further modified to aid the occurrence of color changes due to chemical color reactions on the paper due to the catalytic activity of enzymes and redox in AgNPs for further testing and detection of pesticides.

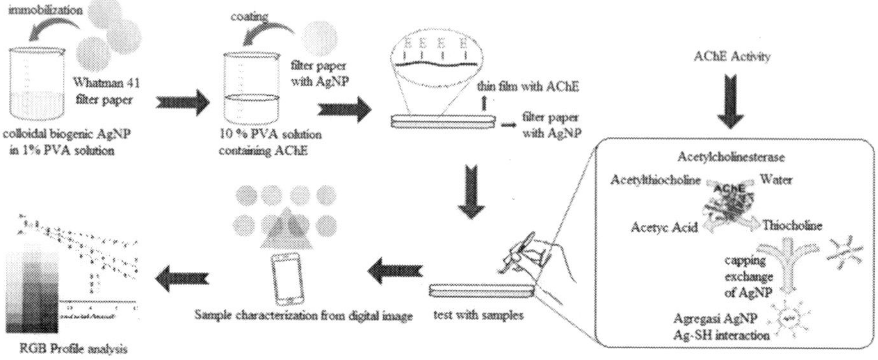

Fig. 1. Biosensor scheme.

2.3 Establishment of profenofos standard curve with paper biosensor

The fabricated paper biosensor is used for the detection approach based on chemical color reaction and integrated with the inspection machine [12]. Chemical color changes are observed as three base colors, R (red), G (green), and B (blue), which are conferred in the appropriate values. Digital cameras (smartphone cameras) and ImageJ image processing software can quantify the resulting image's color and identify differences in image resolution. The inspection machine is used to quantify the RGB values of chemically colored sectors on a paper biosensor under constant temperature and maintained light conditions. The standard curve is made from the change in RGB value (color intensity with increasing concentration series of standard solutions. Then, the sample is determined, and the RGB value obtained from the sample is calculated using the standard curve linear regression equation to obtain the concentration of the test substance in the sample.

2.4 Real sample analysis

Method validation for determining the performance of paper biosensors in pesticide detection, the GCMS method (QP210 Ultra, Shimadzu) was used as a comparison [8]. The RTX®-5MS column and helium gas were stationary and mobile phases. The injection temperature is 250°C. The qualitative is based on the retention time of the chromatogram peaks. At the same time, for the quantitative one, the calibration curve is used by plotting the concentration vs. chromatogram peak area. The sample concentration was determined from the chromatogram area and calculated based on the linear regression equation. Vegetable samples obtained from a local market in Mataram, Indonesia, including cayenne pepper (fruit), long beans (pod), and chinese cabbage (leaves and stem), were used as real samples. 100 g of the fresh sample was mashed, and 50 mL of Tris-HCl buffer pH 7.0 was stirred. The mixture was separated by centrifugation at 8000 rpm for 5 minutes, the supernatant was separated, and the pesticide content was analyzed. Samples were analyzed using both methods (biosensor and GCMS).

3 Results and discussion

3.1 Biosensor scheme

This work uses a paper biosensor for pesticide residue detection in vegetables. It is based on AChE inhibition and utilizes AgNP as a color-changing agent (Figure 1). The AgNP that has been adsorbed onto the filter paper evenly has a brownish-yellow color, and then it is coated with a PVA film that already contains AChE. The paper biosensor surface containing AChE makes it easier for enzymatic reactions to occur first. The ATCh substrate interacts with AChE, and a hydrolysis reaction occurs to produce acetic acid and TCh [13]. TCh has a thiol group, -SH, which further interacts with AgNPs on the paper side. The -SH group tends to interact more easily with Ag to form Ag-SH. Loss of Ag in AgNP to Ag-SH causes a change in the color of the biosensor paper from brownish yellow to pale yellow [14]. Pesticides inhibit the AChE enzymatic reaction, which correlates to the color change of the paper biosensor. Paper biosensors utilize silver nanoparticles for pesticide residue detection, making them easy and simple analytical devices, but they still have high sensitivity in detecting pesticides. To date, there have been very few reports on paper biosensors for pesticide detection in real samples, so testing the performance of paper biosensors has always been an interesting challenge. For this reason, testing on real samples is needed to validate the performance of this paper biosensor.

3.2 Biosensor response analysis

Profenofos is used to control insects (especially Lepidoptera) and mites. The largest agricultural market regarding the total weight of active ingredients uses profenofos pesticides in vegetable crops. Profenofos is a non-systematic insecticide and acaricide that works as a skin contact poison, inhalation poison if it enters the respiratory system, and stomach poison if ingested. Profenofos can biochemically inhibit the action of the acetylcholinesterase enzyme. Profenofos inhibits AChE irreversibly through covalent binding of the serine hydroxyl group of the AChE active site by the phosphoryl group of profenofos, rendering it unable to hydrolyze ACh [15]. Therefore, profenofos was used as a test material for detecting paper biosensors in this study.

Digital image information has been used in analytical applications, showing precision and accuracy in quantitative measurement results data. This procedure determines color information through the chromaticity of the standard red, green, and blue (RGB) color components. These intensities can be collected and provide a basis for quantitative measurements via the ImageJ tool, which is a colorimetric measurement method with a portable format [5]. The standard curve in the quantitative analysis determines the relationship between the concentration of the standard solution and the color intensity (mean RGB). It is used to quantitatively ensure the feasibility of the colorimetric test (Figure 2).

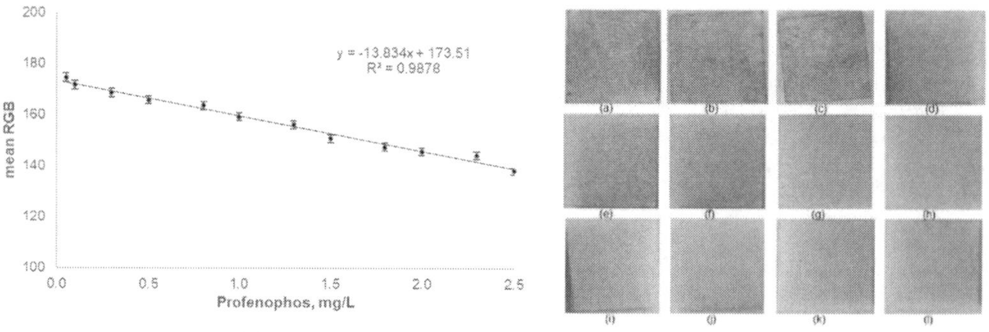

(a) (b)

Fig. 2. Biosensor performance with profenofos standard curve (a) and color of a paper biosensor at various profenofos concentrations, including (a) 2.50; (b) 2.25; (c) 2.00 (d) 1.75; (e) 1.50; (f) 1.25; (g) 1.00; (h) 0.75; (i) 0.50; (j) 0.25; (k) 0.10; (l) 0.05 mg/L.

The paper biosensor shows the sensor response to catalytic activity by changing the color of the paper from brownish yellow to pale yellow. Pesticides inhibit the catalytic activity, which can be seen quantitatively in the standard curve in Figure 2a. Meanwhile, changes in the color of the paper biosensor can be seen in Figure 2b. The corresponding change in the color of the paper biosensor, which can be detected by the naked eye in quantizing several concentrations of pesticides (Fig. 2b), implies the simplicity of paper biosensors for this purpose due to the elimination of time-consuming sampling, sample preparation, and procedures in traditional analytical laboratories [5]. As a result, this feature is appropriate for point-of-care (POC) testing. A more accurate inference about the response of the paper biosensor can be obtained by plotting the standard curve between observed color intensity (mean RGB) vs. pesticide concentration, as shown in Figure 2a. The curve shows a linear response in the region of the desired concentration from 0.05 to 2.5 mg/L of profenofos with a correlation coefficient (r) of 0.994. The paper biosensor's detection limit was 0.04 mg/L of profenofos. Repeating measurements of the profenofos standard solution using a paper biosensor 3 times gives a coefficient of variance (CV) of 0.48%. The lower the CV means the lower level of dispersion around the average value, indicating the more precise biosensor measurement results.

3.3 Biosensor performance

The next stage in the analysis of biosensor performance is method validation, which is carried out by determining pesticide residues in real samples and comparing the results obtained using certified reference methods. In this case, the method used is GCMS, which is a method commonly used in the quantitative analysis of pesticide residues in the laboratory. Here, the samples were spiked with profenofos at a 0.5 mg/L concentration. The results of the quantitation of profenofos in several vegetable summarized in Table 1. Table 1 shows the measured pesticide content values corrected by adding a standard profenofos solution. The results of measuring the pesticide content using the paper biosensor and the GCMS method showed that several samples were detected to contain pesticides. However, they were still below the threshold allowed by the Indonesian authorities, so they were still at a safe level for consumption [16]. Additionally, low levels of pesticides might be caused by several natural factors including degradation by temperature, leaching by water, etc. The paper biosensor developed for this purpose agreed well with the GCMS method. It was not significantly different using the t-test. Therefore, paper biosensors utilizing AgNP can be applied on-site in detecting pesticide residues, offering easy, fast, and inexpensive analysis.

Table 1. Measurement of pesticide concentration in real samples.

Sample	Various vegetable samples	Biosensor, mg/L	GCMS, mg/L
1	Cayenne pepper	0.020±0.003	0.023±0.001
2	Long beans	0.013±0.002	0.010±0.001
3	Chinese cabbage	0.055±0.004	0.048±0.001

4 Conclusions

A paper-based colorimetric biosensor for detecting profenofos has been successfully designed using AgNPs as an indicator. Profenofos in the sample inhibits AChE activity and reduces color fading of the paper (which contains the AgNP), a colorimetric method with a portable format via the ImageJ tool. The sensor response was measured as the average RGB value of the paper and showed a linear relationship with the profenofos concentration in the range of 0.05-2.50 mg/L. The detection limit of the paper biosensor is 0.04 mg/L. The results of biosensor measurements on vegetable samples showed conformity with GC-MS method. Therefore, this paper biosensor can be applied to determine profenofos easily and quickly.

This study was funded by the Kementrian Pendidikan, Kebudayaan, Riset dan Teknologi Republik Indonesia (Kemendikbudristek RI) [decree number: 2349/UN18.L1/PP/2023].

References

1. K.S. Rajmohan, R. Chandrasekaran, S. Varjani, Indian J. Microbiol. **60**, 125–138 (2020).
2. M. Tudi, H.D. Ruan, L. Wang, J. Lyu, R. Sadler, D. Connell, C. Chu, Int. J. Environ. Rsearch Public Heal. **18,** 1112, 1–23 (2021).
3. FAO, WHO, *Pesticide residues in food and the environment* (FAO, Rome, 2016).
4. S.M. Kang, J.H. Won, J.E. Han, J.H. Kim, K.H. Kim, S.H. Sung, Molecules **28**, 8(3343), 1–9 (2023).
5. M. Kavruk, V.C. Özalp, H.A. Öktem, J. Anal. Method Chem. **2013**(932946), 1–8 (2013).
6. M. Lee, K. Oh, H.K. Choi, S.G. Lee, H.J. Youn, H.L. Lee, D.H. Jeong, ACS Sensor **3**, 1, 151–159 (2018).
7. D. Hermanto, M. Mudasir, D. Siswanta, B. Kuswandi, N. Ismillayli, Kuwait J. Sci. **49**, 1, 1–13 (2022).
8. D. Hermanto, N. Ismillayli, H. Muliasari, R. Wirawan, S. R. Kamali, Glob. J. Environ. Sci. Manag. **10**, 1, 1–12 (2024).
9. P. Prosposito, L. Burratti, I. Venditti, Chemosensors **8**, 2, 1–29 (2020).
10. A. Loiseau, V. Asila, G. Boitel-Aullen, M. Lam, M. Salmain, S. Boujday, Biosensors **9**, 2, 1–39 (2019).
11. D. Hermanto, N. Ismillayli, I. Irmawanti, I. Fathulloh, D.N. Cahyani, U.K. Zuryati, H. Muliasari, R. Wirawan, Karbala Int. J. Mod. Sci. **9**, 1, 80–88 (2023).
12. L.M. Fu, M.K. Shih, C.W. Hsieh, W.J. Ju, Y.L. Tain, K.C. Cheng, J.H. Hsu, Y.W. Chen, C.Y. Hou, Biosensors **11**, 12, 1–16 (2021).
13. D.N. Kumar, A. Rajeshwari, S.A. Alex, M. sahu, A.M. Raichur, N. Chandrasekaran, A. Mukhreje,. RSC Adv. **5**, 76, 61998–62006 (2015).
14. G. Garai-Ibabe, L. Saa, V. Pavlov, Analyst **139**, 1, 280–284 (2014).
15. B. Kuswandi, M.A. Hidayat, E. Noviana, Biosensors Bioelectronics **12**, 1–22 (2022).
16. Menkes, Mentan, Join decision by the ministers of health and agriculture RI. (1996).

The effectiveness of biofertilizer biofilm formulation in promoting the growth of Pak choi (*Brassica rapa* L.) plant

Retno Rosariastuti[1*], *Addina* Harir[2], *Sri* Hartati[1], *Evi* Irawan[1], and *Setya* Nugraha[1]

[1]Department of Soil Science, Faculty of Agriculture, Universitas Sebelas Maret, Surakarta, Indonesia
[2]Undergraduate Program of Soil Science Dept., Faculty of Agriculture, Universitas Sebelas Maret, Surakarta, Indonesia

Abstract. One of the climate changes is the El Nino phenomenon, causing the soil to dry out, so it needs to be addressed because it will impact plant growth. Farmers use a lot of inorganic fertilizers (NPK), and in the long term, it can cause soil damage. One alternative treatment is to use organic fertilizers or biological fertilizers. Biofilm biofertilizer is a form of biological fertilizer. This study aims to determine the effect of dosing biofilm biofertilizer on the growth of Pak choi plants so that it can help reduce the use of inorganic fertilizers. This type of research was an experiment in a greenhouse using a completely randomized design (CRD) with six treatments, consisting of A (100% inorganic fertilizer), B (75% inorganic fertilizer and 25% biofilm biofertilizer), C (50% inorganic fertilizer and 50% biofertilizer biofilm), D (25% inorganic fertilizer and 75% biofertilizer biofilm), E (100% biofertilizer biofilm), and F (no inorganic fertilizer and no biofertilizer biofilm). Each treatment was repeated four times. Parameters used in this study included plant height, number of leaves, dry weight and wet weight, and leaf diameter. Data analysis used the ANOVA test with a 95% confidence level, followed by Duncan's Multiple Range Test (DMRT). The results showed that the combination of 25% biofertilizer biofilm treatment with 75% inorganic fertilizer gave the best results compared to the control and other treatments. The treatment of 25% Biofilm biofertilizer with 75% inorganic fertilizer was able to increase plant height by 21.175 cm or 74.24% higher than control, number of leaves by 17.25 strands or 65.62% higher than control, leaf width by 7.125 cm or 61.80% higher than control, wet weight by 28.083 g or 97.29% higher than control, and dry weight by 1.869 g or 95.88% higher than control so that this treatment can be an alternative to reduce the use of inorganic fertilizer.

* Corresponding author: retnobs@staff.uns.ac.id

1 Introduction

Climate changes such as rainfall trends, temperature trends, and changes in average rainfall significantly affect crop production results. One of the phenomena resulting from climate change is El Nino, where deficient rainfall causes the land to become dry. El Nino is a phenomenon of warming sea surface temperatures in the central and eastern Pacific Ocean. Warming sea surface temperatures increase the potential for cloud growth in the Pacific Ocean. This condition causes reduced rainfall in Indonesia. El Nino also causes drought conditions and hot weather in parts of Indonesia [1]. This needs to be addressed because it impacts plant growth and production. One of the leaf vegetable plants that is popular with many people is Pak choi. Pak choi belongs to the Brassicaceae family which originates from China and has high economic value [2]. These vegetables contain high nutritional value the body needs, such as protein, fat, carbohydrates, calcium, phosphorus, iron, vitamin A, vitamin B, vitamin C, minerals, and fiber [3]. The production potential and market demand for this vegetable is very large so it needs to be managed appropriately [4].

Pak choi plants (*Brassica rapa* L.) can reach a 15-30 cm height with oval-shaped, green leaves and light green or white stalks. Pak choi can grow optimally at 100-200 meters above sea level [5]. The air temperature needed for Pak choi growth ranges from 15°C-30°C with air humidity of 80%-90% [4]. These plants can be planted yearly and are even better if planted in loose soil, rich in organic matter, and good drainage with an acidity degree (pH) of 6 to 7 [6].

The availability of nutrients in the form of nitrogen, phosphorus, and potassium is needed in large quantities to form Pak choi biomass. Fertilization is an effort to increase Pak choi productivity by replacing lost nutrients and increasing the supply of nutrients required by plants to increase crop production and quality [7]. However, in the practice of cultivating vegetables, especially Pak choi, they still depend on high doses of inorganic fertilizers [8]. Excessive use of inorganic fertilizers will harm the environment. The structure will harden soil given excessive inorganic fertilizers, making it difficult for roots to grow and absorb nutrients. Applying inorganic fertilizers also leaves chemical residues that can inhibit performance and reduce the diversity of microorganisms beneficial to plants [8]. Therefore, an alternative is needed to reduce the use of inorganic fertilizers. One way is to fertilize using biological fertilizers [9]. Biofilm Biofertilizer is a natural fertilizer that contains various kinds of microbes that can increase the availability of nutrients for plants [10]. Biofertilizers are classified as environmentally friendly fertilizers and can play a dual role by producing phytohormones beneficial to plants [11]. Applying Biofilm Biofertilizer to rice plants (*Oryza sativa* L.) is proven to replace about 50% of inorganic fertilizers while increasing crop yields by 24%. Application of BFBF is an effective method to reduce the use of inorganic fertilizers by half in *C. asiatica* cultivation, thereby minimizing the negative impact of inorganic fertilizers [12]. The addition of biological fertilizers is expected to substitute inorganic fertilizers so that the use of inorganic fertilizers can be reduced.

2 Materials and methods

2.1 Study area

This research was conducted from May to August 2023 in the Greenhouse of the Faculty of Agriculture, Universitas Sebelas Maret (UNS). Plant analysis was carried out at the UNS Soil Biology and Biotechnology Laboratory.

2.2 Materials and equipment

The materials used include Nauli f1 Pak choi plant seeds, Andisol soil, inorganic fertilizer (Urea, Sp-36, and KCl), bacteria and fungi isolated from previous research [13], alcohol, aquadest, spirtus, PDB (Potato Dextrose Broth), NB (Nutrient Broth), YEMA (Yeast Extract Manitol Agar), Pikosvkaya medium, Jansen medium, crystal violet 0,1%, dan PBS (Phosphate Buffer Saline). The equipment needed is a pot, seedling tray, scales, hand colony counter, spectrophotometer, hemacytometer, autoclave, measuring cup, Erlenmeyer, cotton pads, plastic wrap, microscope, rack and test tubes, Petri dishes, tube needles, bunsen, pipette, micropipette, chips, tissue rolls, incubators, shakers, rulers, ovens and vortexes.

2.3 Research methods

The experimental design used was a Completely Randomized Design (CRD) with six treatments, consisting of control (no inorganic fertilizer and no biofertilizer biofilm), A (100% inorganic fertilizer), B (75% inorganic fertilizer and 25% biofertilizer biofilm), C (50% inorganic fertilizer and 50% biofertilizer biofilm), D (25% inorganic fertilizer and 75% biofertilizer biofilm), and E (100% biofertilizer biofilm) with four repetitions.

2.4 Procedure

The fungal inoculum was propagated on PDB (Potato Dextrose Broth*)* media. The bacterial inoculum was reproduced using several media, namely YEMA (Yeast Extract Mannitol Agar) for symbiotic BPN, Pikosvkaya for BPF, and Jensen for non-symbiotic BPN using the slant media method. 6 ml of each, then taken and grown on NB (Nutrient Broth) media. Monocultures are stored in a shaker for three days at 33-37°C. After three days, a micropipette transferred the fungal inoculum to a new PDB. At the same time, each bacterial inoculum was inoculated into a PDB medium containing fungal inoculum and allowed to form a biofilm. Bacterial cell density was measured using a hemacytometer. Biofilm formation was observed qualitatively using the staining method [14]. The bacterial dose given was 10^6 cells.g^{-1} of soil. In its application, the inoculum density is 10^{10} sel.mL^{-1}. The inorganic fertilizers used are 300 kg/ha of Urea fertilizer, 200 kg/ha of SP-36 fertilizer, and 100 kg/ha of KCl fertilizer [15]. The dose of fertilizer is then adjusted to the needs of the Pak choi plant. Based on soil weight, the amount of Urea fertilizer for Pak choi plants was 0.45 g/pot (100%) and adjusted to each amount of 75%, 50%, and 25%, respectively 0.33 g/pot, 0.22 g/ pot, and 0.11 g/pot. The amount of SP-36 fertilizer for Pak choi is 0.3 g/pot (100%) and adjusted to each amount of 75%, 50%, and 25,% respectively, 0.22 g/pot, 0.15 g/pot, and 0.07 g/pot. The dose of KCl fertilizer for Pak choi is 0.15 g/pot (100%) and adjusted to each amount of 75%, 50%, and 25%, respectively 0.11 g/pot, 0.07 g/pot and 0.03 g/pot. Inorganic fertilizer incubation was carried out one day before transplanting. The planting medium used is Andisol soil. The soil is put into heat-resistant plastic and weighed 3 kg. The soil is then sterilized using steam (steamed for 2 hours/day for three consecutive days) [16]. The application of biofertilizer biofilm was carried out 7 HST around the roots. The amount of inoculum given was 25%, 50%, 75%, and 100%, respectively, 3 ml, 6 ml, 9 ml, and 12 ml, with a bacterial density of 10^{10} cells.mL^{-1}. Seeding is carried out for one week in the tray or until the seedlings have an average number of leaves of 3-4. After that, the Pak choi seedlings were transferred to pots. One pot contains 1 Pak choi seed. The application of inorganic fertilizer was carried out one day before transplanting, while the application of biofilm biofertilizer was carried out at 7 HST. Maintenance carried out is watering and spraying organic pesticides. Harvesting is around 30-45 days after transplanting. The observation parameters were plant height, leaf width, number of leaves (direct observation and

https://doi.org/10.1051/e3sconf/202346701027

measurement in the field using a ruler), wet weight, and dry weight (direct weighing including leaves, roots, and stems). Observations are made every week. Data were analyzed statistically using Analysis of Variance (ANOVA) (95% significance level) followed by Duncan's Multiple Range Test (DMRT) (95% significance level). A correlation test was conducted to see the observed parameters' relationship.

3 Result and discussion

3.1 Plant growth

Table 1. Growth of Pak choi plants.

Treatments	Plant Height (cm)	Number of Leaves (strand)	Leaf Width (cm)
Control	5.100a	5.500a	2.750a
A	21.175d	17.250e	7.125e
B	19.800d	16.000de	7.200e
C	17.775c	14.000cd	6.050c
D	19.600d	13.750c	6.300c
E	14.775b	10.250b	3.850b

Notes: Control (no inorganic fertilizer and no biofertilizer biofilm), A (100% inorganic fertilizer), B (75% inorganic fertilizer and 25% biofertilizer biofilm), C (50% inorganic fertilizer and 50% biofertilizer biofilm), D (25% fertilizer inorganic and 75% biofilm biofertilizer), and E (100% biofilm biofertilizer)

3.1.1 Plant height

Based on the ANOVA results, treatment significantly affected plant height (Table 1). The highest average increase in plant height was in treatment A (100% inorganic fertilizer), namely 21.175 cm or 75.91% higher than the control. Treatment A had no significant difference from treatments B and D. The control treatment had the lowest mean with a value of 5.1 cm. Nutrients N, P, and K are essential nutrients for plants. Biological fertilizers act as inoculants made from living organisms' active ingredients, which fix certain nutrients or facilitate the availability of nutrients in the soil for plants. Biological fertilizer can be effective if organic or inorganic fertilizer is added to the application as a substrate to reproduce [17]. Providing input in the form of adding inorganic fertilizers can help metabolic processes in plants run well. Inorganic fertilizers are needed to stimulate the formation of roots, which will support the establishment of plants and the building of plant height [18].

3.1.2 Number of leaves

Based on the ANOVA results, treatment significantly affected the number of leaves (Table 1). The highest average increase in leaf number was in treatment A (100% inorganic fertilizer), namely 17.25 pieces or 68.11% higher than the control. Treatment A had no significant difference from treatment B. The control treatment had the lowest mean with a value of 5.5 strands. Inorganic fertilizers have an advantage compared to biological fertilizers, which dissolve quickly in water so that nutrients are readily available to plants. The addition of inorganic fertilizer will enable plants to grow their vegetative organs optimally from the start of their development because inorganic fertilizer is fast-release (quickly available) [18]. In addition, the number of leaves will be affected by the height of the plant. The higher the plant, the more leaves will also increase because the leaves are located on the stem segments [19].

3.1.3 Leaf width

Based on the ANOVA results, treatment significantly affected leaf width (Table 1). The highest mean increase in leaf width was found in treatment B (75% inorganic fertilizer + 25% biofilm biofertilizer), which was 7.2 cm or 61.80% higher than the control. Treatment B had no significant difference from treatment A. The control treatment had the lowest average, with a value of 2.75 cm. Combining inorganic fertilizers and biofilm biofertilizers can increase leaf width because the amount of nutrients contained is sufficient to supply the nutrient needs of the Pak choi plant. An adequate supply of nutrients will help plants to form protein so that with enough macro and micronutrient requirements for plants, the amount of protein included will increase and will increase the amount of protoplasm in plant cells and will eventually increase the width of leaves that are rich in chlorophyll. The nutrient element that most influences the growth and development of leaves is nitrogen [20].

3.2 Plant weight

Plant weight is differentiated between wet and dry weight. The following is data on the wet weight of Pak choi (Figure 1).

3.2.1 Wet weight

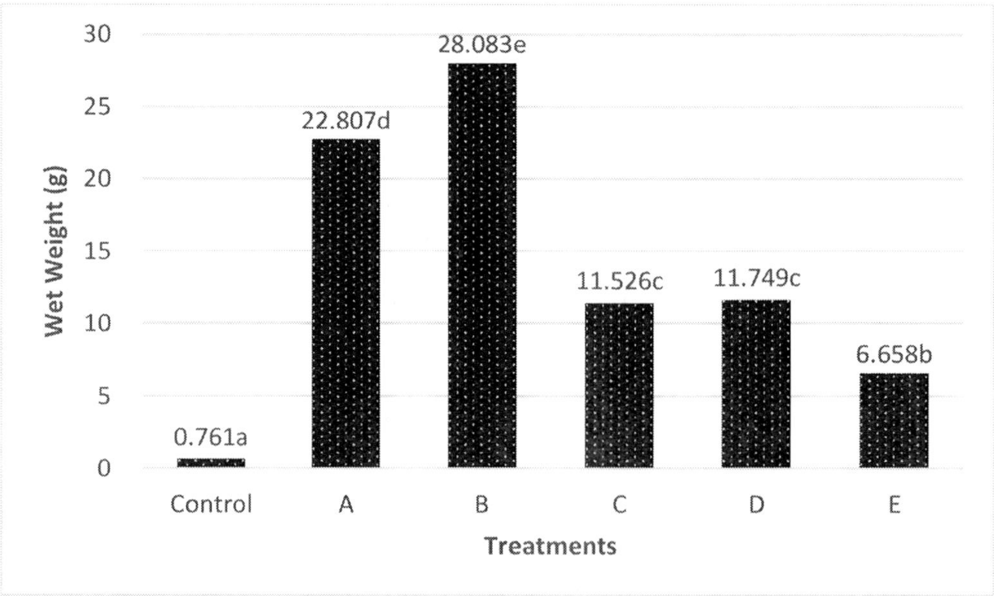

Notes: Control (no inorganic fertilizer and no biofertilizer biofilm), A (100% inorganic fertilizer), B (75% inorganic fertilizer and 25% biofertilizer biofilm), C (50% inorganic fertilizer and 50% biofertilizer biofilm), D (25% fertilizer inorganic and 75% biofilm biofertilizer), and E (100% biofilm biofertilizer)

Figure 1. Wet weight of Pak choi plants.

Based on Figure 1, biofilm can increase plant wet weight. The highest increase was in treatment B (75% inorganic fertilizer + 25% biofilm biofertilizer), which was 28.083 g or 97.29% higher than the control. The difference between treatment B and other treatments was also significant. The control treatment had the lowest result, with a value of 0.761 g. The wet weight content of plants is influenced by the water content in plant tissues or organs, nutrients, and organic matter contained in a plant. Microorganisms included in the biofertilizer biofilm play a role in decomposing organic matter. The more decomposition processes by decomposer microorganisms, the availability of nutrients in the planting medium will increase, affecting crop production. The availability of nutrients in sufficient conditions will allow the photosynthesis process to run smoothly so that assimilation can be translocated to all plant parts. In the end, there will be an increase in plant wet weight [21]. Next is the dry weight data for Pak choi plants (Figure 2).

3.2.2 Dry Weight

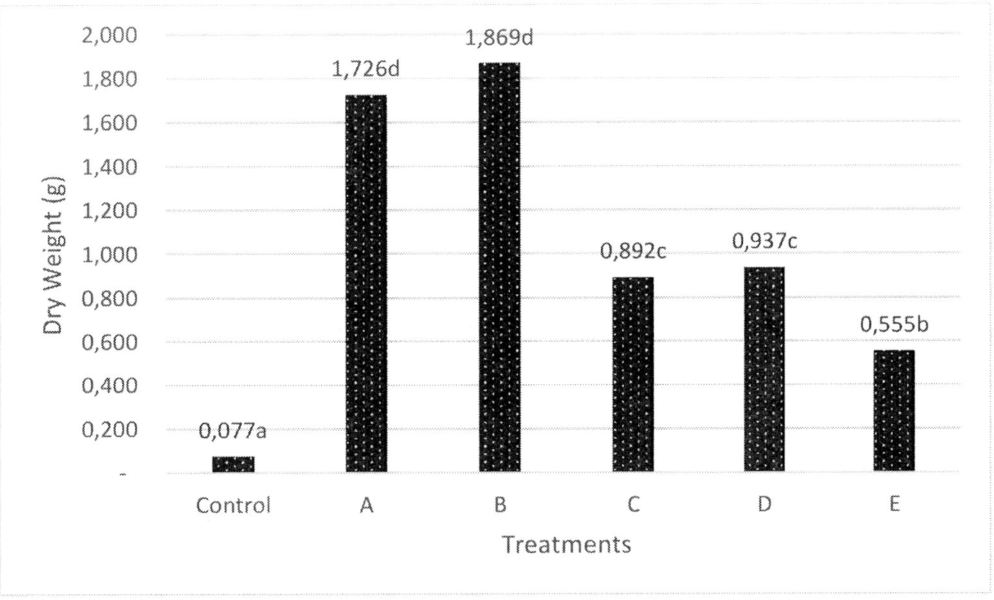

Notes: Control (no inorganic fertilizer and no biofertilizer biofilm), A (100% inorganic fertilizer), B (75% inorganic fertilizer and 25% biofertilizer biofilm), C (50% inorganic fertilizer and 50% biofertilizer biofilm), D (25% fertilizer inorganic and 75% biofilm biofertilizer), and E (100% biofilm biofertilizer)

Figure 2. Dry weight of Pak choi plants.

Based on Figure 2, biofilm can increase plant dry weight. The highest increase was in treatment B (75% inorganic fertilizer + 25% biofilm biofertilizer), which was 1.869 g or 95.88% higher than the control. Treatment B was not significantly different from treatment A. The control treatment had the lowest results, with a value of 0.077 g. Plant dry weight reflects the nutritional status of a plant. It is also an indicator that determines whether or not a plant's growth and development are closely related to nutrient availability. The higher the dry weight value of the plant produced, the better the plant's growth and the more nutrients it will absorb. By adding microbial inoculant, the presence of nutrients in the soil can increase so that it can stimulate plant growth [22]. The results of the correlation analysis between each variable can be seen in Table 2 below:

Table 2. Correlation between observed variables in Pak choi plants.

Variables	Plant height	Number of leaves	Leaf width	Wet weight	Dry weight
Plant height	1	0.902**	0.896**	0.770**	0.782**
Number of leaves	0.902**	1	0.925**	0.826**	0.850**
Leaf width	0.896**	0.925**	1	0.856**	0.853**
Wet weight	0.770**	0.826**	0.856**	1	0.979**
Dry weight	0.782**	0.850**	0.853**	0.979**	1

Notes:**= Very Significant Correlation (Level 0.01); *= Significant Correlation (0.05 Level).

Based on Table 2, it is known that all plant growth variables have a close correlation and are positive. Harvest results show optimal photosynthesis processes. Optimal photosynthesis is supported by leaf color, which has a high level of greenness; leaf green is influenced by the availability of N in the soil or growing media [23]. So that, in the end, it supports all components of growth and optimal harvest results, too. The results of this correlation indicate that the combination of inorganic fertilizers and biofilm biofertilizers can be used as complete fertilizers because these two types of fertilizers can complement each other's nutrient needs in Pak choi plants.

4 Conclusion

The treatment of 25% biofilm biofertilizer with 75% inorganic fertilizer was able to increase plant height by 21.175 cm or 74.24% higher than control, number of leaves by 17.25 stand or 65.62% higher than control, leaf width by 7.125 cm or 61.80% higher than control, wet weight by 28.083 g or 97.29% higher than control, and dry weight by 1.869 g 95.88% higher than control so that this treatment can be an alternative to reduce the use of inorganic fertilizer.

References

1. N. Novitasari, J. Sujono, S. Harto, A. Maas, and R. Jayadi, *J. Disaster Res.*, **147** 939–948 (2019)
2. D. Priadi, H. Wibowo, and E. S. Mulyaningsih, *J. Biodjati*, **4** 175–183 (2019)
3. P. N. Sari, M. Auliya, U. Farihah, and N. E. A. Nasution, *J. Phys. Conf. Ser.*, **1563** (2020)
4. E. H. Pudjiwati and D. D. Asmina, *J-PEN Borneo J. Ilmu Pertan.*, **2**, 26–33 (2020)
5. E. N. Agustina, S. Laili, and L. D. Ratna, *Biosaintropis (Bioscience-Tropic)*, **8**, 122–128 (2022)
6. T. S. Catharina, N. A. Rosadi, And I. A. Nopiari, *Ganec Swara*, **17**, 176 (2023)
7. D. Priadi and F. Nuro, *Biosaintifika J. Biol. Biol. Educ.*, **9**, 217–224 (2017)
8. R. Tao, Y. Liang, S. A. Wakelin, and G. Chu, *Appl. Soil Ecol.*, **96**, 42–51 (2015)
9. M. Mazid and T. A. Khan, *Int. J. Agric. Food Res.*, **3**, 10–23 (2014)
10. Sudadi, Suryono, and E. Triharyanto, *IOP Conf. Ser. Earth Environ. Sci.*, **200** (2018)
11. S. Nosheen, I. Ajmal, and Y. Song, "*Sustain.*, **13** 1–20, (2021)
12. G. Seneviratne, M. L. Kecskés, and I. R. Kennedy, *Biofertiliser Ina.*, **130** (2015)
13. R. Rosariastuti, Sutami, S. Nugraha, and B. S. Amanto, *Biodiversitas*, **24**, 2125–2133 (2023)

14. A. Maurya, P. S. Kumar, and A. Raj, *Chemosphere*, **286**, 1–13, (2022)
15. H. Sunarjono, *Bertanam 36 Jenis Sayur*. Jakarta: Penebar Swadaya, (2013)
16. Sutami, Purwanto, and R. Rosariastuti, *IOP Conf. Ser. Earth Environ. Sci.*, **905** (2021)
17. S. J. Li *Huanjing Kexue/Environmental Sci.*, **43**, 4735–4744 (2022)
18. I. Atmaja, *J. Log.*, **19**, 63–68 (2017).
19. N. Akhtar *Nanomaterials*, **12**, 1–21 (2022)
20. V. Torres-Olivar *J. Agric. Sci. Technol.*, **4**, 29–37, (2014)
21. O. O. Aluko, C. Li, Q. Wang, and H. Liu, *Int. J. Mol. Sci.*, **22**, 1–29 (2021)
22. T. Permatasari dan Nurhidayati, *J. Sains dan Seni POMITS*, **3**, 44–48 (2014)
23. W. Hou *Front. Plant Sci.*, **11,** 1–13 (2020)

Study of sediment transport distribution pattern after normalization in the Majalaya Area, Citarum River

Martha Natalia Pirsouw[1], *Joko* Nugroho [2], *Faizal Immaddudin Wira* Rohmat [2,3*], *Nabila Siti* Burnama [3], *Arno Adi* Kuntoro [2], *Mohammad* Farid [2] and *Hadi* Kardhana [2]

[1]Master Student of Water Resources Management, Faculty of Civil and Environmental Engineering Bandung Institute of Technology, Jl. Ganesa No.10 Bandung 40132, Indonesia
[2]Water Resources Development Center, Bandung Institute of Technology, Jl. Ganesa No.10 Bandung 40132, Indonesia
[3]Water Resources Research Group, Bandung Institute of Technology, Jl. Ganesa No.10 Bandung 40132, Indonesia

Abstract. Majalaya area in West Java, Indonesia, is prone to extreme rainfall events that cause flooding of the Citarum River. The river flows through the area, naturally containing sediments from erosion upstream that disturb the river balance. Some flat areas of the river retain sediments and cause excessive accumulation, resulting in changing the storage capacity of the river. The accumulation disrupts the flow of water and causes flooding in the area. To mitigate this problem, the River Basin Authority has implemented sediment dredging and river normalization with a total length of 5.4 km in 2020, successfully preventing flooding in the first year. However, subsequent floods that affected more than 7,000 people return to occur due to changes in land use, human activities, and unmanaged erosion upstream. This study examines the sediment accumulation pre-2020, immediately after the normalization efforts 2020, and the current sediment profile 2022. The result provides an overview of the river's morphology, sediment distribution patterns, and potential future profiles. The research results can be used to evaluate flood management risk in the area, including the compound flood risk caused by continuous sedimentation and climate change, e.g., heavier and longer rainfall.

1 Introduction

Floods are one of the most frequent natural hazards that worsen because of extreme rainfall due to climate change [1,2]. Reducing greenhouse gas emissions to mitigate climate change has recently become a significant global concern. In March 2023, a synthesis report by the International Panel on Climate Change (IPCC) confirmed that human activities are responsible for a steady increase in greenhouse gas emissions, with the potential to raise the Earth's temperature to 1.5°C by the early 2030s, exceeding the pre-industrial level of 1.1°C since about 1850. These findings indicate that there are currently high concentrations of

* Corresponding author: faizalrohmat@itb.ac.id

greenhouse gases in the atmosphere. This disruption of air temperature significantly impacts the hydrological cycle, resulting in extreme rainfall. As an integral part of the hydrological cycle, rivers are vulnerable to climate change. The natural drainage of watersheds results in rapid increases in inflow due to extreme rainfall, leading to flooding when storage capacity is exceeded [1]. These extreme rains can also contribute to river damage and degradation. Floods brought nutrient-rich sediments [3]. Sediment buildup results in reduced water storage capacity. This is also exacerbated by the presence of neglected critical land in the watershed, as is the case with the Citarum River in West Java, Indonesia.

Citarum River is a source of raw water for the districts and the capital of Indonesia, Jakarta, providing about 50 m3/dt [1]. However, the management of the Citarum watershed is increasingly out of control due to human activities and land use changes that continue to occur continuously, causing high urbanization and critical watersheds, which lead to changes in the upstream capacity of the Citarum River and uncontrolled sediment accumulation along the river, exacerbated by climate change[1,4]. With existing climate change indicators, increased rainfall intensity and concentrated water flow with sediment from land degradation into the river [3,5]. The condition of the Citarum watershed is worsening because of the uncontrolled sedimentation process in the upstream area.

Sediment can be divided into some transportation mechanisms: dissolved, suspended, drift, and bottom sediment. Sedimentation is strongly influenced by the amount of water runoff because the greater the flow in the watershed, the more sediment in the river [6]. The entry of sediment into the river will cause an increase in surface runoff, and deposition will cause the river to silt up [7]. The amount of sediment will greatly affect the volume of the river when the rainy season arrives because it can cause flooding [1,2,8–10]. The deposition process can be influenced by several factors, such as the influence of human activities, such as land use change, and natural sediments, such as high rainfall, which can cause soil erosion [11,12]. The excess sediment in the Citarum River has caused overflow in recent years. Fast-moving water erodes the riverbank, carrying soil and sediment to other areas. The Majalaya district, part of the Citarum River basin, has significant sediment accumulation. This buildup is one of the factors that changed the capacity of the upstream Citarum River, causing flooding in the area. In 2020, the Citarum River Basin Center conducted sediment dredging of the Citarum River by increasing the river's depth to 4 m along a 5.5 km stretch. This dredging was intended to restore the river's storage capacity so that no major overflows would occur, but just two years later, in 2022, the river experienced greater sediment accumulation. This situation must be analyzed carefully. We need to study the distribution of sediment shape and how it is transported to find better ways to prevent sedimentation.

2 Materials and method

Based on the literature review, it was determined that the Majalaya area may experience flooding with larger discharge volumes. Therefore, a comparative model should be used to assess the storage capacity of the Citarum River. This model compares data from each river cross-section to obtain riverbed elevation patterns as a sediment distribution formed by either river discharge or dredging activities that alter the river's capacity.

2.1. Study area

The Citarum River is the longest and the largest river in West Java, 269 km, and is important in providing water for West Java and Jakarta people, irrigating rice fields, and acting as a water source for >2000 factories [13]. The Majalaya Area is part of the Upper Citarum watershed with 207.2 km^2, located in the central part of West Java Province. The topography of the Upper Citarum watershed is relatively flat, and the outer of the watershed is mountains, with an average basin slope of 26.9% and an extension ratio of 0.46 [1,4]. The elevation of the catchment area in the mountains ranges from 2595 m.asl to 664.3 m.asl at the river gauges, with an average of 1393 m.asl. Although the dredging that has been carried out is expected to restore the storage capacity of the river and prevent large overflows, by 2022, it is seen that sediment buildup is increasing. Therefore, it is important to investigate how sediment transportation occurs; the length of the reviewed area is 1.2 km, with the review points shown in Figure 1.

Fig 1. Study area.

2.2. Data description

2.2.1. Cross-section data

River cross-section data were used for sedimentation assessment to understand the transformation of river channel properties concerning sediment movement and accumulation over time. The cross-section data analyzed corresponds to the condition before normalization (in 2009), the condition under construction (in 2020), and the condition after normalization in the Citarum River (in 2022) by reviewing 1.2 km of the 5.4 km normalized river segment. Cross-section data was sourced from direct measurements by the Citarum River Basin Authority (Balai Besar Wilayah Sungai Citarum).

2.2.2. Topographic data

Topographic data is crucial in comprehending river morphology and topographic variations due to sedimentation. It can potentially aid in identifying significant changes in elevation within the river channel caused by erosion or sediment accumulation. The data was procured from a topographic survey by the Citarum River Basin Authority.

2.3 Methodology

Studying the river's topography and examining its cross-sectional data is the method used to analyze how sediment settles in the river. In 2009 (pre-normalization), 2020 (post-normalization), and 2022 (present sediment profile), several measures were taken to gather data. Then, the collected data could be examined by comparing those years' size and height measurements and identifying locations with notable sedimentation alterations.

Topographic data can map the shape and elevation of the land surrounding the river, identifying sections of the river that may have experienced erosion or sediment accumulation and areas prone to soil erosion. Cross-section data describes the size and shape of the river channel at specific points. Collecting cross-section data at intervals enables monitoring of river channel dimension changes over time. For an illustrated explanation, reference the detailed flowchart displayed in Figure 2.

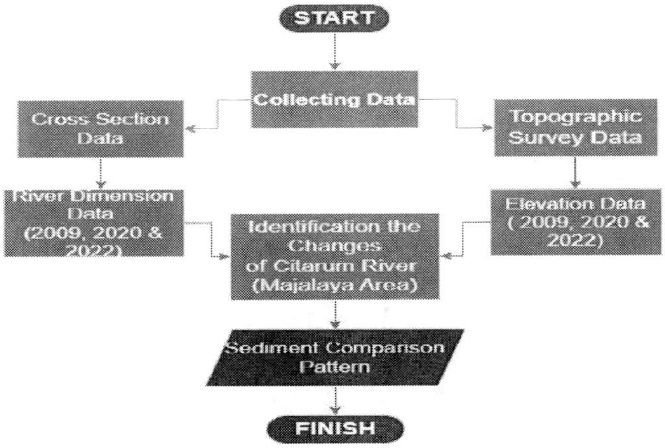

Fig 2. Flowchart study.

3. Result and discussion

Cross-section data from 2009 to 2022 at each point needs to be adjusted based on existing topographic data so that the cross-section data adjustments can be seen in Table 1; 24 data have been adjusted. Data adjustment is carried out from the three data in each year by adjusting each centerline center point on each cross-section data and then calculating the elevation comparison based on existing data between 2009 to 2020 and 2020 to 2022.

Table 1. Cross-sectional data adjustment.

Cross Number	2009	2020	2022
1	CTR 214	M 138 (5+508)	4+200
2	CTR 215	M 137 (5+438)	4+250
3	CTR 216	M 136 (5+393)	4+300
4	CTR 217	M 135 (5+354)	4+350
5	CTR 220	M 129 (5+132)	4+550
6	CTR 221	M 128 (5+091)	4+600
7	CTR 222	M 126 (5+002)	4+650
8	CTR 223	M 125 (4+976)	4+700
9	CTR 224	M 124 (4+916)	4+750
10	CTR 225	M 123 (4+881)	4+800
11	CTR 226	M 122 (4+865)	4+850
12	CTR 227	M 120 (4+798)	4+900
13	CTR 228	M 119 (4+754)	4+950
14	CTR 229	M 118 (4+711)	5+000
15	CTR 230	M 117 (4+668)	5+050
16	CTR 231	M 116 (4+629)	5+100

17	CTR 232	M 114 (4+560)	5+150
18	CTR 233	M 112 (4+492)	5+200
19	CTR 234	M 111 (4+453)	5+250
20	CTR 235	M 110 (4+400)	5+300
21	CTR 236	M 108 (4+349)	5+350
22	CTR 237	M 107 (4+306)	5+400
23	CTR 238	M 106 (4+273)	5+450
24	CTR 239	M 105 (4+210)	5+500

From the adjusted data, we focused specifically on the cross-section data from the as-built normalization plan and performed elevation comparison calculations for the cross-section data from 2009 before normalization through 2020. In addition, we compared the cross-section data for the year 2020 with the stream data that had experienced sedimentation in the year 2022 (Figure 3).

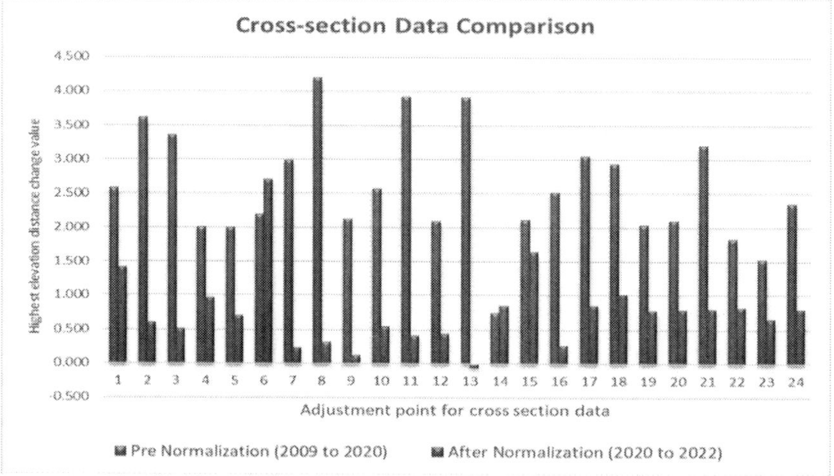

Fig 3. Result of Comparison Cross-section Data.

The graph shows the result of this study, significant changes between the highest elevation comparisons in each section. Elevation changes that occur at the bottom of the river channel that rises or falls are a response to temporal variations in the nature of water flow and sediment transport [7]. Elevation increases that occur in the river under review tend to experience higher elevation increases, indicating the condition of the riverbed that has undergone a process of additional sedimentation.

With a duration of 11 years before normalization, the accumulation is still considered reasonable, with the highest elevation comparison of 4,198 in section no. 8, but after normalization with a duration of only two years, the highest elevation in section no. 6 is 2,704. With such a short duration, the elevation increase has reached >50%, meaning that the pattern of sediment accumulation in the riverbed has changed more than the previous pattern, thus quickly changing the storage capacity of the river. This change in storage capacity due to sediment needs to be addressed appropriately and quickly to prevent negative impacts that will arise [7], especially flooding that occurs in Majalaya [1,9]. The river is basically a very dynamic system and continues to change over time, so the possibility of the centerline point reviewed in the cross-section data is not the same due to the movement of the river addition. However, this elevation monitoring approach will provide a view of sedimentation that continues to accumulate, especially for bedload in sediment transport, which greatly affects the river's capacity [7].

4. Conclusion

This study compares elevations to determine the sediment height in each cross-section before (2009) and after (2020) normalization, as well as from normalization (2020) to sediment accumulation (2022). The comparison spans 11 years before normalization and two years of sedimentation in the river cross-section. The comparison results show that only changes occurring within a 2-year become faster after normalization. This indicates that the changes in the cross-sectional capacity of the riverbed cause a decrease in the river's storage capacity during extreme rainfall conditions, leading to a faster potential for water overflow.

Reference

1. F. I. W. Rohmat, Z. Sa'adi, I. Stamataki, A. A. Kuntoro, M. Farid, and R. Suwarman, Urban Clim **46**, (2022)
2. N. S. Burnama, F. I. W. Rohmat, M. Farid, A. A. Kuntoro, H. Kardhana, and Solehudin, *Proceedings of the 2 Nd ITB Graduate School Conference Strengthening Multidisciplinary Research to Enhance Its Impact on Society Rainfall-Runoff Analysis Using Satellite Data in The Majalaya Watershed* (2022)
3. M. G. Macklin and J. Lewin, Quat Sci Rev **114**, 228 (2015)
4. S. Y. Siswanto and F. Francés, Environ Earth Sci **78**, (2019)
5. H. A. Rahman, International Journal of Malay-Nusantara Studies **1**, (2018)
6. J. Daramola, E. J. Adepehin, T. M. Ekhwan, L. K. Choy, J. Mokhtar, and T. S. Tabiti, Water (Switzerland) **14**, (2022)
7. A. M. Pfeiffer, B. D. Collins, S. W. Anderson, D. R. Montgomery, and E. Istanbulluoglu, Water Resour Res **55**, 6795 (2019)
8. N. S. Burnama, F. I. W. Rohmat, M. Farid, A. A. Kuntoro, H. Kardhana, F. I. W. Rohmat, and W. Wijayasari, Water (Basel) **15**, 3026 (2023)
9. N. S. Burnama, F. I. W. Rohmat, M. Farid, and W. Wijayasari, in *IOP Conf Ser Earth Environ Sci* (Institute of Physics, 2023)
10. I. Dewa Gede A Junnaedhi, E. Riawan, R. Suwarman, T. Wahyu Hadi, A. Lubis, N. Joko Trilaksono, R. Rahayu, P. Kombara, R. Waskito, H. Ekalaya Oktora, R. Supriatna, A. Anugrah, A. Haq Mudzakkir, and W. Setiawan, in *IOP Conf Ser Earth Environ Sci* (Institute of Physics Publishing, 2017)
11. O. S. Areu-Rangel, R. Bonasia, F. Di Traglia, M. Del Soldato, and N. Casagli, Sustainability (Switzerland) **12**, (2020)
12. J. Pandey and R. Singh, Appl Water Sci **7**, 1669 (2017)
13. R. R. Utami, G. W. Geerling, I. R. S. Salami, S. Notodarmojo, and A. M. J. Ragas, Science of the Total Environment **738**, (2020)

Organic fertilizer decomposed with biofilm biofertilizer to nutrients status and yield of shallot on acid and alkaline soil

Sudadi[1*], *Widyatmani Sih* Dewi[1], and Hadiwiyono[2]

[1]Soil Science Study Program, Faculty of Agriculture, Sebelas Maret University, Surakarta, Central Java, Indonesia
[2]Agrotechnology Study Program, Faculty of Agriculture, Sebelas Maret University, Surakarta, Central Java, Indonesia

Abstract. The use of organic fertilizer in organic farming increases carbon sequestration so will reduce CO_2 concentration in the atmosphere and impact of global warming. Research aims to study the effect of biofilm biofertilizer (BiO_2) formulas and organic fertilizer dosages on soil nutrients status and shallot yield in acid (Alfisol) and alkaline (Vertisol), Experiment was conducted in field with two treatment factors says BiO_2 formula (F_0, F_1, F_2, and F_3) and doses of organic fertilizer composted with biofilm biofertilizers i.e 0, 10 and 20 tonha^{-1} (D_0, D_{10}, and D_{20}). As a comparison treatment was NPK at a rate of 200 kg urea, 200 kg ZA, 400 kg SP-36 and 250 kg KCl per hectare). The observed variables include available-P, exchangeable-K and shallot bulb weights. Data analyzed by ANOVA followed by DMRT at 95% level confidence. The dose of organic fertilizer increased significantly of observed variables, while the formula of BiO_2 was not. The application of organic fertilizer at 20 tonha^{-1} increased available-P, exchangeable-K and weight of shallot bulbs on Vertisols and Alfisols. Compared to the NPK fertilizer treatment, the application of 20 tonha^{-1} organic fertilizer provided higher shallot yield on Alfisol soil but lower on Vertisol soil. It need further reseacr to identified the factor causing lower yield on Vertisol.

1 Introduction

The successful integration of a consortium of functional rhizobiota that provide nutrients and biological agents that prevent soil-borne diseases (basal root rot) into biofilm biofertilizer formula that effectively provides nutrients and prevents moler disease attacks in the previous research [17 – 19] has become the basis for further development of the BiO_2, which is more consistent in its ability to be used in the field. There were two formula of biofilm biofertilizer which have similar effectivity in the previous research. Field testing

* Corresponding author: sudadi62@staff.uns.ac.id

will verify which formulas have consistent performance and also which treatments can optimize their performance.

Combining microbes and organic materials in one fertilizer formula will bring the microbes closer to the nutrients they need so that they are better able to develop and be active in the soil. The research that has been carried out has succeeded in combining the P-solubilizing fungus Aspergillus niger and the sulfur-oxidizing fungus Penicillium nalgiovensis together with natural phosphate powder, sulfur and organic materials in one biosulfo fertilizer formula which is able to increase the available P and S simultaneously so that it can increase shallot yields on three types of soil. namely those with acidic, neutral and alkaline reactions [1, 2]. Combining functional microbes that provide nutrients and biological agents in one biological fertilizer formula (biofertilizer) is thought to better ensure their growth and activity in the plant root environment, forming a microbial biofilm so as to better guarantee their performance in increasing nutrient availability and uptake by plants and preventing the development of microbes that cause infectious diseases. land. Various countries have succeeded in developing and utilizing biofertilizers and biopesticides. to increase fertilizer efficiency, increase crop yields, preserve natural resources and food security [4, 5, 6]. According to [7, 8] biofilm is a complex community of various microbial species attached to or close to plant roots. This biofilm can be developed in vitro and at high cell density can be used as a biofertilizer or biocontrol agent in plants.

The inclusion of functional microbes that provide nutrients and biological agents that prevent soil-borne diseases in a fertilizer formula is intended to bring it closer to the nutrients needed so that after being given to the soil it will adapt more quickly and develop into a dominant microbial community in the rhizosphere environment, forming a microbial biofilm that will increase the availability of nutrients and their uptake by plants and prevent the development of disease-causing microbes. Biofilm is a form of association of various microbial species which is believed to have a greater ability to increase nutrient availability, produce growth factors and resist plant disease attacks compared to monoculture or polyculture populations that do not form biofilms [6, 7 9, 10]. Biofertilizer biofilm is more resistant to environmental stress, attacks by predators and antagonists and has the ability to increase crop yields higher than conventional biofertilizers (monoculture, mono inoculums). For example, in rice plants the increase in dry weight of stover by conventional biofertilizer was only 7% and biofilm biofertilizer was 25% compared to control, and biofilm biofertilizer + ½ dose of recommended fertilizer increased 55% from treatment with full doses of NPK fertilization. Existing research results show the advantages of biofertilizer microbes that form biofilms compared to those that do not form biofilms, both in increasing nutrient availability, production of growth-promoting substances (eg IAA) and their role in preventing plant disease attacks [11, 12, 9, 13, 14, 15, 7, 6, 10, 16].

This research is a follow-up to the results of the previous research to verify, optimize and select the best formula, as well as to identify factors that become weaknesses of the 'initial' formula of BiO_2 fertilizer in field so that the quality can be improved. From this research, a 'selected' BiO_2 fertilizer formula will be obtained which can be developed as a BiO_2 fertilizer product that is ready for use by farmers.

2 Material and method

The experiment aims to study and determine the combination of organic material dose treatment and the 'initial' formula for biofilm bio-fertilizer with biological agents which has the highest performance in supporting plant growth, preventing basal root rot attacks and providing shallot yields on acidic and alkaline soils to be selected as the 'selected formula'.

'BiO_2. The experiment was arranged in a randomized completely block design (RCBD) with the experimental treatment plan **as shown in the Table 1 :**

Table 1. The experimental treatment factors of the research

Treatment factor		Soil pH characteristic	
BiO_2 **Formula**	**Dosages of organic fertilizer, tonha^{-1}**	**acidic**	**alkaline**
	D_0 (0)	$F_0 D_0$	$F_0 D_0$
F_0	D_1 (10)	$F_0 D_1$	$F_0 D_0$
	D_2(20)	$F_0 D_2$	$F_0 D_0$
	D_0 (0)	$F_1 D_0$	$F_0 D_0$
F_1	D_1 (10)	$F_1 D_1$	$F_0 D_0$
	D_2(20)	$F_1 D_2$	$F_0 D_0$
	D_0 (0)	$F_2 D_0$	$F_0 D_0$
F_2	D_1 (10)	$F_2 D_0$	$F_0 D_0$
	D_2(20)	$F_2 D_0$	$F_0 D_0$
	D_0 (0)	$F_3 D_0$	$F_0 D_0$
F_3	D_1 (10)	$F_3 D_0$	$F_0 D_0$
	D_2(20)	$F_3 D_0$	$F_0 D_0$
NPK*	NPK*	NPK*	NPK*

* NPK fertilizer dosage was 200 kg urea, 200 kg ZA, 400 kg SP-36 and 250 kg KCl per hectare

Each treatment combination was repeated three times. The variables observed included plant growth and yield as well as the availability of N, P, K and S nutrients, pH and soil organic matter. Data were analyzed using the F test at a confidence level of 95% followed by Duncan's multiple range test if there was a significant effect.

3 Results and discussion

Shallots require soil with a crumb structure, medium to clay texture, good drainage/aeration, contains sufficient organic matter, and the soil reaction is not acidic (soil pH: 5.6 – 6.5). The most suitable soil for shallots is alluvial soil or its combination with Glei-Humus or Latosol soil. Soil that is sufficiently moist and water does not stagnate is favored by shallot plants.

As shown in the Table 2, the Alfisols soil used has a pH of 5.98 (slightly acidic). The C-organic content is 0.07% (very low). The nutrient content of Alfisols is very low, both total and available. Total N is 0.38% (medium). Total P in alfisols soil is 0.08%, which is very low, with available P content analyzed using the Bray method of 0.12 ppm. The total K content was 1.131%, available K was 0.102% (moderate), total S was 15.964 ppm, and water-soluble S was 0.16 ppm (low).

Meanwhile the Vertisols soil used has a loamy loam texture. The pH of Vertisols soil is 7.49, which is slightly alkaline. The total N content in vertisols soil is 0.46%, which is low. The total P in the vertisols soil was 1.212% and the available P was 3.26 ppm, which was very low. The total S value content of vertisols soil is 101.23%, and the available S or water-soluble S is 0.17 ppm, which is low.

Table 2. Some of chemical and physical properties of Alfisol and Vertisol soils used in the experiment

Soil order	Soil properties	Value	Classification*
Alfisols	Soil texture		**Silty loam** - clay 7,14% - silt 57,18% - sand 35,68%
	Volume density	1,89	-
	pH-H_2O	5,98	Slightly acid*
	Organic matter content (%)	0,07	Very low*
	CEC (me%)	18,98	Fair *
	Total-N (%)	0,38	Fair *
	Total-K (mg/100g)	1,131	Low *
	Exchangeable-K (me%)	0,102	Low *
	Total-P (%)	0,08	Very low *
	Available-P (ppm)	2,94	Very low *
	Total-S (%)	15,964	High
	Available-S (ppm)	0,16	Low **
Vertisols	Soil texture		**Clay loam**
	pH-H_2O	7,49	Slightly alkaline*
	Organic matter content (%)	0,96	Very low*
	CEC (me%)	28,45	Fair *
	Total-N (%)	0,46	Low *
	Total-K (mg/100g)	13,6	Low *
	Exchangeable-K (me%)	0,1	Low *
	Total-P (%)	1,212	Low *
	Available-P (ppm)	3,26	Very low*
	Total-S (%)	101,23	Very high
	Available-S (ppm)	0,17	Low **

* Classification according to Soil Research Institute, Bogor, Indonesia (2005)
** Classification according to Soil *Quality Index (*SQI*) for Assessing Forest Soil Health* (2007).

The use of organic fertilizer at a dose of 20 tonha^{-1} provides plant growth as indicated by plant height **(Fig. 5 and 6)** that is not significantly different to the treatment using NPK fertilizer (200 kg urea, 200 kg ZA, 400 kg SP-36 and 250 kg KCl per hectare). This shows that at these dose (20 tonha^{-1}), organic fertilizers is able to meet the nutrient needs of shallot equivalent to the nutrient provided by chemical NPK fertilizers. This can be shown by the higher soil available-P and exchangeable-K of organic fertilizer treatment (20 tonha^{-1}) compared to the treatment using NPK fertilizer **(Fig. 1, 2, 3 and 4)**.

Beside provide plant nutrient, organic fertilizer improve soil physical, chemical and biologial properties to support plant growth better than chemical fertilizer. The increase of organic fertilizer dose will increase nutrient available to plant.

 https://doi.org/10.1051/e3sconf/202346701029

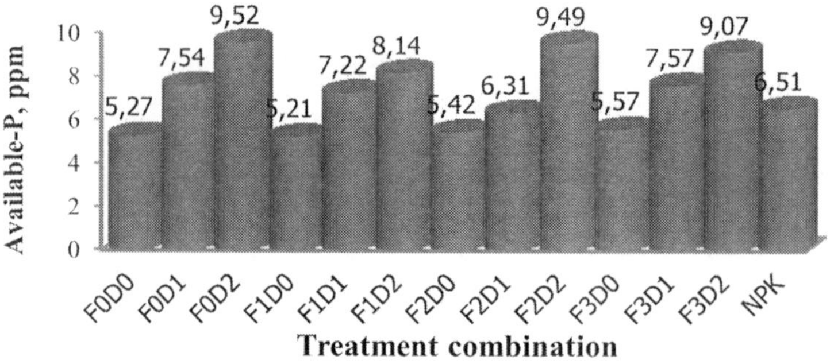

Fig.1. The influence of the treatment on available-P of Vertisol soil.

Both soil available-P (Fig.1) as well as soil exchangeable-K (Fig.2) are increase linearly with the increasing of organic fertilizer doses used. In alkaline soil, phosphorus often fixed by calsium and magnesium that make it not available to the plant.

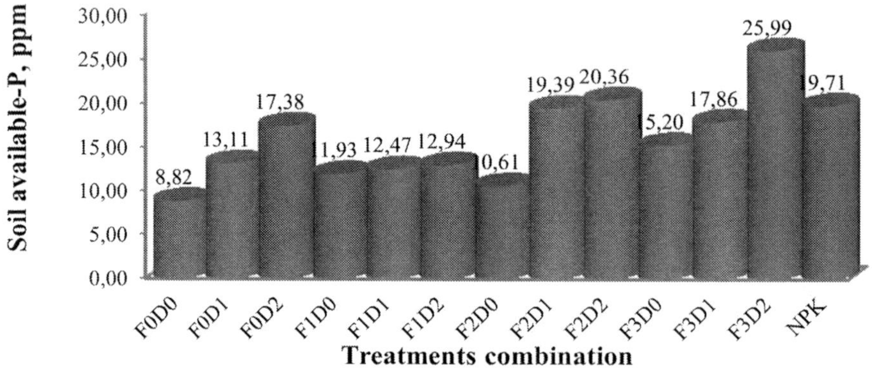

Fig. 2. The influence of the treatment on available-P of Alfisol soil.

While in acid soil P can be fixed by aluminium and iron that make it low availability to the plant. The use of organic fertilizer will improve soil microbial activity and produce organic acid which will increase soil pH and chelating Al and Fe in acid soil or reduce pH and chelating Ca and Mg in alkaline. This will increase soil available-P in acid or in alkaline soil.

Just like soil available-P, the increase of organic fertilizer dose increase exchangeable-K as shown in Fig. 3 and 4. Soil exch-K increase linearly with the increase of organic fertilizer dose. Potassium is the essential nutrient for plant including shallot. It need to fill the sitoplasms, to keep cell turgor and rigidity so it is important to plant resistance againt pest and diseases.

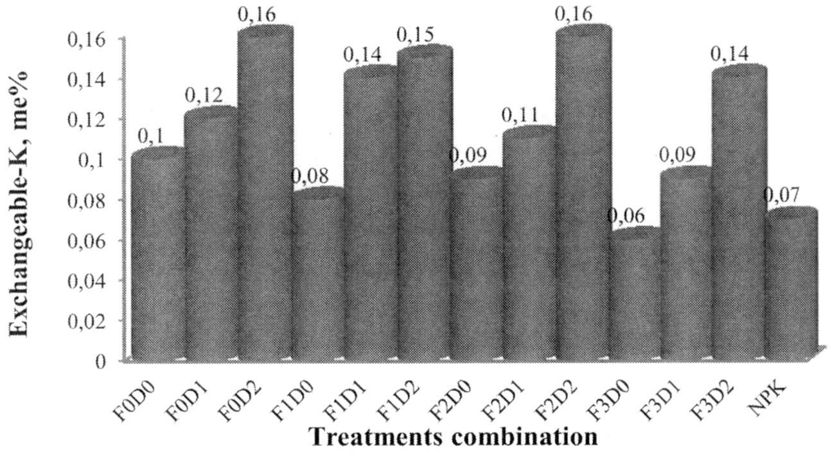

Fig. 3. The influence of the treatment on exchangeable-K of Vertisol soil.

Both in Vertisol and Alfisol soil, the availability of soil P and K was higher for the organic fertilizer treatment than for NPK fertilizer treatment. This maybe caused of the high content of K of the organic fertilizer

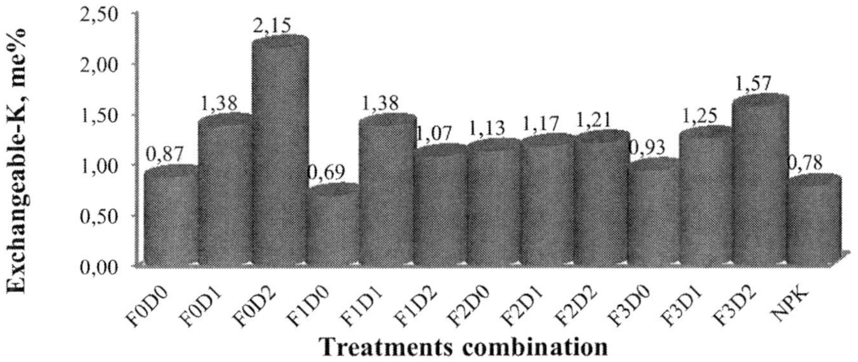

Fig. 4. The influence of the treatment on exchangeable-K of Alfisol soil.

Plant need nutrient and other factor for their growth. Shallot is one of the high economic value horticulture that need high available soil nutrients especially macro nutrient such N, P and K to growth and give optimal yield.

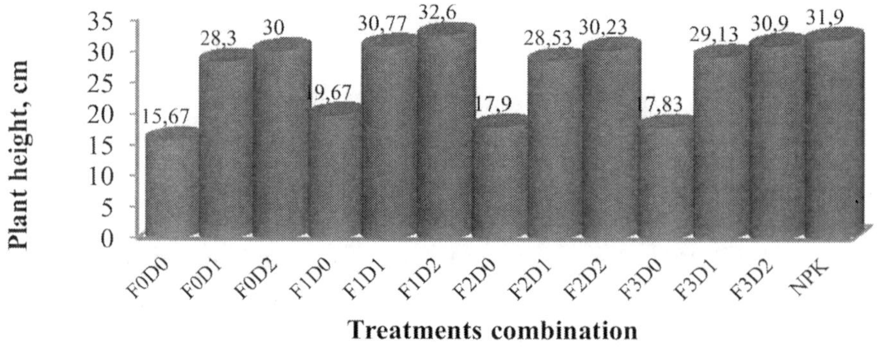

Fig. 5. The influence of the treatment on shallot height at Vertisol soil

From the Fig. 5 and 6 can be shown that the use of organic fertilizer decomposed with biofilm biofertilizer (BiO_2) can provide P and K for shallot growth as well as NPK fertilizer does as indicated by the plant height for the organic fertilizer treatment was slightly higher than NPK fertilizer treatment.

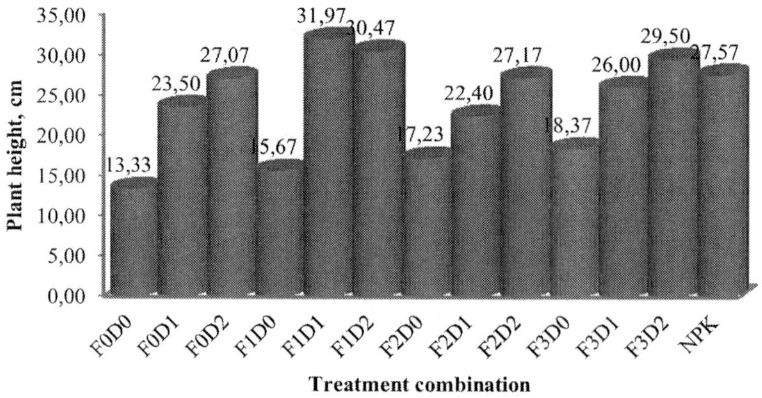

Fig. 6. The influence of the treatment on shallot height at Alfisol soil

Plant yields are generally in accordance with plant growth. This also tend to happens to the yield of shallot bulbs on Vertisol and Alfisol soils as shown in Fig.7 and 8. Increasing the dose of organic fertilizer increases the yield of shallot bulbs, except at a dose of 20 tonha^{-1} with formulas F_2 and F_3 on Vertisol soil which reach the highest (maximum) at a dose of 10 tonha^{-1}. This can occur because at a dose of 10 tonha^{-1} the organic fertilizer used is already able to meet the needs of plants. At a dose of 20 tonha^{-1}, there was a decrease in F_2 and F_3 treatment, it is suspected that there were certain nutrients outside of P and K or decomposition results that caused plant growth to be hampered. NPK fertilizer at recommendations dosage provide the highest yields. It mean that there is something that not yet known in this research that make the shallot yield tend to decrease with higher dose than 20 tonha^{-1}.

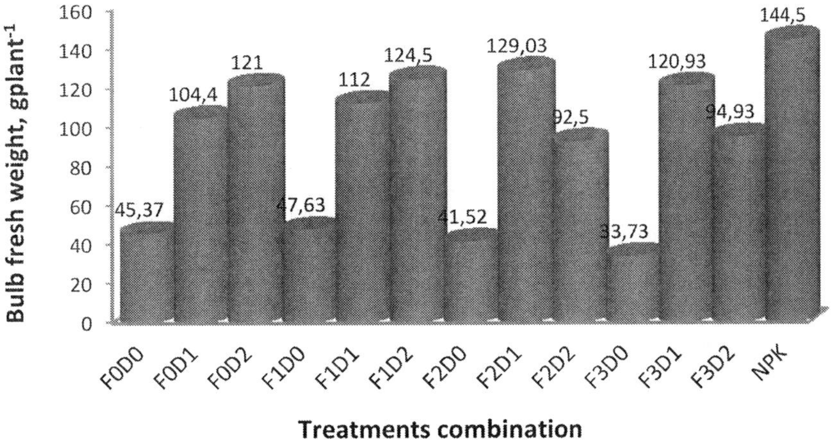

Fig. 7. The influence of the treatment on shallot yield in Vertisol soil

Different result from Alfisol soil, which show higher yield of organic fertilizer treatment than NPK fertilizer as shown in Fig. 8.

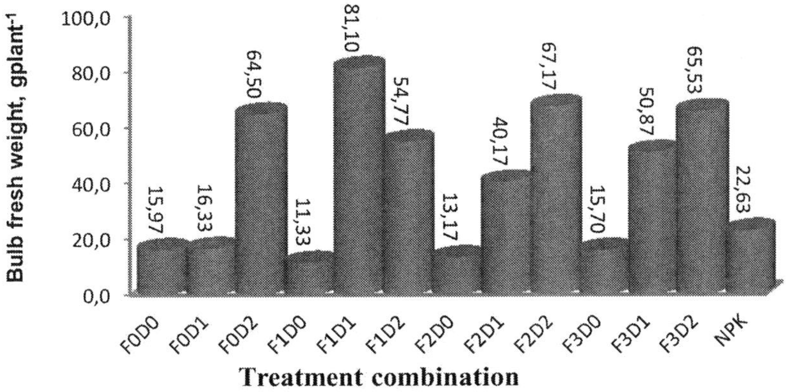

Fig. 8. The influence of the treatment on shallot yield in Alfisol soil.

In Alfisol soil, the yield of shallot bulbs in the treatment with an organic fertilizer dose of 20 tonha-1 gave higher shallot yields compared to the recommended dose of NPK fertilizer treatment, allegedly due to the physical-chemical-biological properties of the soil which became better after applying organic fertilizer, when compared to the effects application of NPK fertilizer. This can happen because the presence of organic and microbial material from biofertilizer biofilms which is more effective than conventional biofertilizers can improve the physical-chemical properties of the soil much better because its ability to provide nutrients, produce growth-promoting substances and stimulate soil aggregation can be much higher [7,9–16].

4 Conclusion

1. The dose of organic fertilizer affect significantly to the available-P, exchangeable-K, plant height and shallot yields, but the BiO_2 fertilizer formula does not. The use of organic fertilizer decomposed with biofilm biofertilizers at the dose of 20 $tonha^{-1}$

provides higher av-P, Exch-K and shallot yields than NPK fertilizer treatment (200 kg urea, 200 kg ZA, 400 kg SP-36 and 250 kg KCl per hectare).

2. The used of organic fertilizer decomposed with Formula BiO_2 1 (F1) is not in optimal dose yet as shown by the plant yield which is still increasing linearly. Meanwhile, for organic fertilizer decomposed with the BiO_2 2 (F_2) and BiO_2 3 (F_3) formulas, the optimal dose is 10 tons per hectare. Lower yields from the NPK treatment may becaused of low nutrients content that could not provided plant nutrients demand optimally for crop production.

This research publication is supported by the Research Group Grant scheme (HGR-UNS Research) A for the 2023 fiscal year with the contract number 228/UN27.22/PT.01.03/2023.

References

1. Sudadi, Sumarno, and J. Widada, Research Report of National Strategis Research Fund Batch II (Second year) (Surakarta, 2010)
2. Sudadi, Sumarno, and J. Widada, Research Report of National Strategis Research Fund Batch II (First year) (Surakarta, 2009)
3. S.-S. Kao, FFTC Ext. Bull **599**, 1 (2007)
4. T. Yokohama and T. Ohyama, FFTC Ext. Bull **600**, 1 (2007)
5. S. Y. Chien, C. C. Young, and C. L. Wang, FFTC Ext. Bull **603**, 1 (2007)
6. G. Seneviratne and H. S. Jayasinghearachchi, J Biosci **28**, 243 (2003)
7. G. Seneviratne, R. M. M. S. Thilakaratne, A. P. D. A. Jayasekara, K. A. C. N. Seneviratne, K. R. E. Padmathilake, and M. S. D. L. De Silva, Microbial Strategies for Crop Improvement 51 (2009)
8. G. Seneviratne, A. P. D. A. Jayasekara, M. S. D. L. De Silva, and U. P. Abeysekera, Soil Biol Biochem **43**, 1059 (2011)
9. U. V. A. Buddhika, G. Seneviratne, and C. L. Abayasekara, International Journal of Scientific and Research Publications **4**, (2014)
10. J. P. Verma, J. Yadav, K. N. Tiwari, and D. K. Jaiswal, Soil Biol Biochem **70**, 33 (2014)
11. S. Gaind, (2011)
12. A. O. Adesemoye and J. W. Kloepper, Applied Microbiology and Biotechnology 2009 85:1 **85**, 1 (2009)
13. T. Danhorn and C. Fuqua, Annu Rev Microbiol **61**, 401 (2007)
14. D. N. Rodríguez-Navarro, M. S. Dardanelli, and J. E. Ruíz-Saínz, FEMS Microbiol Lett **272**, 127 (2007)
15. H. S. Jayasinghearachchi and G. Seneviratne, Soil Biol Biochem **38**, 405 (2006)
16. W. M. M. S. Bandara, G. Seneviratne, and S. A. Kulasooriya, J Biosci **31**, 645 (2006)
17. Sudadi, Suryono, Triharyanto, E. 2018. *IOP Conference Series: Earth and Environmental Science Open Access* Volume 200, Issue 126 November 2018 Article number 012006
18. Sudadi, Setyawan, C.Y. 2021. *IOP Conference Series: Earth and Environmental ScienceOpen Access*Volume 905, Issue 115 November 2021 Article number 012121
19. Sudadi, Nurjanah, M., Minardi, S. 2021. *IOP Conference Series: Earth and Environmental ScienceOpen Access*Volume 637, Issue 18 January 2021 Article number 012092

Are emerging economies a pollution haven? a new era in dirty industrial trade

Ana Shohibul Manshur Al Ahmad[1*] and Sarjiyanto[1]

[1]Economics Department, Faculty of Economics & Business, Sebelas Maret University, Surakarta

Abstract. The hypothesis that international trade causes pollution and environmental damage is still debated and creates heat rather than light. Right extremists argue that trade encourages investment and economic growth while accretion benefits cover the environmental cost. Other extremists claim that international trade is a pollution transmitter (flight industry) from developed countries to countries with laxer policies and low costs. It creates a world pollution haven through the production and trade of dirty industries, especially in countries with abundant factors and low costs. This study aims to investigate the existence of pollution havens as a source of environmental pollution in developing economies. This research method uses a unit root test to estimate the presence of pollutants in each country. This study also employs the dirty industry's trade balance index (TBI) for 44 countries worldwide. The results of this study show that most emerging economies tend to be pollution havens (PH) of developed countries. This finding provides different effects on the pollution haven effect (PHE): full, partial, and clean PH. The implication is that strong environmental policies are necessary for developing economies through a commitment to implementing the SDGs. In addition, global conformity, technological innovation, and strong institutions are required to achieve a clean economy and sustainable development.

1 Introduction

International trade is increasingly globalizing the problem of pollution deaths and has shifted human health impacts from importing countries to producing countries. Thomson Reuters data shows that more than 700,000 premature deaths occur due to pollution generated by export-based industries. Exploiting resources for production activities and trade results in severe environmental pollution and damage without solid environmental policies. Increasing world production and trade will lead to a scaling up in pollution [1, 2] and will damage the environment [3]. The development of exports causes positive scale effects on the formation of emissions and compositional effects through income and structural transformation between economic sectors. Increased exports lead to a rise in the dirtiness of industries [4]. Dirty industries are emissions and pollution-intensive sectors [5] compared to other industries, whether air, water, or soil pollution. It requires environmentally friendly

* Corresponding author: shohibulana@staff.uns.ac.id

technology, clean production, renewable energy, and strong environmental policies to minimize the negative impact of production and trade.

Loose environmental policies in emerging economies lead to a tendency to become a haven for developed countries. This results in more significant degradation and pollution in emerging economies, which become pollution havens. Under free trade, dirty industries shift from countries with high environmental standards to lower environmental stringency. Environmental standards and export performance (dirty products) have an inverse relationship in emerging economies [6, 7]. Environmental stringency causes industry inflexibility, leading to suboptimal production and exports. Weak environmental policies (home country) increase the intensity of multinational companies (home country), causing pollution to increase sharply [8]. In addition, the strength of environmental policies in emerging economies erodes comparative advantage; they are confronted with favorable trade conditions or uncontrolled environmental pollution. If the industry is pollution-intensive (either capital or labor-intensive), then countries with abundant factors will export pollution-intensive products even though they have strong environmental policies [9, 10]. Hence, this automatically leads to a decrease in the comparative advantage for such products due to rejection from other countries.

On the other hand, the pollution effect in emerging economies will be self-limiting [5]. It is due to the balancing and scale effect (benefits) in economic growth, increased income, changes in production techniques, technical expertise, investment in clean technology, and public demands for environmental quality and protection. Increased trade and environmental pollution have a divergent relationship caused by global pollution [11]. Environmental pollution is a human hazard because they want to exploit resources for maximum profit and ignore environmental sustainability. Trade and economic growth lead to more efficient use of resources and technology distribution [12]. The requirements of consumers and the global environment for products that are friendly and provide environmental protection will pressure producers. Innovation in clean production and environmentally friendly energy is compulsory for global manufacturers. In addition, exports and environmental standards positively correlate [13] and encourage improving environmental standards [14]. Meanwhile, the pollution havens effect (PHE) is only partial, and in a few cases, where the trend of PHE is not significant [15].

This study aims to examine the PHE of dirty products trade in emerging economies. This study employs a unit root test on the trade balance index (TBI) variable to check the existence of PH. The contribution of this study is that we empirically document the pollution haven of 39 emerging economies and further establish that dirty industries play an essential role in forming pollution haven in emerging economies. Most studies state that emerging economies are victims of pollution haven for dirty products. This result may contribute to the concentration of dirty industries in the world. On the one hand, dirty industries contribute significantly to growth in emerging economies, but on the other hand, they face strict environmental standards applied globally. Global consumer requirements and the convergence between polluting industries and the Sustainable Development Goals (SDGs) are becoming a new chapter in global trade patterns.

2 Research method

2.1 Data & variables

This study employs the trade balance index (TBI) analysis center. TBI measures the international trade performance for dirty products [14] [16]. Specifically, the construction of TBI is as follows:

$$\mathrm{TBI}_{jt}^{i} = \left[\frac{X_{jt}^{i}-M_{jt}^{i}}{X_{jt}^{i}+M_{jt}^{i}}\right] \tag{1}$$

Where *i* is sector *i* (dirty product group based on SITC Rev 3); *j* refers to country *j*; and *t* is the current year *t*. X^i_{jt} is the export for product group *i* of country *j*, and *M* is the import.

TBI describes how much a country specializes in sector *i*, i.e., if it specializes in exports (net-exporter) or imports (net-importer). This index ranges between -1 and 1 ($-1 \leq TBI \geq 1$). In general, to see whether a country is a net exporter or net importer is from the resulting coefficient if TBI is less than zero/negative ($-1 \leq TBI \geq 0$), then it is a net-importer and vice versa. The data in this study are the export and import values of the dirty products group, then calculate the TBI value. This data is published by The United Nations, namely the United Nations Commodity Trade Statistics Database (UN COMTRADE). This study uses trade data based on the Standard International Trade Classification Revision 3 for the 3-digit code.

In this study, dirty industries consist of six (3-digit) product groups, namely: Metal Mining (281, 283); Primary Nonferrous Metals (681, 682, 683, 684, 685, 686, 687, 689); Pulp and Paper (251, 641, 642) Primary Iron and Steel (671, 672, 673, 674, 675, 676, 677, 678); Chemical (512, 513, 514, 582,583,584,585) [5], and Cement Industry (283, 661) [17] [18].

2.2 The unit root test model

The unit root test is the most popular test to explain whether the model is stationary (or nonstationary) [19] with the Augmented Dickey-Fuller approach [20]. Theoretically, this (stochastic) unit root process can be described as follows:

$$Y_t = \rho Y_{t-1} + \epsilon_t \qquad -1 \leq \rho \geq 1 \tag{2}$$

where ϵ_t is a white noise error term.

Then we include the trade balance index (TB) variable into the model (TB → Y) as the dependent variable, so the estimation model is as follows:

$$TB_t = \rho TB_{t-1} + \epsilon_t \qquad -1 \leq \rho \geq 1 \tag{3}$$

If we regress the model (3), TB_t on the lagged (one period) value of TB_{t-1}, assuming $\rho = 1$, the model becomes a walking model without random drift (nonstationary stochastic process).

For theoretical reasons, we manipulate the model by including TB_{t-1} on both sides (left and right) as a subtractor, resulting in the model:

$$TB_t - TB_{t-1} = \rho TB_{t-1} - TB_{t-1} + \epsilon_t \tag{4}$$
$$TB_t - TB_{t-1} = (\rho - 1) TB_{t-1} + \epsilon_t \tag{5}$$

Then, rewrite the alternative model:

$$\Delta TBI_t = \vartheta TBI_{t-1} + \epsilon_t \tag{6}$$

Where $\vartheta = (\rho - 1)$ and Δ is the first difference operator.

For hypotheses H_0 and H_1, we estimate equation 6 and test the null hypothesis that $\vartheta = 0$. If $\vartheta = 0$, then $\rho = 1$, i.e., it has a unit root, meaning it is not stationary. Note however that if $\vartheta = 0$ then the equation becomes $\Delta TB_t = TB_t - TB_{t-1} = \epsilon_t$. As long as ϵ_t white noise error term, the last equation is the most precise.

2.3 Robustness Test

To test the robustness of the ADF model, we employ The Kwiatkowski-Phillips-Schmidt-Shin (KPSS) test [21]. It attempts to test whether the time series is non-stationary due to a unit root or stationary in linear mean/trend. A stationary time series reflects that the mean and variance are constant over time. This model also combines deterministic trends, random walks, and stationary errors. The equation is as follows:

$$Y_t = \vartheta_t + (r_t + \alpha) + \epsilon_t \tag{7}$$

Where $r_t = r_0 + u_t$ is a random walk with initial value $r_0 = \alpha$ as intercept, t is the time index, and u_t is independent identically distributed $[u_t \sim IID\ (0\sigma^2_u)]$. If $\sigma^2_u = 0$, then $r_t = r_0$, then all t and Y_t are stationary. Furthermore, the hypothesis $H_0 : \sigma^2_u = 0$ is stationary (trend or level), and $H_A : \sigma^2_u > 0$ is a unit root.

3 Result and discussion

Table 1 presents the results of dirty product testing based on unit root test for 39 emerging economies (* indicates stationary and as PH at level). Based on the ADF test, we find three different pollution haven models for dirty product trade. *First*, full pollution havens, i.e., Cambodia, Costa Rica, Mexico, and Kazakhstan, suffer from full PH (all dirty industry groups/DIG), followed by India and Guatemala (5 DIG exposed as PH). Based on the 2022 Environmental Performance Index (EPI) data [22], they are ranked 154, 68, 73, 93, 180, and 167 (out of 180 countries) with scores of 30.10, 46.3, 45.50, 28, and 18.90 (on a scale of 100), respectively. This condition shows that their environmental policies are feeble, especially in India, which occupies the lowest position [22]. They pursue economic growth to exclude sustainable development oriented towards a green economy. Environmental institutions and policies in Cambodia have been established, but their supremacy and implementation are poor. Internal conflicts and suboptimal environmental governance exacerbate this, so businesses only pursue profits without regard to environmental impacts.

Meanwhile, in India, environmental policies are lax, especially in enforcement. India's industrialization has led to industrial waste and particles that are harmful to the environment, including air, soil, water, and human health, and the infrastructure related to their management is inadequate. India is the world's third largest emitter of air pollution, contributing 2.65 billion metric tonnes to the atmosphere [23]. It is due to the country's use of coal, oil, and gas as its primary energy sources. Large environmental legislative bodies have been unable to improve environmental performance. Cambodia, India, and Guatemala's TBI are negative across time and cross-section, indicating a net importer. Imports of dirty products, whether direct (intermediate goods) or investment (FDI), cause domestic production processes to generate pollution in the host country.

Table 1. Stationary test result of pollution haven existence based on TBI.

No	Country	Dirty Industries					
		Metal Mining (MM)	Primary Nonferrous Metals (PNM)	Pulp and Paper (PP)	Primary Iron and Steel (PIS)	Chemical (CHC)	Cement Industry (CEM)
1	Cambodia	-0.448**	-0.679**	-0.434**	-1.115**	-0.757**	1.060**
2	China	-0.175	-0.398**	-0.017	-0.192	0.015	-0.029
3	Indonesia	-0.220	-0.138	-0.141**	-0.106	-0.132*	-0.262
4	Malaysia	-0.015	-0.105	-0.027	-0.018	-0.076	-0.283*
5	Philippines	-0.206	-0.449*	-0.406*	-0.167	-0.473*	-0.365*
6	Thailand	-0.416*	-0.3105*	-0.174	-0.333*	-0.134	-0.174
7	Viet Nam	-0.247	0.014	0.135	-0.013	-0.569*	-0.094
8	Argentina	-0.405*	-0.166	-0.220	-0.097	-0.008	-0.186
9	Bolivia	-0.405*	-0.166*	-0.220	-0.097*	-0.008*	-0.186
10	Brazil	-0.627*	-0.030	-0.074	-0.141	-0.216	-0.071
11	Chile	-0.005	-0.174	-0.460	-0.321	-0.138	-3.705
12	Colombia	-0.176	-0.308*	-0.155*	-0.226	-0.128	-0.059
13	Costa Rica	-0.735*	-0.477*	-0.309*	-0.526*	-0.732*	-0.494*
14	Ecuador	-0.167	-0.040	-0.552*	-0.617*	-0.290*	0.052
15	El Salvador	-0.102	-0.031	-0.515*	-0.519*	-0.409*	-0.192
16	Jamaica	-0.933*	-0.595*	-0.594*	-0.910*	-0.274	-0.438*
17	Mexico	-0.286*	-0.218*	-0.677*	-0.324*	-0.152*	-0.645*
18	Paraguay	-0.170	-0.648*	0.055	-0.261*	-0.410*	-0.231
19	Peru	-0.291*	-0.275*	-0.086	-0.237*	0.000	-0.211
20	Uruguay	-0.066	-0.305*	-0.011	-0.221	-0.107	-0.475*
21	India	-0.281*	-0.230*	-0.034*	-0.236*	-0.300*	-0.065
22	Estonia	-0.388*	-0.784*	-0.309*	-0.407*	-0.269	-0.220
23	Czech	-0.181	-0.304*	-0.181	-0.204*	-0.495*	-0.306*
24	Greece	-0.477*	-0.232	0.061	-0.165	-0.065	-0.110
25	Hungary	-0.678*	-0.455*	-0.114	-0.680*	-0.172	-0.495*
26	Poland	-0.056	-0.152	-0.060	-0.144*	-0.059	-0.124
27	Slovakia	-0.244	-0.282	-0.253*	-0.192	-0.356*	-0.215
28	Bulgaria	-0.370*	-0.936*	-0.446*	-0.148	-0.857*	-0.205*
29	Croatia	-0.326	-0.850*	-0.243*	-0.599*	-0.040	-0.299*
30	Egypt	-0.227	-0.377*	-0.029	-0.282	-0.099	-0.132
31	Guatemala	-0.305*	-0.423*	-0.194*	-0.432*	-0.290*	-0.292*
32	Kazakhstan	-0.653*	-1.052*	-0.568*	-1.169*	-0.323*	-0.353*
33	Pakistan	-0.250	0.487*	-0.104	-0.241	-0.147	-0.204
34	Romania	-0.108	-0.031	-0.538*	-0.034	-0.257*	-0.087
35	Russia	-0.705*	-0.318*	-0.153	-0.180	-0.188	-0.759
36	Ukraine	-0.284	-0.071	-0.375*	-0.153	-0.357*	-0.174
37	Turkey	-0.147	-0.480*	-0.047	-0.841*	-0.017	-0.449*
38	Belarus	-0.133	-0.027	-0.121	-0.540	-0.020	-0.243
39	Armenia	-0.713*	-0.537*	-0.071	-0.454*	-1.083*	-0.223

Note: *) and**) are significant at 5% and 1%, respectively.

Based on Table 1, the dirty industries of Mexico and Kazakhstan show stationary (<1% and <5%), which means that they are PH for all DIG. The pollution in Mexico and Kazakhstan is inseparable from weak environmental policies. Although they have sound policies, they must pay more attention to implementation and enforcement. In the Mexico, disharmony between the federal government and local governments led to chaos in implementing these policies, and the standards were considered a relative failure [24]. For companies, environmental policies are informal, leading to non-compliance by businesses. It is supported by the dominance of business groups over ecological activists. In addition, non-renewable energy (fossil fuels), as a form of "sovereignty," causes enormous environmental pollution. Kazakhstan's mining industry is a leading sector due to endowment factors. Mining as the primary energy source makes the country the largest emitter of carbon dioxide in Central Asia (ranked 14th in the world) [25], where the industrial sector contributes up to 40 percent of pollution. In addition, Kazakhstan is also stalled by the stagnation of environmental technology, especially in producing and treating environmentally friendly residues (solid, liquid, and gaseous waste). Although Kazakhstan is promoting Carbon Capture and Storage (CCS), its heavy dependence on fossil fuels poses a formidable challenge to creating sustainable development. This finding fully confirms the existence of PH for emerging economies (in favor of PHH). Although they have a set of environmental policies, they may need to be stronger in enforcement and policymakers' commitment.

Second, clean pollution haven. Chile and Belarus are the only emerging economies free from PH for all dirty industries. Chile has strong environmental standards. It is reflected in Chile's EPI ranking 16th in 2022 with a score of 6.6 in the total composition of measures [24]. The relatively rapid reform of environmental policy (The Climate Act 2022) and the industry's responsibility for its operations are some factors that reduce pollution and environmental damage, so Chile is not a PH. In addition, the endowment factor of the mining industry requires them to create a clean and sustainable environment, so they invest heavily in technological innovation to control emissions. Belarus's success as a non-PH is the government's commitment to the National Sustainable Development Strategy (NSDS) for the 2021-2030 period, which emphasizes green economy, green innovation, and integration of national development plans with the SDGs.

Finally, Partially pollution haven, 31 countries are exposed as pollution haven (PH) but only for some products. Table 1 (*) illustrates the presence of a unit root for some DIGs ranging from one to four bundles, which explains that they are PH for some DIGs. China, Malaysia, Vietnam, Argentina, Brazil, Greece, Egypt, Pakistan, and Poland are countries with PH for only one product bundle. Referring to the Yale Centre for Environmental Law & Policy 2022 data, the EPI scores are 28.40, 35, 20.10, 41.10, 43.60, 56.20, 35.50, 24.60, and 50.60 respectively, out of a total score of 100 [22]. Asian countries have relatively low environmental policies but tend to have solid policies and commitments to reducing environmental pollution. China has strong fiscal and regulatory policies to reduce energy-related emissions. Carbon and customs taxes are the most effective tools in reducing emissions and premature deaths while maintaining overall domestic welfare. A progressive CO2 tax scenario 2017 of $5 per tonne and increasing to $70 per tonne [26] is an effective solution in curbing pollution in China. The Jing-Jin-Ji program is green financing that supports the 13th Five-Year Plan (2016-2020), and PforR is effective in reducing the use of fossil energy and supporting the use of renewable energy in China [27]. Meanwhile, Poland tends to be restrictive towards the EU's green policy to achieve environmental neutrality by 2050. They argue for the efficient use of coal for the industry and significant investments in infrastructure for using renewable fuels.

It reflects the different readiness and commitment of governments to sustainable development. This finding is also a milestone in the evolution from a dirty economy to a new order in the form of a green and sustainable economy. Global demands incredibly advanced nations require emerging economies to respond quickly to the various environmental and quality management standards they apply. Scale effects must handle the technical effects of pollution arising from the industrialization and trade of emerging economies. Emerging economies must adopt and innovate environment-based technologies to maintain a clean environment.

4 Conclusion

Environmental matters are tricky and complex for some emerging economies facing global environmental pressures and economic growth. Weak environmental policies create PH for them. The findings provide new insights and horizons related to the PHE hypothesis in dirty industries trading in emerging economies. Clean economy and sustainable development have become the dreams and ideals of the global community. Institutions, technological innovation, strong environmental policies, and global synergy are the keys to achieving these goals. Emerging economies must strive for more incredible and still manage steady economic growth. However, emerging economies confront various problems to achieve this, including institutions, technology, commitment, economic complexity, sovereignty issues, geography (endowment factors), and several other fundamental issues. Emerging economies have various environmental regulatory tools, participation in world environmental agreements, and SDGs, but commitment and reluctance are central issues.

References

1. R. Straumann, "Hovedoppgave for cand.oecon-graden Exporting Pollution? Calculating the embodied emissions in trade for Norway," 2003.
2. M. Personal, R. Archive, and A. Sen, "M P RA Intra-Industry Trade in Dirty Goods and Environmental Policies Intra-Industry Trade in Dirty Goods and Environmental Policies," 2013.
3. S. S. Rosenthal and W. C. Strange, "The Determinants of Agglomeration," *J. Urban Econ.*, vol. 50, no. 2, pp. 191–229, 2001, doi: https://doi.org/10.1006/juec.2001.2230.
4. E. Akbostanci, G. I. Tunç, and S. Türüt-Aşik, "Pollution haven hypothesis and the role of dirty industries in Turkey's exports," *Environ. Dev. Econ.*, vol. 12, no. 2, pp. 297–322, 2007, doi: 10.1017/S1355770X06003512.
5. M. Mani and D. W. Prdei, "IN SEARCH OF POLLUTION HAVENS? DIRTY INDUSTRY IN THE WORLD ECONOMY, 1960-1995," 1997.
6. A. Mahajan and K. Majumdar, "Environmental policy stringency and comparative advantage of environmental sensitive goods: a study of textile exports in G20 countries," *Benchmarking*, vol. 29, no. 9, pp. 2924–2951, 2022, doi: 10.1108/BIJ-06-2021-0304.
7. S. Cagatay and H. Mihci, "Degree of environmental stringency and the impact on trade patterns," *J. Econ. Stud.*, 2006, doi: 10.1108/01443580610639884.
8. D. K. Kellenberg, "An empirical investigation of the pollution haven effect with strategic environment and trade policy," *J. Int. Econ.*, vol. 78, no. 2, 2009, doi: 10.1016/j.jinteco.2009.04.004.
9. J. Cherniwchan, B. R. Copeland, and M. S. Taylor, "Trade and the environment: New methods, measurements, and results," *Annu. Rev. Econom.*, vol. 9, pp. 59–85, 2017, doi: 10.1146/annurev-economics-063016-103756.
10. W. Du and M. Li, "Influence of environmental regulation on promoting the low-carbon transformation of China's foreign trade: Based on the dual margin of export enterprise," *J. Clean. Prod.*, vol. 244, p. 118687, 2020, doi: https://doi.org/10.1016/j.jclepro.2019.118687.
11. M. T. Aktaş, "Did Globalization Promote Dirty Industries in Turkey? 1," 2015. [Online]. Available: www.ijbssnet.com
12. Y. Lu, "Do Environmental Regulations Influence the Competitiveness of Pollution-Intensive Products?," *Front. Econ. China*, vol. 5, no. 2, pp. 276–298, 2010, doi: https://doi.org/10.1007/s11459-010-0014-3.
13. C. Tisdell, "Globalisation and sustainability: environmental Kuznets curve and the WTO," *Ecol. Econ.*, vol. 39, no. 2, pp. 185–196, 2001, doi: https://doi.org/10.1016/S0921-8009(01)00234-8.
14. J. S. Wilson, T. Otsuki, M. Sewadeb, and T. World Bank, "Dirty Exports and Environmental Regulation Do Standards Matter to Trade?," 2002.
15. S. Indriya and T. Widodo, "M P RA Munich Personal RePEc Archive Do Pollution Havens Exist? Evidence from East Asia Do Pollution Havens Exist? Evidence from East Asia," 2011.
16. G. Lafay, "The measurement of revealed comparative advantages," *International trade modelling : [the papers published in this book were selected from among the best communications presented at the XXVII Conference of the Applied Econometric Association on International Trade held in Montreal, in September 1989]*. Chapman & Hall, London [u.a.], pp. 209–234, 1992.
17. P. Fennell, J. Driver, C. Bataille, and S. J. Davis, "Going net zero for cement and steel," *Nature*, vol. 603, 2022.
18. S. Gross, "The challenge of decarbonizing heavy transport," *Energy Clim.*, no. October, pp. 1–28,

2020.
19. D. N. Gujarati, *Basic econometrics: {McGraw} {Hill}*. 2004.
20. D. A. Dickey and W. A. Fuller, "Distribution of the Estimators for Autoregressive Time Series With a Unit Root," *J. Am. Stat. Assoc.*, vol. 74, no. 366, pp. 427–431, Oct. 1979, doi: 10.2307/2286348.
21. D. Kwiatkowski, P. C. B. Phillips, P. Schmidt, and Y. Shin, "Testing the null hypothesis of stationarity against the alternative of a unit root. How sure are we that economic time series have a unit root?," *J. Econom.*, vol. 54, no. 1–3, pp. 159–178, 1992, doi: 10.1016/0304-4076(92)90104-Y.
22. et al. Wolf, M. J., Emerson, J. W., Esty, D. C., de Sherbinin, A., Wendling, Z. A., *2022 Environmental Performance Index*. New Haven: CT: Yale Center for Environmental Law & Policy, 2022.
23. M. Igini, "5 Biggest Environmental Issues in India in 2023," 2023. https://earth.org/environmental-issues-in-india/# (accessed Oct. 20, 2023).
24. BertelsmannStiftung, "Environmental Policies," *Sustainable Governance Indicators*, 2022. https://www.sgi-network.org/2022/Sustainable_Policies/Environmental_Policies (accessed Sep. 20, 2023).
25. ITA, "Kazakhstan - Country Commercial Guide," *International Trade Administration*, 2022. https://www.trade.gov/country-commercial-guides/kazakhstan-environmental-technologies (accessed Sep. 02, 2023).
26. I. Parry, B. Shang, P. Wingender, N. Vernon, and T. Narasimhan, "Climate Mitigation in China: Which Policies are Most Effective?," *SSRN Electron. J.*, 2021, doi: 10.2139/ssrn.2882584.
27. WorldBank, "China: Fighting Air Pollution and Climate Change through Clean Energy Financing," 2020. https://www.worldbank.org/en/results/2020/06/21/china-fighting-air-pollution-and-climate-change-through-clean-energy-financing (accessed Oct. 02, 2023).

The accuracy of soil moisture prediction using an RGB camera on maize and peanut plantation

Dwi Priyo Ariyanto[1*], *Isnain Rachmanto* Rustam[2], *Mujiyo*[1], *and Komariah*[1]

[1]Soil Science Department, Faculty of Agriculture, Universitas Sebelas Maret, Jl. Ir. Sutami 36a Surakarta, Central Java, Indonesia
[2]Bachelor Degree of Soil Science Department, Faculty of Agriculture, Universitas Sebelas Maret, Jl. Ir. Sutami 36a Surakarta, Central Java, Indonesia

Abstract. Forest observation to use soil moisture meters for monitoring soil moisture are still relatively inefficient if done on very large land because it still relies on relatively much energy and requires a long time. This study aims to determine the accuracy of the use of aerial photographs from the Red-Green-Blue (RGB) camera Drone to estimate soil moisture levels. Aerial photographs were analyzed with the ImageJ application to find digital numbers, and actual soil moisture measurements by the gypsum block method. Soil moisture and digital numbers are then analyzed by regression and correlation. The results show that this method has a low accuracy. However, the comparison between the actual soil moisture and the soil moisture from the prediction doesn't show a significant difference. The accuracy of the estimation depends on the camera settings and the weather, so further calibration and testing are necessary if used under different conditions.

1 Introduction

Water is a vital component of plants. Required by plants in greater numbers than others. Water content in the soil or commonly referred to as soil moisture is a very important factor for plant growth and growth [1]. An efficient tool used today to find out the current soil content is the Soil Moisture Meter. Soil Moisture Meter is useful to indicate the amount of water content from soil samples. This information is very useful for people involved in irrigation system management and for other researchers who need to measure groundwater content [2]. The effort is still relatively inefficient if done on a very large land because it still relies on a relatively large amount of energy and requires a long time so there is a need to find more practical soil moisture.

The use of satellite imagery to determine soil moisture levels has been studied by several researchers in the world. The use of remote sensing with microwaves originating from satellites has succeeded in estimating the dielectric properties of the soil based on ground surface emissivity which leads to the estimation of soil moisture [3]. Access to satellite imagery is quite expensive. Satellite imagery is an image obtained from shooting an area using satellite vehicles operated from space so that access is quite limited and often

* Corresponding author: dp_ariyanto@staff.uns.ac.id

constrained by cloud cover [4]. The use of satellites to take images or images is inefficient if used only in a small area.

According to [5] Unmanned Aerial Vehicle (UAV) is a technology that can conduct data acquisition with low altitude and produce high-resolution images so that it can be a solution in providing data for small island mapping. The camera with a complete sensor has a very expensive price. However, there is a drone with more affordable prices that generally use RGB camera sensors. According to [6], The estimation of soil moisture with an RGB digital camera has moderate accuracy. Supported by research from [7] that digital cameras that compose gray images can estimate soil moisture with high accuracy.

Digital photo analysis can be used as an alternative method of characterization and color quantification that is objective, accurate, and applicable. One of the software to process digital photos is ImageJ software. ImageJ software aids measure and classify some qualitative color quantitatively based on the standard digital value of RGB per pixel [8]. Each pixel is arranged by three color components: R (red), G (green), B (blue). The combination of these three RGB colors produces a distinctive color for each pixel. RGB is a color model consisting of red, green, and blue, combined in a wide array of colors [9]. The purpose of this study is to determine the accuracy of the use of aerial photographs from the RGB camera Drone to estimate soil moisture levels.

2 Research methodology

2.1 Time and location

The research was conducted in the Sukosari, Jumantono District, Karanganyar Regency, Central Java, Indonesia with a geographical location of -7° 37' 49.814" latitude and 110° 56" 52.408" longitude. The type of soil in this research area is Alfisols, this is following the statement of [10], who previously researched the site. The research began in March to August 2019. Tools used in this research include drone, gypsum blocks, and laptops. The land was designed as an RCGD (Randomized Complete Group Design) with 2 treatment factors. The first treatment factor was the amendment with P0 (control), P1 (manure 5 tons / ha), P2 (biochar 5 tons / ha), and P3 (zeolite 5 tons / ha), and the irrigation volume treatment factor I1 (50% x Evapotranspiration actual), I2 (100% x Evapotranspiration actual), and I3 (150% x Evapotranspiration actual). The Evapotranspiration actual (ETa) value is obtained from the daily evapotranspiration value of maize and peanut plants. Each plot or plot with an area of 2 m x 2 m, within a block of 0.25 m then the distance between the blocks is 0.4 m. The treatment plot aims to obtain soil moisture diversity for observation. Test analysis of this study uses linear and multiple regression statistical analysis. Pearson correlation test and T-test to find out whether there is a significant difference between the measurement of the actual soil moisture with the prediction of soil moisture.

2.2 Aerial photoshoot on land

Aerial photographs of the land were taken from May to July 2019. Aerial photo taking was done using a Drone/UAV (Unmanned Aerial Vehicle) model of the DJI Mavic Pro Platinum with camera settings used including ISO 100, Exposure time 1/1000 second, Aperture f / 2.2, with a flying height of a 10-meter drone. Photos were taken on May 7 May 2019, 14 May 2019, 24 May 2019, 29 May 2019; June 3 June 2019, 12 June 2019, 19 June 2019, 21 June 2019; until July 3 July 2019, 13 July 2019, 21 July 2019, 30 July 2019. Aerial Photography was taken at 10:00 a.m. - 11:30 a.m. to anticipate the shadow of plants that covered the soil. The soil moisture measurements carried out simultaneously while taking aerial photos. The aerial photo is then processed into a digital number with ImageJ software. ImageJ is an image processing application developed by NIH (National Institutes of Health) which is often used in biology to measure cells, bacterial colonies to leaf area. This application can calculate the statistics of the specified area and pixel values [11]. A digital number is obtained from the number indicated by the pixel in the photographic image (Figure 1).

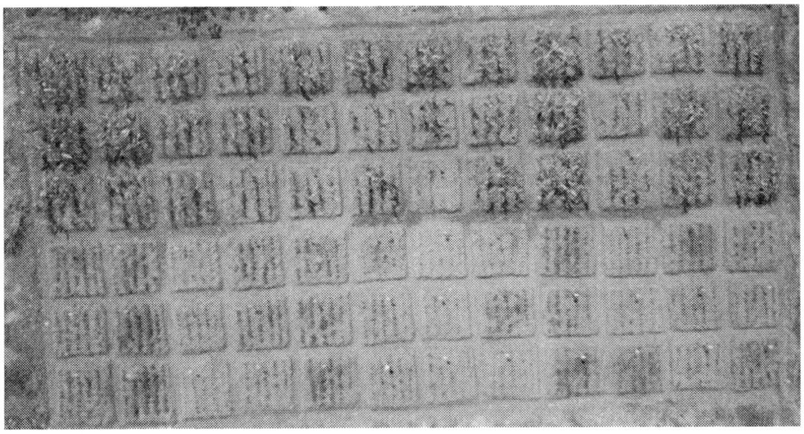

Fig. 1. The Aerial photo on experimental land (13 July 2019).

2.3 Measurement of actual soil moisture content

The actual soil moisture content is measured by the gypsum block method [12]. The gypsum block is then planted in the field and what is measured is the amount of resistance on the brass plate using a multimeter. Soil moisture is adjusted and selected by groundwater retention to normalize the actual soil moisture value. Soil moisture data obtained in this study after adjusting and selecting with groundwater retention amounted to 236 data (138 in maize land and 98 in peanut land).

3 Result and discussion

3.1 Quantitative data on aerial photographs

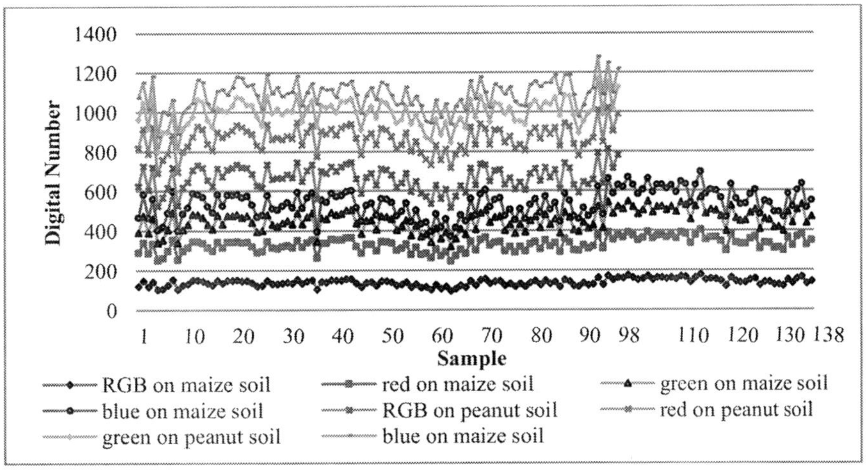

Fig. 2. Digital numbers of the aerial photographs of maize and peanut land.

The aerial photographs obtained are then processed into digital numbers consisting of composite RGB, red, green, and blue. Digital numbers of the aerial photos are shown in Figure 2. The results obtained in Figure 2 on both plantations show that the red digital number has the highest value, then the RGB digital number, green digital number, and blue digital number. The digital numbers used this time are the RGB composite (DN RGB = 0.2989·R +

0.5870·G + 0.1140·B), Band Red, Band Green, and Band Blue. The results obtained in Figure 2 show that the red digital number has the highest value, then the RGB digital number, green digital number, and blue digital number. The red band gets the highest value possible because of the dominant red color in the soil. According to [13] Spectral or Image Characteristics in general state that the ground reflects more light with red wavelengths. Other than that, according to [14] said that Alfisol soil in Jumapolo, Karanganyar had a color of 2.5 YR 3/6 (dark red) on a layer of 0-20 cm. The location of Jumapolo and Jumantono are close to each other and have the same topography, so it can be concluded that the color of the land is close to the same.

3.2 The actual soil moisture

Soil moisture analysis was measured using gypsum block which was then converted using calibration regression that had been done before observation so that the actual soil moisture value was obtained in maize and peanut land. The moisture data is then correlated with the digital number data to find out the moisture relationship with the digital number (Figure 3).

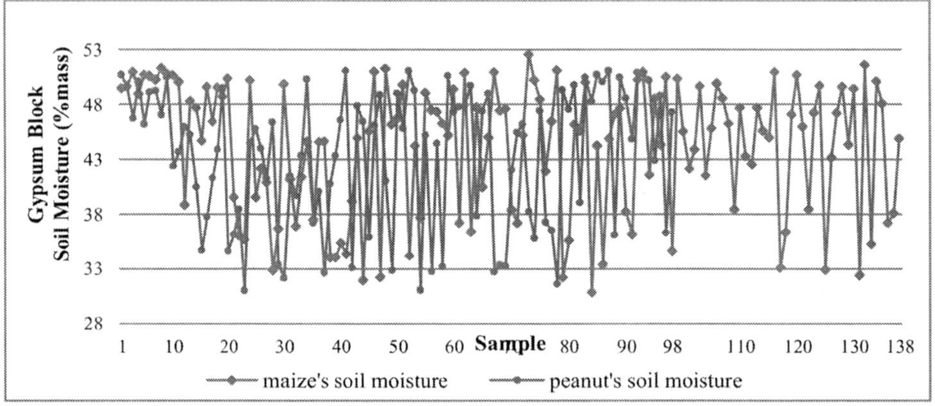

Fig. 3. Soil moisture data from May to July on maize and peanut land.

3.3 Correlation test of digital numbers with actual soil moisture

Table 1. Correlation tests for RGB, Red, Green, and Blue with moisture.

Digital Number	R	P-value
RGB	-0.146	0.02
Red	-0.188	0.00
Green	-0.137	0.03
Blue	-0.066	0.31

Test correlation of RGB, red, green, and blue with moisture on the land of maize and peanuts aims to determine the close relationship between digital numbers and moisture. The RGB, red, green, and blue correlation tests with moisture on the planting of maize and peanuts show that the correlation between moisture with digital numbers is negative or inversely proportional which means the greater the value of digital numbers, the smaller the soil moisture, and vice versa, the smaller the value of digital numbers the soil moisture will get bigger (Table 1). The results obtained are following research conducted by [15] which obtained a negative correlation between the value of red, green and blue with the value of soil moisture. The study showed the soil became darker when soil moisture increased.

3.3 Actual moisture regression with digital numbers

Table 2. Results of actual soil moisture regression analysis with digital numbers

Regression	Digital number	R^2	Y
linear regression	Composite RGB	0.021	51.949 - 0.0569 RGB
linear regression	Red	0.035	59.026 - 0.0781 Red
linear regression	Green	0.018	50.046 - 0.0491 Green
linear regression	Blue	0.004	46.35 - 0.0278 Blue
multiple regression	Composite RGB, Red, Green, and Blue	0.073	57.2 + 1.69 RGB - 0.54 Red - 1.27 Green + 0.12 Blue

The regression in moisture with the RGB digital number yields $R^2 = 0.021$; moisture with a red digital number of yields $R^2 = 0.035$; moisture with a green digital number of yields $R^2 = 0.018$; moisture with a blue digital number of yields $R^2 = 0.004$; and Multiple regression between RGB, red, green, and blue composites with soil moisture yields $R^2 = 0.073$ (Table 2,). The regression with the highest R^2 to the lowest in the soil moisture multiple regression with the RGB, red, green and blue; linear regression soil moisture with red digital numbers; soil moisture with RGB digital number; soil moisture with a green digital number; soil moisture with blue digital numbers. The results of the analysis of these data prove that the multiple regression formula between moisture with the composite digital numbers RGB, red, green, and blue is the most accurate among the other regressions with $R^2 = 0.073$. According to [6], the regression results between moisture and digital numbers have moderate to high accuracy by identifying the camera and settings used, but the results of this study indicate that the use of outdoor camera RGB is less accurate in terms of a very low R^2 value. The use of a drone with RGB cameras is greatly influenced by outside factors such as light and weather. Besides, differences in soil moisture on Figure 3, which is very volatile also cause low R^2 values. The next highest-accuracy regression formula is a linear regression between moisture with digital red numbers. The red color has the longest wavelength in visible light which is around 620 - 750 nm. Supported by research [16], states that the spectral reflection in the soil tends to increase with increasing wavelengths of light so that the red color which incidentally has a high wavelength following the spectral reflection of the soil.

3.4 Comparison of actual soil moisture value and digital number prediction soil moisture

Digital number data obtained on observations, then used in the multiple regression formula with the highest R^2 value. The multiple regression formula produces a predictive value of the digital number of aerial photographs. The predictive value of the digital aerial photograph number is then compared with the actual value of the soil moisture on the land.

Multiple Regression between soil moisture with digital composite numbers RGB, red, green, and blue has a low R^2 but it is expected that the soil moisture values from the regression and the actual are not much different (Table 2,). The experiment compared with the Correlation, T-test, and linear regression between the actual soil moisture obtained from measurements in the field with the predicted soil moisture obtained by the multiple regression formula, yielding T-value = 0.00 and P-value = 1. These results can be concluded that there is no difference between the soil moisture when the measurement is on the field and the soil moisture is predicted from aerial photographs. According to [17] states that multi-spectral imagery which has a high resolution, then combined with soil sampling, is sufficient to provide information on modelling approaches to accurately estimate spatially distributed surface soil moisture. This statement supports the results of this study which states that there is no difference between actual soil moisture measurements and predicted soil moisture measurements.

4 Conclusion

The conclusion that can be drawn from this research is the method of estimating soil moisture by using aerial photography of RGB on maize and peanut plantations have low accuracy. However, the comparison between actual soil moisture on the soil and soil moisture from the prediction of aerial photographs does not show a significant difference so this method still needs to be increased and improved. The accuracy of the estimation is very dependent on the camera settings and the weather so that further calibration and testing is necessary if used under different conditions.

References

1. E. Suharto, J. Ilmu-Ilmu Pertan. Indones. **8**, 44 (2006)
2. S. B. Onoja, J. A. Enokela, and G. O. Ebute, ARPN J. Eng. Appl. Sci. (2014)
3. B. P. Mohanty, M. H. Cosh, V. Lakshmi, and C. Montzka, Vadose Zo. J. **16**, (2017)
4. R. Shofiyanti, **20**, 58 (2011)
5. Y. H. Ramadhani, Rokhmatulloh, A. P. K, and R. Susanti, Maj. Ilm. Globe **17**, 125 (2015)
6. M. B. G. Sakti, Komariah, D. P. Ariyanto, and Sumani, in *IOP Conf. Ser. Earth Environ. Sci.* (2018), pp. 1–6
7. Y. Zhu, Y. Wang, M. Shao, and R. Horton, Can. J. Soil Sci. **91**, 69 (2011)
8. R. V. Kusumah, E. Kusrini, S. Murniasih, A. B. Prasetio, and K. Mahfudz, J. Ris. Akuakultur **6**, 381 (2011)
9. A. H. Hasugian, **III**, 1 (2013)
10. Suntoro, H. Widijanto, Sudadi, and E. E. Sambodo, J. Ilmu Tanah Dan Agroklimatologi **11**, 69 (2014)
11. T. Ferreira and W. Rasband, Image J User Guid. **1.46r**, (2012)
12. T. M. W. Sir and M. E. Bolla, J. Tek. Sipil **1**, 61 (2012)
13. R. O. Kuchenbuch and K. T. Ingram, J. Plant Nutr. Soil Sci. **165**, 573 (2002)
14. A. Wijanarko, Sudaryono, and Sutarno, **2**, 214 (2007)
15. J. F. C. Dos Santos, H. R. F. Silva, F. A. C. Pinto, and I. R. De Assis, Rev. Bras. Eng. Agric. e Ambient. **20**, 1051 (2016)
16. T. Lillesand, R. W. Kiefer, and J. Chipman, *Remote Sensing and Image Interpretation* (John Wiley & Sons, New York, 2007)
17. L. Hassan-Esfahani, A. Torres-Rua, A. Jensen, and M. McKee, Remote Sens. **7**, 2627 (2015)

Biochar as bioretention systems for water quality improvement in Malaysia

Anurita Selvarajoo[1*], *Darshan* Balakrishnan[1], *Aan Mohammad Nusrat* Aman[1], and *Tuck Wai* Yeong[1]

[1]Department of Civil Engineering, University of Nottingham Malaysia, Jalan Broga, Semenyih 43500, Selangor, Malaysia

Abstract. Rapid urbanization and constant development of infrastructure in the civil engineering community has accelerated in the past decades. However, there has been a reduction of permeable surfaces for rainwater and surface runoff to escape, raising concerns regarding flooding and water quality especially in Malaysia where the tropical climate and heavy rainfall applies pressure on the bioretention systems. To lessen the issue, this paper will be looking into biochar's benefits in bioretention systems focussing on the type of biochar which will be most suited in improving water quality. In particular, two types of biochar were compared such as rice husk biochar and palm biochar to be tested as bioretention system and different water samples were run through them. Various water quality parameter tests were conducted to evaluate biochar's performance in reducing pollutants and contaminants from the water samples and thus enable to draw a conclusion on which is the most effective for bioretention use. With a deeper understanding of biochar's capabilities and limitations, stormwater management strategies can be improved to make the construction industry more sustainable.

1 Introduction

With rapid urbanization in Malaysia, the frequent flash floods experienced over the years is a clear indication that the infrastructure is unable to cope with the volume of surface runoff from impermeable surfaces [1]. This has a butterfly effect as fewer impermeable surfaces leads to greater runoff volume which in-turn means more harmful pollutants are being washed into primary waterways, a great concern for water quality. Bioretention systems are at the forefront of stormwater management as they can tackle most issues brought on by excess runoff, however, with rainfall volume increasing and permeable surfaces decreasing, most bioretention systems are running at full capacity and need further enhancing to improve performance. Biochar is made through a process called pyrolysis which is essentially burning organic biproducts with the absence of oxygen [2]. Biochar structurally has a large surface area making it well versed in capturing heavy metal particles and pollutants, essential for improving water quality in a bioretention system. However, even with these benefits there is

[*] Corresponding author:: Anurita.Selvarajoo@nottingham.edu.my

still a lack of understanding on biochar's long-term effects and its performance in bioretention systems. A study was conducted on the role of biochar of stormwater management that would promote green infrastructure in major metropolitan areas worldwide [3]. It has been noted that the hydrophobic nature of the biochar and the electrostatic are the primary mechanisms for adsorption of organic containment so the higher the biochar production temperature the more effective its use [4,5]. It has been noted that biochar also supports the growth of plants and improves aesthetic value of the surrounding areas [3]. Another review has been focused on using biochar based bioretention systems focusing of the chemical and microbial pollutants from stormwater [6]. The review has highlighted the key mechanisms of the removal different impurities: of nutrients is biological, adsorption, anion exchange and precipitation, for heavy metals is sorption, cation exchange, complexation, and plant update, for organic compounds is hydrophobic interaction, hydrolysis, oxidation, and biodegradation, for microbial pollutants is filtration, adsorption and hydrophobic interactions. A study conducted on surface modified biochar reported that with modification by H_2SO_4 excellent results have been obtained with 98% removal efficiency of pollutants [7]. Another study conducted on woodchips and biochar reported that it has effectively removed nitrates, metals and organic contaminants [8].

2 Materials and methods

The biochar samples used in this investigation are rice husk and palm waste. The types of water used are tap water, lake water and sewer treatment water. The water purification system has been conducted within 48 hours. Prior to the experiment the water parameters have been through different tests; BOD (Biological Oxygen Demand), COD (Carbon Oxygen Demand), pH and turbidity. Additionally, the experiment has been repeated twice for reliability of the data obtained. The equipment has been cleaned with distilled water prior to the experiment and the water samples were kept at a constant temperature throughout. Figure 1 shows the experimental set up used for this experiment, a hole was drilled into the top and bottom of plastic jars that are placed on top of the transparent box and a water tube was installed on the top and a valve was installed at the bottom for the water to filter out. Each jar has 6 layers of material; layer 1 is the mesh filter, layer 2 is the plastic netting, layer 3 is the gravel (3cm), layer 4 is the sand (1.5cm), layer 5 is the biochar (5cm) and layer 6 is the sand (1.5cm) mimicking a typical bioretention system [3]. The mesh and the plastic netting are stop the debris from flowing through while the gravel and the sand serve as a drainage and screening layer. The biochar layer is compacted to not allow any presence of air gaps. Each jar was connected to a storage tank inside of the transparent box with a pump (Dophin P200) was used to control the flow rate of the water. A faucet was installed through every storage tank to obtain the water samples. The water samples tap, lake and sewer were all collected from around the University of Nottingham Malaysia campus.

Figure 1. Bioretention experimental system set up.

3 Results and Discussion

3.1 Biochar effect on BOD

The results for each test are given in Table 1, The BOD test measures the quantity of the soluble oxygen required by the microorganisms to decompose organic matter in a certain time frame. This test is important as eutrophication in lakes are contributors to high levels of BOD [1], and combating the BOD levels will have a net positive impact on the ecosystem which inhabits these lakes as they won't be starved of oxygen, which ultimately leads to a higher water quality. Both biochar's performed similarly in tap water samples with palm biochar averaging a BOD reduction of 29.28% across both tests, against rice husks reduction of 32.60%. This trend continued for both lake water and sewer treatment water with palm biochar reducing BOD content by 23.11% and 25.82% reduction achieved by rice husk biochar. The variation in results between each biochar was minute with percentage variation being under 3.5% for every test with the most important water parameter of sewer treatment water returning a result of 20.14% BOD reduction done by palm biochar and 23.64% by rice husk biochar. According to the Taiwanese EPA river pollution index, all tap water results fell in between the range of 1.05 and 1.55 which classifies as non/mildly polluted water samples, while both river and sewer treatment water fell in the moderately polluted water category; each with a BOD range from 5.11 to 5.41 and 8.24 to 9.48, respectively.

3.2 Biochar effect on BOD

The COD test is a parameter which measures the amount of oxygen required to oxidize a particular specimen (conversion from CO_2 to H_2O). Palm biochar was twice as effective in COD content reduction across both tap water tests versus rice husk biochar, however the inverse result was observed for the lake water and sewer treatment water samples. Palm biochar reduced COD content in tap water by 28.57% while rice husk managed to only achieve a 14.28% reduction in the same 48-hour interval. It is difficult to explain this result as rice husk was able to remove 41.94% of COD content in lake water while palm only managed 24.95%, and 52.06% in sewer treatment water, while palm here only managed to remove 46.11% of COD content. Both water samples with much higher levels of pollutants and contaminants than tap water. A material's ability to remove COD is highly dependent on its ability to absorb organic pollutants. Rice husk contains a higher silica content than its palm counterpart. Silica is able to bind with heavy metal ions and trap them when in samples of water. Additionally, palm biochar has a greater surface area to volume ratio over palm biochar, resulting in a greater ability to sweep waste particles. These factors can explain rice husks' superior performance in reducing COD levels of lake water and sewer treatment water.

3.3 Biochar effect on Turbidity

The presence of suspended particles, including sediment, mud, organic materials, and other tiny or decomposing creatures, causes turbidity. As for the analysis of the turbidity in a water sample, it will determine the performance of biochar's filtration capabilities, a vital aspect of water filtration and purification. Suspended solids vary in size and can put heavy pressure and in some circumstances even damage water filtration systems. Therefore, removing these particles from runoff before they are treated for municipal use is an important aspect of maximizing efficiency of water treatment.

The results show that over both test runs, rice husk biochar outperformed palm biochar by lowering the TSS values in both samples. In total, rice husk biochar outperformed palm biochar by an average of 4.51% in the initial test run and 5.33% in the second test run. Although this discrepancy may seem rather insignificant, the large volume of water filtered through the bioretention systems daily means every incremental difference in filtration performance will have significant impacts on efficiency. The average bioretention system filters around 500 L of runoff water per square meter daily, given the average area of Malaysian bioretention system at 48 square meters each bioretention system moves through a volume of roughly 24,000 L/day [9].

This difference can be accounted for by its larger surface area, allowing a greater volume of solids to be absorbed. Another factor for rice husk's superior ability to absorb comes from its manufacturing process where rice husk is heated at temperatures higher than palm during the pyrolysis process giving it a greater porosity. The pyrolysis of rice husk biochar is typically conducted at a temperature of 550°C to 600°C while palm biochar's pyrolysis was done at 400-450°C, which accords with the literature [4,5].

Tap water for this test acted as more of a controlled variable as it has a low turbidity value with insoluble suspended particles. The lake water and sewer treatment water on the other hand contained larger particles as they are untreated and exposed to many external contaminants, therefore the turbidity changes in these tests are particularly important. Both biochar's performed exceptionally well in sewer treatment conditions as they lowered turbidity value by over 50%. Rice husk biochar outperformed palm biochar across both tests, a reason for this could be the much greater surface area it has over its palm biochar counterpart. Visually at the end of the test, the lake water and sewer treatment water samples were far less murky to the naked eye at the end of the 48 hours versus before.

3.4 Biochar effect on pH

Based on the results obtained, both biochar's increased the pH levels of all 3 samples rendering them more alkaline at the end of the 48-hour cycle than their initial conditions. Biochar is generally more alkaline by nature due to its rich mineral content which is heated and burned in pyrolysis. When in a bioretention system, through the process of hydrolysis when biochar reacts with the water flowing, It releases OH^- by its basic cations ($Ca2+$, $K+$, and $Mg2+$) that increase the soil pH From the results shown, palm biochar was able to increase the alkalinity of each sample more than its rice husk counterpart by an average of 72.03% over both tests. Biochar's ability to increase the pH level of samples is a great plus point in water filtration especially for bioretention systems as heavy metals and toxins are more soluble in acidic conditions therefore by increasing the alkalinity of water its surrounding environment is more habitable.

The tap water sample was found to have an initial pH value of 9.06, this is relatively high given the average pH levels of tap water in Malaysia are typically in range of pH7-pH8.5. Both biochar's still worked to increase the alkalinity in their respective tap water samples, notably a 5.41% increase in palm while only a 0.77% increase was found in the rice husk sample during the first round of testing. This increase is consistent with both test runs. Palm biochar's mix contains palm whereas rice biochar's contains rice husk. This difference in organic material means varying properties; palm biochar has a greater mineral content than rice husk biochar, it is more alkaline as a result. When compared to rice husk, palm in itself has a greater concentration of calcium and potassium minerals which help give palm biochar its alkalinity.

Moreover, both biochar's were concurrently purifying lake water and sewer treatment samples. These two samples contained a lot more contaminants and impurities and thus the 48-hour results of these tests would show clear indications on which type is more effective. In lake water across both tests, palm biochar increased alkalinity in its sample by 37.23% more than rice husk biochar did. This trend continued along with sewer treatment water as palm biochar outperformed rice biochar by a staggering 48.11%. Alkaline water conditions promote healthy development of aquatic life and plants as well as soil health. In practical usage, rainwater is more acidic due to the presence of carbon dioxide as well as pollutants which mix into the water. Therefore, biochar's ability to make a sample more alkaline is an important factor to consider when purifying water.

Table 1. Results of all the tests conducted in this experiment after 48 hrs. ([a]Palm waste biochar, [b]Rice husk biochar.)

Water Sample	**Tap**			**Lake**			**Sewer**		
Hours	0	48		0	48		0	48	
Biochar		P[a]	R[b]		P[a]	R[b]		P[a]	R[b]
BOD (mg/L)	1.81	1.28	1.22	7.01	5.39	5.2	11.25	8.99	8.59
COD (mg/L)	14	10	12	62	44.5	37.5	194	104.5	93
Turbidity (mg/L)	22.8	13.95	10.75	79.4	49.7	46.9	472.1	216.2	208.2
pH	9.06	9.5	9.13	6.43	7.0	6.8	5.28	6.07	5.81

4 Conclusion

In conclusion, both biochars were effective at purifying water as they were able to increase alkaline levels, reduce turbidity, suspended solid content as well as BOD and COD levels which are all crucial parameters in determining the quality of a water sample. Palm biochar is more readily available in Malaysia due to the mass palm oil production done locally; however, rice husk biochar was able to outperform palm biochar in most water parameter tests.

References

1. Sharip Z and Zakaria S 2022 Lakes and Reservoir in Malaysia: Management and Research Challenges 1349–55
2. Selvarajoo A and Oochit D 2020 Effect of pyrolysis temperature on product yields of palm fibre and its biochar characteristics *Mater Sci Energy Technol* **3** 575–83
3. Premarathna K S D, Biswas J K, Kumar M, Varjani S, Mickan B, Show P L, Lau S Y, Novo L A B and Vithanage M 2023 Biofilters and bioretention systems: the role of biochar in the blue-green city concept for stormwater management *Environ Sci (Camb)*
4. Ippolito J A, Cui L, Kammann C, Wrage-Mönnig N, Estavillo J M, Fuertes-Mendizabal T, Cayuela M L, Sigua G, Novak J, Spokas K and Borchard N 2020 Feedstock choice, pyrolysis temperature and type influence biochar characteristics: a comprehensive meta-data analysis review *Biochar* **2** 421–38
5. Yang G, Wu L, Xian Q, Shen F, Wu J and Zhang Y 2016 Removal of Congo Red and Methylene Blue from Aqueous Solutions by Vermicompost-Derived Biochars *PLoS One* **11** e0154562

6. Biswal B K, Vijayaraghavan K, Tsen-Tieng D L and Balasubramanian R 2022 Biochar-based bioretention systems for removal of chemical and microbial pollutants from stormwater: A critical review *J Hazard Mater* **422** 126886
7. Lau A Y T, Tsang D C W, Graham N J D, Ok Y S, Yang X and Li X dong 2017 Surface-modified biochar in a bioretention system for Escherichia coli removal from stormwater *Chemosphere* **169** 89–98
8. Ashoori N, Teixido M, Spahr S, LeFevre G H, Sedlak D L and Luthy R G 2019 Evaluation of pilot-scale biochar-amended woodchip bioreactors to remove nitrate, metals, and trace organic contaminants from urban stormwater runoff *Water Res* **154** 1–11
9. Azizan N A, Goh H W, Faudzi L M, Juiani S F and Kiat Chang C 2023 Bioretention System as a Stormwater Best Management Practices (BMPs): A Case Study in Malaysia *IOP Conf Ser Earth Environ Sci* **1238** 012023

Usability analysis of environmental disaster mitigation efforts in Lampung Province: case study of *Jaring Bencana* application

Simon Sumanjoyo Hutagalung[1*], and *Himawan* Indrajat[2]

[1]Departement of Public Administration, Universitas Lampung, Indonesia
[2]Departement of Government Science, Universitas Lampung, Indonesia

Abstract. Disaster management requires a variety of innovations to intervene or prevent the adverse effects of the disaster, including through various applications made for various purposes and uses. One of the applications is the *Jaring Bencana*, to strengthen the mitigation aspects of natural and non-natural disasters, including environmental disasters. This study aims to examine the level *of usability* in disaster applications by measuring the variables of effectiveness, efficiency, and user satisfaction. The method used in this study is *usability testing* with technique *performance measurements* and *retrospective think-aloud*. Respondents in this study 30 individuals as test participants, were selected purposively based on criteria of interest and responsibility in disaster management. Results from testing *usability* show that mark component effectiveness is significant at 89%, mark efficiency component of 0.24 *goals/second,* and satisfaction score of 68. It can be concluded that the *Jaring Bencana* application has been functionally effective and efficient, although it must be optimized to achieve maximum user satisfaction. The results of this research have implications for the development of disaster applications that are in accordance with disaster needs.

1 Introduction

Indonesia has many natural disasters, such as earthquakes, tsunamis, floods, and landslides. The natural disasters that occurred in Indonesia in 2022 have caused enormous losses. Material losses reached 1.5 trillion rupiah [1]. These losses are not only in the form of physical damage but also economic and social losses. The Indonesian government continues to strive to reduce the risk of natural disasters. These efforts include building disaster-resistant infrastructure, increasing community preparedness, and strengthening early warning systems [2]. However, these efforts are still insufficient to overcome Indonesia's high risks of natural disasters. Communities need to know the types of natural disasters that often occur in their area and ways to reduce the risk of these disasters [3]. By increasing awareness and vigilance of natural disasters, Indonesian people can reduce the risk of life

* Corresponding author: simon.sumanjoyo@fisip.unila.ac.id

and material losses due to natural disasters. One of the efforts of various parties is the development of various disaster-related technologies, such as development of various digital applications.

Based on these efforts, researchers see that several disaster applications in Indonesia have features and functions that seek to contribute to strengthening disaster. Several experts have studied disaster applications, and the study provides an analysis of existing disaster risk assessment and response tools to enable the user community [4]. Mobile applications can potentially assist disaster response by providing an avenue for distributing relevant and time-critical information to the public [5]. Various disaster apps are already in the app market. However, engaging users in maintaining disastrous smartphone applications is a challenge [6]. The research presents the author's experience implementing a chatbot system in Taiwan government departments for disaster response [7]. Existing research has investigated various mobile applications that engage with crowds during disaster situations. The findings show that disaster applications can be general-purpose or specific for disaster purposes. This review further focuses on built-for-disaster-purpose applications and shows the various interactions these applications foster with the public and the added value contributions of applications throughout the disaster lifecycle [8]. There are three areas of need for future research: app engagement prior to the disaster response stage, public behavior and motivation towards app use, and mobile app usability.

Therefore, based on the existing problems, the *Jaring Bencana* application needs the usability test, which is crucial and can be used as material in future application development plans [9]. Several studies regarding usability have been carried out, and research conducted by Tan's research shows that five usability factors significantly influence continuance intentions. The main positive influences are (1) the user's perception of whether the application delivers its functionality (application utility), (2) whether it is reliable, and (3) whether it presents information that can be easily understood. Furthermore, too much focus on (4) user interface graphics and (5) user interface inputs can discourage continuation intentions. The challenge of finding essential usability factors for m-government applications and selecting the right factors for a particular m-government application [10]. The results include six usability factors: learnability, simplicity, satisfaction, security, privacy, and memorability [11]. Several other studies have compared user perceptions collected using a final questionnaire based on the System Usability Scale [10]. This paper intends to use a similar approach to analyze the prototype of the *Jaring Bencana* application, which is to be introduced to disaster management groups in the regions. So, the problem in this paper is, how is the achievement of the usability value of the *Jaring Bencana* application? Moreover, what are the problems in testing the *Jaring Bencana* application faced by user groups? This study aims to examine the level of usability in disaster applications by measuring the variables of effectiveness, efficiency, and user satisfaction.

2 Research method

The research methodology contains the stages or description of the research conducted. The methodology is applicable in every research to make it easier. This research was conducted using a mixed method approach, which combined survey instruments and interviews with the target group of application users—adjusted research stages with stages on analysis and evaluation with the usability testing method on the *Jaring Bencana* application.

2.1. Data collection

Data collection consisted of three stages: user identification by testing 30 participants divided into three age groups; teens (13-17 years), college students / young adults (18-24 years), and adults (25-64 years). It is because the target user of the *Jaring Bencana* application is the public aged 15 - 64 years old. After identification of the user, create a task scenario by creating a set of tasks for, the list oftask scenarios can be seen in the table 1 below.

Table 1. Task Description

Job description	Task Description
Task 1 (T1)	Log in to the application
Task 2 (T2)	Try the report disaster info menu around
Task 3 (T3)	Try to find disaster response volunteers near
Task 4 (T4)	Trying to respond to the disaster report menu that appears
Task 5 (T5)	Find educational information features for disaster risk reduction (environmental conditions that have the potential for disaster or activities)
Task 6 (T6)	Find notifications of disaster location positions (earthquakes, floods, etc.)

The final activity is to conduct interviews to gather the necessary in-depth information. At the same time, type data testing, which may be used to evaluate usability, contains all of the application's functionality. When users utilize the *Jaring Bencana* application, the performance measurement technique is employed to assess its efficacy and efficiency. Effectiveness measurement can be calculated based on the success and failure of each respondent's tasks. Measurement of effectiveness can be calculated based on the success and failure of the task scenario that each respondent has done. Effectiveness is calculated by setting the number binary "1" for respondents used in the study. It consists of quantitative data and qualitative data. Qualitative data results from *Usability Testing* using the *Retrospective think-aloud* (RTA) technique.

2.1.1. Analysis and Data Processing

The information gathered during the data gathering stage is subsequently processed. The processed data is then evaluated to produce findings. The following is how data is processed in usability study on the *Jaring Bencana* application: (a). Usability testing is a technique for evaluating the quality and usability of a service or product by directly participating application users to find out problem information on tested mobile device applications. In the usability success complete tasks scenario, there are ten ways and a "0" for respondents who failed. Equation (1) is used to compute the respondent's success rate, whereas equation (2) is used to calculate the respondent's failure rate.

$$Success = \frac{Number\ of\ tasks\ successfully\ performed}{Total\ tasks} \times 100\% \qquad (1)$$

$$Failed = \frac{Number\ of\ tasks\ failed\ to\ perform}{Total\ tasks} \times 100\% \qquad (2)$$

Respondent success level is calculated by dividing the number of successfully completed scenario tasks by the total number of given task scenarios, and then multiplying the result by 100%. The failure rate of the respondent is computed by dividing the number of failed task scenarios by the total number of offered task scenarios and multiplying the result by 100%. Based on the average time to finish each work from all existing responses, the calculation results will highlight which tasks are most important to accomplish [11]2w. Meanwhile, the level of efficiency can be measured from the time of completion of the task with the success or failure of carrying out the task using the equation (3).

$$Time\ based\ Efficiency = \frac{\sum_{j=1}^{R}\sum_{i=1}^{N}\frac{n_{ij}}{t_{ij}}}{NR} \qquad (3)$$

Whereas:

N	:	task totals
R	:	total participants
Nij	:	Result of task i by user j; if the user completes the task successfully, then Nij = 1. If it does not work, then Nij= 0.
Tij	:	Time taken by user j to complete task i; If the user does not complete the task, the time is measured until the user quits the task.

The System Usability Scale (SUS) will be used to assess the user satisfaction component. SUS is a well-known usability testing tool that is dependable, popular, efficient, and cost-effective. When calculating the System Usability Scale (SUS) score, there are various rules to follow. The rules for calculating questionnaire scores according to the user's score decreased by one; and for each odd-numbered question received. Each question has an even number; the final score is determined by subtracting the value five from the user's question score. The SUS score is obtained by multiplying the sum of each question's scores by 2.5. Further calculations seek each respondent's SUS score by averaging all scores and dividing by the number of respondents. The average of SUS score from all respondents was derived, and the score was then adjusted to the SUS assessment. The average SUS score from the number of studies is 68, hence scores above 68 are considered above average, while scores below 68 are considered below normal.

Retrospective think-aloud analysis is also used to test a system that incorporates users, or end users, by verbalizing what users feel and think when utilizing the system in a sustainable manner. Verbalization is accomplished by conducting interviews with respondents to learn about their experiences and opinions while using the *Jaring Bencana* program. The verbalization comprised the respondent's impression of the message or challenges conveyed, as well as the comments made after the testing session.

3 Result and discussion

The usability testing was implemented on 30 people who had fulfilled the test requirements. The requirements are that test participants are of productive age; aged 15 – 64 years, use the application on Android smartphones, and have never used *Jaring Bencana*. The usability test conducted in this study aims to calculate the effectiveness, efficiency, and satisfaction components using performance measurement and retrospective think-aloud techniques. This test is carried out by giving a series of tasks to the test participants to complete. The results of the data processing effectiveness and efficiency in the tests carried out are as follows.

3.1. Component *effectiveness*

The purpose of measuring the level of effectiveness is to provide an overview of the application's ability to support the convenience of test participants in using the application based on success in carrying out tasks. The results of the success rate are shown in form percentage. All data successful test takers tasks and test participants who failed to carry out the task are shown in Table 2.

Table 2 shows results from successful test takers in doing the task. Data in the table is represented using binary numbers 0 and 1 to simplify calculations. A value of '1' will be given if the participant can complete the assigned task correctly and mark '0' if the participant test failed the task. Most of the results from completing tasks are not happening or much failure. There are several anomalies in tasks T2 and T3. Some test takers experienced difficulties in finishing tasks T2 and T3. After the interview, the participants tested had an understanding of task 2. However, the problem is that some test participants are less focused and less aware of the location of the hazard info button and its recommendations. The result of calculating the success rate of using this application is 89% above average success, i.e., 78%. This figure concluded that the application is practical and has a good level of learnability.

Table 2. Results of participant test (%)

Category	T1	T2	T3	T4	T5	T6	Average
Success	100	80	76.7	100	86.7	90	89
Fail	0	20	23.3	0	13.3	10	11.1

3.2. Component efficiency

To measure component efficiency, the time needed by the test to complete the assigned task is recorded and recorded. The time spent completing the task is counted from when the researcher finishes giving the assignment, and the test participants start using the application. The time calculation will be stopped when the test participant states that the task has been completed or the test participant gives up on completing the task. Table 3 shows the times needed by every participant test to complete the task.

Table 3. Results for participant test in completing tasks (units/seconds)

Task Category	T1	T2	T3	T4	T5	T6	Average/Participant
Average Per Task	11.5	27.6	30.3	19.1	12.5	6.8	18.0

If seen from component efficiency, the teens' category performs more efficiently than other categories. In comparison, the adult category performs better than the *college students category*. Efficiency calculation is based on calculated time using the success data of participants in completing the tasks listed in Table 4 and the time data needed by participants to complete the tasks in Table 5. The calculation results show the time-based efficiency value of the *Jaring Bencana* Application is 18.0 second or 0.24 goals /sec.

3.3. Component Satisfaction

The satisfaction component in this study was measured by filling out an online questionnaire given to 30 users of the *Jaring Bencana* application. SUS is a measurement of the level of user satisfaction by giving questionnaires to users after the application (Brooke, 2013). Respondents were asked to fill in data self and ten statements from System Usability Scales (SUS) with a scale Likert 1- 5. Based on the calculation, the SUS obtained from the respondents was 68. Score SUS on 68, so it would be considered above average, and anything below 68 below average. The SUS score on the *Jaring Bencanas* application is still below the average and has received *a grade scale* D with *adjectival ratings* "OK" and the *acceptability range* entered on the category "*marginally acceptable.*"

3.4. Retrospective Think Aloud

Retrospective think-aloud data was obtained by conducting interviews with participants test. An interview was conducted to find out test participants' experiences and opinions while carrying out scenario assignments using the Jaring Bencana application. The list of questions used follows the interview framework that has been made before. Questions were further filed by the answers of participants to explore the problem. From the results of interviews with all test participants, several problems were found, and the test participants gave suggestions. A list of problems can be seen in Table 4.

Table 4. Problems and suggestions for all test participants

No	List of Usability Issues
1	Some are still having trouble finding features and menus in the application.
2	Need instructions for use for new users.
3	I found a button whose name/label is unclear.
4	Interface design and similar buttons make it confusing
5	The application requires a stable internet network because it has to load GPS
6	Application loading feels heavy when using low HP specifications.
7	I found some bugs in some menu features.
8	There is no offline feature for loading maps yet.

Based on Table 4, it is necessary to improve and strengthen several functions in the application. A more concise guide that can be effectively understood by new users is needed. Improvement and strengthening of this feature will later strengthen the benefits of this application for its user group.

4 Conclusion

Based on the analysis, it can be concluded that the results of the evaluation of the *Jaring Bencana* application have complied with the criteria of usability in effectiveness and efficiency, as well as user satisfaction in using the application. The results evaluation used performance measurement techniques that measure components' effectiveness and efficiency to 30 test participants, resulting in an effectiveness component value of 89% and an efficiency of 0.24/sec. The value shows that the level of learnability of the application *Jaring Bencana* is on average. While the value of the satisfaction component is 68, it is included in the grade scale category D with the information "OK." Then, the retrospective think-aloud technique was conducted through interviews with test participants to produce several identifications of usability problems applications.

References

1. BNPB Indonesia, (2023)
2. M. T. Sinaga, B. Asyik, and D. Miswar, JPG (Jurnal Penelitian Geografi) **7**, (2019)
3. A. Setiawati and S. S. Dewi Cahyani Puspitasari, (2017)
4. H. T. Ferguson, S. Gesing, and J. Nabrzyski, in *2016 49th Hawaii International Conference on System Sciences (HICSS)* (2016), pp. 2872–2881
5. W.-Y. Lin, T.-H. Wu, M.-H. Tsai, W.-C. Hsu, Y.-T. Chou, and S.-C. Kang, Automation in Construction **91**, 182 (2018)
6. J. Basak, P. Bhaumik, S. Roy, and S. Bandyopadhyay, in *Proceedings of the 21st International Conference on Distributed Computing and Networking* (Association for Computing Machinery, New York, NY, USA, 2020), pp. 1–6
7. M.-H. Tsai, C.-H. Yang, J. Y. Chen, and S.-C. Kang, KSCE J Civ Eng **25**, 503 (2021)
8. A. Yaneri, Jurnal Papatung: Jurnal Ilmu Administrasi Publik, Pemerintahan Dan Politik **3**, 12 (2020)
9. T. A. Purnomo, R. A. Widyanto, A. Setiawan, P. Hendradi, and P. Suksmasetya, J. Phys.: Conf. Ser. **1517**, 012089 (2020)

https://doi.org/10.1051/e3sconf/202346702001

10. P. Kureerung, L. Ramingwong, S. Ramingwong, K. Cosh, and N. Eiamkanitchat, Information **13**, 470 (2022)
11. M. L. Tan, R. Prasanna, K. Stock, E. E. H. Doyle, G. Leonard, and D. Johnston, International Journal of Disaster Risk Reduction **50**, 101874 (2020)

Assessing the ecological impact of the mass entry of Rohingya refugees in Ukhia, Cox's Bazar, Bangladesh: A remote sensing-based analysis

Toerhthangquor Shuvrangshu Joyotu[1*], Komariah[2], and *Agung* Hidayat[3]

[1]Master's program, Department of Environmental Science, Graduate School, Universitas Sebelas Maret, Surakarta, Indonesia
[2]Faculty of Agriculture, Universitas Sebelas Maret, Surakarta, Indonesia
[3]Department of Environmental Science, Graduate School, Universitas Sebelas Maret, Surakarta, Indonesia

Abstract. The massive influx of Rohingya refugees to Bangladesh from Myanmar in August 2017, impelled the government of Bangladesh to clear thousands of acres of forestland in Ukhia and Teknaf sub-districts to allow their accommodation. Subsequently, the refugees also cut down trees thus resulting in rapid deforestation in that region. Recently, however, the Rohingya refugees in collaboration with UNHCR, have undertaken a massive tree plantation project. In this study, we calculated a Normalized Difference Vegetation Index (NDVI) using remote sensing data in the Ukhia sub-district to analyze changes that occurred in the years following the 2017 influx and evaluated the effectiveness of the reforestation effort. After calculating the NDVI, we found that there was an implication of a slight increase in vegetation. The results show that the total area of vegetation coverage in February, 2017 was 215 square kilometers and in 2019 it had gone down to 197.92 and by 2023 it had recovered to 235 square kilometers. Thus, we can conclude that the reforestation process should be continued as the changes will become more visible. As Bangladesh is highly vulnerable to climate change, continued deforestation will only aggravate the situation and thus this raises the importance of reforestation.

1 Introduction

Although Rohingya refugees have been coming to Bangladesh for a long period of time, [1] even before Bangladesh got its independence in 1971, [2–4] it was August 2017, when more than 700,000 [5] Rohingya refugees from the neighboring country Myanmar came to Bangladesh to seek living facility [5,6]. To ensure adequate facilities, the government of Bangladesh had to make environmental changes [5,7] in 2 registered camps namely Kutupalong under the Ukhia sub-district and Nayapara under the Teknaf sub-district, [8] within which, Kutupalong is the largest refugee camp in the world [9]. Now, the number of Rohingya refugees has increased to nearly a million [10,11] and outnumbered the local

[*] Corresponding author: joyotu16@gmail.com

population of Ukhia and Teknaf sub-district in the Cox's Bazar district [7]. With the increasing number of refugees, a lot of unregistered camps, both with and without the host community were built. Apart from Bangladesh, Rohingya refugees went to seek shelter in different countries for example India, Indonesia, Malaysia, Pakistan, and Thailand [4,12].

Right in 2017, when Bangladesh made the decision to open its border to accept Rohingya refugees, it started facing a multidimensional problem as the country itself is a small and densely populated country. To deal with the huge refugee entry within a short period of time, the Government of Bangladesh allocated 2,500 hectares of protected forest land which was then swept for further expansion of the settlement. [13] Moreover, refugees cut many of the remaining trees for firewood due to a lack of alternative cooking fuels. What had been a sanctuary for wildlife, including endangered Asian elephants, quickly became a bare and refugee settlement. [5,14] Bangladesh itself is a country with the possibility of becoming a worse possible victim of climate change, this huge swipe out of green land puts a huge question mark on the policy of the Bangladesh government to address the Rohingya issue.

Previous empirical studies implied that deforestation might have several negative impacts on the environment which include desertification, loss of wildlife habitat, [15] soil erosion, and as a forest works as a carbon storage, a decrease in the forest cover might disturb the local climate by affecting the day-time temperature variation which might increase the risk of global climate change. [16,17] After the Rohingya influx, several remote-sensing-based analyses were performed right after the settlements took place in the sensitive ecological region that assessed the land use and land-cover change, forest degradation, and loss of environmental sustenance. [14]

A couple of researches have been done regarding the overall change in LULC (land use land cover) pattern. [6,7,11,18,19]. They have found that the loss of forestland was the highest within the time period of 2010-2020 in comparison with other decades varying from 1990 to 2020. In this period, refugee camps increased by 21.29 sq. km., (582%). However, they have found significant losses for the vegetation by 51.81 sq. km. (17%), and water bodies by 10.39 sq. km (15%) [14].

Several analyses till December 2017 came up with quantified values about the loss of vegetation cover. Hassan [8] showed the data of the refugee camp expansion along with the forest loss both pre and post-refugee influx. He found that the refugee settlement went from 1.75 sq. km. to 15.30 sq. km. within the year 2017 which were constructed mostly by replacing forest land. At the same time, the forest cover encircling those camps was decreased by 22.83 sq. km. While Hassan was focused on the overall situation, Imtiaz [7] was centered mainly on the Teknaf sub-district and Teknaf Wildlife Sanctuary. According to his findings, these places have lost 12.84 sq. km. and 1.03 sq. km. of vegetation respectively. This drastic change in the vegetation cover within solely 2017 also clearly indicates a link between the Rohingya influx and the deterioration of the forest cover. Analysis over a longer period of time showed some more drastic fall in the vegetation cover. Braun [19] showed that 15 sq. km. of vegetation cover was cleared due to the expansion of refugee camps within a few years. The refugee camp, which accounted for only 0.36 sq. km. rocketed to 18.50 sq. km. within just 2016 to 2019 within which the expansion was the most in July 2017.

The primary gap among these studies is a couple of studies were conducted with the data until 2017 or early 2018. Therefore, these analyses are just a partial picture of the drastic effect that has taken place. Other studies have discussed the LULC pattern but none of them have assessed the validity of the reforestation process which started in 2021. The main goal of this research is to assess the LULC using the Normalized Difference Vegetation Index or NDVI) due to the establishment of infrastructure for the Rohingya refugees. Also, due to the reforestation efforts taken by the Rohingya people, we would like to assess whether the reforestation worked well to make the region greener than before or not.

2 Research methodology

To understand the ecological impact of Rohingyas on the land of Bangladesh, we used remote sensing technology in this quantitative research. The primary idea was to calculate NDVI to validate the observation of news that has come across the mainstream media as well as the impact of the reforestation program. To get satellite data, primarily we thought of working with Sentinel-2 because of its better spatial resolution. However, the launching period of Sentinel-2 was just before the Rohingya influx and it was not fully operational until 2018. Thus, we could not use Sentinel-2 data.

Because of that, we tried to work with sentinel-1 which was fully functional before the event took place but it had another problem which is called blank image. It was acknowledged by the ESA (European Space Agency) and they have suggested using other satellites to take data from that particular period due to faulty pictures. As for our last option, we chose Landsat-8 satellite data which worked perfectly fine but the only problem it had was its spatial resolution. In comparison with Sentinel-1/2, it had a little worse spatial resolution of 30 meters. But seemingly, it did not create any problem to calculate the NDVI and validate the problem that we were working on.

With an area of 261.8 square kilometers, the Ukhia sub-district within the Cox's Bazar district is our study area which is located in the south-eastern part of Bangladesh with coordinates 21.2833°N, 92.1000°E. The Kutupalong refugee camp (21.2126°N, 92.1634°E) is situated in this sub-district along with 25 other refugee camps. [20] Geographically, Ukhia possesses a subtropical monsoon climate with an average temperature between 17.3° C to 32.1° C.

Multispectral satellite imagery from usgs.gov was used to analyze the LULC over eight different years; two before the influx, and 6 consecutive years after the Rohingya influx. We have taken data from different dates in the month of February to get low cloud coverage. The cloud coverage was set below 40% to get a clear picture to do the NDVI calculation.

Table 1: Date of satellite data acquisition for years 2014 and 2017-2023.

Year	Acquisition date	Year	Acquisition date	Year	Acquisition date
2014	19th February, 2014	2018	14th February, 2018	2021	22nd February, 2021
2017	11th February, 2017	2019	1st February, 2019	2022	09th February, 2022
2018	14th February, 2018	2020	20th February, 2020	2023	28th February, 2023

To calculate the NDVI, we used the NDVI equation which is,

$$\text{NDVI} = \frac{NIR - RED}{NIR + RED} \quad (1)$$

Here, NIR is near infra-red band and RED is the red band that we get from raw data.

As we have worked using Landsat 8, we needed only band 5 and band 4 as band 5 is the NIR and band 4 is the RED band. So for Landsat 8, our final equation to calculate NDVI was,

$$\text{NDVI for Landsat 8:} \quad \frac{Band\ 5 - Band\ 4}{Band\ 5 + Band\ 4} \quad (2)$$

We have calculated the satellite data using ArcGIS 10.8 and we have found that the NDVI has changed over the period of time which clearly indicates that there was a change in the greenery of that area.

3 Results

We have taken 8 different satellite images of the years 2014, and 2017-2023. The result of the consecutive years shows signs of positive vegetation. We have classified the NDVI result into 6 different classes namely water, built-up area, barren land, shrub and grassland, sparse vegetation, and dense vegetation as shown in Figure 2.

As it is seen from the figure 1, in the year 2014, long before the Rohingya influx, the Ukhia sub-district had healthy vegetation with 26.1 square kilometers of dense vegetation which was nearly 10% of the total area of the sub-district. Gradually the dense vegetation area decreased to only 1.6 square kilometers months before the Rohingya influx in 2017 which was only 0.6% of the total area.

As seen from Table 2, Ukhia had 91.09% of overall vegetation in 2014 which decreased to 81.42% months before the Rohingya influx in 2017. Whereas in 2019, Ukhia had the lowest percentage of vegetation which was 74.95% despite having nearly 20 square kilometers of dense vegetation. But from 2019, the overall vegetation has started to climb up and showed a gradual increase in the total vegetation area up until now.

Table 2: The total area by percentage of 4 different classes that we have found after the LULC classification.

Year	Area (Percentage)				
	Built-up area	Shrub and grassland	Sparse vegetation	Dense vegetation	Total vegetation
2014	3.05	35.38	45.83	9.88	91.09
2017	9.26	51.78	29.03	0.60	81.42
2018	12.19	44.52	31.67	1.41	77.59
2019	15.68	31.92	35.53	7.50	74.95
2020	11.17	45.29	31.73	1.47	78.48
2021	5.83	47.52	35.78	2.58	85.89
2022	4.65	45.99	39.28	2.71	87.98
2023	4.07	42.03	39.31	7.65	88.99

This year, Ukhia had 89% of total vegetation which is just slightly lower than that of the total vegetation in 2014. The incremental area of total vegetation has an implication that given the current processes are ongoing, the total vegetation cover will increase further.

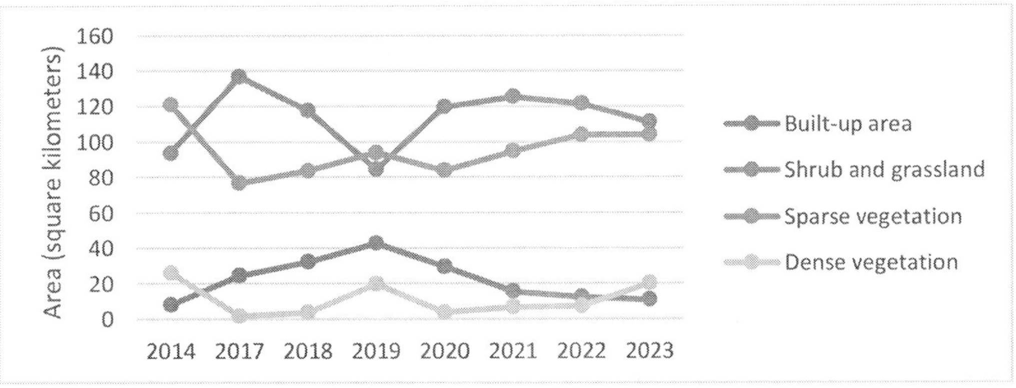

Figure 1: LULC classification based on four different classes shows the variation of the area over the years 2014 and 2017-2023. The data were found by calculating the NDVI of the years mentioned.

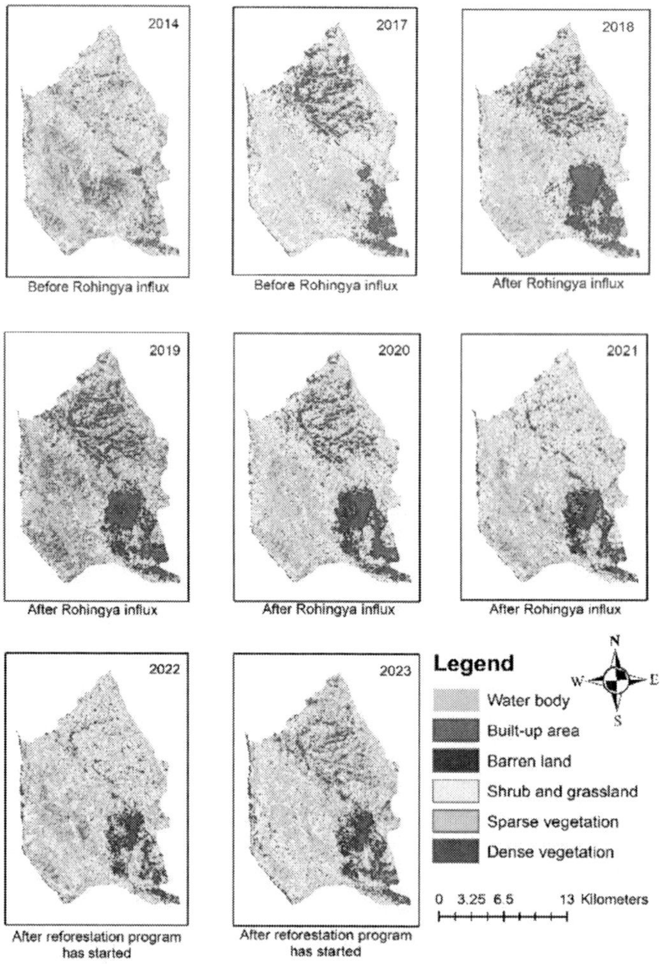

Figure 2: N classification of the Ukhia sub-district starting from 2014 and 2017-2023.

4 Discussion

In this research, we have categorized the timeline into three separate branches namely before the refugee influx, during the influx, and after the reforestation process begins. Before the Rohingya refugee influx, we have seen typical deforestation that is commonly happening in Bangladesh. From 2014 to 2017, Ukhia lost about 10% of vegetation cover. Sparse vegetation at that time was reduced by 44.36 square kilometers. On the other hand, during the influx, we have found deterioration in vegetation cover around the southern and northern parts of the Ukhia sub-district. In 2014, the total vegetation cover was 91.09% which in 2019 dropped to 74.95%. From 2020 to 2021, we have found a good improvement in the means of positive vegetation in Ukhia. By this time, the total vegetation cover had improved from 78.48% to 85.89%.

After the official reforestation program was started, we found a gradual increase in the total vegetation, and in the year 2023, it had 88.99% of the total vegetation. These results show and strengthen the other research regarding the ecological impacts of building camps for Rohingya refugees that the drastic loss of vegetation was linked to the refugee influx. [7,21] But most importantly, our result shows that the reforestation project is actually working well despite vegetation being something that requires time to show up. Thus, we believe if this project can be continued at this pace, then the vegetation cover will be increased and more visible.

One of the key limitations of our research is that we did not analyze data regarding the Teknaf sub-district which also has some refugee camps. Research with a broader study area might come with a better understanding of the overall vegetation cover for the year 2023.

5 Conclusion

In this research, we have found that after the Rohingya influx in 2017, deterioration of vegetation cover took place until 2019 straight. But from 2020, we have seen that the vegetation has started increasing at a gradual pace. Our analysis also implies that after the reforestation project was taken, the total vegetation increased over a short period of time. As the reforestation process has been running since 2021, which is not a long time in these cases of vegetation, we strongly believe that with time, we will be able to see better vegetation cover if the reforestation process continues. So, the continuation of this reforestation program might be able to overcome the loss that has been taking place since the Rohingya influx in 2017.

Reference

1. M. R. Habib, Asian Polit. Policy **13**, 577 (2021)
2. F. Exch. 62 **62**, 24 (n.d.)
3. Amnesty International, Amnesty Int. (2016)
4. J. S. Sidhu and S. N. Parnini, J. Int. Stud. (2020)
5. S. Bepler, Pacific Geogr. **28**, 4 (2018)
6. M. Rahman, M. Islam, and T. Chowdhury, J. Environ. Sci. Nat. Resour. **11**, 9 (2019)
7. S. Imtiaz, Ecocycles **4**, 16 (2018)
8. M. M. Hassan, A. C. Smith, K. Walker, M. K. Rahman, and J. Southworth, Remote Sens. **10**, 1 (2018)
9. N. Sultana, S. Sultana, R. Saha, and M. M. Alam, Southeast Asia A Multidiscip. J. (2023)
10. M. N. Momem, *The Rohingya Refugee Crisis: Implications for Regional Security* (Springer International Publishing, 2021)
11. M. M. Hassan, M. Duveneck, and J. Southworth, Ecol. Inform. **74**, 101966 (2023)
12. T. Joarder, I. Sutradhar, M. I. Hasan, and M. M. I. Bulbul, Cureus **12**, 8 (2020)
13. R. D. la Portilla, Unhcr (2021)
14. F. Hossain and D. M. Moniruzzaman, Environ. Challenges **2**, 100024 (2021)
15. J. Liu, M. Linderman, Z. Ouyang, L. An, J. Yang, and H. Zhang, Science (80-.). **292**, 98 (2001)
16. R. G. Newell, Enviornmental Econ. Manag. **40**, 211 (2000)
17. D. Morales-Hidalgo, S. N. Oswalt, and E. Somanathan, For. Ecol. Manage. **352**, 68 (2015)
18. M. Rahman and T. A. Chowdhury, (2019)
19. A. Braun, F. Fakhri, and V. Hochschild, Remote Sens. **11**, (2019)
20. ISCG, 2018 (2018)
21. N. Akhtar, M. K. Uddin, and Y. Tan, GeoJournal **87**, 4269 (2022)

Analysis of economic and social vulnerability to disaster risk in Tawangmangu and Kejajar Districts

Lusia Advenia Dwi Rahayu[1*] and Suryanto[2]

[1]Master Economics and Development Studies, Postgraduates, Sebelas Maret University, Surakarta, Central Java, Indonesia
[2]Faculty of Economics and Business, Sebelas Maret University, Surakarta, Central Java, Indonesia

Abstract. Climate change is a major concern. Climate change causes changes in the seasons and hydrological systems. Landslides are one of the most common hydrometeorological disasters in Indonesia. This study identifies indicators of economic and social vulnerability to the risk of landslides in the Tawangmangu and Kejajar sub-districts. The method used is scoring and weighting of each economic indicator namely the Gross Domestic Regional Product at Current Prices (PDRB ADHB) of the sub-district and the number of poor people, as well as social indicators used, including population growth rate, sex ratio, population density data, the number of children at risk, the number of elderly at risk which are analyzed using the SoVI index referring to the Regulation of the National Disaster Management Agency No. 2 Year 2012 and research conducted by Sulma, 2012 to determine the vulnerability of the two sub-districts to landslides. Tawangmangu and Kejajar sub-districts are classified as areas with moderate vulnerability, so the economic and social conditions of the Kejajar sub-district are more vulnerable to the risk of landslides. If a region has a high SoVI index due to landslides, it can affect agricultural productivity because landslides affect agricultural land. The government can use disaster studies as a basis for determining broader regional policies and accelerating regional economic recovery through the improvement of economic facilities and infrastructure in disaster-prone production sectors.

1 Introduction

Climate change is a phenomenon of primary concern. The IPCC states that the impacts of climate change in the Southeast Asian region, including Indonesia, are expected to increase threats to food security, human health, water availability, biodiversity, and sea-level rise. Climate change leads to seasonal shifts and alterations in the hydrological system, increasing flood intensity and rising sea temperatures [1]. Disasters will always impact all aspects of human life, including political, economic, social, and cultural aspects. Economic and social

* Corresponding author: lusia.advenia@student.uns.ac.id

aspects are very vulnerable to the effects of a disaster because both aspects are closely related to people's lives. The community can directly feel the impact of changes in these two aspects. Both aspects must have a strong foundation because when an area is affected by a disaster, the likelihood of that area being in a high vulnerability level is very low. The vulnerability level of the economic and social aspects of the community affected by a disaster can be seen based on the amount of losses experienced by the community and the number of people affected by the disaster.

One of the hydro-meteorological disasters that frequently occurs in Indonesia, especially in Central Java Province, is landslides. The occurrences of landslides in Central Java from 2010 to 2019 reached 1,811 incidents [2]. This disaster has negative impacts, including decreased forest and land productivity due to environmental imbalances, leading to land degradation that can trigger landslides. Areas in Central Java prone to landslides include the Regency of Wonosobo and Karanganyar. Based on data from the Disaster Risk Assessment of Central Java from 2016 to 2020, the Wonosobo Regency has a high hazard index for landslides, evident from the potential hazard area of 44,149 ha, compared to the regency's total area of 98,468 ha. The Karanganyar Regency also has a high hazard index for landslides, with a potential hazard area of 17,827 ha, against the Regency's total area of 77,378.64 ha. Several areas in both Karanganyar and Wonosobo Regencies are also vulnerable to landslides, such as the Tawangmangu Sub-district in Karanganyar, as depicted in the following Figure 1:

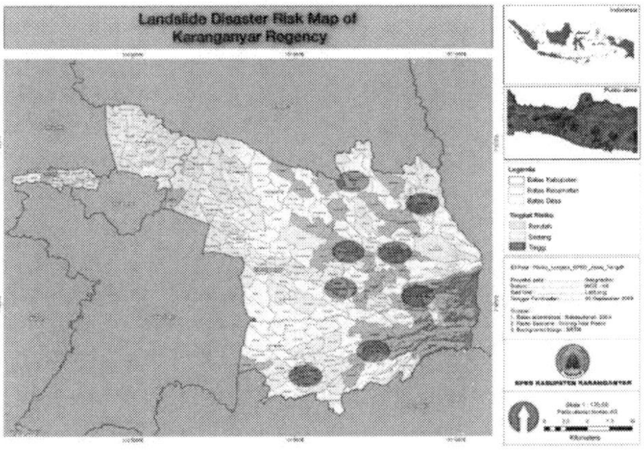

Fig. 1. Landslide Vulnerability Map of Karanganyar Regency.

And also the Kejajar Sub-district in Wonosobo Regency, where the highland area of Dieng is located. The vulnerability of the area to landslide disasters in the Kejajar Sub-district can be seen in the following Figure 2 :

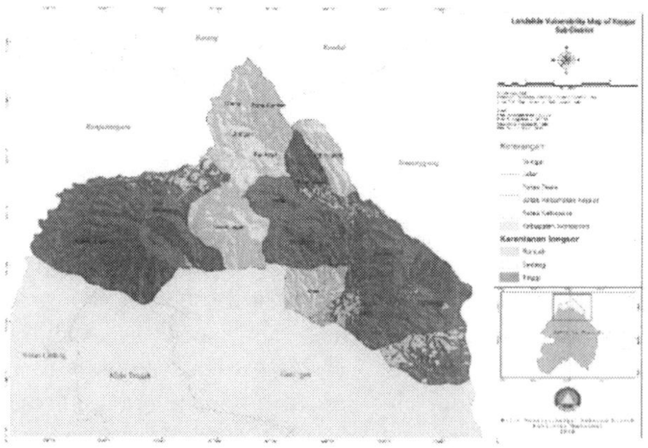

Fig. 2. Landslide Vulnerability Map of Kejajar Sub-district.

Research on the level ability to landslides in the Tawangmangu Sub-district and Kejajar Sub-district has been conducted frequently, but it has mostly focused on physical aspects. Mapping landslide-prone areas in the Kejajar Sub-district using Geographic Information System (GIS) technology with an overlay method was done to provide an overview of the landslide disaster threat in the Kejajar Sub-district [3, 4]. Physical aspect research on landslides in the Tawangmangu Sub-district has also been conducted [5]. This study is important to study the economic and social vulnerability levels in Tawangmangu and Kejajar Sub-districts to assess the community's ability to face landslide disasters.

2 Method

This research is conducted using a quantitative approach. The quantitative data in this study includes information related to regional disaster data, Gross Regional Domestic Product at Current Prices (GRDP CP) of sub-districts, population growth rate, gender ratio, population density, the number of vulnerable children, the number of vulnerable elderly population, and the number of impoverished population. This study uses secondary data, a data source that indirectly provides data to the data collector [6]. The secondary data required for this research comes from various government agencies or Local Government Work Units (SKPD) in the Karanganyar and Wonosobo Regencies. The research analysis method employs scoring and weighting methods for each indicator using a quantitative descriptive approach. The scoring and weighting methods are modified using the Social Vulnerability Index (SoVI), based on Regulation of the National Disaster Management Agency No. 2 Year 2012 and research conducted by Sulma, 2012, to determine the vulnerability of the two sub-districts to landslides. The related research aims to determine the sensitivity level of the Tawangmangu and Kejajar sub-districts to landslide disaster risks.

3 Result and discussion

3.1 Indicators of Economic and Social Vulnerability in Tawangmangu and Kejajar Sub-districts

Economic and Social Vulnerability Indicators in Tawangmangu and Kejajar Sub-districts. In this study, the economic vulnerability indicators used are the Sub-district's Gross Regional Domestic Product (GRDP) and the number of people living in poverty, while the social vulnerability indicators used include population density, population growth rate, gender ratio, and vulnerable groups such as children and the elderly [7]. Based on data analysis, it is known that the GRDP of Tawangmangu Sub-district in 2019 reached IDR 1,903,764.11 million, approximately 5.4% of the total GRDP of Karanganyar Regency, while the GRDP of Kejajar

Sub-district in the same year reached IDR 1,034,353.14 million, around 5.49% of the total GRDP of Wonosobo Regency. Based on data obtained from the Integrated Social Welfare Data (DTKS) of the Central Java Social Service, the number of people living in poverty in Tawangmangu Sub-district in 2019 was 3,617 individuals, accounting for 7.9% of the total population in Tawangmangu Sub-district. In Kejajar Sub-district, the number of people living in poverty reached 5,987 individuals, representing 13.9% of the total population.

The social vulnerability indicators related to population density in Tawangmangu Sub-district increased from 2017 to 2018, while in 2019 the population density in Tawangmangu Sub-district experienced a decrease. In 2019, the population density was 651.11 individuals per square kilometer. Meanwhile, in Kejajar Sub-district, the population density increased from 2017 to 2019. In 2019, the population density was 753 individuals per square kilometer. Regarding the social vulnerability indicator of population growth rate until 2019, the population of Tawangmangu Sub-district reached 45,598 individuals with a population growth rate of 0.74%. In contrast, the population of Kejajar District in 2019 was 43,368 individuals with a population growth rate of 0.53%. Another social vulnerability indicator is the gender ratio. The gender ratio in Tawangmangu Sub-district is 98.43%, while in Kejajar Sub-district, it is 104.73%.

Social vulnerability indicators for vulnerable groups consist of the population of children and the elderly. In Tawangmangu Sub-district in 2019, there were a total of 17,586 individuals in the children age group and 7,964 individuals in the elderly age group who fall into the vulnerable category. In contrast, in the Kejajar Sub-district, there were 16,382 individuals in the children age group and 4,533 individuals in the elderly age group who were considered vulnerable.

3.2 Level of Economic and Social Vulnerability in Tawangmangu and Kejajar Sub-districts

Classification of Economic and Social Vulnerability Levels is determined based on the scoring results assigned to each indicator. The placement of each indicator refers to the predetermined classifications of low, moderate, and high vulnerability, as shown in the following Table 1:

Table 1. Scores of Economic and Social Vulnerability Indicators.

Indicator	Unit	Score Low (1)	Medium (2)	High (3)
Standardized Values		**< 0,33**	**0,33 – 0,66**	**> 0,67**
Economic Aspect				
Sub-District Gross Regional Domestic Product	Percent	< 3	3 – 6	> 6
Population Below the Poverty Line	Individuals	< 2.600	2.600 – 4.910	> 4.910
Social Aspect				
Population Density	Individuals /Km^2	< 420	420 – 2.234	> 2.234
Population Growth Rate	Percent	< 0,5	0,5 - 1	> 1
Gender Ratio	Percent	> 99,86	99,86 – 104,06	< 104,06
Child Population	Individuals	<6.500	6.500 – 18.295,2	> 18.295,2
Elderly Population	Individuals	< 2.200	2.200 – 5.711,2	> 5.711,2

Source: Author, 2020. (Modified from Regulation of the National Disaster Management Agency No. 2 Year 2012 and Sulma, 2012)

The economic vulnerability level based on the Gross Regional Domestic Product (GRDP) indicator in Tawangmangu Sub-district is classified as class one with a low category, while Kejajar Sub-district is also classified as class one with a low category. As for the indicator of the population living in poverty, the number of impoverished residents in the Tawangmangu Sub-district is lower than in the Kejajar Sub-district. Therefore, the poverty indicator in the Tawangmangu Sub-district is classified as class one with a low category, while Kejajar Sub-district is classified as class two with a moderate category. In summary, information regarding the scores of the economic vulnerability indicators for Tawangmangu and Kejajar Sub-districts can be seen in Table 2:

Table 2. The score of the Economic Vulnerability Indicator in Tawangmangu Sub-district.

Indicator	Unit	Score		Proportion
		Tawangmangu Sub-district	Kejajar Sub-district	
Sub-District Gross Regional Domestic Product	Percent	Low (1)	Low (1)	40 %
Population Below the Poverty Line	Individuals	Low (1)	Medium (2)	60 %

Source: Primary Data Processing, 2020.

Based on the calculation of the economic vulnerability index using the SoVI equation, it can be determined that the economic vulnerability index in Tawangmangu Sub-district is 0.33. Thus, the level of economic vulnerability in the Tawangmangu Sub-district falls into the category of moderate economic vulnerability. Meanwhile, the economic vulnerability index in Kejajar Sub-district is also calculated using the SoVI equation, resulting in a value of 0.53. Therefore, the level of economic vulnerability in the Kejajar Sub-district is classified as moderate economic vulnerability. In summary, information regarding the scores of social vulnerability indicators in both Tawangmangu and Kejajar Sub-districts is presented in Table 3:

Table 3. Scores of Social Vulnerability Indicators in Tawangmangu and Kejajar Sub-districts.

Indicator	Unit	Score		Proportion
		Tawangmangu Sub-district	Kejajar Sub-district	
Population Density	Individuals/Km^2	Low (1)	Low (1)	30 %
Population Growth Rate	Percent	Medium (2)	Medium (2)	25 %
Gender Ratio	Percent	Low (1)	High (3)	15 %
Child Population	Individuals	Low (1)	Low (1)	15 %
Elderly Population	Individuals	Medium (2)	Low (1)	15 %

Source: Analysis Result, 2020.

Based on the calculation of the Social Vulnerability Index (SoVI), it is found that the social vulnerability index in Tawangmangu Sub-district is 0.47. Therefore, the level of social vulnerability in Tawangmangu falls into the category of moderate social vulnerability.

Meanwhile, the social vulnerability index in Kejajar Sub-district is 0.51. According to the established classification, the level of social vulnerability in Kejajar is categorized as moderate social vulnerability.

3.3 Comparison of Economic and Social Vulnerability Levels in Tawangmangu Sub-district and Kejajar Sub-district

Based on the calculations using the SoVI method, it is evident that the economic-social vulnerability index in Tawangmangu Sub-district is 0.40, falling into the low category and the medium category for one of the social vulnerability indicators, namely the gender ratio. For the Kejajar Sub-district, the economic-social vulnerability index is 0.52, classifying it into the medium category. Therefore, it can be concluded that the economic and social conditions in the Kejajar Sub-district exhibit a higher level of vulnerability to the occurrence of landslide disasters. This is also consistent with previous research findings that the Tawangmangu Sub-district is an area with a moderate to high potential for landslide disasters in residential areas, especially during the rainy season [5]. The vulnerability level of the Tawangmangu Sub-district falls into the category of sub-district/area with a moderate level of vulnerability to potential landslide disasters that may occur throughout the Karanganyar Regency [8]. Research related to vulnerability to landslide disasters in the Kejajar Sub-district has also been conducted by the Regional Disaster Management Agency (BPBD) of Wonosobo Regency [9]. The results indicate that the Kejajar Sub-district has a moderate to high potential threat of landslides. In conclusion, both of these Sub-districts exhibit a medium level of vulnerability to landslide disasters.

3.4 The Impact of Landslide Disasters on Food Vulnerability

Landslides pose significant environmental risks, threatening human and livestock lives, as well as destroying land use systems and agricultural production. These impacts have severe repercussions on the livelihoods of those affected, their economic conditions, and food security [10]. Landslide disasters occurring in various countries have resulted in serious damage across multiple aspects of human life. The consequences of landslide disasters include land degradation for farmers, as agricultural land becomes unusable for years, crops and seeds are destroyed, and losses occur in terms of livestock and food stocks [11]. Research conducted by Mertens et al [12] found evidence of income losses among Ugandans working in the agricultural sector due to landslides. As a result, they are unable to meet their basic needs, leading to hunger and the necessity to seek alternative income sources. Adaptive and diversified land use planning and rural livelihood improvement can reduce the vulnerability of mountain communities to climate change and food insecurity [13]. Efforts that can be made to prevent landslides and maintain food security in landslide-prone areas by utilizing sloping land by planting forestry and agricultural species can provide benefits in terms of soil and water conservation and economic aspects [14].

4 Conclusion

Based on the analysis conducted using the SoVI method, the results show that Tawangmangu Sub-district has lower economic and social vulnerability levels to landslide risks compared to Kejajar Sub-district, with a percentage of 0.40, while Kejajar Sub-district has a percentage of 0.52. It can be concluded that the economic and social conditions in Kejajar Sub-district are more vulnerable to the risk of landslide disasters. When an area has a high socio-economic vulnerability due to landslides, it can lead to a decrease in the productivity of farmers because farmland is affected by landslides. Additionally, such disasters can result in challenges to the resilience of the local population. Mitigation efforts to address these issues may include regional government policies prioritizing disaster risk assessments as the basis for development and investment decisions. Furthermore, focusing on improving economic infrastructure in sectors vulnerable to disasters can help the regional economy recover more quickly. The results of this research can be used as considerations for the government in formulating policies that are in line with the local conditions and needs when disasters occur, ensuring that the well-being of the community does not decline.

This research supported by UNS research funding (The Impact of Climate Change to Food Security and Economic Growth: Panel Error Correction Model and Arima)

References

1. R. K. Pachauri, A. Reisinger, L. Bernstein, Intergovernmental Panel on Climate Change. Core Writing Team, and Intergovernmental Panel on Climate Change, *Climate Change 2007 : Synthesis Report* (n.d.)
2. BNPB (2020)
3. S. B. Nugraha, Wahid Akhsin, and Andi Irawan Benardi, Jurnal Geografi **12**, 202 (2015)
4. E. M. Astuti, Djarot Shadarto Widyatmoko, and Sudibyakto Sudibyakto, Jurnal Universitas Gajah Mada: Majalah Geografi Indonesia **25**, (2011)
5. H. S. Naryanto, Jurnal Sains Dan Teknologi Indonesia **13**, 74 (2011)
6. P. Sugiyono, *Metode Penelitian Bisnis: Pendekatan Kuantitatif, Kualitatif, Kombinasi, Dan R&D* (CV. Alfabeta , 2017)
7. Riyanto, (2014)
8. M. Cholil, I. Hardjono, and Rudiyanto, in *Prosiding Seminar Nasional Pendidikan Geografi FKIP UMP.* (2018)
9. BPBD Kab. Bantul, *Tanah Longsor* (2014)
10. T. Hofer, in *Landslide Science and Practice: Global Environmental Change* (Springer Science and Business Media Deutschland GmbH, 2013), pp. 141–143
11. H. L. Aalbers, H. A. J. Moll, I. D. Brouwer, M. Van Dorp, C. B. Houtman, M. L. Brouwer, M. M. M. Zuurbier, and E. C. M. Hagenaars, *FOCUS ON FOOD INSECURITY AND VULNERABILITY A Review of the UN System Common Country Assessments and World Bank Poverty Reduction Strategy Papers* (2003)
12. K. Mertens, L. Jacobs, J. Maes, C. Kabaseke, M. Maertens, J. Poesen, M. Kervyn, and L. Vranken, Science of the Total Environment **363**, 1032 (2016)
13. P. C. Tiwari and B. Joshi, in *Landslide Science and Practice: Global Environmental Change* (Springer Science and Business Media Deutschland GmbH, 2013), pp. 183–189
14. Karyati, S. Sarminah, Karmini, A. M. Akbar, and R. Hermansyah, Sains Tanah **18**, 98 (2021)

Analyzing lead time for flood early warning system in the upstream Citarum River

Aristi Ayuningsi Ode Asri[1], *Faizal Immaddudin Wira* Rohmat[2,3*], *Hadi* Kardhana[2,3], *Arno Adi* Kuntoro[2,3], and *Mohammad* Farid[2,3]

[1] Master Student of Water Resources Engineering, Faculty of Civil and Environmental Engineering, Bandung Institute of Technology, Ganesa Street, Number 10, Bandung 40132, West Java, Indonesia
[2] Faculty of Civil and Environmental Engineering, Bandung Institute of Technology, FTSL Building, Ganesa Street, Number 10, Bandung 40132, West Java, Indonesia
[3] Water Resources Development Center, Bandung Institute of Technology, Ganesa Street, Number 10, Bandung 40132, West Java, Indonesia

Abstract. The Citarum River is a staple water source for the people of West Java, irrigating rice fields in West Java, generating electricity in Java and Bali, supplying water to Jakarta, and operating hydroelectric dams. West Java is prone to flooding because of the concave topography of the Upper Citarum. Floods in Majalaya and Sapan are typically of short duration, with water levels increasing rapidly. Therefore, implementing a predictive early warning system is essential for effective flood control. This study examines lead times utilizing Global Satellite Mapping of Precipitation (GSMaP) data, water levels, and discharge information to provide Majalaya and Sapan communities with early flood warnings and enable timely evacuations. The analysis reveals that Majalaya's lead time is often less than an hour (with 76% incidence and 24% risk), while Sapan's lead time is commonly less than four hours (with 86% occurrence and 14% risk). Utilizing effective lead times enhances the preparation before a flood in the study area. Given the importance of pre-flood evacuation lead times, this research can inform the development of early warning systems in Indonesia, particularly in Majalaya and Sapan.

1 Introduction

The Citarum River, located in West Java Province, Indonesia is a staple water supplier for the province's people. The river supplies the majority of the province's water needs, irrigates 250,000 hectares of rice field in West Java's North Coast, generates 1700 MW of electricity for Java and Bali, 80% of Jakarta Capital Region's water supply, as well as providing energy source for the Cirata, Saguling, and Jatiluhur cascade hydropower dams [1 – 4]. Aside from its water availability, during the rainy season, natural disasters such as floods and landslides often occur in the Citarum Watershed, especially in the Upper Citarum River (UCR) watershed, which is the area upstream to the first of the cascade reservoirs (Saguling) [5].

* Corresponding author: faizalrohmat@itb.ac.id

The geographical and morphological characteristics of the UCR affect the high incidence of flooding, with the Majalaya area being one of the hotspots of flooding [5, 6].

Floods in Majalaya occur almost every year since 2008 [6]. Usually, floods occur in a relatively short period with depths ranging from ankle-high to 1 m [7]. A flood incident in Majalaya on 22 February 2018 inundated the Majalaya area of 5.35 km^2 with a peak discharge of 155.82 m^3/s. The time between rain events and peak discharge is less than 8 hours [8]. Flood water rises quickly with a fairly heavy flow, which can be categorized as a flash flood [5] . Local governments and communities have already carried out structural and non-structural approaches to overcoming floods. Structurally, the government has managed floods on the Citarum River by dredging and straightening the river with the most recent large-scale river normalization conducted in 2019 – 2021 [2, 8] . The normalization proved effective for a short while, given that two years after the normalization had been completed, the floods continued to reoccur. In response to this, combined with the morphological location of Majalaya being that of an inflection point from a steep river slope to a mild one. A non-structural approach is needed to reduce flood risk, in addition to the structural measures. One example is the implementation of the flood early warning system (FEWS) in the area.

An important aspect of FEWS is lead time, defined as the time between the peak rainfall (when the warning is issued) and the peak discharge and water level [9]. A thorough understanding of the lead time characteristics of an area can significantly improve the safety of communities in Majalaya and Sapan from potential flood hazards by facilitating timely evacuation and reducing flood losses. The study uses GSMaP satellite rainfall data, water level, and discharge data from a discharge gauge to determine the lead time for flood emergencies in The Upstream Citarum River.

2 Study sites and methods

2.1. Overview of the study area

The UCR is located in West Java, Indonesia. Topographically, the UCR is a basin surrounded by mountains, with the center being a flat alluvial area, which is partly responsible for making the area prone to flooding. The highest elevation of this watershed is 2,500 m above sea level [4]. The short duration of the rain event encouraged the community in UCR to continue developing a simple early warning system that allows for early anticipation. There are three rain stations on the UCR: Majalaya, Sapan, and Dayeuhkolot. However, only Majalaya and Sapan gauges are used in this research. The catchment areas of Majalaya and Sapan are 205.49 km^2 and 621.21 km^2 respectively, as shown in Fig.1.

2.2. Methods

The research methodology of the study is shown in Fig. 2. The process begins with a literature study, data collection, data correction, and lead time analysis. An explanation of the detailed steps taken is presented below:

1. Rainfall observation data, hourly rainfall data, water level, and discharge are shown in Table 2.
2. Rainfall correction using bias correction with satellite data from GSMaP and observational data with the Thiessen method.
3. Correction of water level and discharge using a Jasa Tirta Corporation 2 (PJT2) stream gauge data by deleting data that has a significant increase and decrease in water level above 2 m in 1 hour.
4. Determine 70 lead time representative data from rainfall, water level, and discharge that have been corrected.
5. Analyze and determine Majalaya and Sapan lead time.

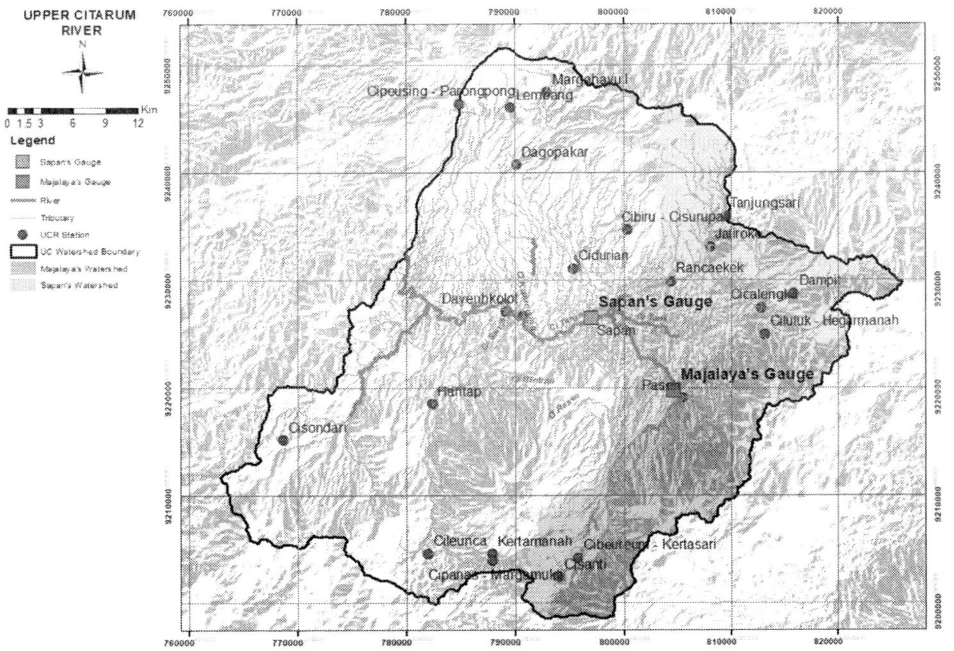

Fig. 1. Study area of the Majalaya and the Sapan.

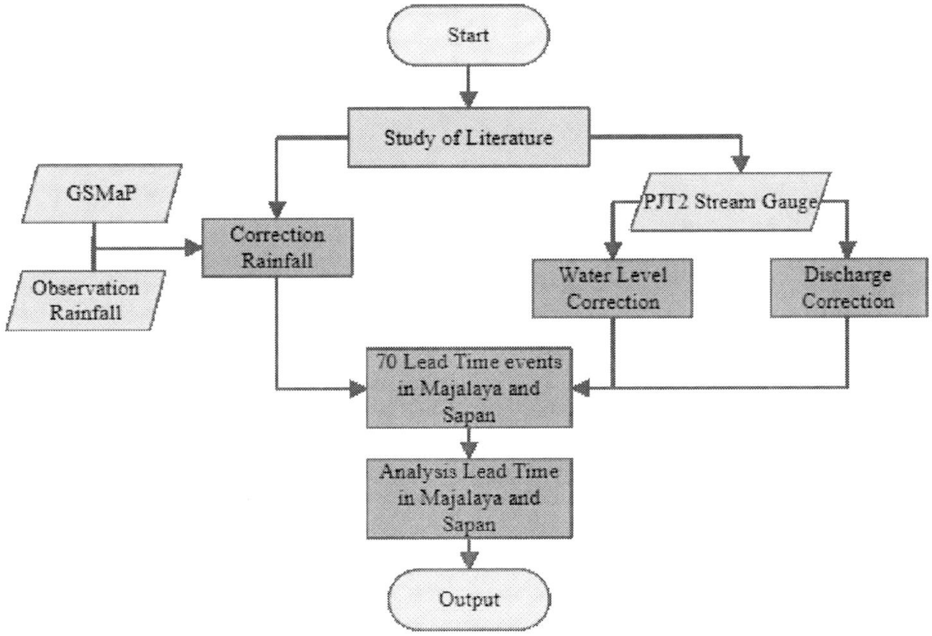

Fig. 2. Methodology flowchart of the study.

Table 1. Data used in this study.

Category	Data	Specification	Data Source
Rainfall gauge	Gauge data from Rancaekek, Sapan, Cicalengka, Paseh, Jatiroke, Tanjungsari, Cibereum, Cisanti, Ciluluk, Dampit, Kertamanah, Cipanas, Kertasari, Bojongsoang, Hantap.	Daily since 2014 until 2021	BBWS Citarum
Hourly rainfall	GSMaP	Version 7, since 2014 until 2022	JAXA
Water level	Majalaya and Sapan gauge	Hourly since 2018 until 2022	Jasa Tirta Corporation II
Discharge	Majalaya and Sapan gauge	Hourly since 2018 until 2022	Jasa Tirta Corporation II

2.2.1. Bias correction rainfall

The utilization of satellite rainfall data necessitates correction and validation against surface observed rainfall data from Majalaya and Sapan. One bias correction method uses the quantile mapping from the "qmap" package in R-software, which increases the R^2 statistical criterion by one level. The quantile mapping adjusts rainfall data to match or resemble observed rainfall measurements, accomplished by converting the data into percentiles and quantitatively using the e-CDF (empirical cumulative distribution function) [10].

2.2.2. Water level and discharge correction

The PJT2 website provided data on water levels and discharge data with a resolution of one hour. However, specific errors occurred during the data recording process, which caused some of the data to be unsuitable for analysis due to significant water level fluctuations over particular periods. To address this issue, data were revised by eliminating entries that signaled a rise or a drop in water level greater than two meters within one hour.

2.2.3. Representative lead time events

Representative 70 events were selected from 42,293 discharge and water level data. This data is consistently selected based on several criteria:

1. Data on the highest water level and the highest discharge each year from 2018-2022 are representative events.
2. The rise in water level and discharge in Majalaya occurred earlier than the incident in Sapan.
3. Rain events cause an increase in water level and discharge.
4. The highest rainfall event causes a peak discharge and water level.

To simplify the determination of representative events, plotting by RStudio is initially performed to observe each occurrence. As a result, a total of 70 representative data points were obtained.

3 Result and discussion

From the 70 representatives data, we determined the lead time for each event. As shown in Fig. 3, the range of lead time events in Majalaya is from 1 to 5 hours, with the majority of lead time events occurring within 2 hours, while the least frequent events, from rain onset to peak runoff, occur within 5 hours. Meanwhile, lead time variations in Sapan range from 1 to

12 hours, with peak discharge events from the highest rainfall taking place within 5 to 6 hours. The lead time events with the least frequency occur within the time frames of 3, 9, 10, 11, and 12 hours, with only one occurrence each.

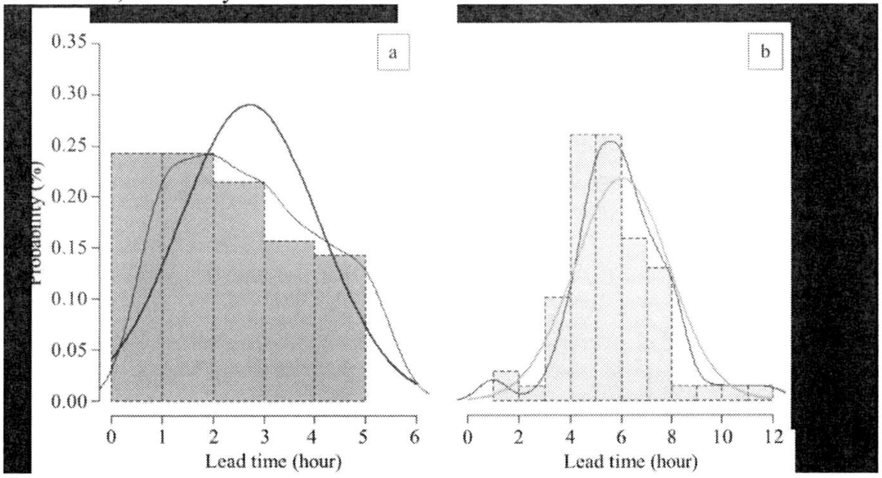

Fig. 3. Number of events and probability of occurrence based on lead time in (a) Majalaya and (b) Sapan Watershed.

Based on the number of lead time events, the probability of lead time occurrences is determined using the Probability Density Function (PDF) through R-software. The probability of lead time events in the Majalaya Watershed to occur beyond 2 hours is 52%, with lead times that often occur within 1 hour and 2 hours. However, the risk of several events with less than 2 hours lead time is 48%. Thus, almost the same possibility makes the community unsure in determining the early warning and evacuation time. So, the probability of a more convincing lead time for the community is more than 1 hour, which is 76%. The safe time in Majalaya to evacuate before flooding is 1 hour. Nevertheless, the possibility of events occurring under 1 hour cannot be disregarded, accounting for 24% of the total occurrences in Majalaya. This implies that there might be instances where the time between receiving a warning and peak discharge, or even flooding, is less than 1 hour.

The highest probability of lead time events occurring in the Sapan region is beyond 4 hours, at 84%. The most frequent lead time events occur within 5 and 6 hours. Within 4 hours or more, the local community in the Sapan area has more opportunities to carry out early evacuation using the early warning system. However, there is a risk of several incidents of rising peak discharge after the highest rain in less than 4 hours lead time of 14%. In conclusion, the lead time for flood evacuation and early warning readiness in the Sapan area is 4 hours. However, the potential for certain events causing flood risk in the Sapan area in less than 4 hours cannot be disregarded.

Majalaya's lead time is shorter than Sapan's because Majalaya is located the most upstream on the UCR, where similar rainfall occurs. Consequently, when water levels rise in Majalaya, Sapan's community receives more information to be prepared.

Practically, lead times in Majalaya and Sapan can be determined using real-time data, although additional lead times are not yet determined. Predicting and providing extra time on lead time requires further study by considering forecasting data.

4 Conclusion

This study determines the lead time based on rainfall, water level, and discharge data. Lead time signifies the duration for executing evacuation following the highest rainfall after a warning is issued up to the time of peak discharge. The Majalaya community requires 1 hour for early evacuation, while Sapan necessitates 4 hours. Due to Majalaya' s position of the

UCR, its lead time is shorter, and flood events are initiated there. The lead time in Majalaya is shorter than in Sapan, where the rainfall starts in the Majalaya watershed, the farthest upstream section of the UCR. Determining the lead time can allows the Majalaya and Sapan communities to prepare, evacuate early, and develop an effective flood early warning system to mitigate flood impacts.

References

1. Fulazzaky, M. A. *Challenges of integrated water resources management in Indonesia. Water (Switzerland)* **6**, 2000–2020 (2014).
2. Putuhena, W. M. *Restorasi Citarum Hulu: Mewujudkan Sungai Citarum yang Sehat untuk Kesejahteraan Masyarakat. Puslitbang Sumber Daya Air* (ITB Press, 2018).
3. Enung, Kusuma, M. S. B., Kardhana, H., Suryadi, Y. & Rohmat, F. I. W. *Hourly Discharge Prediction Using Long Short-Term Memory Recurrent Neural Network (Lstm-Rnn) in the Upper Citarum River. Int. J. GEOMATE* **23**, 147–154 (2022).
4. Ratri, D. N. *et al. Calibration of ECMWF SEAS5 based streamflow forecast in Seasonal hydrological forecasting for Citarum river basin, West Java, Indonesia. J. Hydrol. Reg. Stud.* **45**, 101305 (2023).
5. Burnama, N. S., Immaddudin, F., Rohmat, W., Farid, M. & Kuntoro, A. A. *The Utilization of Satellite Data and Machine Learning for Predicting the Inundation Height in the Majalaya Watershed*. 1–19 (2023).
6. Junnaedhi, I. D. G. A. *et al. Majalaya Flood Early Warning System: A Community Based Approach. IOP Conf. Ser. Earth Environ. Sci.* **71**, 0–12 (2017).
7. Sa'adi, Z. *et al. Spatiotemporal Rainfall Projection in Majalaya basin, West Java, Indonesia under CMIP6 Scenarios*. 1–32 (2022).
8. Rohmat *et al. Urban Climate Flood modeling and baseline study in urban and high population environment : A case study of Majalaya , Indonesia. Urban Clim.* **46**, 101332 (2022).
9. Kardhana, H., Valerian, J. R., Rohmat, F. I. W. & Kusuma, M. S. B. *Improving Jakarta's Katulampa Barrage Extreme Water Level Prediction Using Satellite-Based Long Short-Term Memory (LSTM) Neural Networks. Water (Switzerland)* **14**, (2022).
10. Burnama, N. S., Rohmat, F. I. W., Farid, M. & Wijayasari, W. *Utilization of quantile mapping method using cumulative distribution function (CDF) to calibrated satellite rainfall GSMaP in Majalaya watershed. IOP Conf. Ser. Earth Environ. Sci.* **1165**, (2023).

Nickel mining in Obi Island: problem or solution for mitigating climate change effects?

Ignatius Agung Satyawan[1*], *Togu Marisi* Pardede[2], *Muhammad Sultan* Kevin[3]

[1]Department of International Relations and ASEAN Study Center, Universitas Sebelas Maret, Surakarta, Indonesia
[2]Department of International Relations, Universitas Sebelas Maret, Surakarta, Indonesia
[3]Department of International Relations, Universitas Sebelas Maret, Surakarta, Indonesia

Abstract. Climate change has become a global issue that threatens human life. According to Intergovernmental Panel on Climate Change (IPCC), since 1850 there has been an increase in temperature and sea level so that small islands will sink. As the largest archipelagic country in the world, Indonesia is committed to reduce carbon emissions as a means for mitigating the effects of climate change. The current program is to replace fossil energy with renewable energy. Nickel is a mineral for basic materials to make batteries that can be used as an energy source for electric vehicles which are environmentally friendly. Indonesia has the world's largest nickel reserves and contributes to 37% of the world's total nickel production in 2021. One of the areas that produces nickel is Obi Island in North Maluku. Massive nickel production in this island has negative on environment as well as social-economic on local community. The purpose of this study is to analyse the profit and loss impact of nickel mining in Obi Island. Using qualitative descriptive method through library research this study shows that nickel mining is able to increase regional income but is accompanied by environmental degradation and increased social problems for the local people.

1 Introduction

Global warming and climate are the biggest challenges facing the world today. The impacts have been felt all over the planet, starting from rising global temperatures, extreme weather that is increasingly occurring, to ecological changes that affect human life and the earth's ecosystem. One of the main causes of climate change is the emission of greenhouse gases, and the energy sector is the main source of these emissions. Therefore, many countries, including Indonesia, have committed to reduce greenhouse gas emissions and tackle climate change.

Indonesia, as a country with a large population and rapidly growing economy, has an important role to play in the global effort to address climate change. The Minister of Environment and Forestry, Dr Siti Nurbaya signed the Paris Agreement on Climate Change on 22 April 2016. The Paris Agreement is a monumental global agreement to

* Corresponding author: agungsatyawan@staff.uns.ac.id

combat climate change. National commitment is demonstrated through the Nationally Determined Contribution (NDC) for the 2020-2030 period, as well as actions before 2020. In addition, the Paris Agreement has been supported by 195 countries.

One of the efforts made by Indonesia to fulfil its Paris Agreement commitments is to introduce electric vehicles as an alternative to fossil fuel vehicles. However, in the development of electric vehicle batteries, the raw material for making them is nickel. Indonesia, as the owner of the largest nickel reserves in the world, certainly has an important role in making the transition from conventional vehicle energy to electric fuelled vehicles. Based on data collected from the Ministry of Energy and Mineral Resources, it is recorded that Indonesia has 52% of the world's nickel reserves with a total of 72 million tons [1]. By realizing this advantage, the Indonesian government issued a downstream policy as an effort to increase the selling value of nickel.

One of the locations that has the largest nickel reserves is on Obi Island in North Maluku Province. Obi Industrial Estate has been designated as a national strategic project based on Presidential Regulation Number 109. 2020. Harita Nickel, one of the largest nickel mining companies in the world, manages a nickel mine on Obi Island. This article analyses the positive and negative impacts of Nickel mining on Obi Island. Below is the figure example of Obi Island:

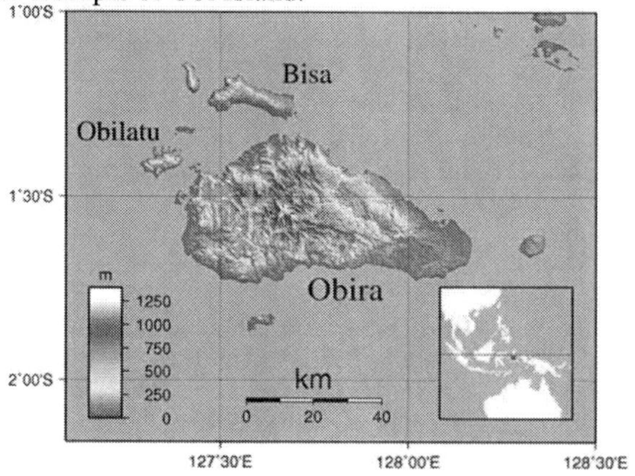

Fig.1. Obi Island [2].

2 Discussion

As a result of the downstream policy by the Indonesian government for nickel mining products, the construction of smelters is important because of their use to process nickel into content that has added value. However, of course this nickel processing has both positive and negative impacts.

2. 1 Positive impacts

The positive impact of nickel mining and processing is employment. The quite massive addition to the workforce is employment in the mining and industrial sector. This occurs in line with the increasing number of smelters operating and the number will continue to increase until the end of the year to process raw nickel ore in North Maluku, as well as the increasing number of contract workers/laborers to accelerate the construction of new smelters for 2022 and the following year. President Joko Widodo claims that before downstream, there were only 500. After downstream mining workers in North Maluku has increased to 46,500 workers.

Apart from absorbing labor, the presence of nickel processing plants and its smelters provides state revenue from taxes. President Joko Widodo revealed that the value of nickel exports before downstream was only IDR 30 trillion, but as a result of downstream the value of nickel exports increased to IDR 510 trillion. Tax revenues from the nickel downstream sector are of course increasing. In 2022 alone state revenue will reach IDR 17.96 trillion [3]. This revenue has increased 10.8 times compared to 2016 which was only IDR 1.66 trillion. Meanwhile, corporate income tax revenue in 2022 was recorded at IDR 7.36 trillion or an increase of 21.6 times compared to 2016 which was only IDR 0.34 trillion [3].

2.2 Negative impacts

Before nickel smelter company existed, Obi Island's air quality was relatively good. Local residents can breathe easily and cases of severe respiratory illness are rare. However, after the establishment of a nickel smelter company, emissions of toxic gases such as sulfur dioxide (SO2) and nitrogen oxides (NOx) have increased significantly. Nickel smelters produce these dangerous gases during the process of smelting nickel ore at high temperatures. Over time, the impact of this air pollution has become more pronounced, with residents experiencing respiratory problems such as asthma, bronchitis and throat irritation more frequently. The increase in cases of respiratory diseases also increases the public health burden, including increasing medical costs [4].

Nickel smelter industry requires large amounts of water for processing and cooling purposes. Before the nickel smelter company existed, Obi Island's aquatic ecosystem functioned well. However, after the establishment of the smelter company, liquid waste containing toxic chemicals, such as heavy metals and hazardous compounds, was discharged into the surrounding waters [5]. The presence of nickel smelter processing companies produces toxic waste which pollutes the waters around the island. This waste pollutes sea water and has a direct impact on marine biota, including skipjack tuna. This pollution threatens the survival of skipjack tuna and makes it unfit for export. Thus, the economy of local fishermen who rely on catching skipjack tuna suffered heavy losses. Agriculture is the backbone of the economy and the main livelihood for the people of Obi Island. Before the establishment of the nickel smelter company, crop yields were quite stable and people could rely on agriculture as a source of income. However, with soil and water pollution by nickel smelter waste, crop yields have decreased significantly. Worsening soil quality and contaminated water result in decreased quality of crop yields. Farmers face difficulties in maintaining local food security, and dependence on outside resources for basic food is increasing. The Indonesian government has an ambitious target to green cities by achieving zero emissions in electric vehicles. However, on the other hand, pollution of the marine environment by nickel smelter waste contradicts these efforts. Ironically, the government is promoting a clean environment on land while destroying the marine ecosystem which is so important to the people of Obi Island [6].

The existence of nickel smelter companies has created complex social conflicts between local residents and mining companies. The 'Harika' Nickel Company is also privileged because it has the status of a vital state object, making it exclusive and untouchable [7]. As a result, the Indonesian government strictly protects this island with the presence of army personnel. This hampers access to transparency for the public and journalists who want to investigate the impact of nickel smelter companies. This limited information makes it difficult for the public to fully understand the negative impacts caused by these companies.

The establishment of the nickel smelter company has brought major changes to the urban planning and lifestyle of the people of Obi Island. Massive industrial facilities and urbanization have changed the face of the island permanently. This has threatened the island's cultural heritage, including its unique local traditions, language and culture. As the industry grows, many elements of local culture are marginalized or even forgotten, threatening the sustainability of cultural heritage that has existed for centuries. The fact that 21 of the 23 nickel smelters in Indonesia are owned by Chinese companies is a serious problem. This illustrates China's dominance in the Indonesian nickel industry, including its mines. Regulations such as the President Law on nickel exports and plans to downstream the nickel industry are becoming less relevant because most of these assets are owned by foreigners. This has an impact on state control of Indonesia's natural resources and increases the risk of unsustainable exploitation [8].

2.3 The way out

Nickel is the main raw material for lithium batteries as an energy source for electric vehicles, which is very important for fulfilling Indonesia's commitment to reducing carbon emissions. Therefore, currently nickel is a very strategic mining material and automatically invites investors to invest so that nickel becomes a mining material that has high economic value. However, nickel mining and its processing has an impact on environmental damage and causes social, economic and cultural problems for communities living around the mine. Therefore, enforcing legal regulations is an absolute requirement to achieve maximum use value of nickel while reducing negative risks to downstream nickel mining.

3 Conclusion

Processing and mining of nickel commodities on Obi Island can indeed have a positive impact on society, such as opening up employment opportunities, income from the tax sector, resulting in an increase in the economy in North Maluku. The construction of a nickel factory and smelter on Obi Island is also an effort to support the downstream of nickel processed products such as electric vehicle batteries which will later become a government initiative in converting gasoline-fueled vehicles into electric fuel. This is a commitment to fulfill Indonesia's agreement and concern for the environment. However, to fulfill this commitment, Indonesia has to sacrifice the environment on Obi Island. Behind the government's mission to realize the mission of zero vehicle emissions in urban areas, in fact there are various kinds of environmental exploitation violations on Obi Island which are of course caused by the Nickel company's neglect of environmental safety. This can be seen from the pollution of drinking water due to factory waste, the worsening air quality for breathing, and the decline in the quality of fish caught by fishermen due to sea water pollution. The only way to overcome this is to enforce and regulate the law in a transparent and just manner.

References

1. www.esdm.go.id/id/booklet/booklet-tambang-nikel-2020
2. https://id.wikipedia.org/wiki/Pulau_Obi
3. Mediatama, Grahanusa, *Kontan.co.id*, 13 Aug. 2023
4. https://www.mongabay.co.id/2023/05/30/munculkan-banyak-masalah-nikel-indonesia-untuk-siapa/
5. S. Yudo and T. Hernaningsih, Kondisi kualitas dan kebutuhan air Desa Kawasi di kawasan pertambangan nikel Pulau Obi, *Jurnal Rekayasa Lingkungan*, **14**, 1 (2021), 1–13
6. T. Sudinda, *Analisa potensi sumber daya air Pulau Obi Halmahera, Jurnal Rekayasa Lingkungan,* **13**, 2 (2021), 166–81
7. https://www.antaranews.com/berita/3458874/harita-nickel-bantah-tudingan-merusak-lingkungan-di-pulau-obi
8. R. Qothrunnada, *peran tenaga kerja asing china pada proyek smelter di Indonesia pada tahun 2018-2021*, *Journal of International Relations*, **8** (2022), 766–79

Mitigating the impact of environmental disasters on food systems in Indonesia: strategies and implications

Armi Susandi[1,2*], *Aristyo R.* Wijaya[2], *Mustafid* Ihsan[3], and *Ahmad W.* Nugroho[2]

[1]Sekolah Tinggi Intelijen Negara, Bogor, Indonesia
[2]Meteorology, Faculty of Earth Technologies & Sciences, Institut Teknologi Bandung, Bandung, Indonesia
[3]Bioengineering, School of Life Technologies & Sciences - Engineering, Institut Teknologi Bandung, Bandung, Indonesia

Abstract. This study investigates the impact of environmental disasters on food systems in Indonesia and proposes strategies for environmental disaster reduction. The study utilizes a case study approach, analyzing the impact of environmental disasters on food systems in selected regions of Indonesia and proposing a mathematical function using an inverse function to map the food systems vulnerability index to the disaster capacity index. The study also considers social, economic, and cultural dimensions in the analysis to ensure the strategies are contextually appropriate. The analysis focuses on identifying successful strategies for environmental disaster reduction and quantitatively assessing their impact on food systems.The research highlights the importance of investing in adaptive and coping capacity to enhance food security and mitigate the impact of environmental disasters, with the potential to increase production by 3-5% in local areas in Indonesia. The study's findings have important implications for policy and practice, emphasizing the importance of integrating environmental disaster reduction strategies into national and regional development plans to ensure the resilience and sustainability of the food systems in Indonesia.

1 Introduction

Indonesia is a country that is highly vulnerable to the impacts of climate change, particularly in the agricultural sector. The country's food systems are heavily dependent on natural resources, making them highly susceptible to the effects of climate change [1, 2]. The impacts of climate change on food systems in Indonesia include changes in rainfall patterns, increased frequency and severity of extreme weather events, and rising sea levels. These impacts have led to reduced agricultural productivity, crop failures, and food insecurity in many parts of the country [1, 2]. The impact of climate change on food systems in Indonesia is a significant concern, as the country is home to over 270 million people, and food security is a critical issue [3].

* Corresponding author: armi.susandi@itb.ac.id

Therefore, due to the inevitable conditions of climate change, the urgency of strategies for environmental disaster reduction is paramount in Indonesia. The country is prone to various environmental disasters, including floods, landslides, droughts, and forest fires, which have a severe impact on food systems [4]. These disasters can cause crop damage, loss of livestock, and destruction of infrastructure, leading to food shortages and increased food prices [2]. Therefore, it is essential to develop effective strategies for environmental disaster reduction to ensure the resilience and sustainability of the food systems in Indonesia.

A review of existing literature on the impact of climate change on food systems in Indonesia and strategies for environmental disaster reduction reveals that there are significant gaps and limitations in current research and strategies [5, 6]. While some studies have focused on the impact of climate change on food systems, few have examined the relationship between climate change and environmental disasters and their impact on food systems. Additionally, there is a lack of comprehensive and contextually appropriate strategies for environmental disaster reduction that consider social, economic, and cultural dimensions.

Hence, this study aims to examine the impact of environmental disasters on food systems in Indonesia and propose strategies for environmental disaster reduction. The research utilizes a case study approach, analyzing the impact of environmental disasters on food systems in selected regions of Indonesia. The study incorporates successful strategies for environmental disaster reduction that could potentially increase production in local areas in Indonesia. The analysis considers social, economic, and cultural dimensions to ensure the strategies are contextually appropriate.

2 Methodology

To determine and examine the impact of environmental disasters on the food system in Indonesia, this study followed the method by L. Wen, et al. [7] to develop an indicator to measure the impact of environmental disasters on the food system by superimposing a climate hazard index with agricultural vulnerability index (see Figure 1).

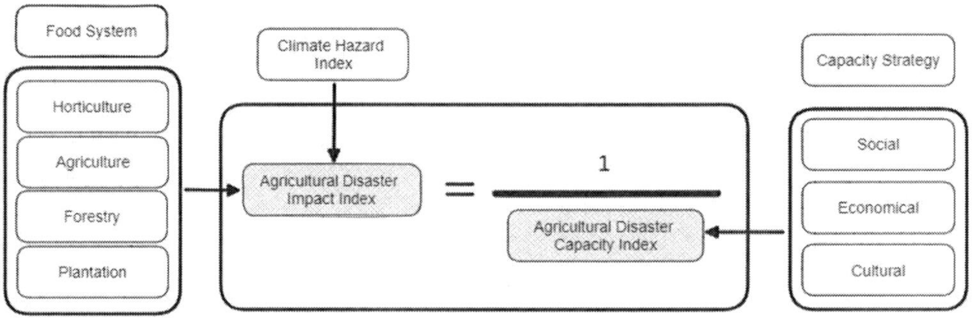

Fig. 1. The workflow of the study.

Following the impact indicator, this study also develops an adaptive and coping capacity index that considers social, economic, and cultural aspects as shown in Table 1. The consideration of the proposed capacity index was based on the arguments that the existing one only measures the capacity to prevent, mitigate, and respond to the disaster with a government perspective which is not related to the end user as intended [8]. Hence, this study tried to explore several indicators to address this issue.

Table 1. Capacity Index Indicators.

Dimension	Indicators
Social	Community awareness and preparedness for disasters
	Access to information and communication during disasters
	Availability of social support networks during disasters
Economy	Economic losses due to disasters
	Availability of financial resources for disaster response and recovery
Culture	Cultural practices and knowledge related to disaster response and
	Cultural sensitivity in disaster response and recovery efforts

Later, by considering the hypothesis that increasing the adaptive and coping capacity, will decrease the negative and destructive impact of environmental disasters on the food system in Indonesia, the study will create a mathematical function by using an inverse function. Hence, assuming that x was a food systems vulnerability index due to climate-related hazards and y was a disaster capacity index, if the function $f(x)$ maps x to y, then the inverse function maps y back to x. The relationship between food systems vulnerability and disaster capacity can be represented by the following equations.

$$f(x)=f^{1}\,(y) \qquad (1)$$

$$f^{1}\,(y)=\frac{1}{y!} \qquad (2)$$

In this case, this study can use the inverse function to map y to x in a way that satisfies the condition. As y gets bigger, $1/y$ gets smaller, so this function satisfies the condition. Later, the value of $f(x)$ will be compared to the value of $f^{1}(y)$ to test the effectiveness of the mitigation strategy.

3 Results

3.1 Analysis of the impact and mitigation of environmental disasters on the food system

Following the methodology, this section will explain in detail the findings of this study. To identify the impact of environmental disasters on food systems in Indonesia and propose strategies for disaster risk reduction and food security, the study will highlight the vulnerability of Indonesia to climate disasters and strategies to mitigate their adverse effects on food systems. Figure 2 shows the result of the map for each index.

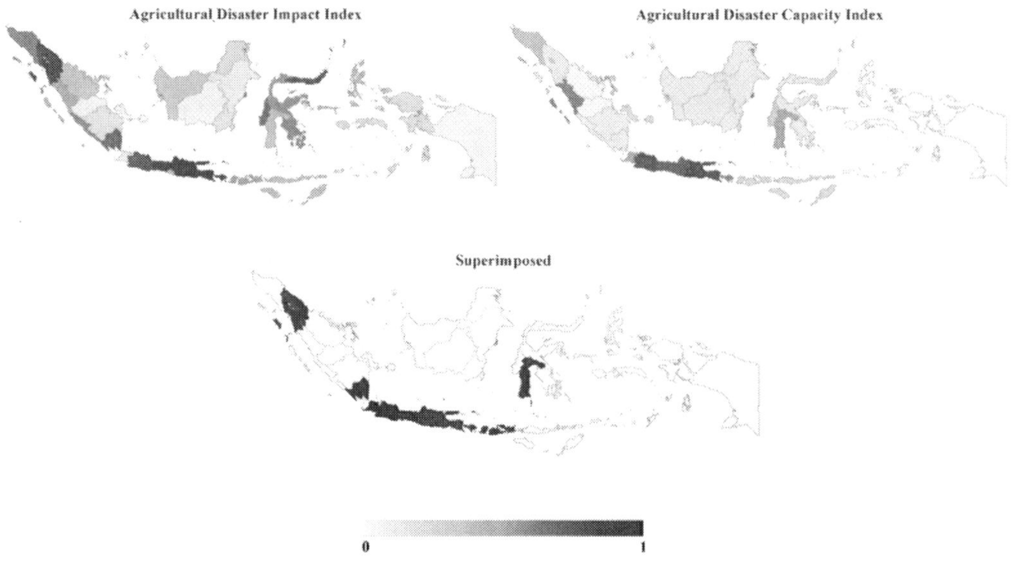

Fig. 2. Map of each Agricultural Disaster Impact Index and Agricultural Disaster Capacity Index and the superimposed for each province in Indonesia.

The Agricultural Disaster Impact Index measures the impact of environmental disasters on agricultural production in each province, while the Agricultural Disaster Capacity Index measures the capacity of each province to mitigate the impact of environmental disasters on agricultural production. The scale is from 0 to 1, with a higher index meaning the resilience of the food system from the impact of the disaster.

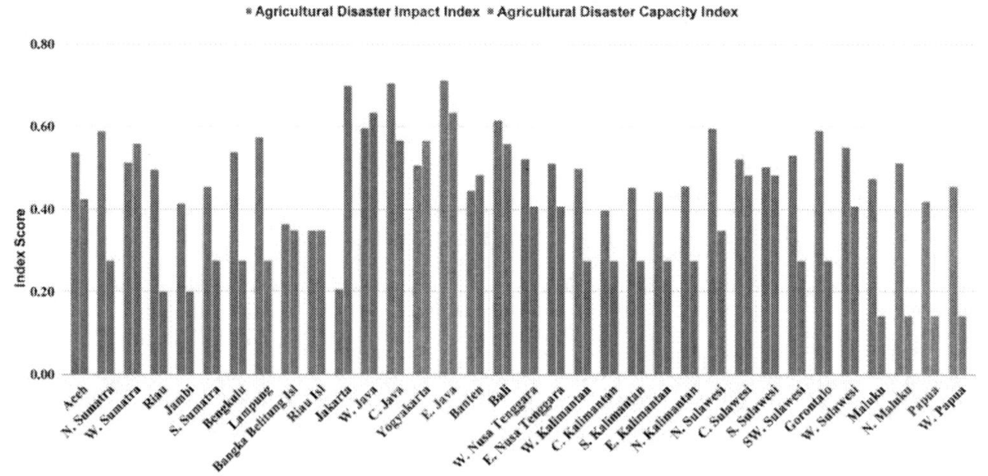

Fig. 3. Comparison of Agricultural Disaster Impact Index and Capacity Index.

Figure 3 shows that Central Java has the highest Agricultural Disaster Impact Index 0.71, indicating that it is highly vulnerable to environmental disasters. On the other hand, Jakarta has the highest Agricultural Disaster Capacity Index of 0.70, indicating that it has the highest capacity to mitigate the impact of environmental disasters on agricultural production. The figure also shows that several provinces, such as Riau, Jambi, and Maluku have low Agricultural Disaster Capacity Index values, indicating that they have limited capacity to mitigate the impact of environmental disasters on agricultural production.

3.2 Discussions

As shown in the result, the Agricultural Disaster Impact Index and Agricultural Disaster Capacity Index for each province in Indonesia can be used to identify the provinces most vulnerable to environmental disasters and have limited capacity to mitigate their impact on agricultural production. It is known that Jakarta has the highest Agricultural Disaster Capacity Index which can be attributed to the city's well-developed infrastructure, disaster management systems, and access to resources. On the other hand, Central Java indicates that it is highly vulnerable to environmental disasters which can be attributed to the province's location in the middle of Java Island, which is prone to natural disasters such as floods, landslides, and earthquakes.

Meanwhile, several provinces, such as Riau, Jambi, and Maluku, have low Agricultural Disaster Capacity Index values, indicating that they have limited capacity to mitigate the impact of environmental disasters on agricultural production. This can be attributed to the provinces' limited resources, infrastructure, and disaster management systems. The proposed disaster capacity index based on social, economic, and cultural dimensions can provide valuable information on the key factors contributing to the success of the strategies for disaster risk reduction and food security in each province that depend on several key factors. Henceforth, based on the findings, the following strategies can be proposed to mitigate the impact of environmental disasters on agricultural production in Indonesia:

1. Develop and implement disaster management systems that are tailored to the specific needs of each province.
2. Invest in infrastructure development, such as irrigation systems, roads, and bridges, to improve access to resources and markets.
3. Promote community-based approaches to disaster risk reduction and food security, such as the establishment of farmer cooperatives and the use of traditional knowledge and practices.
4. Encourage the adoption of new technologies and innovations, such as renewable energy-powered water pumps, to improve water management and increase agricultural productivity.
5. Strengthen the capacity of local governments and stakeholders to develop and implement effective strategies for disaster risk reduction and food security.

Overall, the result of this study highlights the need for a comprehensive and context-specific approach to disaster risk reduction and food security in Indonesia. The proposed disaster capacity index based on social, economic, and cultural dimensions can provide valuable information on the key factors contributing to the success of the strategies and inform the development of effective strategies for disaster risk reduction and food security in each province which could increase the production up to 3-5%.

4 Conclusions

The study provides valuable insights into the relationship between adaptive and coping capacity and the impact of environmental disasters on food security in Indonesia. The research is based on the hypothesis that increasing adaptive and coping capacity will decrease the negative and destructive impact of environmental disasters on the food system in Indonesia. The study proposes the creation of a mathematical function using an inverse function to map the food systems vulnerability index to the disaster capacity index, with the aim of improving food security in each province, and potentially increasing production by 3-5%.

The findings of the study underscore the critical role of adaptive and coping capacity in mitigating the impact of environmental disasters on food security. The proposed mathematical function using an inverse function provides a robust framework for analyzing the relationship between the food system's vulnerability and disaster capacity. The potential increase in production by 3-5% in each province highlights the significant impact that enhancing adaptive and coping capacity can have on food security in Indonesia.

In conclusion, the study's findings have important implications for policy and practice. The research highlights the importance of investing in adaptive and coping capacity to enhance food security and mitigate the impact of environmental disasters. The proposed mathematical framework provides a valuable tool for policymakers and practitioners to analyze and address the relationship between food security and environmental disasters, with the potential to significantly improve food security and resilience in Indonesia. The proposed set instrument can be used to measure the food system's vulnerability after a disaster. Successful strategies for environmental disaster reduction could potentially increase production by 3-5% in local areas in Indonesia. Further research is needed to develop more comprehensive and accurate methods and models for measuring potential food production loss after a disaster.

References

1. A. Mirzabaev, L. Olsson, R. B. Kerr, P. Pradhan, M. G. R. Ferre, and H.Lotze-Campen, in *Science and Innovations for Food Systems Transformation*, edited by J. von Braun, K. Afsana, L. O. Fresco, and M. H. A. Hassan (Springer International Publishing, Cham, 2023), pp. 511–529
2. A. Murwani, CIPS Indonesia (2022)
3. Z. Rozaki, in *Advances in Food Security and Sustainability*, edited by M. J. Cohen (Elsevier, 2021), pp. 119–168
4. T. Takama, in *Non-Traditional Security Challenges in Asia* (2017), pp. 93–133
5. Y. V. Markhof, G. Ponzini, and P. R. Wollburg, *Measuring Disaster Crop Production Losses Using Survey Microdata: Evidence from Sub-Saharan Africa* (The World Bank, 2022)
6. A. Munirah and N. A. Norfarizan-Hanoon, Food Res **6**, 304 (2022)
7. L. Wen, H. Liu, J. Chen, B. Fakieh, and S. M. Shorman, Applied Mathematics and Nonlinear Sciences **7**, 275 (2022)
8. R. Bali, RESEARCH REVIEW International Journal of Multidisciplinary **7**, 40 (2022)
9. A. B. Setiawan, A. Nugroho, D. Wiyanti, M. Yusuf, A. Maslikhatun, and P. Bowo, JEJAK: Jurnal Ekonomi Dan Kebijakan **16**, (2023)

Adaptation and mitigation management of drought, wildfire, and climate change toward the development of natural tourism in Jatirejo District, Mojokerto Regency, Indonesia

Eko Budi Santoso[1*], *Muhammad Ifaldy Amrila* Zulkarnain[1], *Azura Calista Shafa* Kamila[1], *Gde Abhicanika Pranata* Dyaksa[1], *Shinta Novia* Vera[1], and *Annisa Fadhilah* Ashari[1]

[1]Department of Urban and Regional Planning, Sepuluh Nopember Institute of Technology, Sukolilo, Surabaya 60111, Indonesia

Abstract. Global climate change, drought, and the threat of wildfires become major challenges in tourism development. Jatirejo District has a unique natural landscape and is stated as an Environmental Strategic Area in the form of Raden Soerjo Forest Park, which is frequently hit by drought, wildfire, and climate change. This study aims to identify those disasters that threaten natural tourism, see how far the gap with ideal conditions and formulate adaptation and mitigation strategies. The methods used in this research are spatial analysis to identify the disaster risks and qualitative method by gap analysis to formulate the strategies. The findings show that wildfires in 2019 at R. Soerjo Forest Park had a coverage area of 64.7 hectares. Moreover, 59% of the total area has been identified as a very high-risk index of wildfire and drought. There is also a significant increase in temperature by 0.7°C-1°C. This study shows the large gap between the severity of the disaster and the ideal conditions according to the guidelines. The potential of tourism destinations, as well as the threat of disasters, shows the need for adaptation and mitigation strategies for tourism development which contain three phases, specifically prevention and preparation, response, and recovery.

1 Introduction

Global climate change, drought, and the threat of wildfires are potentially major challenges to developing the nature tourism sector [1]. Nature-based tourism relies on ideal climatic and environmental conditions, making it vulnerable and sensitive to climate change and environmental degradation [2]. Several studies refer to these three phenomena as having affected many natural tourism destinations worldwide, including from the supply side of tourism management and the demand side related to the number, preferences, and timing of tourist visits [3]. Climate change, drought, and forest fires, in correlation with the tourism sector, have a major impact on employment, economic development, and national income [4,5]. Thus, shaping tourism that is resilient to global climate change, drought, and fire threats

* Corresponding author: santosoeb@gmail.com

is very urgent. Furthermore, it is essential under the United Nations-International Strategy for Disaster Reduction (UN-ISDR) to reduce the disaster risk [6].

Jatirejo district, Mojokerto regency, represented in Fig. 1 has 8,5 thousand hectares of total area and ten natural attractions including the R. Soerjo Grand Forest Park (Tahura). Although it has been stated as a strategic environmental area, there was an increase in wildfires from 2017 to 2023 by converting 126.45 ha of forest area in Jatirejo [7]. Furthermore, in 2022, there was a drought disaster due to the depletion of water sources [8]. Those disasters disrupted nature tourism activities in Jatirejo, such as climbing, recreation, and other natural tourism [9]. A similar phenomenon has been studied by [10], which shows the significant impact of climate change and drought on natural tourism in the Caribbean, offers adaptation and mitigation efforts through green building implementation, and talks about the economic effect which there is no spatial analysis using there to show the disaster risk condition. In addition, similar research about landslide disaster mitigation plans in Bantul district [11], focuses on the land risk spatial analysis instead of giving the specific process to formulate the mitigation strategies. We have not found a comprehensive study that discusses drought, wildfires, and climate change risk by using spatial analysis tools and validate it through gap analysis to compare with the actual condition, and also formulate the adaptation and mitigation strategies in a single publication. Hence, this study wants to identify the disasters and formulate the adaptation and mitigation strategies by seeing the gap between the real disaster condition with the ideal condition as the standards, to find the suit best practices that can be implemented in the study area according to its characteristics.

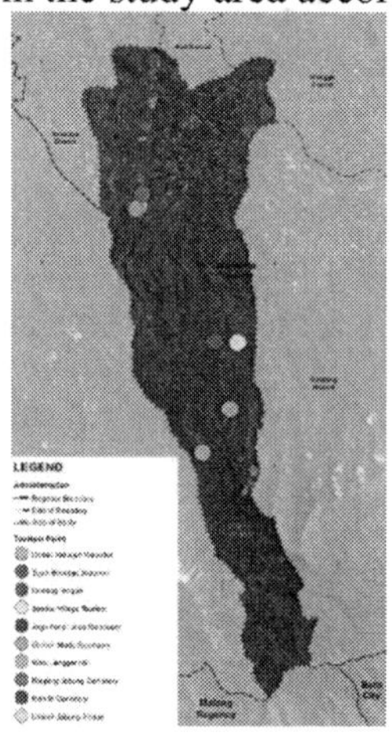

Fig. 1. Administrative and Tourism Attraction Map

2 Material and methods

2.1 Climate change and disaster risk analysis

This analysis was conducted to determine climate change conditions, drought, and wildfire disaster risks. Climate change can be seen from Land Surface Temperature (LST) and rainfall index. LST was identified through the remote sensing method, using Landsat 8 image from USGS, then processed into temperature data using spatial Geographic Information System (GIS) with ArcGIS program. The rainfall index is identified through the GIS interpolation method using data from the Climate Hazards Group InfraRed Precipitation with Station data (CHIRPS). Information on drought and fire risk was obtained through GIS overlay and interpolation methods based on Digital Elevation Model (DEM) data from InaRISK. Land risk assessment is based on three parameters according to the formula agreed upon in the Hyogo Framework for Action [12]:

$$R = \frac{H}{C} \times V \quad (1)$$

Where R is the risk; H is a hazard; C is capacity, and V is vulnerability.

Those parameters are then overlaid to obtain a disaster risk map. Drought and fire risk index values are divided into 5 classes: 1) very low, 2) low, 3) medium, 4) high and 5) very high.

2.2 Adaptation and mitigation management formulation

The disaster management and adaptation strategies will be formulated through gap analysis and best practices by comparing several literature studies and using qualitative descriptive methods to formulate the strategies. Gap analysis is either a tool or a process to identify where gaps are and what differences exist between the current situation and "what ought to be" in place [13]. In this research, we compare between the ideal condition and the current condition to identify the gaps. The existing conditions data were obtained by a field survey in the study area and the ideal condition is based on the standardization or regulation that we get from many study literature processes and has been adjusted to the study area characteristics.

This research simulates the disaster management cycle of wildfire, drought, and climate change through three phases adaptation and mitigation. Those are: 1) prevention and preparation, 2) response, and 3) recovery [14].

3 Results and Discussion

3.1 Wildfire, drought, and climate change disasters

The analysis results show an increase in temperature of 0.7°C-1°C in Jatirejo, as presented in Fig. 2. In addition, there was also a decrease in rainfall intensity during the dry season by 62%. The decline was not matched by good water management, causing the water content in the soil to decrease or even disappear and affecting the water supply for the community and tourism activity. Furthermore, the analysis results show that the risk of wildfires and drought is dominantly at a very high level as presented in Fig. 3 for wildfire risk and Fig. 4 for drought risk which is symbolized by the color red in maps. By Fig. 3 we find that Jatirejo has a very high wildfire risk index with a coverage area of 5110 ha (59%) while Fig. 4 shows 5178 ha (60%) of very high drought risk index of the total area of Jatirejo.

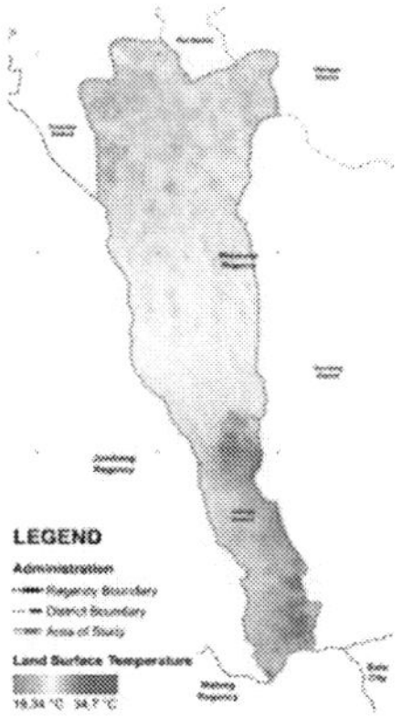

Fig. 2. Land Surface Temperature Map

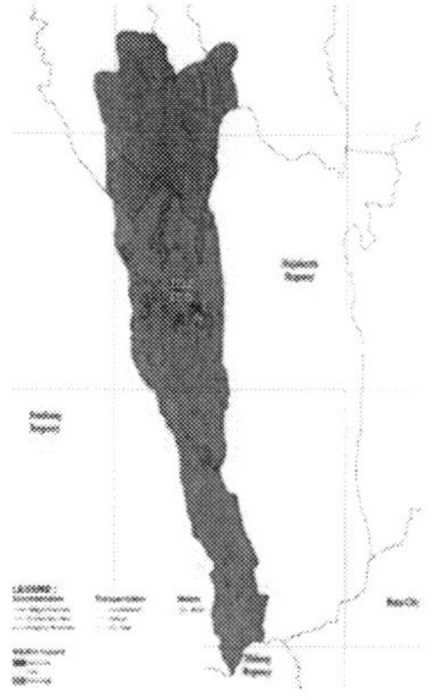

Fig. 3. Wildfire Risk Map

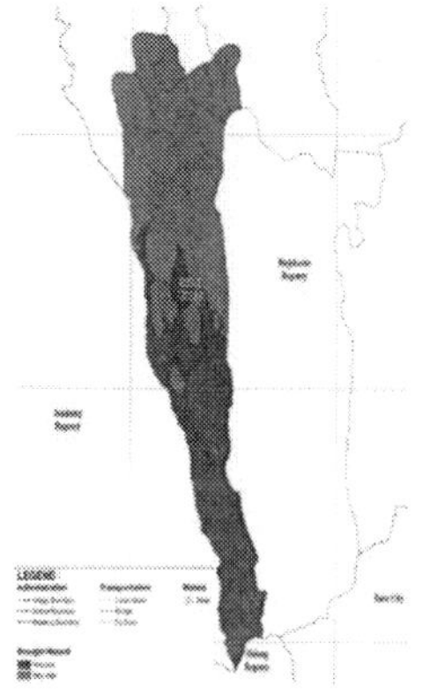

Fig. 4. Drought Risk Map

3.2 Gap Condition

There is a large gap between the disaster conditions in the study area and the ideal condition according to the guidelines, as shown in Table 1. By the gap analysis, this research finds that the study area still lacks adaptation and mitigation efforts according to the standards. Through the gap analysis, It can be seen which best practices are suitable for existing conditions which are then elaborated through strategies.

Table 1. Gap analysis table.

Ideal conditions	Existing Conditions
• Hold global average temperature rise below 1.5°C by the end of this century. • Increasing resilience and adaptive capacity • Implementing the "Climate Village" program *Guidelines: Suppression of Average Temperature Rise Under the Paris Agreement To The UN Framework Convention On Climate Change*	• Have temperature increases of 0.7- 1°C in 2040. • There is no disaster evacuation route or risk map. • Lack of community and stakeholder engagement.
• Provide shelters/posts in certain locations equipped with fire extinguishers. • Give more attention to buildings in fire-prone areas. *Guidelines: Optimization of Standard Operating Procedures for Land and Forest Fire Control Activities (Perdirjen Perubahan Iklim No. P.12/PPI /SET/KUM.1/12/2020)*	• Lack of availability of shelters with fire extinguishers • There is no standardization of building and residential development in fire-prone areas
• The area allowed to construct nature tourism facilities is at most 10% of the area stipulated in the license. • Provide early warning system facilities. • Implementing cooperation schemes between the government and the tourism industry. *Guidelines: Botanical Forest Park Tourism Management is carried out by the principles of conservation of natural resources and ecosystems. (Permen LHK No. P.8/MENLHK /SETJEN/KUM.1/3/ 2019)*	• Built-up nature tourism areas below 10% • Lack of early warning system facilities • Lack of stakeholder cooperation in Disaster Rapid Response Team capacity

3.3 Adaptation and mitigation strategies

Based on the comparison between the ideal condition and the existing condition on the gap analysis, it was found that the study area still lacks adaptation and mitigation efforts by the standards. Therefore, through these gaps and best practices, disaster adaptation and mitigation strategies can be formulated, as follows.

1. Prevention and Preparation Phase by reduced risk exposure [15]. Hazard maps help inform this phase by predicting or identifying potential damage [16]. This phase can be implemented by establishing disaster risk maps and publishing periodic information on LST and rainfall.
2. Response Phase, by land use planning [17], and involving the role of the community [18]. Land use planning can be implemented by managing watersheds and water reservoirs, including water conservation, creating evacuation routes and shelters, and providing infrastructure based on disaster risk reduction. While, involving the role of the community can be implemented by preserving forest and land fire fighting units, conserving landscapes and nature tourism by optimizing action plans, conducting

stakeholder cooperation in the capacity of the Disaster Rapid Response Team, and conducting training and socializing on disaster action plans.

3. Recovery Phase, by policymaking [10]. Policymaking can be implemented such as retrofitting and investing in environmentally friendly buildings for drought disasters and wildfires, policies contain the standardization of building and residential development in fire-prone areas and involve the role of the community.

4 Conclusion

The results analysis shows that Jatirejo is affected by climate change, drought, and wildfire disasters. The spatial analysis results show an increase in temperature of 0.7°C-1°C and the risk of wildfires and drought is dominantly at a very high level covering 39% of the Jatirejo area that has disrupted nature tourism activities. Those data were then validated with the actual conditions through the gap analysis that shows the large gap between the existing and ideal conditions which need serious strategies against the disasters. In response to these problems, disaster mitigation and adaptation management were formulated in three phases, including 1) prevention and preparation, 2) response, and 3) recovery. The overall form of mitigation and adaptation includes preparing a map of disaster-prone locations, water reservoir construction, watershed management, water conservation, evacuation routes, assembly points, land use planning, conducting stakeholder cooperation, and policy development. The formulation of strategies in this research is a mitigation and adaptation effort in minimizing the impact of disasters which is considered adequate and able to answer the problem.

References

1. S. Ma, ScholarWorks @ UARK Managing Climate Change and Weather Extremes for Nature-based Tourism Organizations in the United States, (2021)
2. H. Kilungu, R. Leemans, P. K. T. Munishi, S. Nicholls, B. Amelung, Forty Years of Climate and Land-Cover Change and its Effects on Tourism Resources in Kilimanjaro National Park, *Tour. Plan. Dev.*, **16**, 2, pp. 235–253, (2019)
3. S. M. De Urioste-Stone, L. Le, M. D. Scaccia, E. Wilkins, Nature-based tourism and climate change risk: Visitors' perceptions in Mount Desert Island, Maine, *J. Outdoor Recreat. Tour.*, **13**, (2015)
4. W. L. Hambira, J. Saarinen, Policy-makers' perceptions of the tourism–climate change nexus: Policy needs and constraints in Botswana, *Dev. South. Afr.*, **32**, 3(2015)
5. K. Dube, G. Nhamo, Climate variability, change and potential impacts on tourism: Evidence from the Zambian side of the Victoria Falls, *Environ. Sci. Policy*, **84**, (2018)
6. UN-ISDR, What is the Sendai Framework for Disaster Risk Reduction?, *United Nations Officer for Disaster Risk Reduction*, (2020)
7. BPBD, *Disaster History of Mojokerto District*, (2020)
8. M. Romadoni, *Musim Kemarau, Warga Jatirejo Mojokerto Dapat Bantuan Pengadaan Air Bersih Dari Polda Jatim Artikel ini telah tayang di Surya.co.id dengan judul Musim Kemarau, Warga Jatirejo Mojokerto Dapat Bantuan Pengadaan Air Bersih Dari Polda Jatim, Surya.co.id*, (2023)
9. M. Romadoni, *Dampak Kebakaran Hutan, Pendakian Kawasan Taman Hutan R. Soerjo Mojokerto Ditutup Sementara Artikel ini telah tayang di Tribunjatim-timur.com dengan judul Dampak Kebakaran Hutan, Pendakian Kawasan Taman Hutan R. Soerjo Mojokerto Ditutup Sementara*, *Tribun Jawa Timur*, (2023)
10. K. Dube, G. Nhamo, Vulnerability of nature-based tourism to climate variability and change: Case of Kariba resort town, Zimbabwe, *J. Outdoor Recreat. Tour.*, **29**, (2020,)
11. M. Teguh, S. Aminatun, W. Erlangga, Landslide disaster mitigation plan in Karang Tengah Village, Bantul district, Yogyakarta, *E3S Web Conf.*, **156**, (2020)

12. S. Bachri, S. Sumarmi, L. Y. Irawan, Y. E. Aldianto, K. S. B. Utomo, A. K. Wardhani, Coping Mechanism in the Face of Landslide Hazard in East Java Indonesia, *Geoj. Tour. Geosites*, **40**, 1 (2022)
13. S. Kim, Y. Ji, Gap Analysis, (2018)
14. QDMC, Queensland State Disaster Management Plan, *Brisbane*, (2015)
15. L. Gannon, Bushfire Hazard Management Plan, (2016)
16. P. Deckers, W. Kellens, J. Reyns, W. Vanneuville, P. De Maeyer, Geospatial Techniques in Urban Hazard and Disaster Analysis, *Geospatial Tech. Urban Hazard Disaster Anal,* (2010)
17. C. Gonzalez-Mathiesen, A. March, Nine design features for bushfire risk reduction via urban planning, *Aust. J. Emerg. Manag.*, **29**, 3, (2014)
18. A. Rom, I. Kelman, Search without rescue? Evaluating the international search and rescue response to earthquake disasters, *BMJ Glob. Heal.*, **5**, 12(2020)

The role of Weather Modification Technology for forest and land fire disaster mitigation in the perspective of carbon emission reduction in Indonesia

Budi Harsoyo[1,2*], *Rizaldi* Boer[2], *Edvin* Aldrian[1], *Lailan* Syaufina[2], *M. Bayu Rizky* Prayoga[1], *Muhamad Djazim* Syaifullah[1], *Ari* Nugroho[1], and *Chandra* Fadlilah[1]

[1]National Research and Innovation Agency, Jakarta, Indonesia
[2]IPB University, Bogor, Indonesia

Abstract. This paper reviews the effectiveness of implementing Weather Modification Technology (WMT) in efforts to control forest and land fire disasters in Indonesia. Analysis is carried out on several parameters in the field that can be measured and observed. WMT has been proven to be able to increase rainfall intensity by 12.9% and shorten the duration of Consecutive No Rain Days, which in turn can reduce the number of hotspot events in the target area. The implementation of WMT during the rainy season transition period has also been proven to be able to increase the peatland groundwater level, which ultimately shortens the drought period in the target area. The WMT implementation program, which has become increasingly well-planned in the last 4 years (2020-2023), has proven to be able to reduce the number of hotspot incidents, the extent of forest and land fires, and the amount of carbon emissions in Indonesia quite significantly. Based on updated data until September 2023, the number of hotspots, burnt area, and total carbon emissions in Indonesia were "only" 22%, 38.9%, and 5.3% respectively compared to the numbers in 2019, which was both an El Nino year.

1 Introduction

Indonesia is a country that is vulnerable to the impacts of climate change and has a reasonably high-risk index for natural disasters [1]. Based on data on disaster events in Indonesia from the National Agency for Disaster Countermeasures (BNPB), disaster events in Indonesia are dominated by the type of hydrometeorological disaster. Apart from being triggered by human-caused environmental damage, the increasing frequency of hydrometeorological disasters is also caused by climate change. An example is the forest and land fires in Indonesia every year. This disaster has historically tended to recur yearly and increase during the dry season. The intensity of fires can worsen when El Nino occurs, resulting in a prolonged drought in Indonesia [2].

* Corresponding author: budi042@brin.go.id

The phenomenon of forest and land fire disasters has always been a concern of the Government of Indonesia. Not only causing losses in the social and economic sectors, these disasters also often give a negative image due to the transboundary haze problem. Forest and land fires are also closely related to carbon emissions, which significantly affect climate change in the long term. Therefore, referring to the Nationally Determined Contribution (NDC), which the Government of Indonesia has prepared as a follow-up to the ratification of the Paris Agreement on the 2015 UN Work Convention on Climate Change, the Government of Indonesia is committed to setting a total target of reducing carbon and Greenhouse Gas (GHG) emissions of 2,869 MTon CO2e in 2030. Forestry and Other Land Uses (FOLU) is the sector with the most significant emission reduction contribution, which is 17.2% of the total target of 29% with an unconditional mitigation scenario or 24.1% of the real target of 41% with conditional mitigation scenarios. This amount has considered the emissions from forest and land fires that occur every year in Sumatra and Kalimantan [3].

One of the efforts to reduce carbon emissions is to reduce the incidence of forest and land fires. The total carbon content in Indonesia is estimated at 14 gigatonnes from the forest sector [4]. Forest and land fires in 1997 coinciding with El Nino were estimated to release as much as 1 gigatonne of carbon [5]. Thus, if the intensity of forest and land fires can be reduced, this will automatically reduce carbon emissions in Indonesia.

The Ministry of Environment and Forestry (KLHK), as the National Climate Change Focal Point in Indonesia, has made many efforts to control forest and land fire disasters. One of them is by utilizing Weather Modification Technology (WMT). The Government of Indonesia first used WMT to deal with massive fires during El Nino in 1997. In subsequent implementations since 2009, WMT began to be routinely used in efforts to deal with forest and land fires that have occurred almost every year in Sumatra and Kalimantan. El Nino in 2015, which also resulted in severe forest and land fires, has provided lessons and has become a starting point for the paradigm reform of using WMT in the disaster management scheme for forest and land fires in Indonesia. Previously WMT was carried out when the disaster was already massive for fire suppression. Since 2016 WMT has been carried out earlier for prevention purposes with the aim of re-wetting peatlands so that they do not catch fire easily. With earlier implementation during the rainy season transition period, the rain yields obtained from WMT implementation can be more optimal and shorten the drought period on peatlands prone to fire [6].

This article reviews the evaluation of the results of implementing WMT in efforts to prevent forest and land fires disasters which have been programmed by the Indonesian Government to anticipate the impact of El Nino in 2023. The output resulting from this research is expected to provide an overview of the effectiveness, contribution and value of the benefits of implementing TMC in efforts controlling forest and land fire disasters in Indonesia.

2 Materials and methods

Several research samples were taken from a number of WMT implementations that have been carried out for the year El Nino 2023. A number of data analyzed included data on rainfall, peatland groundwater level (GWL), number of hotspots, burnt area and total carbon emissions. The rainfall data analyzed uses data from the Tropical Rainfall Measuring Mission (TRMM) Satellite taken from the JAXA Realtime Rainfall Watch sites: *https://sharaku.eorc.jaxa.jp/GSMaP/*. Data on SM and peatland GWL used in this study were obtained from the SIPALAGA sites: *https://sipalaga.brg.go.id/* and *https://prims.brg.go.id/*, which are managed by the Peatland and Mangrove Restoration Agency and (BRGM). Data on number of hotspots, burnt area and total carbon emissions taken from Ministry of Environment and Forestry (KLHK) – SIPONGI sites: *https://sipongi.menlhk.go.id*.

 https://doi.org/10.1051/e3sconf/202346703001

3 Results and discussion

The implementation of WMT has proven to increase the intensity of rainfall in the target area. Figure 1 shows the daily rainfall intensity during WMT implementation in Riau Province from May 17 – June 9, 2023. The figure shows that relatively low rainfall occurred during the week before the implementation of the WMT in Riau Province (7 – 16 May 2023). The intensity is much lower than the historical average and the actual value for the previous year (2022). This condition then improved during the WMT implementation period from 17 May to 9 June 2023, where the intensity of daily rainfall was generally higher than the previous year's historical and actual average values. Please note that 2023 is an El Nino year while 2022 is a La Nina year.

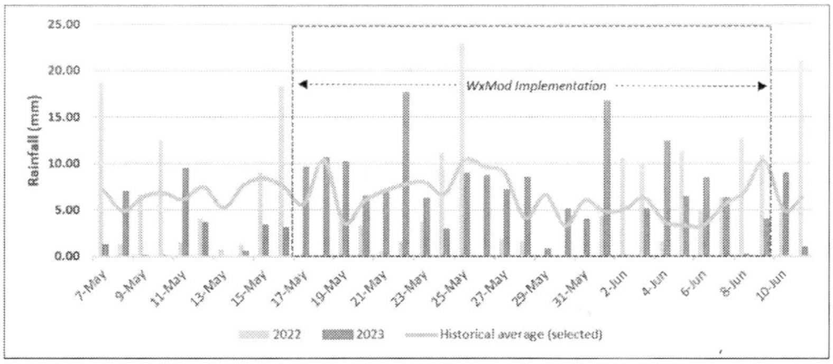

Figure 1. Rainfall in Riau Province during the WMT implementation period.

Furthermore, Figure 2 compares the spatial distribution of rainfall before and during the WMT implementation period in Riau Province. To facilitate comparison, it is made in 3 periods with a duration of about one decade each. In the left figure (period 6-16 May 2023), about one decade before the TMC implementation period, the rainfall intensity in Riau Province was relatively the same as in other nearby provinces. Contrast this with the description of the first decade of the implementation of WMT in the period 17-27 May 2023 (middle figure) and the second decade in the period 28 May – 9 June 2023 (right figure). This figure shows that the rainfall intensity in Riau Province is higher than in other nearby provinces. Areas with high rainfall intensity in Riau Province appear to be concentrated around cloud-seeding areas. The line drawn on the map records the track of cloud seeding during the WMT implementation.

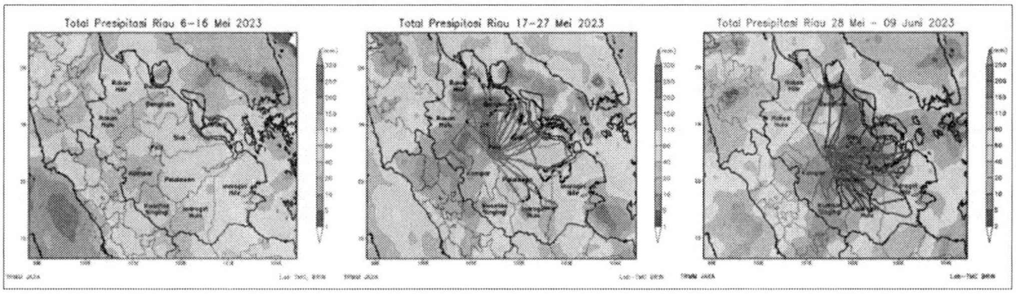

Figure 2. Spatial distribution of rainfall in Riau Province before (left) and during the WMT implementation period (middle and right).

The value of the increase in rainfall in Riau Province during the WMT implementation period from 17 May to 9 June 2023 was calculated using the Target Only Method. The calculation compares the actual rainfall in the target area obtained during the WMT implementation period with historical rain at the same time interval. From the calculation results, the value of increasing rainfall is 12.79%. In addition, the volume of rains obtained from the effects of cloud seeding activities during the WMT implementation period was also calculated. The calculation is based on the precipitation thickness approach measured from the TRMM satellite during the WMT implementation period multiplied by the area affected by cloud seeding activity in the target area. The total volume of rainfall resulting from cloud seeding during the WMT implementation in Riau Province was 208.1 million m^3.

The results of the WMT implementation have also proven to be able to shorten the duration of Consecutive No Rain Days (CNRD). Figure 3 shows the analysis results from BMKG (updated May 31, 2023) on CNRD monitoring in all regions of Indonesia. For the Riau Province, which has received WMT implementation support since May 17, 2023, CNRD is dominant for a relatively short period (1-5 days).

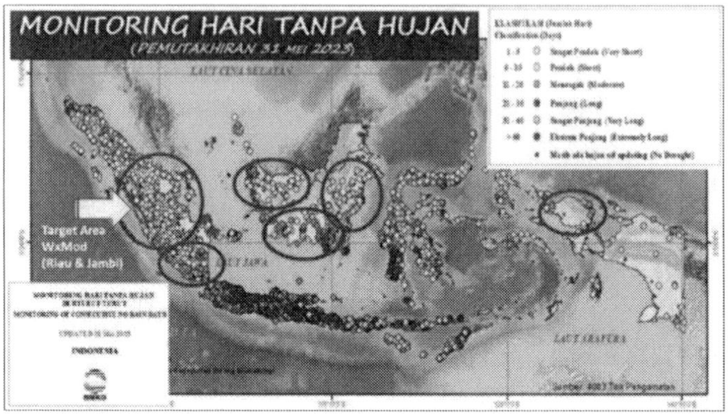

Figure 3. Monitoring of Consecutive No Rain Days, Updated 31 May 2023 (Source: BMKG).

WMT implementation aims to keep the peatland groundwater level (GWL) above a threshold of not less than -40 cm. This is because peatland areas will have a high fire hazard vulnerability if the GWL is below 40 cm, and burning areas only occur when the GWL is below 60 cm [7]. The fire risk can be reduced by up to 30% if rewetting is prioritized in areas that are most prone to burning [8].

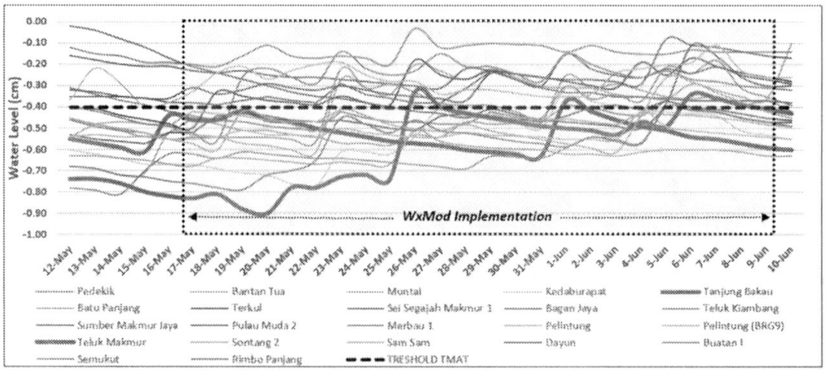

Figure 4. Peatland Groundwater Level from 22 observation stations in Riau Province during the WMT implementation period.

Figure 4 shows the peatland GWL conditions of 22 observation stations in Riau Province from May 12 - June 10, 2023. During the WMT implementation (May 17 - June 9, 2023), as many as 18 stations experienced an increase, and 4 stations experienced a decrease in peatland GWL. The highest increase occurred at Teluk Makmur Station, which experienced a rise of +0.42 m (initial GWL = -0.83 m; final GWL = -0.41 m). Meanwhile, the highest decline occurred at Tanjung Bakau Station, which decreased by -0.13 m (initial GWL = -0.46 m; final GWL = -0.59 m).

The WMT implementation paradigm reform, which has been implemented since 2016, has proven to be very significant in reducing the number of hotspots in Indonesia. Figure 5 shows how the number of hotspots in Sumatra and Kalimantan has decreased significantly since 2016, when WMT was implemented when a massive forest and land fire disaster occurred for fire suppression purposes and starting in 2016, WMT was implemented earlier as an anticipatory effort for peatland rewetting. For comparison, 2015 and 2019 were both El Nino years. However, the number of hotspots in 2019 was reduced to almost half of the number in 2015. The number of hotspots until September 2023 is 5,501, "only" 22% compared to the number in 2019.

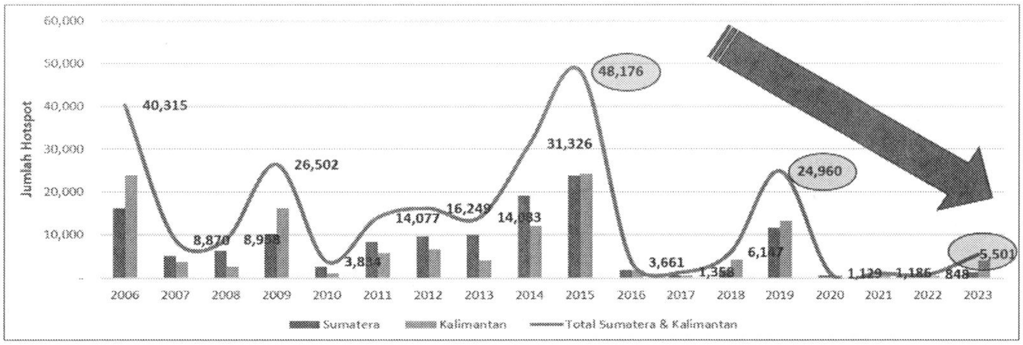

Figure 5. Number of hotspots in Sumatra and Kalimantan during the period 2006 - September 2023.

Figure 6 shows the annual rainfall pattern and the accumulated hotspot events in Sumatra (left) and Kalimantan (right) during El Nino years. During El Nino in 2014, an increase in the number of hotspots in Sumatra had started to occur since the beginning of the year during the dry period in February. This condition was able to be improved in 2019 when the increase in the number of hotspots began in the second dry period in August – September. In 2015, the increase in the number of hotspots occurred in June, while in 2019 the spike started in August and reached a peak in September (left image). A similar condition also occurs for Kalimantan. The increase in the number of hotspots generally occurs since the July period. However, in 2019, the spike started in August and reached its peak in September (right image).

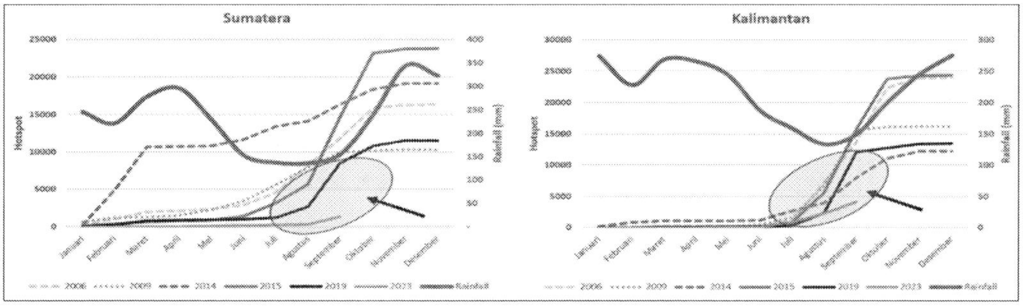

Figure 6. Annual rainfall patterns and accumulated number of hotspots in Sumatra and Kalimantan in several El Nino years.

Based on the description above, it can be seen that there has been a delay in the period of the surge in the number of hotspots on Sumatra and Kalimantan, which has been delayed by around 1-2 months. This indicates that WMT's efforts to rewet peatlands have succeeded in shortening the drought period and have been very significant in reducing the number of forest and land fire incidents.

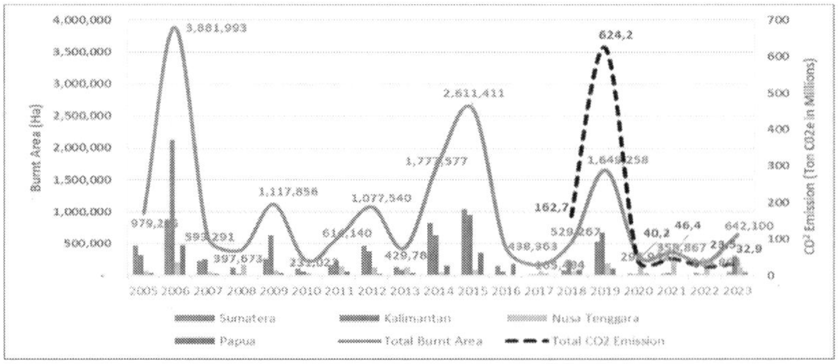

Figure 7. Total burnt area (2005-September 2023) and total carbon emissions (2018-August 2023) in Indonesia.

The decrease in the number of hotspot incidents is directly proportional to the decrease that also occurred in the burnt area and total carbon emissions, as presented in Figure 7. El Nino in 2006 burned 3.8 million hectares of forest and land. While in El Nino the following years the numbers continued to decrease. Each area is 2.6 million hectares (2015) and 1.6 million hectares (2019). Based on data from the KLHK's SIPONGI website, until September the area of forest and land fires in 2023 "only" reached 642.1 thousand hectares or only around 38.9% of the amount in 2019. Likewise for total carbon emissions originating from fire incidents. In El Nino in 2019, total carbon emissions in Indonesia were 624.2 million tons of CO_2e, while in 2023, which is also El Nino, data until August 2023 was "only" 32.9 million tons of CO_2e or only around 5.3% of the amount in 2019. Even though it hasn't been a complete year, at least this number is still very far compared to other El Nino years or several previous years.

4 Conclusion

An interesting fact that can be obtained from the description above is that the total burnt area and also the total carbon emissions in Sumatra and Kalimantan after 2019 were able to be reduced very significantly. During period 2020-2022, the highest burnt area came from Nusa Tenggara, not from Sumatera or Kalimantan. At least this indicates that the TMC efforts, which have been routinely carried out every year since 2009 in Sumatra and Kalimantan and since 2016 have become increasingly well-planned in their implementation, are starting to show significant results. This can be proven by the number of areas where forest and land fires in Sumatra and Kalimantan were able to be suppressed. Precisely the Nusa Tenggara region which has never received the TMC assistance program has even increased the area of forest and land fires and is even higher than Sumatra and Kalimantan. This fact can at least become input and concern for the Government of Indonesia to start considering whether or not the Nusa Tenggara region needs support for WMT implementation as has been routinely carried out in Sumatra and Kalimantan.

References

1. V. Thomas, J. R. G. Albert, and C. Hepburn, "Contributors to the frequency of intense climate disasters in Asia-Pacific countries," *Clim. Change*, vol. **126**, no. 3–4, pp. 381–398, (2014).
2. S. Nurdiati, A. Sopaheluwakan, and P. Septiawan, "Spatial and Temporal Analysis of El Niño Impact on Land and Forest Fire in Kalimantan and Sumatra," *Agromet*, vol. **35**, no. 1, pp. 1–10, (2021).

3. KLHK, "Updated Nationally Determined Contribution Republic of Indonesia," (2021).
4. S. S. Saatchi *et al.*, "Benchmark map of forest carbon stocks in tropical regions across three continents," *Proc. Natl. Acad. Sci. U. S. A.*, vol. **108**, no. 24, pp. 9899–9904, (2011).
5. S. E. Page, F. Siegert, J. O. Rieley, H. D. V. Boehm, A. Jaya, and S. Limin, "The amount of carbon released from peat and forest fires in Indonesia during 1997," *Nature*, vol. **420**, no. 6911, pp. 61–65, (2002).
6. B. Harsoyo and I. Athoillah, "Paradigma Baru Pemanfaatan Teknologi Modifikasi Cuaca dalam Upaya Penanganan Bencana Kebakaran Hutan dan Lahan di Indonesia," *J. Sains dan Teknol. Modify. Cuaca*, vol. **23**, no. 1, pp. 1–9, (2022).
7. M. Taufik, M. Tri, A. Sulaiman, D. Murdiyarso, I. P. Santikayasa, and B. Minasny, "Agricultural and Forest Meteorology An improved drought-fire assessment for managing fire risks in tropical peatlands," *Agric. For. Meteorol.*, vol. **312**, no. May 2021, p. 108738, (2022).
8. D. Murdiyarso, I. Lestari, B. B. Hanggara, M. Saragi-Sasmito, I. Basuki, and M. Taufik, "Managing Water Regimes," pp. 355–369, (2021).

Utilization of biomass wastes: coconut and *Pangium edule* shells as activated carbon for energy storage device material

Diah Susanti[1*], *Haniffudin* Nurdiansah[1], *Azza*h *Dyah* Pramata[1], *Tubagus Noor* Rohmanuddin[1], *Rindang* Fajarin[1], *Yeny Widya* Rakhmawati[1], *George Endri* Kusuma[2], and *Fathan* Bahfie[3]

[1]Materials and Metallurgical Engineering Department, Faculty of Industrial Technology and System Engineering, Institut Teknologi Sepuluh Nopember (ITS), Surabaya, Indonesia
[2]Mechanical Engineering Department, Surabaya State Shipbuilding Polytechnique, Jl. Teknik Kimia, PPNS, Surabaya, Indonesia
[3]Research Unit for Mineral Technology, Indonesian Institute of Sciences, Jl. Ir. Sutami km. 15 Tanjung Bintang, Lampung Selatan 35361, Indonesia

Abstract. In this research, activated carbon (AC) was synthesized from biomass wastes of coconut and *Pangium edule* shells and utilized as a material for electrochemical double-layer capacitors (EDLC), which are eco-friendly energy storage devices. This research was intended to bridge the need for greenhouse gas-free energy storage device and the handling of abundant biomass wastes. These efforts would undoubtedly contribute to mitigating climate change. To begin the research, the coconut and *Pangium edule* shells underwent carbonization at varying temperatures of 600°C and 700°C for 2 hours. Subsequently, they were subjected to chemical activation using KOH and physical activation at varying temperatures of 110°C and 600°C. Some characterization techniques, including SEM, XRD, TGA/DSC, BET, Iodine number, and proximate analysis, were employed to analyze the materials. The capacitive properties of EDLC electrodes were assessed through cyclic voltammetry (CV). After carbonization at 700°C and subsequent physical activation at 600°C, the coconut and *Pangium edule* shells exhibited the highest active surface area of 548.542 m^2g^{-1} and 333.4 m^2g^{-1}, respectively. Notably, the EDLC demonstrated a maximum specific capacitance of 364.5 Fg^{-1} at 2 mVs^{-1} scan rate. These findings indicate the viability of utilizing AC from biomass waste as a promising material for EDLC applications.

1 Introduction

Coconut holds a prominent position in Indonesia's agricultural sector. In 2021, the country achieved a substantial output of coconut products, totaling 2.85 million tons from both public and private plantations. This figure saw a further uptick in 2022, reaching 2.86 million tons. Notably, there has been a consistent average annual growth rate of 0.14% in production up

* Corresponding author: santiche@mat-eng.its.ac.id

to the year 2026 [1]. As coconut production increases, it is inevitable that there will also be a rise in biomass waste, including coconut shells and fruit husks

Furthermore, *Pangium edule*, known as 'kluwek' in the Indonesian language, is a fruit product commonly used as an ingredient in traditional dishes. Similar to coconut shells, the shells of *Pangium edule* also generate biomass waste. The coconut ad *Pangium edule* shells are organic materials consisting of various components, including cellulose, hemicellulose, and lignin [2]. Lignin, a high-molecular-weight polymer, acts as a binder, lending strength to plant cells [3].

Activated carbon, a type of charcoal, is created by subjecting it to activation processes involving CO_2 gas, water vapor, or chemicals. These processes create open pores, greatly improving its capacity to absorb substances such as color and odor. Activated carbon typically contains 5 to 15 percent water, 2 to 3 percent ash, with the remaining composition being carbon. It is commonly available in granular or powdered forms [4].

The adsorption capacity of activated carbon is heavily dependent on its pore structure, consisting of micropores (< 20 Å), mesopores (20 - 500 Å), and macropores (> 500 Å) [5]. Different raw materials and processing methods result in diverse pore distributions. Activated carbon rich in micropores excels in adsorbing small molecules, such as gases with low contamination levels. In contrast, activated carbon with a predominance of macropores is well-suited for adsorbing larger molecules and is especially effective in decolorization processes [6].

In general, the production of activated carbon comprises three primary stages:

a. Dehydration: This initial step involves heating raw materials to approximately 170°C to remove moisture. b. Carbonization: Organic matter is transformed into carbon during this phase, typically occurring at temperatures ranging from 400 °C to 900 °C. The resulting carbon is then cooled, washed, and processed to remove activating chemicals. Temperatures exceeding 170°C can lead to the formation of CO, CO_2, and acetic acid. c. Activation: During this stage, pores in the carbon structure expand as tar decomposes. Activation can be achieved using either steam or CO_2 as activators. The activation process can be achieved through either chemical or physical (thermal) treatments [7].

The electric double-layer capacitor (EDLC) functions by establishing electrical double layers between activated carbon and the electrolyte, effectively acting as a dielectric. During charging and discharging cycles, ion absorption and desorption processes occur in both activated carbon electrode layers. When voltage is applied to the electrodes, ions migrate to the electrode surfaces during charging and move away during discharging. [8].

The research on activated carbon from biomass wastes have been widely reported with various results. However, the usage of *Pangium edule* was rarely reported. In addition, the research on the activated carbon-based EDLC have been reported by many papers. For example, Deng et.al (2022) reported EDLC from coconut shell-based activated carbon which resulted in the highest active surface area of 2410 m^2g^{-1} and the highest specific capacitance of 395 Fg^{-1} [9].

In this research, we synthesized activated carbon through both physical and chemical activation processes, employing coconut and *Pangium edule* shells as raw materials. Subsequently, we utilized this activated carbon to create EDLC electrodes and quantified their specific capacitances. These endeavors aimed to establish a connection between harnessing biomass from coconut and *Pangium edule* shells and fulfilling the requirements of environmentally sustainable energy storage devices.

2 Experimental Details

The coconut and *pangium edule* shells were initially washed and crushed into flakes sized between 0.1-3 mm. Carbonization was conducted in a horizontal furnace at temperatures of 700°C and 800°C, with a 2-hour holding time. The samples were placed in alumina crucible during this process, and nitrogen gas was introduced to ensure uniform pore size in the activated carbon. Following carbonization, the carbon samples were pulverized using a blender until they reached a consistency that could pass through a 120-mesh sieve.

In the chemical activation process, charcoal and KOH were combined in a glass beaker, with distilled water added in a 1:1:4 ratio of water, carbon, and KOH. This mixture was then heated and stirred on a magnetic stirrer hot plate at 80°C for 4 hours, with a stirring speed of 200 rpm. Afterward, sedimentation and washing steps were carried out until the pH approached neutrality.

For physical activation, the carbon deposits resulting from the chemical activation process underwent hydrothermal heating at varying temperatures between 110°C and 600°C, each held for 4 hours. This involved placing the still-moist activated carbon mixture into a crucible, which was then inserted into an autoclave and heated in a furnace.

The samples were characterized using X-Ray Diffractometer (XRD) Panalytical X'pert, Scanning Electron Microscope (SEM) FEI Tecnai S50, Brunauer-Emmet-Teller (BET) Quantachrome, Iodine number, and proximate analysis (ASTM D 1762-84). The activated carbon powder was impregnated into Nickel foam as electrodes for EDLC. The capacitance was measured using Cyclic Voltammetry (Solartron Electrochemical Measurement) at 2, 5, 10, 20 and 50 mVs^{-1} scan rates using 1 M of Na_2SO_4 electrolyte at potential windows of -0.7 to -0.2 V vs. Ag/AgCl saturated KCl.

3 Result and Discussion

Table 1 presents the proximate analysis results of raw materials from coconut and *Pangium edule* shells, following ASTM D 1762-84 standards. Notably, both samples exhibited similar fixed carbon contents exceeding 74%, signifying their high carbon content.

Table 1. Proximate Analysis of Raw Material (ASTM D 1762-84).

Shell	Fixed Carbon (%)	Volatile Matters (%)	Ash (%)
Pangium edule	74.59	23.8	1.61
Coconut	74.62	22.8	2.58

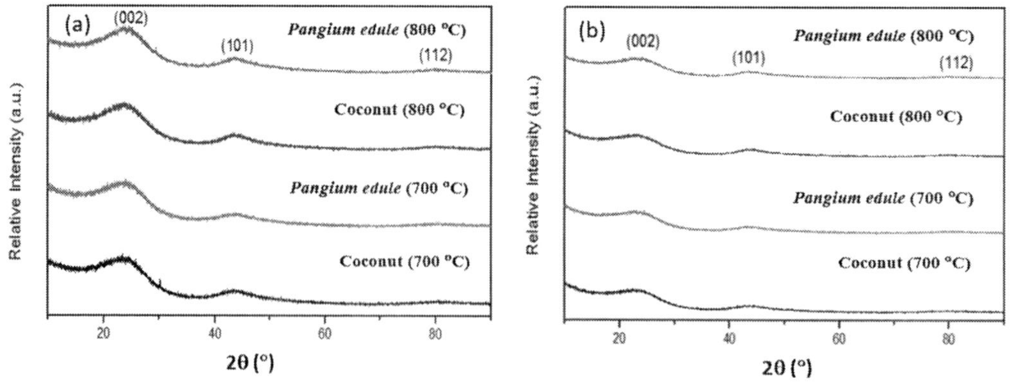

Fig. 1. The XRD patterns of the activated carbon resulted from (a) activation temperature of 160 °C and (b) activation temperature of 600 °C for coconut and *Pangium edule* shells which carbonized at 700 °C and 800 °C, respectively.

Fig. 1 illustrates a noticeable distinction between the activated carbon samples produced at different activation temperatures, namely 160°C (a) and 600°C (b). It is evident that the samples activated at 600°C exhibit a more amorphous structure, as indicated by their less intense peaks. These XRD peaks align with the JCPDF 75-1621 pattern for carbon, which is characterized by a hexagonal structure. Specifically, the three peaks observed at approximately 2θ = 26°, 45°, and 81° correspond to the (002), (101), and (112) crystallographic planes, respectively.

Fig. 2 displays SEM images of activated carbon obtained from coconut and *Pangium edule* shells. These materials were activated at two different temperatures, namely 160 °C and 600 °C, and subsequently carbonized at 600 °C and 800 °C. The corresponding pore sizes can be found in Table 2. Notably, carbonization at 800°C yielded a combination of mesopore and macropore structures. In contrast, carbonization at 700°C resulted in a macropore structure.

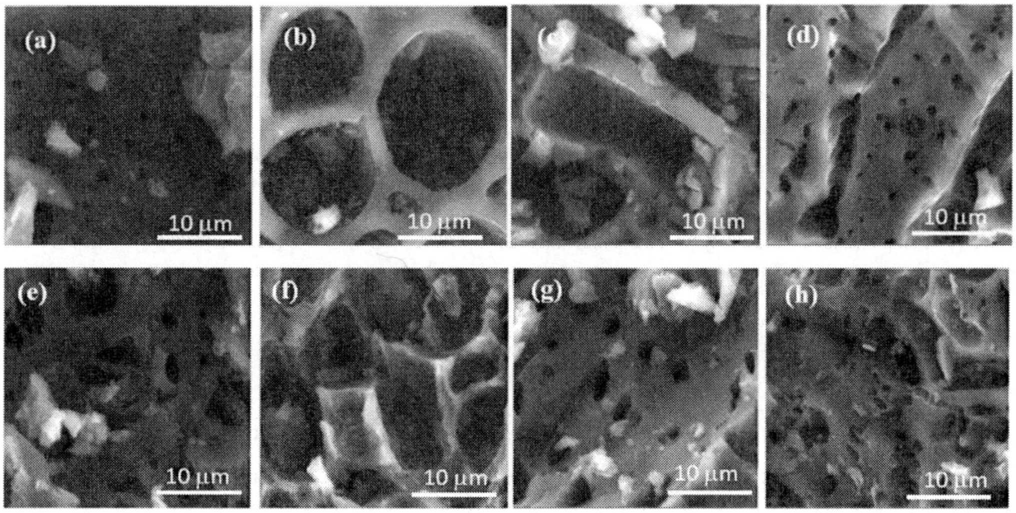

Fig. 2. SEM Results for Coconut Shell Activated Carbon (a, b, c, d) and *Pangium edule* Shell (e, f, g, h). SEM test results for activated carbon derived from both coconut shell and *Pangium edule* shell under various carbonization and physical activation conditions:
Carbonization Temperature 700°C, Physical Activation Temperature 110°C (a, e) and 600°C (b, f)
Carbonization Temperature 800°C, Physical Activation Temperature 110°C (c, g) and 600°C (d, h).

Table 2. Pore diameter, Iodine number, BET total volume, and BET active surface area.

No	Shell	Carbonization Temp. (^{0}C)	Physical Activation Temp. (^{0}C)	Pore Diameter (μm)	Iodine No (mg g^{-1})	BET Total Volume (cm^3g^{-1})	BET Active Area (m^2/g)
1	Coconut	700	110 (a)	3.02 - 12.08	826.4	95.5	275.9
2			600 (b)	4.56 - 17.32	1122.9	189	548.5
3		800	110 (c)	0.204 - 0.70	823.8	79.6	230.03
4			600 (d)	0.39 - 1.19	1036.2	104.3	303.65
5	*Pangium edule*	700	110 (e)	1.23 - 6.75	778.4	90.4	230.57
6			600 (f)	5.72 - 10.76	968.8	115.3	333.4
7		800	110 (g)	0.64 – 4.0	756.4	56.5	143.8
8			600 (h)	0.4 - 2.02	957.2	106.4	307.01

Table 2 provides data on the iodine number, BET total volume, and active surface area for all activated carbon samples. Generally, samples activated at 600°C exhibited higher iodine numbers, BET total volumes, and BET active surface areas compared to those activated at 110°C. Furthermore, activation at 700°C led to higher iodine numbers, BET total volumes, and BET active surface areas compared to activation at 800°C, due to consistently larger pore diameters. The highest BET active surface areas recorded were 548.542 m^2/gram and 333.4 m^2/gram, observed in the case of coconut shell and *Pangium edule* shell, respectively, carbonized at 700°C and activated at 600°C.

According to SNI-06-3730-1995, activated carbon must adhere to specific parameters. These include a maximum water content of 15%, a maximum ash content of 10%, a maximum weight loss at 950°C of 25%, and an Iodine number of 750 mg/gram. Additionally, activated carbon is defined by an active surface area ranging from 300 to 3000 m^2/gram. Based on the data presented in Table 1 and Table 2, it can be concluded that the activated carbon materials were obtained through a physical activation process at a temperature of 600°C (b, d, f, h in Fig. 2 and Table 2).

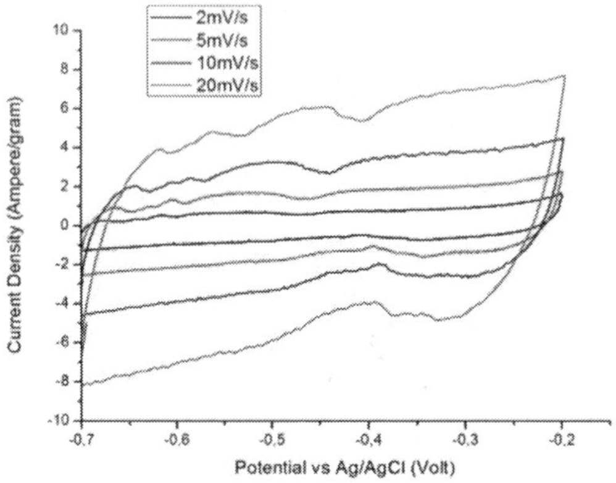

Fig. 3. Cyclic voltammogram of activated carbon – based EDLC from coconut shell.

Table 3. The specific capacitances of EDLC from coconut shell – based activated carbon.

Scan rate (V/s)	**Specific Capacitance (F/gram)**
	Activated Carbon from Coconut Shell
0.002	364.50
0.005	329.30
0.01	302.22
0.02	271.56
0.05	209.22
0.1	159.59

Figure 3 illustrates the cyclic voltammogram (CV) of the EDLC constructed with activated carbon derived from coconut shells. The presence of a rectangular, mirror-like CV curve is a distinctive hallmark of EDLC behavior. In Table 3, we present specific capacitance values at various scan rates. Notably, an increase in scan rate correlates with a reduction in specific capacitance. Slower scan rates afford more time for charges to be stored and released from the less accessible electrode material sites, resulting in higher specific capacitance values. At a scan rate of 2 mV/s, the highest specific capacitance was achieved, measuring 364.5 F/g. This figure is comparable to the result reported by Deng et al. (2022), which stood at 395 F/g, despite their significantly larger active surface area of 2410 m^2/g. [9].

4 Conclusion

Activated carbon material has been successfully synthesized from both coconut shell and *Pangium edule* shell, which underwent carbonization at varying temperatures of 700°C and 800°C. Subsequently, they were chemically activated using KOH and physically activated at 600°C. Among these processes, the highest values for key parameters defining activated carbon were achieved by the coconut shell material carbonized at 700°C and physically activated at 600°C, resulting in an active surface area of 458.542 m^2/gram. In terms of its electrochemical performance, the highest specific capacitance of the electric double-layer capacitor (EDLC) electrode constructed from coconut shell-activated carbon reached 364.5 F/g at a scan rate of 2 mV/s. While this value may appear moderate, it remains suitable for specific applications.

These authors acknowledge express the sincere gratitude for the financial support provided by the Department of Materials and Metallurgical Engineering (DMME) at Institut Teknologi Sepuluh Nopember (ITS) through the Department Research Fund Batch 2, under Contract No: 2417/PKS/ITS/2023. Authors also appreciate the invaluable research facilities made available by DMME for this study.

References

1. R. Zikria, Jakarta, 2022. Accessed: Sep. 13, 2023. [Online]. Available: https://satudata.pertanian.go.id/assets/docs/publikasi/Outlook_Kelapa_2022.pdf.
2. Q. Wang, J. Sarkar, Int. J. Energy Prod. Manag. **3 (1),** 34–43 (2018) doi: 10.2495/EQ-V3-N1-34-43.
3. O. E. Babatunde, J. M. Yatim, M. Y. Ishak, R. Masoud, R. Meisam, J. Teknol. **77(12),** 23–30 (2015) doi: 10.11113/jt.v77.6304.
4. K. Sartova, E. Omurzak, G. Kambarova, I. Dzhumaev, B. Borkoev, Z. Abdullaeva, Diam. Relat. Mater. **91,** 90–97 (2019) doi: 10.1016/J.DIAMOND.2018.11.011.
5. T. Zhang, Y. Yang, X. Li, N. Wang, Z. Zhou, Environ. Sci. Pollut. Res. **26**, 9161–9173 (2019) doi: 10.1007/s11356-019-04392-y.
6. M. A. Tadda, A. Ahsan, A. Shitu, M. ElSergany, T. Arunkumar, B. Jose, M. A. Razzaque, N. N. Nik Daud, J. Adv. Civ. Eng. Pract. Res. **2(1)**, 7–13 (2015).
7. S. J. Zhang, H. Q. Yu, H. M. Feng, Carbon N. Y. **44(10)**, 2059–2068 (2006) doi: 10.1016/J.CARBON.2005.12.047.
8. B. E. Conway, *Electrochemical Supercapacitors: Scientific Fundamentals and Technological Applications*, Springer (1999).
9. M. Deng, J. Wang, Q. Zhang, Mater. Lett. **306**, 130934 (2022) doi: 10.1016/j.matlet.2021.130934.

Does carbon emission disclosure and environmental performance increase firm value? Evidence from highly emitted industry in Indonesia

Ari Kuncara Widagdo[1], *Siti Rochmah* Ika[2]*, *Maria Febiana* Neni[2], *Handoko Arwi* Hasthoro[2], and Widiawati[2]

[1]Department of Accounting, Sebelas Maret University, Surakarta, Indonesia
[2]Department of Accounting, Janabadra University, Yogyakarta, Indonesia

Abstract. Whether investors take into account the company's information related to carbon emission mitigation and the company's environmental ranking in their investment decisions is an interesting research avenue. The objective of this study is twofold. First, it tests whether the level of carbon emissions disclosure differs by industry sector. Second, it examines whether carbon emission disclosures as reported by companies and their environmental performance affect the value of the firm. This study utilizes 102 companies in 2022 that are included in highly polluting industries as a sample, which is divided into 4 different sectors: consumer goods, energy, basic industry, and miscellaneous industry. The results of the analysis of variance (ANOVA) test reveal that the extent of carbon emissions reporting is different across industry sectors. Meanwhile, the multivariate regression results reveal that carbon emission disclosures positively influence the value of the firm. Environmental performance, however, does not impact firm value. This study informs the company's management that promoting higher levels of carbon emission mitigation and reporting would boost the company's reputation, which would in turn increase its value.

1 Introduction

The issue of climate change and carbon emissions by corporations has received widespread attention in recent years. Over the past few years, a sizable number of businesses throughout the world have implemented programs to track, manage, and ultimately cut their carbon footprints [1]. However, companies may incur additional expenses as a result of their efforts to report and reduce their carbon footprint. An increased likelihood of costly environmental liabilities is associated with higher levels of carbon emissions [2]. A company's value will be affected by its emission level to the degree that the market adjusts stock prices to account for the costs associated with carbon emissions both presently and in the future [3]. In this

* Corresponding author: ika@janabadra.ac.id

research, we investigate if a company's stock price changes as a result of the disclosure of its carbon emission levels. This research also investigates whether environmental performance achieved by a company affects its value as reflected in its stock price. Besides, this study examines first whether the level of carbon emission disclosures differs by industry sector.

Numerous studies have explored the company-value effect of greenhouse gas emissions and greenhouse gas emission disclosures in single-country settings and cross-country settings. Using a cross-country dataset, Hassan and Romilly [4] and Choi and Luo [3] find that an increase in carbon emissions would decrease firm value. The repercussions are more pronounced for companies operating in nations that have implemented a domestic carbon emission trading system and have strict environmental rules in place [3]. Similarly, a study in the Shanghai and Senzhen Stock Exchanges [5], in the US [1], and in Japan [6] confirms that a reduction in carbon emission volume would enhance firm value.

Studies of the relationship between the amount of carbon emission reporting and the value of a company show that in China, the amount of carbon emission reporting tend to enhance firm value [2], and this relationship is stronger in non-state-owned companies [7]. In Indonesia, some studies find a favorable association between greenhouse gas emission disclosures and firm value using different industry samples. These studies cover the observation of manufacturing companies [8,9], firms receiving awards in the Indonesian Sustainability Reporting Award [10], companies in Sharia Stock Index [11] and non-financial companies [12,13]. A study that took all companies listed on the country's stock exchange as a sample, however, exhibits contradictory results, indicating carbon emission disclosure has an adverse effect on firm value [14].

The present study will test whether carbon emission reporting affects firm value in highly polluted industries in Indonesia. The industry includes four different sectors: consumer goods, energy, basic industry, and miscellaneous industry. The observed sample is different from previous studies done in the country. This study also examines whether different industry sectors produce different levels of carbon emission disclosures. Using the most recent data, i.e., the year 2022, to avoid the COVID-19 effect, this study contributes to the extant literature by providing a comparison test of greenhouse gas emission disclosure among the four industry samples in the post-COVID-19 period.

Since there will always be an inherent asymmetry of information between companies and investors, signaling theory is often discussed in relation to the stock market [15]. However, management signals to investors may help reduce the impact of information asymmetry. Some examples of signals are carbon disclosure and carbon performance reporting and financial performance, both of which inform investors about management's environmental vision for the company's future [16]. Hence, it is expected that investors will value more companies that obtain ISO 14001 certification and a high PROPER (Company Performance Rating Assessment Program in Environmental Management rating by Ministry of Environment). Previous studies confirm the value and relevance of greenhouse gas emission disclosures [2,7,10–13] and environmental performance [17,18]. Thus, this study posits that the extent of carbon emission disclosures and environmental performance are likely to increase firm value.

2 Material and methods

2.1 Sampling and Data

The research population consists of companies listed on the Indonesia Stock Exchange (IDX) in highly emitting industries, namely in the energy, manufacturing, and transportation sectors. The classification of the industry follows previous studies [19–21]. As of December 31, 2022, there were 112 companies listed in this sector.

https://doi.org/10.1051/e3sconf/202346704002

The sample selection was based on the following criteria: (1) The company discloses carbon emissions reporting and policies to reduce carbon emissions in both the annual report and sustainability report; (2) the company follows PROPER or ISO 14001; and (3) complete research data is available. The data is taken from the company's annual report and sustainability report on the IDX website and company website. At the time this research began (June 2033), 10 company annual reports were not yet available, including seven in the transportation sector. Therefore, the research sample did not include the transportation sector. The final research sample consisted of 102 companies.

2.2 Variables and data analysis

The dependent variable in the research is firm value (FV). FV is identified as the monetary amount that potential purchasers are willing to offer in the event of the company's sale. The valuation of a firm reflects the degree of investor confidence and the impression of the management's ability to effectively oversee the company [13]. This study employs Tobin's Q formula from Chung and Pruitt [22] to measure firm value, as presented in Equitation (1).

$$Tobin's\ Q = \frac{MVE+DEBT}{TA} \tag{1}$$

Where,
MVE = Multiplying the stock's closing price on the last day of the fiscal year by the number of shares outstanding
DEBT = Total Debt
TA = Total Assets

The independent variables in the study are carbon emissions disclosures (CAED) and environmental performance (EP). In line with previous researches [10,11,13,20], to measure the extent to which information about carbon emissions is disclosed, this study adapts an indicator developed by Choi et al. [19]. The index's 18 components address the following areas of interest related to CAED: climate change (threats and opportunities), carbon dioxide emissions (accounting), use of energy (accounting), greenhouse gas reduction (costs), and carbon emission accountability. The scoring of the index is unweighted; hence, a statement about the company's carbon emission reporting receives a score of one if it directly pertains to an item in the index and a score of zero otherwise. A company's CAED level is calculated by dividing its index score by 18, the maximum achievable index score.

EP was evaluated using the ISO 14001 and/or PROPER environmental performance ratings for businesses published by the Ministry of the Environment. Proper ratings are color-coded, from "black" (the least) to "gold" (the utmost). Black and red are indicators of subpar quality. The color blue indicates compliance with applicable environmental management laws and regulations by the company. Green and gold stars are above-average rating indications. In contrast to "green," which indicates mere compliance, "gold" indicates environmental excellence in both manufacturing and service, as well as an ethical and socially responsible approach to business. There are no "black" or "red" ratings given to any of our samples in this study. Therefore, we use blue as our standard indicator color. In this study, a nominal variable is used to assess environmental performance. A "blue" rating is equivalent to a zero score. A score of one is given to the corporation depicted in green. A corporation earns a 2 if it has ISO 14001 certification and a 3 if it earns the highest possible gold rating. Some businesses have both ISO 14001 and a "green" grade. Whichever is higher will be used for scoring.

Profitability (Profit) and firm size (SIZE) are the control variables of the study. Profit is the ratio of net income to total assets, while SIZE is the natural logarithm of a company's total assets. To test whether different industries produce different levels of CAED, this study utilizes the Anova test. Meanwhile, to analyze whether CAED, EP, Profit, and SIZE affect FV, the following regression model is formulated, as depicted in Equitation (2).

$$FV = \alpha + \beta 1CAED + \beta 2EP + \beta 3Profit + \beta 4SIZE + \varepsilon \quad (2)$$

Where,
FV is the firm value, CAED is the corporation's carbon emission disclosure, EP is the environmental performance, profit is the profitability of a company, and SIZE is the firm's size. The measurement of each variable has already been explained in the above paragraphs.

3 Results and discussion

3.1 Descriptive statistics of CAED and the anova test

Table 1 exhibits the descriptive statistics of the CAED score that are categorized into four different industries. Table 1 also displays the results of the comparison test among industries using the Anova test.

Table 1. Descriptive statistics of CAED per industry sector and results of Anova test.

Industry	N	Mean	Min	Max	SD	F	Sig
Consumer goods	26	64.5	38.8	94.4	17.31	3.573	0.009
Energy	24	75.4	44.4	100	17.63		
Basic industry	30	65.5	38.8	94.4	15.53		
Miscellaneous industry	13	56.8	33.3	94.4	18.86		
All sample	102	67.4	33.3	100	17.31		

Table 2. Descriptive statistics of the research variables.

Variables	Mean	Min	Max	SD
FV	1.555	0.358	8.996	1.306
CAED	67.4	33.3	100	17.31
SIZE	29.90	26.36	33.54	1.573
Profit	0.044	-0.447	0.520	0.135
EP	1.29	0	3	1.001
Frequency of EP	0=33	32.4%		
	1=13	12.7%		
	2=49	48.0%		
	3=7	6.9%		
N	102	100.0		

As depicted in the table, the average score of CAED for highly emitting industries is 67.4%. The number is higher than those in nonfinancial companies [12,13], 17.8%, and in the Sharia Stock Index [11]. The score, however, is lower than those in the same industry in previous studies [20,23], at 72%.

The highest average score of CAED is in the energy industry (75.4%), while the lowest is in miscellaneous industries (56.8%). The CAED mean score in the consumer goods industry is 64.5%, ranking third after the basic industry (65.5%). For the consumer goods industry, the figure is higher than that of the same industry as reported in a previous study (47.6%) [24].The table shows the results of the Anova test: the F value is 3.573 and the sig value is 0.009, far above 0.05, indicating the extent of carbon emissions reporting is different across industry sectors.

Table 2 shows the descriptive statistics of the variables used for the regression analysis. The table displays the EP category range from 0 to 3 based on the color of the PROPER rating and ISO certification. The blue PROPER rating (32.4%) comes in second place to the ISO certification category (48%), which has the highest percentage of EP. The lowest percentage is in the gold PROPER rating (6.9%).

2.2 Results of the linier regression analysis

Table 3 presents the results of the multiple regression test. As shown in the table, CAED is positively associated with firm value at the 5% level. It indicates that CAED tends to enhance a firm's reputation in the eyes of investors, which in turn increases its value. The results are in line with signaling theory, where CAED is good news from the management to the shareholder, and shareholders react to the news with an increase in share price. The results are consistent with previous studies in China [2,7] and Indonesia [8], [10–13]. EP does not influence firm value. It implies that investors ignore the environmental management rating and certification achieved by the company. The results are consistent with previous studies in Indonesia, which used ISO certification only as a proxy for EP [8,10].

Table 3. Results of multiple regression analysis.

	B	t	Sig	TOL	FIV
Constant	6.473	2.554	0.012		
CAED	0.110	2.392	0.019	0.750	1.334
EP	0.006	0.044	0.965	0.874	1.145
Size	-0.213	-2.338	0.021	0.755	1.324
Profit	2.409	2.502	0.014	0.913	1.096
F value & sig	3.177		0.017		
Adjusted R^2		0.116			
Kolmogorov Smirnov sig		0.082			

For the control variables, profitability positively affects firm value, while firm size negatively affects firm value. The negative relationship between firm size and firm value indicates investors perceive that firms with high assets possess high risk, which in turn reduces the value of the firm. The findings are consistent with a previous study that took companies receiving sustainability reporting awards as a sample [10].

The adjusted R^2 value of 0.116 indicates that CAED, EP, Profit, and SIZE may predict only 11.6% of the variance that affects firm value, while the other 88.4% are affected by other variables that were not captured in the research model. The F sig suggests that the regression model is fit to the research data.

4 Conclusion

This research aims to test whether the level of carbon emissions disclosure differs by industry sector and to examine whether carbon emission disclosures as reported by companies in highly emitting industries and their environmental performance affect the value of the firm. Data analysis reveals that different industries produce different levels of greenhouse gas emission reporting. Carbon emission disclosures and profitability are good news for investors because they increase firm value. Meanwhile, environmental performance does not affect firm value. Since the adjusted R^2 is relatively low (11.6%), future studies may add variables to examine factors that impact on firm value. This article may provide insight to the management of a company that reporting policies and activities related to carbon emissions may increase the firm's reputation, which in turn increases its value.

References

1. E. M. Matsumura, R. Prakash, and S. C. Vera-Muñoz, Account. Rev. **89**, 695 (2014)
2. H. Yan, X. Li, Y. Huang, and Y. Li, Financ. Res. Lett. **37**, 101680 (2020)
3. B. Choi and L. Luo, Br. Account. Rev. **53**, 100909 (2021)
4. O. A. . G. . Hassan and P. Romilly, Bus. Strateg. Environ. **27**, 0 (2018)
5. Z. Y. Sun, S. N. Wang, and D. Li, Environ. Sci. Pollut. Res. **29**, 60189 (2022)
6. C. Saka and T. Oshika, Sustain. Accounting, Manag. Policy J. **5**, 22 (2014)
7. Y. Yang, J. Wen, and Y. Li, Int. J. Environ. Res. Public Health **17**, (2020)
8. A. Noor and Y. L. Ginting, Int. J. Contemp. Account. **4**, 151 (2022)
9. S. Rachmawati, Int. J. Contemp. Account. **3**, 133 (2021)
10. M. Hardiyansah, A. T. Agustini, and I. Purnamawati, J. Asian Financ. Econ. Bus. **8**, 123 (2021)
11. M. Hardiyansah and A. T. Agustini, J. Ekon. Dan Bisnis Islam (Journal Islam. Econ. Business) **7**, 51 (2021)
12. M. Monica, F. E. Daromes, and S. Ng, J. Ilm. Akunt. Dan Bisnis **16**, 343 (2021)
13. S. Ng, R. Jao, F. E. Daromes, and Monica, J. Akunt. **26**, 426 (2022)
14. G. I. Muhammad and Y. A. Aryani, J. Din. Akunt. Dan Bisnis **8**, 1 (2021)
15. R. D. Morris, Account. Bus. Res. **18**, 47 (1987)
16. R. R. Datt, L. Luo, and Q. Tang, Account. Res. J. **32**, 417 (2019)
17. S. Sarumpaet, M. L. Nelwan, and D. N. Dewi, Soc. Responsib. J. **13**, 817 (2017)
18. K. Khanifah, U. Udin, N. Hadi, and F. Alfiana, Int. J. Energy Econ. Policy **10**, 96 (2020)
19. B. B. Choi, D. Lee, and J. Psaros, Pacific Account. Rev. **25**, 58 (2013)
20. A. K. Widagdo, B. A. Rahanyamtel, and S. R. Ika, IOP Conf. Ser. Earth Environ. Sci. **1016**, (2022)
21. K. H. Rupley, D. Brown, and R. S. Marshall, J. Account. Public Policy **31**, 610 (2012)
22. K. H. Chung and S. W. Pruitt, Financ. Manag. **23**, 70 (1994)
23. S. R. Ika, Yuliani, A. Okfitasari, and A. K. Widagdo, IOP Conf. Ser. Earth Environ. Sci. **1016**, 1 (2022)
24. W. Anggita, A. A. Nugroho, and Suhaidar, J. Akunt. **26**, 464 (2022)

The performance of room air conditioner with Cooling Seasonal Performance Factor (CSPF) in Jakarta and Bandung, Indonesia

Andriyanto Setyawan[1*], *Luga Martin* Simbolon[1], *Hafid* Najmudin[2], *Annisa Ayu* Julyanti[1], *Annisa S.* Kurniasetiawati[1], and Susilawati[1]

[1]Department of Refrigeration and Air Conditioning Engineering, Politeknik Negeri Bandung, Jl. Gegerkalong Hilir, Ciwaruga, Bandung, Indonesia
[2]Balai Besar Bahan dan Barang Teknik, Bandung, Jl. Sangkuriang 14 Bandung, Indonesia

Abstract. Cooling Seasonal Performance Factor (CSPF) is one of methods for measuring the performance of a room air conditioner to assess its energy efficiency and, therefore, its effect on climate change and environmental impact. A set of bin temperatures from a certain region was used in this method to compare the performance of an air conditioner. Due to the variety of hourly local ambient temperature distribution, the performance of an air conditioner could be different in various cities in Indonesia. In this paper, a room air conditioner was tested on the basis of ISO 5151 and analyzed by using ISO 16358-1. The CSPF calculation is then modified based on the ambient local temperature distribution for Jakarta and Bandung. The CSPF for Jakarta City is 1.45% lower and the CSPF for Bandung is 1.49% higher than that of ISO 16358-1-based calculation. It means that for more realistic performance of an air conditioner, the modification of measurement standard could be applied based of local temperature distribution.

1 Introduction

Cooling seasonal performance factor (CSPF) or often denoted as ISO 16358-1 is a part of ISO 16358 for measuring the energy efficiency of an air conditioner (AC) [1]. Another standard for measuring the performance of a room air conditioner is ISO 5151 [2]. Instead of single outdoor temperature (35°C or 46°C) used in ISO 5151, ISO 16358-1 evaluates the performance of an AC in an ambient temperature range of 20°C to 35°C. This gives the more realistic method in calculating the performance of an AC.

The use of ISO 16358-1 has been encouraged to be implemented in Southeast Asia [3] and even worldwide [4,5]. In Indonesia, the CSPF has been applied to study the performance of fixed-speed and variable-speed AC [6]. The seasonal performance factor of 4 cities in Indonesia were compared and it was reported that AC with inverter get more benefits with the application of CSPF. A study to evaluate the performance of inverter air conditioner using CSPF has been conducted in 10 cities in Vietnam [7]. A range of CSPF from 4.6 to 7.4 has

* Corresponding author: andriyanto@polban.ac.id

been reported in this study. This method has also been combined to a cloud-based AI (Artificial Intelligence) to improve the performance of air conditioners [8].

In this paper, the performance of AC calculated by ISO 5151 and ISO 16358-1 will be compared. In addition, the results will also be compared with the modified CSPF for Jakarta and Bandung that represent cities with high and low daily average temperatures. The data of temperature bin distribution is taken from previous study concerning the use of CSPF in four cities of Indonesia [6].

2 Methodology

The test was conducted on a 9000 Btu/hr AC in a test facility that contains two compartments (indoor and outdoor) with different climate conditions, as can be seen in Fig. 1. ISO 5151 was used as a standard for testing the unit. A dry-bulb temperature of 27°C and a wet-bulb temperature of 19°C were maintained in the indoor compartment. Whereas 35°C dry-bulb and 24°C wet-bulb temperature was maintained in the outdoor compartment. Both compartments were equipped with a heater, a cooling coil, and a humidifier to maintain the desired conditions.

During the tests, the indoor unit of the air conditioner was placed in the indoor compartment while the outdoor unit was in the outdoor compartment. This method has been applied in various studies to test the performance of an AC [9–13].

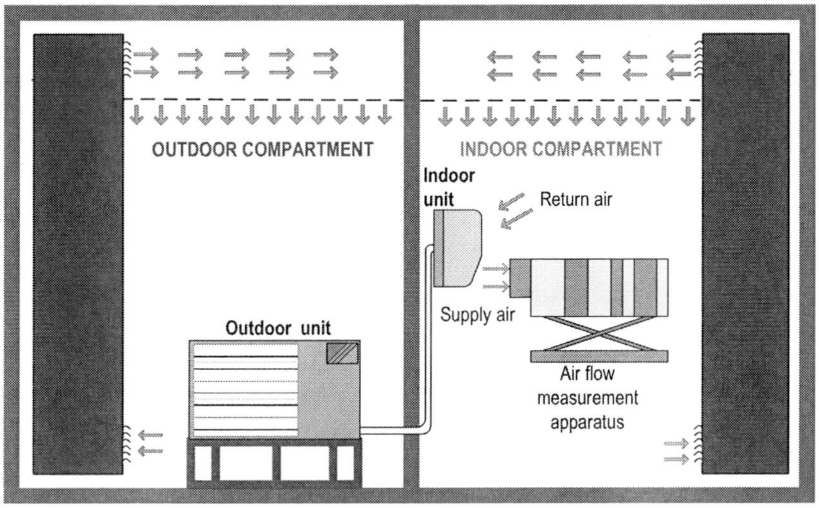

Fig. 1. Setup of air conditioner test.

Thirty-two parameters were measured and calculated in the test and the most desired values are: power consumption, cooling capacity, and coefficient of performance. Once these parameters are obtained, the CSPF of the air conditioner can be calculated by using Eq. (1).

$$F_{CSP} = \frac{L_{CST}}{C_{CSE}} \quad (1)$$

F_{CSP} represents the cooling seasonal performance factor for hot climate, L_{CST} represents the cooling seasonal total load, and C_{CSE} denotes the cooling seasonal energy consumption. The cooling seasonal total load can be expressed by Eq. (2).

$$L_{CST} = \sum_{j=1}^{m} L_c(t_j) \times n_j + \sum_{j=m+1}^{n} \phi_{ful}(t_j) \times n_j \quad (2)$$

L_c represents the cooling load for each bin temperature, t_j denotes ambient temperature for each bin temperature, and n_j is bin hour. L_c could be calculated using Eq. (3).

$$L_c(t_j) = \phi_{ful}(t_{100}) \times \frac{t_j - t_0}{t_{100} - t_0} \quad (3)$$

This equation is valid if $L_c(t_j) \leq \phi_{ful}(t_j)$ $(j = 1 \; to \; m)$.

To calculate the cooling capacity as a function of outdoor temperature, Eq. (4) can be used.

$$L_c(t_j) = \phi_{ful}(35) - \frac{\phi_{ful}(35) - \phi_{ful}(29)}{35 - 29} \times (35 - t_j) \quad (4)$$

This equation is valid if $L_c(t_j) > \phi_{ful}(t_j)$ $(j = m + 1 \; to \; n)$.

The cooling seasonal energy consumption (C_{CSE}) in Eq. (1) can be calculated using Eq. (5).

$$C_{CSE} = \sum_{j=1}^{n} X(t_j) \times P_{ful}(t_j) \times \frac{n_j}{F_{PL}(t_j)} \quad (5)$$

In this equation, $X(t_j)$ is the operation factor, P_{ful} is the power input characteristics, and F_{PL} is the part load factor. The operation factor can be expressed as the ratio between L_c and ϕ. The value of $X(t_j)$ is equal to 1 if $L_c > \phi$.

The power input characteristics is defined using Eq. (6).

$$P_{ful}(t_j) = P_{ful}(35) - \frac{P_{ful}(35) - P_{ful}(29)}{35 - 29} \times (35 - t_j) \quad (6)$$

Using degradation coefficient (C_D) with a default value of 0.25, the part load factor can be calculated by Eq. (7).

$$F_{PL}(t_j) = 1 - C_D(1 - X(t_j) \quad (7)$$

To apply the CSPF for assessing the performance of the air conditioner, the following definitions are used:

- The cooling load is defined as zero when the load is zero and defined as $\phi_{ful}(t_{100})$ at 100% load.
- The temperature is denoted as t_0 when the load is zero and defined as t_{100} at 100% load.

During cyclic operation, indicated by $L_c(t_j) \leq \phi_{ful}(t_j)$, the parameter $X(t_j)$ should be calculated by Eq. (8).

$$X(t_j) = L_c(t_j)/\phi(t_j) \text{ and } \phi(t_j) = \phi_{ful}(t_j) \quad (8)$$

Whereas if the unit is operated at full load, indicated by $L_c(t_j) > \phi_{ful}(t_j)$, then Eq. (9) can be used.

$$(t_j) = F_{PL}(t_j) = 1\cdot \quad (9)$$

3 Results and Discussion

3.1 The conditions of the compartments

As described in the Methodology section, the outdoor compartment was maintained at 35°C dry-bulb and 24°C wet-bulb, whereas the indoor compartment was kept at 27°C dry-bulb and 19°C wet-bulb [9–13]. These conditions were achieved by using a cooling coil, a heater, and a humidifier.

As can be seen in Fig, 2, the conditions of the compartments are very close to that of predetermined condition. The dry-bulb temperature of the outdoor compartment can be maintained at an average of 34.98°C with a range of 34.72°C to 35.22°C. The wet-bulb temperature is in the range of 23.85 to 24.30°C, with an average of 24.01°C. The indoor compartment dry-bulb temperature is in the range of 26.74 to 27.16°C with an average of 26.98. Meanwhile, the indoor wet-bulb temperature ranges from 18.82 to 19.12°C with an average of 10.01. Fig. 2 shows the typical conditions of the indoor and outdoor compartments during the test.

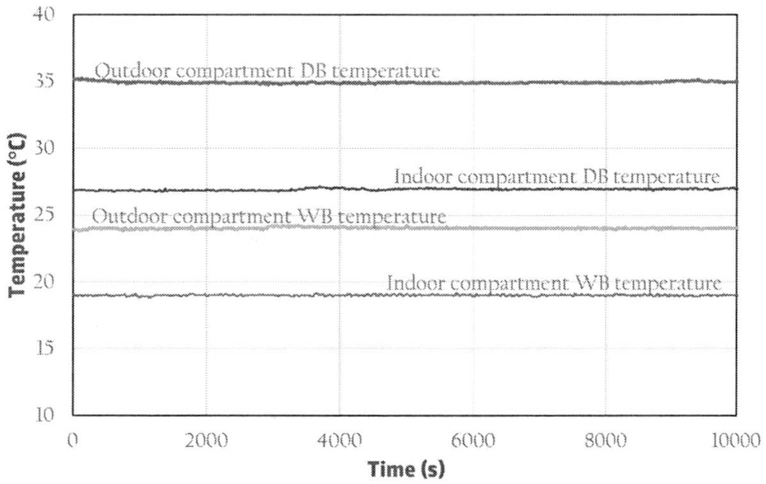

Fig. 2. Sample conditions of indoor and outdoor compartments during the tests.

3.2 Input power

The profile of the power drawn by the air conditioning unit is shown in Fig. 3. Using ISO 5151 test procedure, the average power consumption of the air conditioning unit is 724 Watts. With this method, the power is higher than that of CSPF ISO 16358-1 method, that is 631 Watts. As Bandung has a lower daily average temperature than that of used in ISO 16358-1, the power consumption of the air conditioner is lower (608 Watts). Jakarta, with somewhat higher daily temperature than that of ISO 16358-1, gives a slightly higher power consumption of 656 Watts. It similar to that reported in the previous study for non-inverter AC [6].

In terms of percentage, the power consumption calculated using ISO 5151 is higher than that of ISO 16358-1, modified Jakarta CSPF, and modified Bandung CSPF with differences of 6.19%, 4.66%, and 7.32%, respectively. This means that an air conditioner installed in Bandung will give the lower energy consumption compared to that of Jakarta and most other cities in Indonesia. This confirms that the higher outdoor temperature results in the higher power consumption [12].

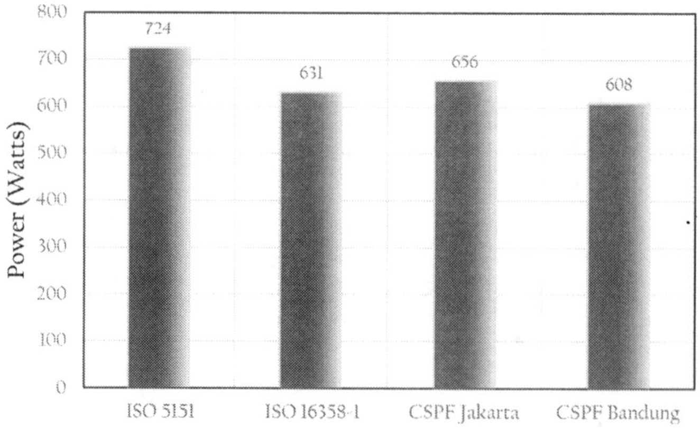

Fig. 3. Power consumption of air conditioner under test for different test conditions.

3.3 Cooling capacity

Contrary to power consumption, the cooling capacity of the test results with ISO 5151 gives the lowest value compared to the results of calculations using ISO 16358-1 and its modifications for Jakarta and Bandung. As can be seen in Fig. 4, the cooling capacity under ISO 5151 test is 2635 W. The ISO 16358-1 gives a cooling capacity of 2938 W. This is higher than that of Jakarta (2857 W) but lower than that of Bandung (3013 W). If the test result of ISO 5151 is used as the basis, then ISO 16358-1 gives 11.48% higher cooling capacity. The calculation based on Jakarta temperature distribution provides 8.40% higher cooling capacity and the highest is given by Bandung, which is 12.86% higher than that of ISO 5151. These results are in agreement with [12].

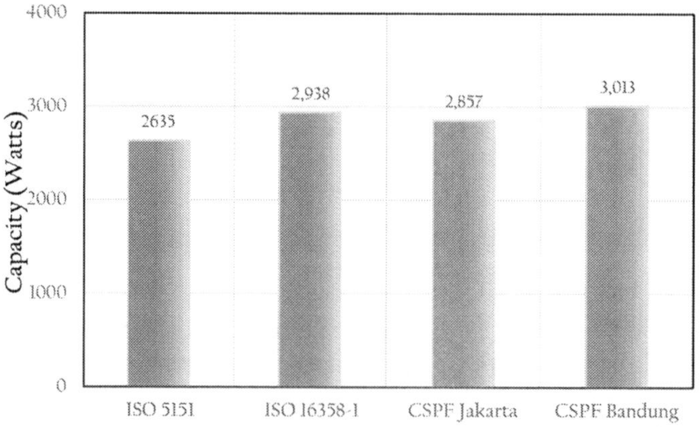

Fig. 4. Cooling capacity of air conditioner under test for different test conditions.

3.4 Cooling seasonal performance factor (CSPF)

The power consumption and the cooling capacity affect the COP of the air conditioner. Under ISO 5151, the unit has a COP of 3.64, that means each 1 W of work given to the unit will give 3.64 W of cooling capacity. This performance is lower than those of Indonesia CSPF (3.87) and its modified CSPF for Jakarta (3.81) and Bandung (3.92). In terms of percentage, Indonesia CSPF gives 6.19% higher COP than that of ISO 5151. Jakarta and Bandung CSPF provide COP which are 4.66% and 7.32% higher than that of ISO 5151. The profile of COP under different test conditions is presented in Fig. 5. Because this study using fixed-speed AC, the CSPF is then lower than that of study in Vietnam, in which inverter air conditioners were used [7].

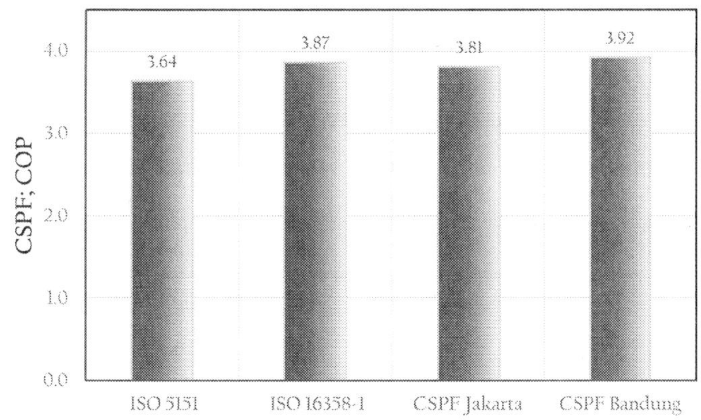

Fig. 5. COP of air conditioner under test for different test conditions.

4 Conclusion

The tests of AC performance using ISO 5151 and ISO 16358-1 (CSPF) on a 9000 Btu/hr nominal cooling capacity have been accomplished. Based on these tests, the following conclusions can be drawn. The highest power consumption of the AC was reached by the test using ISO 5151, followed by the standard CSPF, modified Jakarta CSPF, and modified Bandung CSPF. In terms of cooling capacity, the highest was reached by Bandung (3013 W) and the lowest was recorded by the test using ISO 5151 (2635 W). If ISO 5151 is used as a basis for the overall performance of the AC, then the ISO 16358-1 testing produces performance of 106.2%, while the modified CSPF for Jakarta and Bandung produce performance of 102.9% and 107.3%, respectively. For more realistic performance of an air conditioner, the modified standard could be applied based on the local temperature distribution.

References

1. Á. Andrade, Á. Restrepo, and J. E. Tibaquirá, Energy Reports **7**, 537 (2021)
2. ISO, *ISO 5151:2010 Non-Ducted Air Conditioners and Heat Pumps — Testing and Rating for Performance* (2010)
3. W. Y. Park, N. Shah, V. Letschert, and P. Blake, *Harmonizing Energy-Efficiency Standards for Room Air Conditioners in Southeast Asia* (2021)
4. W. Y. Park, *Assessment of Seasonal Energy Performance for Room Air Conditioners in Multiple Climates* (2022)
5. B. Holuj, W. Y. Park, N. Shah, N. Horowitz, and A. Hillbrand, *Energy-Efficient and Climate-Friendly Air Conditioners* (2019)
6. A. R. Irsyad, A. D. Pasek, and E. Philander, IOP Conf. Ser. Earth Environ. Sci. **927**, (2021)
7. C. D. Huy, T. Q. Dung, N. V. Dzung, and T. M. Thang, IOP Conf. Ser. Mater. Sci. Eng. **1145**, 012076 (2021)
8. D. Lee and F. P. Tsai, Energies **13**, (2020)
9. A. Setyawan, AIP Conf. Proc. **2248**, 7 (2020)
10. A. Setyawan, J. Teknol. **82**, 23 (2020)
11. A. Setyawan and A. Badarudin, IOP Conf. Ser. Mater. Sci. Eng. **830**, (2020)
12. A. Setyawan, S. Susilawati, T. Sutandi, and H. Najmudin, Int. J. Heat Technol. **39**, 1483 (2021)
13. W. H. Mitrakusuma, A. Badarudin, Susilawati, H. Najmudin, and A. Setyawan, J. Adv. Res. Fluid Mech. Therm. Sci. **90**, 42 (2022)

Performance of a room air conditioner under varied duration of system evacuation

Luga Martin Simbolon[1], *Hafid* Najmudin[2], *Andriyanto* Setyawan[1*], and *Risma* Septiany[1]

[1]Department of Refrigeration and Air Conditioning Engineering, Politeknik Negeri Bandung, Jl. Gegerkalong Hilir, Ciwaruga, Bandung, Indonesia
[2]Balai Besar Bahan dan Barang Teknik, Bandung, Jl. Sangkuriang 14 Bandung, Indonesia

Abstract. Refrigeration system evacuation is one of the most important steps in achieving refrigeration or air conditioning system quality. This procedure is used to remove water vapor and non-condensable gas from the pipeline and the heat exchangers that can be dangerous for a vapor compression refrigeration system. In this paper, the performance of an air conditioner (AC) was experimentally tested under varied duration of system evacuation from 0 to 60 minutes with an increment of 15 minutes. Cooling capacity, power consumption, and coefficient of performance were tested in accordance with ISO 5151 and ISO 16358-1. Generally, the ISO 16358-1 gives the lower power consumption and higher cooling capacity and coefficient of performance. ISO 16358-1 test results in 12.8% lower power consumption than that of ISO 5151. On the contrary, ISO 16358-1 test gives higher cooling capacity and coefficient of performance by 11.5% and 6.2%, respectively. It is also observed from this study that the longest evacuation period results in the most efficient operation, thus providing less effect on climate change and less environmental impact. However, it should be observed in a longer period of operation.

1 Introduction

The demand for air conditioners (AC) in the world tends to increase due to the increasing standard of living and the need for thermal comfort. In addition, global warming is also driving the increasing sales of air conditioners around the world. In developed countries, the amount of energy consumed by buildings is about 40% of the total energy demand [1]. Consumption of electrical energy for refrigeration and air conditioning systems reaches 45% in the residential and commercial sectors. This, of course, significantly contributes to emission of greenhouse gas. Due to the rapid growth of the economy and human population, energy consumption is projected to increase by around 71% from 2003 to 2030. For hot regions such as in the Middle East, air conditioning machines can even consume as much as 70% of the total energy demand [2]. Therefore, efforts to save energy are needed so that natural resources can be used more efficiently, and CO2 emissions and the issue of climate change can be anticipated better [3].

---

* Corresponding author: andriyanto@polban.ac.id

One of the important processes that affects durability and energy efficiency of an air conditioner is the evacuation process. This process is carried out to remove water vapor and air from the cooling unit. Otherwise, the water vapor will melt, freeze, and block the flow of refrigerant, which can reduce the efficiency of the system or even damage the compressor [4,5]. The evacuation process could also minimize the presence of air or non-condensable gas can also reduce the capacity of the cooling unit.

So far, no research publications have been found that discuss the effect of system evacuation duration on AC performance. Instead of system evacuation, the studies of the effect of refrigerant charge have been conducted for decades [6–9]. In textbooks on refrigeration and air conditioning, it is only explained that an evacuation system up to a certain vacuum level is important to ensure the refrigeration system works well.

This paper discusses the effect of the system evacuation duration on the performance of an air conditioner. Three main parameters, i.e., power consumption, cooling capacity, and coefficient of performance (COP) are analyzed on the basis of ISO 16258-1 [10] as well as ISO 5151 [11].

2 Methodology

A residential air conditioner with a nominal cooling capacity of 9000Btu/hr was used in this test. The indoor unit of the AC was installed in the indoor compartment and the outdoor unit was installed in the outdoor compartment. The tests were carried out based on ISO 5151 and have been applied in previous studies [12–16]. The analysis and calculation were performed on the basis of ISO 5151 and ISO 16358-1.

During the tests, the outdoor compartment was kept at a dry-bulb temperature (DBT) of 35°C and wet-bulb temperature (WBT) of 24°C. Whereas, the DBT and WBT of the indoor compartment were maintained at 27°C and 19°C, respectively. Detailed method and experimental setup can be found in previous studies [15,16].

The testing of the AC unit was started by installing indoor and outdoor units and checking for leaks. If there is no leakage in the unit, the evacuation process can be carried out. After the evacuation process, the valve on the manifold gauge connected to the vacuum pump is closed, the vacuum pump is turned off, and wait for about 15-30 minutes. If there is no pressure rise, the system is ready to be charged with refrigerant. The amount of refrigerant charged into the system is in accordance with the manufacturer's recommendations. The duration of the evacuation process in this study was varied using 5 variations, i.e., 60 minutes, 45 minutes, 30 minutes, 15 minutes, and 0 minute.

From this test, 32 parameters can be measured and calculated. Three main quantities, namely input power, cooling capacity, and COP can be derived from these data based on ISO 5151. The calculation of Cooling Seasonal Performance Factor (CSPF), Cooling Seasonal Total Load (CSTL), and Cooling Seasonal Energy Consumption (CSEC) were accomplished in accordance with ISO 16358-1. Detailed calculations for CAPF, CSTL, and CSEC have been investigated [17].

3 Results and Discussion

3.1 Power consumption

Figure 1 shows the effect of evacuation duration on the power required to operate the AC unit. The data were obtained from 7 runs of experiment. When the AC unit is operated without prior evacuation, the power consumption appears to be higher than if it is evacuated first. An average of 760 W and a range of 759 – 761 W of power was recorded in this experiment. If the piping system was evacuated for 15-minute duration, the average power decreases to 738 W. Evacuation durations of 30, 45, and 60 minutes result in almost the same power consumption with an average of 736.7 W and a maximum deviation of only about 0.21%. Meanwhile, when compared to conditions without evacuation, the difference in energy consumption can reach 3.25%. This range of power consumption is similar to that of previous study [13].

In the experiment without previous evacuation, it is possible that there is still water vapor left in the piping, causing a disturbance in the refrigerant flow. In the expansion device the water vapor can melt and freeze, so that it can block the flow of refrigerant and create a pressure drop in the piping. Probably, this is what causes the increase in power consumption in the AC unit.

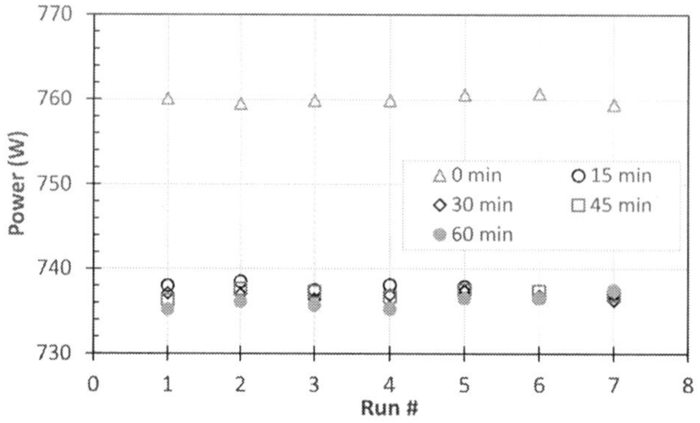

Fig. 1. The power drawn by the AC for various evacuation duration.

3.2 Cooling capacity

The profile of cooling capacity as a function of evacuation duration is shown in Fig. 2. Cooling capacity has an average value of 2648 W with a range of 2645 to 2650 W when the system operates without evacuation. The evacuation process for 15 minutes gave an average cooling capacity of 2658 W with a range between 2654 to 2661 W. Furthermore, the evacuation duration of 30 minutes resulted in an average cooling capacity of 2659 W. The cooling capacities of 2662 and 2665 W were given by experiments with an evacuation duration of 45 and 60 minutes, respectively. This range of cooling capacity is slightly higher than that of reported in previous study in an air conditioner using various distance of condenser air obstruction [13].

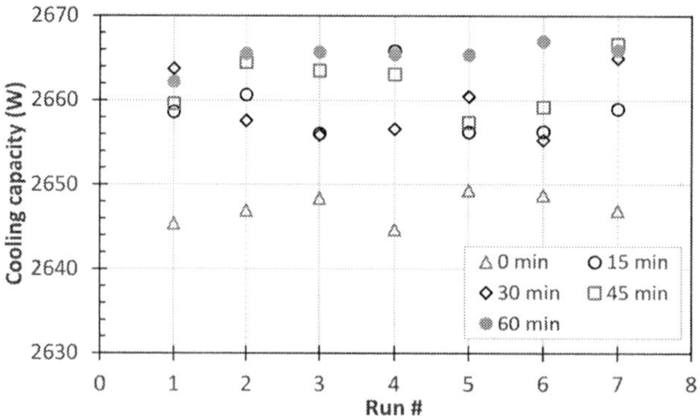

Fig. 2. The cooling capacity for various evacuation duration.

Detailed examination of Fig. 2 reveals that evacuation duration of 15 minutes to 60 minutes result in almost the same cooling capacity. The maximum deviation cooling capacity is only 0.26%. This indicates that duration of evacuation of 15 minutes can be considered as enough to evacuate water vapor and non condensable gas from the AC system as the difference in cooling capacity is very small. However, this should be validated using longer test duration in order of weeks or months, or even longer.

3.3 COP

COP is a parameter of AC performance that depends on the cooling capacity and power required to operate the AC and produce the cooling capacity. It can be expected that the COP increases as the cooling capacity increases and input power decreases. The profile of the COP for different durations of evacuation process is depicted in Fig. 3.

Detailed inspection of Figure 3 divulges that the COP of the AC unit without evacuation (3.48) is 3.94% lower than that of 60 minutes evacuation duration (3.62). The range of COP is similar to that reported in a study using R22, R410A, R290, and R32 [18]. Another finding from this study is the insignificant difference in COP between vacuum durations of 15, 30, 45, and 60 minutes. So that from a performance standpoint, it appears that the duration of the evacuation from 15 minutes to 60 minutes does not seem to have a significant impact. Evacuation duration of 15 minutes, for instance, has a COP of 3.60. This is only 0.56% lower than that of the longest evacuation duration. However, again, longer duration of test should be carried out to evaluate its effect on the long-term performance as well as the lifespan of the AC unit.

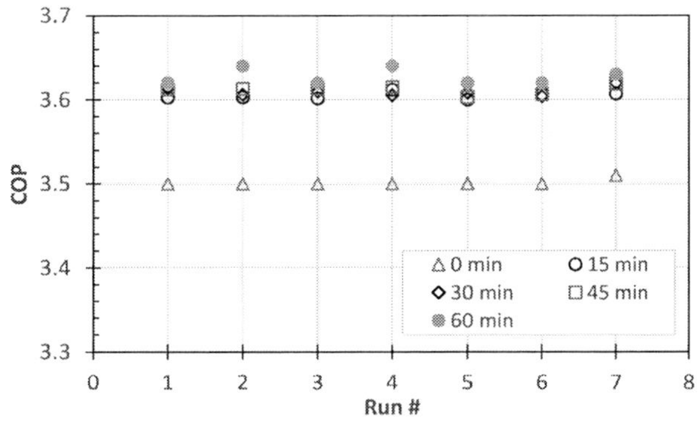

Fig. 3. The COP for various evacuation duration.

3.4 Performance calculation using ISO 16358-1

In ISO 16358-1, the performance of the air conditioner can be expressed in terms of CSPF (Cooling Seasonal Performance Factor). This quantity is defined as the ratio of Cooling Seasonal Total Load (CSTL) and Cooling Seasonal Energy Consumption (CSEC). Generally, CSPF is expressed as F_{CSP}, while CSTL and CSEC are expressed as L_{CST} and C_{CSE}, respectively, as written in Eq. (1-3).

$$F_{CSP} = \frac{L_{CST}}{C_{CSE}} \tag{1}$$

L_{CST} and C_{CSE} can be calculated using the following equations:

$$L_{CST} = \sum_{j=1}^{m} L_c(t_j) \times n_j + \sum_{j=m+1}^{n} \phi_{ful}(t_j) \times n_j \tag{2}$$

and

$$C_{CSE} = \sum_{j=1}^{n} X(t_j) \times P_{ful}(t_j) \times \frac{n_j}{F_{PL}(t_j)} \quad . \tag{3}$$

The complete calculation techniques for L_{CST} and C_{CSE} is available in [17].

By using equations (2) and (3), the total load and the power consumption of the air conditioner for 1 year period can be determined. Therefore, the CSPF can be obtained. The profile of the power and cooling capacity calculated by ISO 16358-1 are depicted in Fig. 4 and 5. From Fig. 4, it can be seen that the power calculated by ISO 16358-1 is lower than that of ISO 5151. The reason is that ISO 5151 tests the air conditioner and measures the power consumption at a constant outside air temperature of 35°C. Meanwhile, ISO 16358-1 calculates power when the unit operates in a temperature range of 21 to 35°C. In general, air conditioners that operate at lower ambient temperatures will also consume less power.

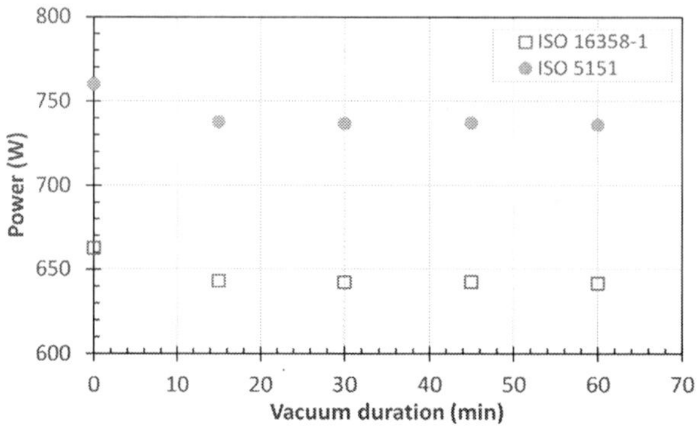

Fig. 4. Power consumption calculated by ISO 5151 and ISO 16358-1.

Conversely, an air conditioner that operates at a lower ambient temperature will produce a higher cooling capacity. This can be seen from Fig. 5, where the calculation with ISO 16358-1 gives a higher cooling capacity compared to the test results with ISO 5151. With ISO 16358-1, the average cooling capacity during the tests is 2963 W. This is higher than the test results with ISO 5151, which is 2658 W.

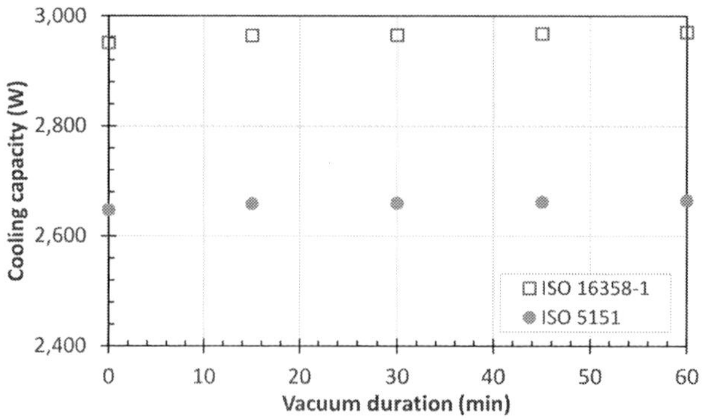

Fig. 5. Cooling capacity calculated by ISO 5151 and ISO 16358-1.

The results of the CSPF calculation with ISO 16358-1 are higher than the COP resulted by the test with ISO 5151. As shown in Figure 6, the results of the ISO 16358-1 give COP values of about 6.2% higher than that of ISO 5151 due to higher cooling capacity and lower power consumption. The CSPF of this study is similar to that of study using non-inverter air conditioners conducted in four main cities in Indonesia [17].

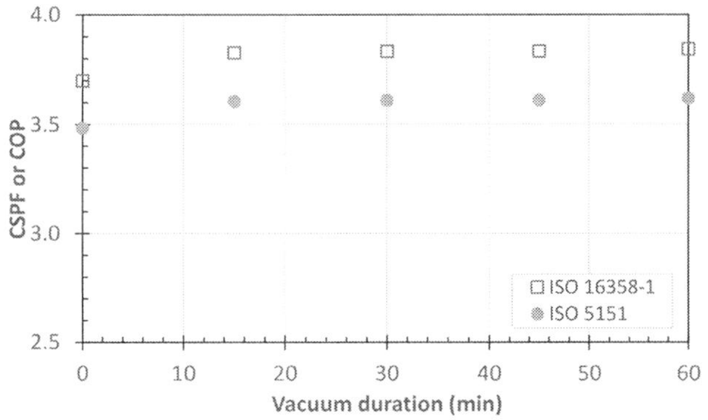

Fig. 6. CSPF and COP calculated by ISO 5151 and ISO 16358-1

4 Conclusion

A test of an AC with a nominal cooling capacity of 9000 Btu/hr or 2.64 kW under 5 different evacuation durations based on ISO 5151 has been accomplished. The highest performance in terms of COP is resulted from the test with longest evacuation duration with an average COP is 3.62. In terms of CSPF, the performance of the AC is 6.2% higher than that of COP calculated by ISO 5151. One of the important findings of this study is the insignificant difference of system performance at evacuation durations of 15, 30, 45, and 60 minutes. This seems to indicate that the evacuation duration of 15 minutes is enough to remove moisture and air from the AC system. However, it is necessary to carry out further studies in a longer test period to determine the effect of evacuation duration on the performance and durability of the system in the long term.

References

1. D. Kumar, S. Kumar, B. A. Bhayo, K. Harijan, and M. A. Uqali, J. Therm. Eng. **6**, 141 (2020)
2. Y. Al Horr, B. Tashtoush, N. Chilengwe, and M. Musthafa, Case Stud. Therm. Eng. **18**, 100574 (2020)
3. T. Ahmad and D. Zhang, Energy Reports **6**, 1973 (2020)
4. Y. Hu and D. P. Yuill, Int. J. Refrig. (2022)
5. Y. Hu, D. P. Yuill, and Y. Chen, Int. Refrig. Air Cond. Conf. (2021)
6. E. Björk and B. Palm, Appl. Therm. Eng. **26**, 829 (2006)
7. K. Li, J. Lan, G. Zhou, Q. Tang, Q. Cheng, Y. Fang, and L. Su, J. Therm. Sci. **28**, 294 (2019)
8. F. Y. Nugraha, A. S. Yatim, and M. I. Alhamid, SSRN Electron. J. 2 (2021)
9. L. Palmiter, J. H. Kim, B. Larson, P. W. Francisco, E. A. Groll, and J. E. Braun, Energy Build. **43**, 1802 (2011)
10. Á. Andrade, Á. Restrepo, and J. E. Tibaquirá, Energy Reports **7**, 537 (2021)
11. ISO, *ISO 5151:2010 Non-Ducted Air Conditioners and Heat Pumps — Testing and Rating for Performance* (2010)
12. A. Setyawan, AIP Conf. Proc. **2248**, 7 (2020)
13. A. Setyawan, J. Teknol. **82**, 23 (2020)
14. A. Setyawan and A. Badarudin, IOP Conf. Ser. Mater. Sci. Eng. **830**, (2020)
15. A. Setyawan, S. Susilawati, T. Sutandi, and H. Najmudin, Int. J. Heat Technol. **39**, 1483 (2021)
16. W. H. Mitrakusuma, A. Badarudin, Susilawati, H. Najmudin, and A. Setyawan, J. Adv. Res. Fluid Mech. Therm. Sci. **90**, 42 (2022)
17. A. R. Irsyad, A. D. Pasek, and E. Philander, IOP Conf. Ser. Earth Environ. Sci.

927, (2021)
18. R. I. Tritjahjono, K. Sumeru, A. Setyawan, and M. F. Sukri, J. Adv. Res. Fluid Mech. Therm. Sci. **55**, 1 (2019)

Scenario of renewable energy transition from fossil energy resources towards net zero emission in Indonesia

Vicky Firmansyah[1], *Made Krisna* Adinarayana[1], *Rizky* Tetrisyanda[1], and *Gede* Wibawa[1*]

[1]Chem-Eng. Dept, Sepuluh Nopember Institute of Technology (ITS), 60111 Sukolilo Surabaya, Indonesia

Abstract. In 2011, Indonesia set a 26% reduction goal for greenhouse gas emissions by 2030 to mitigate the climate change. Based on data from BPS, Indonesia's renewable energy mix in 2021 is 12.16% with a target of 23% in 2025. This indicates that there are challenges faced by Indonesia in many sectors, especially the upstream oil and gas industry as one of the largest emitters of greenhouse gases, in achieving the energy transition target. In this study, trend analysis and data forecasting were carried out using trend analysis of time series data on oil and gas energy supply and consumption data as baseline to propose scenarios for both consumption and utilization energy to achieve net zero emission (NZE) in 2060. This study found that NZE may be achieved by applying energy consumption scenarios including the use of electric vehicles by 10% in 2030, and 90% in 2060 and the use of electric stoves by 25% in 2030, and 90% in 2060. Renewable energy utilization scenarios include geothermal (50%), hydro (50%), mini hydro (50%), solar (80%), and wind (15%) of the existing potential. In addition, early retirement for coal-fired power plants is needed.

1 Introduction

The rapid population growth, economic growth, energy prices, and technological developments have led to increased energy demand worldwide [1]. Its also increases carbon emissions which causes climate change to get worse if the energy sources still coming from fossil energy base. In Indonesia, the energy mix still heavily dominated by fossil energy which is oil and gas contribute 51% and 38% from coal in 2021, which still significantly cause increasing of greenhouse gases (GHGs) emission and other environmental problems [2].

In 2011, Indonesia has a target of reducing GHGs emission production by 26% by 2030, which is also stated in Presidential Decree No. 61 of 2011. To achieve this plan, the implementation of Regional Action Plan for Reducing GHGs (RAN-GRK) are required. In addition, in 2016 Indonesia has signed the Paris Agreement and set new targets for reducing GHGs emission. In accordance with the Nationally Determined Contribution (NDC), the Government of Indonesia seeks to reduce emissions by 29% (without assistance) or 41%

[*] Corresponding author: gwibawa@chem-eng.its.ac.id

(with international assistance) in accordance with the Business as Usuas (BaU) scenario by 2030. The Government of Indonesia also initiated a Low Carbon Development plan for the integration of climate change regulatory mitigation that will be adapted to the National Development Plan [3].

Based on data from the Ministry of Energy and Mineral Resources (MEMR), the utilization of national renewable energy in 2022 only reached 12.3% of the target of 23% by 2025 [2]. It need new more strategy to accelerating GHGs reduction to achieve the target whether from demand side and supply side. Therefore, one solution that can be done to achieve targets in accordance with the Paris Agreement is to implement the use of technology based on the context of renewable energy on a large scale. The concept of using 100% new and renewable energy has also become a global ambition that is planned to be applied from a corporate scale to a larger scale. Even as a country with a total energy contribution of 36% to the total energy supply in ASEAN, Indonesia's contribution to reducing the amount of emissions both on an ASEAN scale and on a global scale is something that really needs to be considered [4].

This study aims to provide an overview of carbon emission reduction using government strategies and additional strategies suggested by the author to achieve Indonesia's net zero emission in 2060 that presented as several scenario.

2 Methodology

2.1 Data sources

This study used secondary data to prepare mitigation scenario and calculate CO_2 emission for 2023 until 2060. The main data used in this study are supply and demand of energy in Indonesia that obtained from BP Statistical Review 2022 and Handbook of Energy & Economic Statistic of Indonesia (HESSI 2021) [2]. Then to compare the result of forecasting supply and energy demand data will be validated with RUEN and publication from Institute for Essential Services Reform (IESR) [5]. HESSI also consist of econometric and demographic essential data to create energy modelling and planning. Besides, the emission factor data obtained from MEMR [6] and another essential data i.e electrification ratio, power plant efficiency, and power plant construction planning were obtained from PLN [7].

2.2 Proposed scenario

Several scenarios that will be proposed in this study are an accumulation of several mitigation strategies to reduce GHGs emission nationally, as follows.

1. Business as Usual (BaU) – Existing scenario from *Rencana Umum Energi Nasional* (RUEN) that released in 2017 and realization data for 5 years after RUEN was released will be projected as the *baseline* scenario.
2. Government Mitigation Scenario (GMS) – Based on electric vehicle (EV) target used, city gas network, and biodiesel usage.
3. High Optimistic Consumption Scenario (HOCS)– Maximizing demand of EV, usage of electric stove, biodiesel usage, and several energy supply strategy i.e cofiring hydrogen on Gas Turbine Combined Cycle (GTCC) power plant, and biomass cofiring on Coal Fired Power Plant (CFPP)
4. Middle Optimistic Consumption Scenario (MOCS) – Reduce EV usage target and the rest is same as HOCS.
5. Middle Consumption and Renewable Energy Scenario (MCRE) - Periodic CFPP retirements, maximizing Indonesia's renewable energy potential, and the rest is same as MOCS.

To see the different strategies used in each scenario, the strategy table is shown in Table 1.

https://doi.org/10.1051/e3sconf/202346704005

Table 1. Table of Mitigation Scenario.

Strategy	BAU	GMS	HOCS	MOCS	MCRE
Motorcycle (EV Based)	(2030;0,07%), (2060;0,187%)	(2025;2%), (2050;50%)	(2030;50%), (2060;80%)	(2030;5%), (2060;60%)	(2030;5%), (2060;60%)
Car (EV Based)	(2030;0,047%), (2060;0,107%)	(2025;2%), (2050;50%)	(2030;30%), (2060;60%)	(2030;20%), (2060;60%)	(2030;20%), (2060;60%)
Electric Stove	(2030;2,32%), (2060;6%)	-	Urban (2060;80%) Rural (2060;70%)	Urban (2060;80% Rural (2060;70%)	Urban (2060;80%) Rural (2060;70%)
City Gas Network	(2060; 10%)	(2024;4%), (2050;50%)			
Biodiesel Mix	Based on historical data	B100 (2030)	B30(2025) B50(2040) B100(2050)	B30(2025) B50(2040) B100(2050)	B30(2025) B50(2040) B100(2050)
Ammonia Cofiring	-	-	(2050;50%)	(2050;50%)	(2050;50%)
Biomass Cofiring	(2022;5%)	(2022;5%)	(2030;20%)	(2030;20%)	(2030;20%)

In addition, for MCRE scenario there is strategy to substitute the electricity supply decrease by CFPP retirement until 2050 with renewable energy included 70% geothermal, 80% hydro, 50% mini hydro, 80% solar, and 15% wind.

2.3 Energy Modelling and Simulation

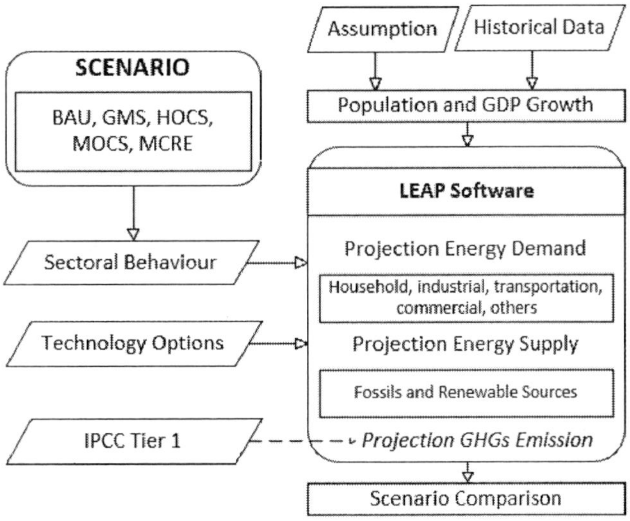

Figure 1. Energy Model for GHGs Emission Mitigation.

GHGs emission calculated from energy supply and demand forecasting result. Low Emission Analysis Platform (LEAP) software. The following energy model is used in Figure 1. Software used to obtained integrated result for supply and demand analysis, GHGs emission analysis that measured in CO_2 equivalent, and made comparison for every single scenario. The LEAP model is the de facto standard for countries to undertake energy planning, integrated resources, greenhouse gas mitigation assessments, and low-carbon development strategies, the LEAP model is a model that meets the criteria for regional energy planning [1].

3 Result and discussion

Energy modelling simulations can be used to calculate the amount of emissions generated in terms of energy demand and energy supply. The increase in energy supply is in line with energy demand, while energy demand grows with the increase in population (assumed to grow at 1.22% annually), GDP growth, and urbanization rate. The energy demand is derived into 5 main sectors namely household, industry, transportation, commercial, and others. Based on demand forecasting that has been done, the largest increase in energy demand is in the transportation sector, which increases from 381 MBOE in 2021 to 1058 MBOE in 2060 with an average increase of 2.65% per year. Thus projection has almost the same comparison with the results of research by Edwaren et al, which grew by around 2.8 times from 2019 to 2060 [8]. Then from the transport sector, the use of gasoline is the largest, which is around 57.86% of all vehicle fuels. This is one reason why it is necessary to increase the usage of electric vehicles.

Based on the results of energy modelling, the emissions produced will still be large if the reduction is carried out only in the transportation sector, so the strategy also seeks to reduce emissions in all sectors that produce carbon emissions. The others strategy are usage of electric stoves to reduce emissions from the gas stoves, and the blending of biodiesel in diesel for industrial activities The following are the results of carbon emissions from the baseline scenario and each mitigation scenario as shown in Table 1.

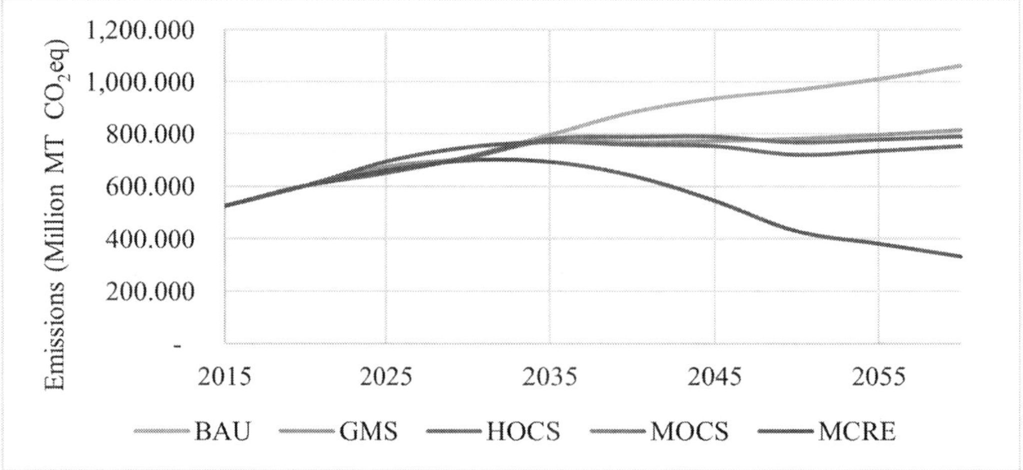

Figure 2. GHGs Emission from Several Scenario.

BAU scenario gives GHGs emission around 1.08 billion MT CO_2eq. Where the amount of emissions is about half of the projected results by BPPT in Indonesia's energy outlook [9], but the realization of the increase in emissions in the study is too high if evaluated in 4 years after the study is made. The GMS scenario provides a 23.25% reduction in GHGs emissions from the BAU scenario. One of the additional mitigations simulated in the HOCS scenario is the use of electric stoves and also an increase in the target of using electric vehicles in 2050 and resulting a 29.02% reduction in GHGs emissions. However, the overly optimistic target use of EV in HOCS causes greater carbon emissions until 2034 than the BAU scenario. This is due to the increasing demand for electricity which is not accompanied by the use of renewable energy sources in electricity generation. Based on the simulations conducted in the MOCS scenario, the target of EV usage was lowered in the early period due to the large amount of electricity generated from coal. Therefore, in MOCS, the target of electric vehicle usage was lowered to 5% in 2030 and 25% in 2040. This scenario provides a 25.54% reduction in GHGs in 2060. The reduction of the electric vehicle usage target also considers how the development of supporting infrastructure and increased public interest in EV usage. To see the effectiveness of each strategy in reducing emissions from the energy demand side, the following table shows the reduction of each strategy in MOCS scenario.

Table 2. GHGs Emission Reduction each Strategy in MOCS.

Strategy	MOCS	GHGs Emission Reduction
Motorcycle (EV Based)	(2030;5%), (2060;60%)	16.73%
Car (EV Based)	(2030;20%), (2060;60%)	
Electric Stove	Urban (2060;80%) Rural (2060;70%)	3.52%
Biodiesel Mix	B30(2025), B50(2040) B100(2050)	3.29%
Ammonia Cofiring in GTCC	(2050;50%)	0.98%
Biomass Cofiring in CFPP	(2030;20%)	3.65%

Based on table 2, the usage of EV provides the largest reduction in emissions from the demand side. While the use of electric stoves is much smaller. This is because gas stoves produce lower emissions when compared to motorized vehicle emissions. Besides, blending biodiesel does not have a significant impact because the carbon emission factor of B50 is 0.77 gram/km when compared to diesel fuel which is 0.88 gram/km [10]. Moreover, the availability of biodiesel needs to be reviewed to be able to meet the demand as a mixed fuel in 2050. Likewise, the cofiring of Ammonia is not significant because the share of GTCC is fewer than CFPP. Cofiring 20% biomass on CFPP has a slightly significant impact because the amount of CFPP is very large in Indonesia. Biomass cofiring has several limitations, including the availability of biomass and the effect of biomass burning in the boiler [11].

GMS, HOCS, and MOCS scenarios lead to an increase in electricity demand. Therefore, the electricity supply must come from New and Renewable Energy (NRE). By utilizing the potential for electricity generation from NRE in Indonesia, MCRE scenario reduce carbon emissions around 68.74%. But a deeper study is needed related to how to take advantage of the potential of such a large NRE in Indonesia. Then, if we trace the remaining carbon emissions in the MCRE scenario, Table 3 displays emissions that still need to be reduced.

Table 3. Amount of CO_2 Emission Remain in 2060 (MTon CO_2e).

Energy Demand Side		Energy Supply Side	
Household	5.167	Power Plant	95.29
Industry	32.136	Oil Refinary	17.78
Transportation	131.778		
Commercial	3,55	Total	287,481

Reducing carbon emissions from energy modeling can only be mitigated through patterns of energy consumption and energy supply [12]. However, from these two aspects, emission reductions are projected to be unable to reach Net Zero Emissions by 2060 for the reasons mentioned earlier. Therefore, other mitigation steps outside these two aspects are needed to reduce carbon emissions. One of them is by capturing, storing, and utilizing carbon emissions generated from the industrial sector, especially the oil and gas industry or commonly referred to as Carbon Capture Storage / Carbon Capture Utilization and Storage (CCS / CCUS) technology [13].

Based on the NRE energy transition scenario, emissions generated from oil refinery activities in 2060 will be 17.78 MTon CO2e and in the power generation sector it is predicted to be 95.29 MTon CO2e. Emissions at this power plant come from the continued use of natural gas, diesel, and biodiesel in their energy sources. Meanwhile, in the aspect of energy demand, there are 173.56 MTon CO2e consisting of sectors as shown in Table 3.

So if the need to absorb carbon emissions by 2060 to reach NZE is 285,701 MTon CO2e. For CCU/CCUS, it can be applied to the energy consumption sector in industry and the energy supply sector in this case is power plants and oil refineries. So that the total need for CCU / CCUS in 2060 is 145,206 MTon CO2e. To reduce carbon emissions in transportation sectors such as buses, trucks, ships, and planes rely on the development of technology to reduce emissions and sequester CO2 naturally such as preserving forest [14]. That way, by 2060 by using mitigation measures in the NRE scenario and the construction of CCU / CCUS with capacity above will be achieved Net Zero Emission.

4 Conclusion

Net Zero Emission in Indonesia may be achieved by applying energy consumption scenarios including the use of electric vehicles by 10% in 2030, and 90% in 2060 and the use of electric stoves by 25% in 2030, and 90% in 2060, biodiesel mix by 2050 for B100, ammonia cofiring in GTCC, and biomass cofiring in CFPP. Renewable energy utilization scenarios include geothermal (50%), hydro (50%), mini hydro (50%), solar (80%), and wind (15%) of the existing potential. In addition, early retirement for coal-fired power plants is needed.

References

1. S. H. Suharyati, J. L. Pambudi, N. I. Wibowo, and Pratiwi, *Indonesia Energy Outlook 2019 : The Impact of Increased Utilization of New and Renewable Energy on the National Economy* (BPPT, Jakarta, 2019)
2. A. Cahyono Adi and F. Lasnawatin, *Handbook of Energy & Economic Statistics of Indonesia. Ministry of Energy and Mineral Resources Republic of Indonesia 2021* (Jakarta, 2021)
3. M. T. Sambodo, C. I. Yuliana, S. Hidayat, R. Novandra, F. W. Handoyo, A. R. Farandy, I. Inayah, and P. I. Yuniarti, Heliyon **8**, (2022)
4. N. Reyseliani and W. W. Purwanto, Renew Energy **176**, 305 (2021)
5. A. P. Tampubolon, (2020)
6. R. Boer, R. G. Dewi, U. Siagian, and M. Ardiansyah, *Pedoman Penyelenggaraan Inventarisasi Gas Rumah Kaca* (2012)
7. PT PLN, *Rencana Usaha Penyediaan Tenaga Listrik PT Perusahaan Listrik Negara (Persero) Tahun 2021 Sampai Dengan Tahun 2030* (Jakarta, 2021)
8. E. Liun, Suparman, Sriyana, D. Dewi, and J. S. Pane, International Journal of Energy Economics and Policy **12**, 467 (2022)
9. E. Siregar, I. Raharjo, Yudiartono, E. Hilmawan, and J. Santosa, *Indonesia Energy Outlook 2021 : Indonesia Energy Technology Perspective* (BPPT, Tangerang, 2021)
10. G. Bieker, *A Global Comparison of the Life-Cycle Greenhouse Gas Emissions of Combustion Engine And Electric Passenger Cars* (Berlin, 2021)
11. P. Adhiguna, *Indonesia's Biomass Cofiring Bet Beware of the Implementation Risks* (2021)
12. S. Solaymani, Energy **168**, 989 (2019)
13. Q. Deng, X. Ling, K. Zhang, L. Tan, G. Qi, and J. Zhang, Front Energy Res **10**, (2022)

14. P. C. Buotte, B. E. Law, W. J. Ripple, and L. T. Berner, Ecological Applications **30**, (2020)

The application of the analytical hierarchy process (AHP) in determining organic waste process technology-oriented-reducing greenhouse gas

I Made Gunamantha[1*], *Made Vivi* Oviantari[2], and *Ni Wayan* Yuningrat[2]

[1]Environmental Management, Universitas Pendidikan Ganesha, Jl. Udayana Singaraja, Bali, Indonesia

[2]Applied Chemistry, Faculty of Mathematics and Natural Sciences, Universitas Pendidikan Ganesha, Jl. Udayana Singaraja, Bali, Indonesia

Abstract. The existence of waste in the environment, which is not managed well, could contribute to global warming and cause climate change on the earth. This research aimed to determine the suitable management preference for organic waste treatment produced in small cities. The methods used are based on pair comparison in hierarchy decisions developed through the analytical hierarchy process. The hierarchy consisted of goals, criteria, sub-criteria and alternatives applied to waste treatment engineering. The expert suggestion was used in a pair comparison matrix to determine the level of technology. The comparison was used to get the significance level of decision criteria and the relative performance of the options. The city's waste managed by the Buleleng government was used to demonstrate the application of the analytical hierarchy process in that region. The result showed that the important factor in deciding on waste treatment for the small city is environment and engineering, with each eigenvector priority (0.28), sociocultural (0.24) and economy (0.20). According to the recruitment preference for Singaraja waste treatment, the analytical hierarchy process showed that controlled landfill and composting are the most suitable treatment, followed by incineration, anaerobic digestion, and mechanical and biological treatment.

1 Introduction

According to data from The National Waste Management Information System (SIPSN), for 2022, 67.23% of the composition of waste generated in urban areas in Indonesia is organic waste [1]. Among them, 40.7% are food waste, 13% are wood/twigs, and 11.31% are paper/cardboard. High organic waste production should be managed well to prevent greenhouse gas emissions. Biodegradable organic waste landfilled at the final processing location will release methane gas. To manage the waste well, the government must provide good infrastructure. Today, most of the waste is managed by landfilling system [2]. However, this method should be improved and replaced by other processes due to limited land area and

* Corresponding author: md_gunamantha@yahoo.com

several environmental problems, such as leachate production besides methane. On the other hand, there are various possible waste processing technology options. Various criteria must be considered in determining the appropriate choice. Multi-criteria analysis is one way to achieve that goal. In this multi-criteria decision maker (MCDM), the aims and criteria wanted were combined in the decision framework provided. This method has been widely used in ecology evaluation by considering complex and interactive environmental problems, socio-cultural, technical, monetary and economic. Besides that, the interchange of the goals and competitive criteria was also considered. The AHP is a multi-criteria analysis which supports the decision-maker in solving and making complex decisions [3,4,5]. The AHP is suitable for waste treatment because it can assist in choosing the specific and effective waste treatment [4,5]. Principally, the AHP is analytical, which supports the process in various criteria and complex problems into easy problem solvers. This study examines municipal waste treatment options using AHP in determining the best and appropriate technology for the organic fraction of municipal solid waste (OFMSW) management.

2 Materials and method

The MCDM framework based on the AHP approach was used to compile and analyze the problem with the decision-maker view hypothesis. The criteria and alternatives used were from experts, waste management regulation and literature. The sub-criteria was confirmed from the literature, interaction and discussion with the TPA manager.

2.1. Research area

The research area was the waste produced in Singaraja, northern Bali. Singaraja is on 8°3'40" - 8°23'00" LS and 114°25'55" -115°27'28" BT geographically, have 82527 people. Administratively, Singaraja consists of 18 districts and one village. The waste produced is estimated at 58000 kg per day with 0.7 /capita/day. The waste comprised more than 68% organic fraction, including 35% part of the plant, 15% food waste and 10% cardboard, respectively. The waste produced was filled by controlled landfill technology at the Bengkala End of Waste Processing (TPA), Kubutambahan, Buleleng Regency, Bali. The TPA has been open since 2006 with an area of 4.8 Ha.

2.2. Data collection

The primary and secondary data were used in this research. The primary data was collected by questionnaire spreading with a non-probability sampling technique. The respondents with knowledge and experience in waste management were chosen, including two academics and three practitioners. Besides the questionnaire, the knowledge acquisition process for structuring the criteria, sub-criteria and waste management technology option was obtained by interviewing the municipal solid waste manager, literature, and other secondary information. The criteria included environment, sociocultural, technique and economic. The environmental criteria included air pollution, pathogenic exposure, and water and soil pollution. Sociocultural criteria included acceptability, community engagement, and susceptibility. The technical criteria consisted of possibilities and sustainability, ease and effectiveness in operation and energy and nutrition recovery. The economic criteria included investment cost and operation and maintenance cost.

2.3. Identification of waste management technology option

This research analyzed four alternatives of OFMSW options: sanitary landfills, composting, incineration, anaerobic digestion, and mechanical and biological treatment. These methods were relevant to waste management in developing countries [5,6]. According to the AHP method, alternatives, sub-criteria, and criteria were distributed into a multilevel hierarchy structure, as presented in Figure 1.

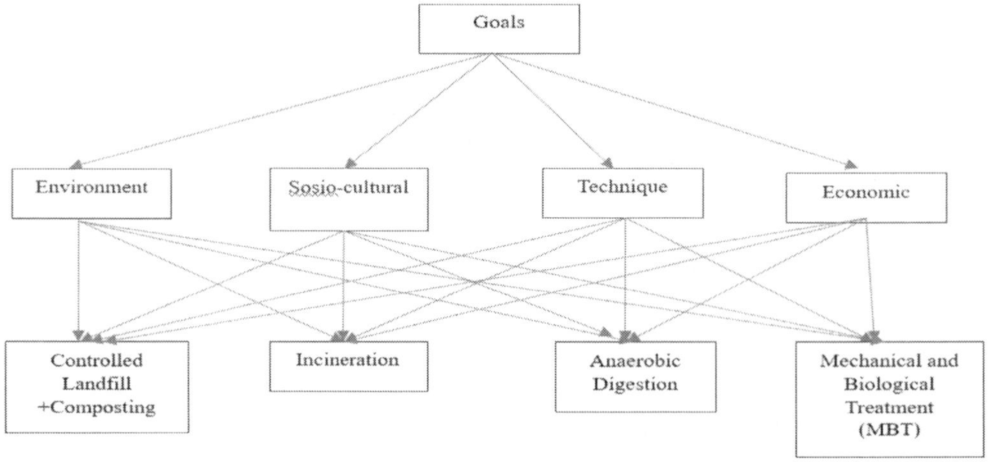

Figure 1. Construction of Hierarchy Analysis.

2.4. AHP steps

Based on the questionnaire responses by the experts, the first step was to create a decision matrix to identify priority weights, which consisted of objectives and sub-criteria, and the expert's scores were collected using the geometric mean. Second, every element was compared in pairs to know one important relationship level to get the goals. The comparison was input into the matrix. The equation determined the priority:

$$AW = \lambda max \times W \quad (1)$$

A is the comparison matrix, λmax is the eigenvalue, and W is the priority vector. Third, the feedback into the decision-making according to the value input was determined through consistency measurement ratio (CR):

$$CR = CI / RI \quad (2)$$
$$CI = \lambda max - n / (n - 1) \quad (3)$$

with CI as the consistency index, n as the dimension of the comparison matrix, λmax as the primary eigenvalue, and RI as the index ratio. If the consistency ratio is less than 0,1 (<10%), the matrix is assumed to be consistent. Otherwise, the measurement is continued by modifying comparison to minimize the inconsistency.

3 Results and discussion

The pairwise matrix comparison between the main criteria: environment, technique, social acceptance and economic to goal with the geomean value from all experts was shown in Table 1.

Table 1. Pair comparison matrix of main criteria related to the goals.

Criteria	Environment	Sociocultural	Economic	Engineering
Environment	1	0.45	1.50	1.73
Sociocultural	0.45	1	0.61	1.00
Economic	1.50	0.61	1	0.92
Engineering	1.73	1.00	0.92	1

To produce a priority vector or eigenvector (right column), the vector in each matrix column (Table 1) was normalized (every column element was divided with the total column value), and the mean of every row of the matrix was the result, as presented in Table 2. The mean of every row was an eigenvector or priority vector of each criterion.

Table 2. Calculation of priority vector.

Criteria	Environment	Sociocultural	Economic	Engineering	Priority Vector
Environment	0.21	0.15	0.37	0.37	0.28
Sociocultural	0.10	0.33	0.15	0.21	0.20
Economic	0.32	0.20	0.25	0.20	0.24
Engineering	0.37	0.33	0.23	0.22	0.28

The following procedure was adopted to check the consistency manually. First, the eigenvalue (λmax) was calculated using the equation (1):

$$\begin{bmatrix} 0.21 & 0.15 & 0.37 & 0.37 \\ 0.10 & 0.33 & 0.15 & 0.21 \\ 0.32 & 0.20 & 0.25 & 0.20 \\ 0.37 & 0.33 & 0.23 & 0.22 \end{bmatrix} \text{x} \begin{bmatrix} 0.28 \\ 0.20 \\ 0.24 \\ 0.28 \end{bmatrix} = \lambda\text{max} \begin{bmatrix} 0.28 \\ 0.20 \\ 0.24 \\ 0.28 \end{bmatrix}$$

$$\text{Matrix of table 2} \times \text{priority vector} = \lambda\text{max} \times \text{priority vector} \quad (1)$$

The mean value (λmax) is obtained from the matrix operation given above, (λmax) average = 4.1913. So, by using equation (3), the consistency index (CI) was calculated by:

$$CI = (\lambda max - n)/(n-1) = (4.1913 - 4)/(4-1) = 0.064 \quad (3)$$

Therefore, the consistency ratio (CR) is calculated by equation (2):

$$CR = CI/RI = 0.064/0.90 = 0{,}07 < 0{,}1 \quad (2)$$

The RI value of 0.90 was obtained from Table 2 for n = 4 (matrix size). Because the CR value (0.07) was less than 0,1 (maximum limit permission), the assessment of the matrix (Table 2) was assumed to be consistent and satisfying logically.

The next assessment involved comparing the effect of sub-criteria on the main criteria. The assessment on economics showed that the effect of investment cost, operation and maintenance costs had the same role on waste management treatment options. Pathogenic exposure had the biggest effect, followed by air, water and soil pollution, considering the main criteria of the environment. The main criteria of the matrix technique showed that the sub-criteria of energy and nutrition recovery were the highest influenced in deciding than easy and effectiveness, possibility and sustainable operation. The effect of acceptability on the main sociocultural criteria had more effect than community engagement and area vulnerabilities. All the assessments in the comparison matrix were consistent because the third matrix's CR value was lower than 0.1.

Furthermore, a comparison value is made to the waste treatment technology options based on each criterion and sub-criteria. The rank of waste management option based on the effect of all the criteria and sub-criteria was presented in Figure 2. Figure 2 showed that anaerobic digestion and incineration were suitable for waste technology based on air, water and soil pollution sub-criteria. Anaerobic digestion was the best choice based on the impact of pathogenic exposure potential and vulnerability. From the land usage view, mechanical and biological treatment was chosen. Incineration was assumed to be the best choice based on air, water and soil pollution, technology sustainability and energy and nutrition recovery potential. According to acceptability, community engagement, investment, operation and maintenance cost, and easy and effective operation, the controlled landfill and composting were assumed to be the proper waste treatment technology.

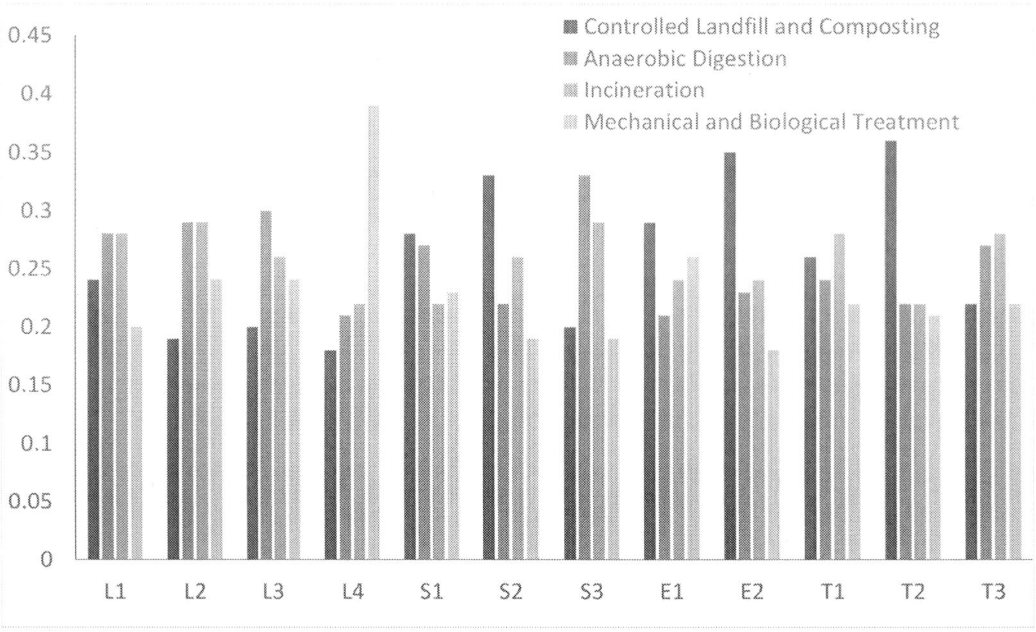

Where: L1 = Air pollution, L2 = Water and soil pollution, L3 = Pathogenic exposure, L4 = Utilization, S1 = Acceptability, S2 = Community engagement, S3 = Susceptibility area, E1= Investment cost, E2= Operating and maintenance cost, T1= Probability and sustainability, T2= ease and effectiveness in operation, dan T3= Energy recovery and nutrients.

Figure 2. Arrangement of waste management alternative based on sub-criteria.

According to the analysis obtained from this research, the suitable waste management technology for Singaraja municipal solid waste was the controlled landfill and composting or incineration followed by anaerobic digestion and mechanical biological treatment. The environment and technique were the factors which informed the waste management option for small cities. This result differed from [7], which reported that anaerobic digestion was the best waste management technology using the AHP method.

Controlled landfill and composting were suitable for the overall aim, sociocultural acceptance criteria, economics and technique [2,8,9]. Besides that, those technologies were cheap and had minimized negative impacts. Contrary to environmental criteria, controlled landfill was not the first rank. Incineration was assumed to be the suitable option for the environmental criteria. Overall, incineration was also equal to controlled landfill and composting. Incineration was not on the first rank in sociocultural, economic and technique, although it was suitable for sustainability sub-criteria and energy recovery in technique criteria. Incineration was considered less easy and more effective than controlled landfill and composting.

The next rank was anaerobic digestion, which was suitable considering to environment criteria. It was unsuitable for comparing the other three criteria to controlled landfill and incineration. Although anaerobic digestion had the potential for energy and nutrition recovery, it was unsuitable for technique criteria compared to incineration. The sustainability and effectiveness of the operation were lower than incineration. The anaerobic digestion processed biodegradable waste. About 80% of the Singaraja waste composition was managed by anaerobic digestion or composting. Incineration was the most suitable technology for quantity and quality. The incineration could process more than 99% of total waste except metal and glass (0.34%).

4 Conclusion

The case study was the quantity and quality of Singraja waste managed by Buleleng municipalities. AHP achieved the preference of criteria and sub-criteria. The result showed that the most decision-making for small cities' waste management was environment and technical, with each eigenvector priority (0.28), followed by sociocultural (0.24) and economic (0.20). Based on the technology options, the controlled landfill and composting were relevant, followed by incineration, anaerobic digestion, and mechanical and biological treatment. Therefore, the Buleleng government should apply the composter technology for the organic waste fraction of municipal solid waste (food, garden, and organic waste) from sources or households, by community groups or centralized composting. Besides, the garbage bank should be managed effectively to minimise the amount of plastic and organic waste in the controlled landfill. This option technology was good not only in economic, socio-cultural and technology but also in environmental aspects. To get high biodegradable waste quality, the government should invest the waste collection and separation from door to door.

References

1. SIPSN, National Waste Management Information System, sipsn.menlhk.go.id/sipsn/ (2020)

2. I. M. Gunamantha, M. V. Oviantari, N. W. Yuningrat, Journal of Environment and Earth Science, **13**, 5 (2023)

3. A. Guillermo, P. Macoun, Center for International Forestry Research(CIFOR) (1999)

4. M. A. Babalola, Environments, **2**, 4, 471–488 (2015)

5. N. Tsydenova, A. V. Morillas, A. A. C. Salas, Recycling, **3**, 3 (2018)

6. N. H. Hoang, C. Fogarassy, Sustainability (Switzerland), **12**, 3, 1–20 (2020)

7. W. A. Qazi, M. F. M. Abushammala, M. H. Azam, Waste Management and Research, **36**, 7, 594–605 (2018)

8. N-S.Toan, D.H. Hanh, T.T.Y Anh, H.K. Thi, H.T.D. Uyen, N-S. Tran, Journal of Soil Science and Agroclimatology, **20**, 1, 94-99 (2023)

9. K. Utami, E. Hanudin, M. Nurudin, Journal of Soil Science and Agroclimatology, **17**, 2, 152-160 (2020)

Energy balance and methane emission prediction in Simmental Cross-breed Cattle Feeding different types of forages and concentrates

Azka Nuruzzaman, *Yuli* Yanti[1]*, *Sulistyo* Sulistyo[1], and *Susi* Dwi Widyawati[1]

[1]Animal Science Study Program, Faculty of Animal Science, Universitas Sebelas Maret., Jalan Ir. Sutami No. 36A Kentingan, Surakarta, Jawa Tengah, 57126, Indonesia

Abstract. This study aims to examine the energy balance of Simmental Cross-breed Cattle given different forage and different proportion of concentrates. This research was conducted at Izzah Farm, Boyolali Regency, Central Java. The material for this research consisted of nine bulls with an average initial body weight of 506±80.09 kg. The research design used a completely randomized design with three treatments and three replications each treatment. The feed proportion was 70:30 (concentrate:forage). The treatments were T1=pollard bran+rice bran (1:4) and kolonjono grass, T2 = pollard bran+rice bran (1:4) and fermented rice straw, and T3 = pollard bran+commercial concentrate (1:4) and fermented rice straw. Data were analysed using the Anova test. The results of this study showed that there was no difference ($P<0.05$) in digested energy and metabolized energy with an average of 1.01 MJ/BB$^{0.75}$/day and 0.77 MJ/BB$^{0.75}$/day. Energy intake was 1.65 MJ/BB$^{0.75}$/day. Energy digestibility and methane emission emission were significantly different with an average of 60.22% and 7.40 KJ/100 KJ GEI. The conclusion of this study is that ransom containing pollard bran+commercial concentrate (1:4) and Fermented Rice Straw has low methane emission.

1 Introduction

The feed aspect plays an important role in fattening cattle, so attention must be paid to aspects of the availability and adequacy as well as the nutritional content of the feed. Apart from that, the economic efficiency aspect of farmers must be considered. This is because almost 70% of the costs incurred to increase livestock productivity are mostly for procuring animal feed. The use of quality and sufficient feed to meet livestock requirements will increase beef cattle production parameters such as daily weight gain, carcass and meat production, feed intake, feed conversion and feed efficiency can be maximized. Feed plays an important role in livestock fattening sector. One important factor that influences daily weight gain of livestock is feed energy intake.

* Corresponding author: yuliyanti_fp@staff.uns.ac.id

Traditional farmers generally use natural resources for feed that is nearby. One of the food resources that easy to reach is kolonjono grass. Kolonjono grass has several advantages such as high productivity, sufficient nutritional content and a palatable [1]. However, the growth area for kolonjono grass is getting narrower, so replacement feed ingredients are needed which are cheap and easy to obtain. One substitute feed ingredient that can be used by farmers is fermented rice straw [2]. The use of fermented rice straw is due to the increase in feed quality due to the involvement of microorganisms in degrading crude fiber, reducing lignin levels and anti-nutritional compounds, so that the digestibility value of feed from waste can increase [2] [3].

The rice straw used for feeding livestock in single form cannot meet the nutritional requirements of livestock. So additional feed in the form of concentrate is needed. The concentrate can be formulated from pollard bran, rice bran or commercial concentrate. However, when using these feed, it is necessary to have the proper proportions. Based on the description above, research was carried out to obtain energy balance in simmental crossbreed cattle given different types of forage and different proportion of concentrates.

2 Materials and methods

This research was conducted from September-December 2022 in Izzah Farm, Sambi District, Boyolali Regency, Central Java. Chemical analysis of feed ingredients was carried out at the Animal Feed and Nutrition Laboratory, Faculty of Animal Science, Sebelas Maret University. Energy test analysis at the Faculty of Animal Husbandry, Gadjah Mada University.

The materials used in this study included nine Simmental cross-breed cattle with initial body weight of 506 ± 80.09 kg, kolonjono (*Brachiaria mutica*) grass, calfeed C-130 commercial concentrate, fermented rice straw, rice bran and wheat bran.

The research design used a completely randomized design (CRD) with three types of treatment (T1, T2, T3) and three replications each treatments.

T1: C1 (rice bran+wheat bran) 70%+ kalanjana grass 30%
T2: C2 (rice bran+wheat bran) 70%+ fermented rice straw 30%
T3: C3 (commercial concentrate+wheat bran) 70%+ fermented rice straw 30%

The preparation stage consists of pen preparation, cattle preparation, forage preparation and concentrate feed preparation. The pen preparation stages include cleaning and repairing. The next preparation stage is the cattle procurement stage, this stage includes administering medicine and vitamins, providing adapted feed for 14 days. The forage preparation stage includes providing forage in the form of kolonjono grass. The concentrate preparation stage includes mixing pollard bran and rice bran until homogeneous and commercial concentrate and rice bran until homogeneous. The nutritional content of feed ingredients is presented in Table 1 and the proportion content of concentrate rations is presented in Table 2.

Table 1. Nutrition content of ration ingredients.

Ration	DM (%)	CP (%)	CF (%)	Ash (%)	Energy (kkal)
Kolonjono	17.4	8.6[c]	1.3[c]	12.8[c]	3798.2
Fermented rice straw	56.9	7.3[a]	22.4[a]	6.1[b]	3574.2
C1	90.3	7.9	24.5	12.8	3547.4
C2	90.3	7.4	23.4	14.9	3581.3
C3	86.4	6.1	28.3	17.6	3319.3

DM= dry matter, CP= Crude Protein, CF= Crude Fiber, [a] [4], [b] [5], [c] [6]

The next stage includes weighing the cattle first to determine the body weight of the cattle. This stage includes feeding, 3% of body weight, carrying out the total collection model, and collecting data. Feeding was carried out according to the treatment, which can be seen in Table 3. Data collection or total collection of faeces and urine was carried out for 7 days. Faeces were collected in the morning and then weighed. The last day of the total collection, all samples were mixed per animal and then 500 grams of samples were taken. The faeces were then dried and ground for proximate analysis. Urine is collected every morning and then weighed. All urine that has been collected for seven days is then mixed evenly per cattle, then a sample of 250 ml is taken. The methane emission was calculated according to Blaxter and Clapperton [7].

Table 2. Feed proportion of concentrate.

Feed Materials	Proportion (%)		
	C1	C2	C3
Pollard bran	66.7	66.7	-
Rice brand	33.3	33.3	20
Concentrate	-	-	80
Total	100	100	100

Table 3. Energy content of feed ration for cattle.

Feed	Energy (cal/g)	Energy content (cal/g)		
		T1	T2	T3
C1	3547.4	2483.2	-	-
C2	3581.3	-	2506.9	-
C3	3319.3	-	-	2323.5
Kolonjono	3798.2	1139.5	-	-
Fermented Rice Straw	3574.2	-	1072.3	1072.3
	Total	3622.6	3579.19	3395.78

Source: Laboratory analysis.

The data obtained were analyzed using analysis of variance (ANOVA). If there are differences between treatments or groups, they will be tested further using the Duncan Multiple Range Test (DMRT) to determine the differences between treatments. All data obtained were analyzed using R studio software version R 4.2.0.

3 Result and discussion

Based on the research that has been carried out, data obtained in the form of average energy consumption, digested energy, energy digestibility, metabolized energy and methane energy are presented in Table 4.

3.1 Gross Energy intake

Energy consumption is the amount of energy consumed by livestock through dry matter consumption. The more rations consumed, the greater the energy consumed by livestock [8]. Energy consumption among treatments was found to be different ($P<0.05$). This is because the energy content in the ration among the treatments was found to be different, namely 3622.6, 3579.19, and 3395.78 kcal for T1, T2, T3, respectively. This research is in line with [9] statement which states that the high or low energy content in feed influences the amount of feed consumed. Thus, the energy content of the three feeds causes relatively different feed consumption. The highest energy consumption in this study was at T3. It is suspected that feed T3 has a higher palatability than T2. High palatability can increase feed intake which can increase energy consumption.

Table 4. Energy balance of Simmental Crossbreed.

Parameters	Treatments			*P* Value
	T1	T2	T3	
Energy intake ($MJ/BB^{0,75}/day$)	$1.67{\pm}0.15^{ab}$	$1.37{\pm}0.19^{b}$	$1.92{\pm}0.10^{a}$	0.014^{*}
Digested Energy ($MJ/BB^{0,75}/day$)	1.14±0.12	0.85±0.17	1.03±0.07	0.078
Digestible energy (%)	$67.27{\pm}3.27^{a}$	$60.2{\pm}4.88^{b}$	$53.18{\pm}0.98^{c}$	0.007^{**}
Energy metabolism ($MJ/BB^{0,75}/day$)	0.88±0.14	0.61±0.16	0.82±0.03	0.079
Methane (KJ/100 KJ GEI)	$7.84{\pm}0.20^{a}$	$7.40{\pm}0.30^{b}$	$6.97{\pm}0.06^{c}$	0.008^{**}

gei : gross energy intake, a, b, c = superscript different letters indicates different
T1: C1 (rice bran+bran) 70%+ kalanjana grass 30%
T2: C2 (rice bran+wheat bran) 70%+ fermented rice straw 30%
T3: C3 (commercial concentrate+wheat bran) 70%+ fermented rice straw 30%

3.2 Digestible Energy

Digestible energy is energy obtained from digested nutrients [10]. Digestible energy is obtained by calculating the difference between the energy consumed by livestock and the energy released through the faeces. The digested energy in this study shows that there is no difference (P>0.05) and has a positive value, which means that there is digested energy that is used for production. The factor causing no significant difference in digested energy was caused by the equation of the speed of the digesta flow. The speed of digesta flow is influenced by several factors, namely the type of feed given, the amount of feed consumed, and the physical ability of the rumen to consume the ration. T1 used forage kolonjono grass while T2 and T3 used fermented rice straw. It is suspected that although the energy consumption at T2 and T3 is high, the digestibility is low. So as to produce the same digestible energy in the three treatments. The digestible energy in Kacang goat fed different composition of concentrate did not show differences [11].

3.3 Energy Digestibility

Energy digestibility is the percentage of feed consumed by livestock and not excreted in the faeces. Energy digestibility in cattle between treatments showed significantly different results (P<0.05). This is presumably due to relatively high energy consumption and energy expended from faeces. This assumption is strengthened by the opinion of [12] who states that high digestibility reflects the large contribution of certain nutrients to livestock, while feed with low digestibility indicates that the feed is less able to supply nutrients for basic living and livestock production purposes.

The average result of the energy digestibility analysis in this study was 60.22%. The results of this study are lower than the results of [13] with an average energy digestibility of 81.6%.

The low digestibility value of fermented rice straw can cause energy digestibility to also be lower than the digestibility value of kolonjono grass. This is because fermented rice straw contains lignin and silica which makes this feed ingredient of low quality [14].

3.4 Metabolized Energy

Metabolized energy in simmental crossbreed cattle among treatments did not show significantly different results (P>0.05). Metabolized energy is obtained from the amount of energy consumption minus the energy contained in urine, the energy contained in faeces, and the energy of methane. Feeding in the treatment of cattle did not show a significantly different effect on metabolized energy. Factors that influence metabolized energy are the physical or chemical properties of the feed which include digestibility, adequacy or balance of substances in the feed ingredients used, feed content, and the intensity of feed fermentation in the rumen.

The results of the analysis of metabolized energy in this study showed an average of 0.77MJ/BB$^{0.75}$/day. The results of this study are higher than the results of research by [15] which showed an average value of metabolized energy of 0.48 MJ/BB $^{0.75}$/day. The results of this study are in line with the opinion of [16] which states that metabolized energy is affected by energy digestibility.

3.5 Methane Energy

Methane energy in cattle between treatments showed significantly different results (P<0.05). Methane gas energy is the energy that comes out of the fermentation process of carbohydrates in the rumen by methanogenic bacteria [17]. It is suspected that this is influenced by high feed consumption resulting in high energy methane gas. This assumption is strengthened by the opinion of [18] that the energy expenditure of methane gas is influenced by feed quality, feed consumption, and feed digestibility.

The average results of methane gas energy analysis show a figure of 7.40 KJ/100 KJ GEI. The average results of methane gas energy analysis show a figure of 4.2%. This is lower than the research results of [19]which showed results of 6.10% and 4.80%. This amount of expenditure is in the range of opinion that methane gas expenditure is around 2-12% [20]. This low percentage indicates that the feed provided is of good quality.

Based on the results of the research conducted, it can be concluded that the provision of different types of forage and concentrates can provide the same results in terms of energy consumption, digested energy, energy digestibility, metabolized energy, and methane energy.

References

1. I.B. Kurnianingtyas, p. 1–38. Thesis. Sebelas Maret University (2012)
2. Y. Yanti, Surahmanto, A. Purnomoadi and Y. Kawamoto, J. Indones. Trop. Anim. Agric. **37**, 3 (2012)
3. D. Riayatsyah, Program Studi Peternakan, Fakultas Pertanian, Universitas Syiah Kuala. Banda Aceh. (2015)
4. N. Suningsih, W. Ibrahim, O. Liandris, R. Yulianti, J. Sain Peternak. Indones. **14**, 2 (2019)
5. M. Firdaus, M. Kamal, *J anim Sci*. (2022)
6. D.A. Lubis, *Ilmu Makanan Ternak* Jakarta: PT. Pembangunan (1992)
7. K.L.C.J.L. Blaxter, 1965 *Rumin. Br. J. Nutr* p. 511–522.
8. H.U. Dewi, Liman, Y. Widodo, *J. Ilm. Peternak. Terpadu* **4**, 2 (2016)
9. B. Saldaña, C.E. Gewehr, P. Guzmán, J. García, G.G. Mateos, *Anim. Feed Sci. Technol.* **221** (2016)
10. E Purbowati, C.I. Sutrisno, E. Baliarti, S.P.S. Budhi, W. Lestariana, E. Rianto, Kholidin, *Semin. Nas. Kebangkitan Peternak*. p. 130–138 (2009)
11. Y. Yanti, W.A. Pamungkas, F. Hermanu, S. Putranto, S.D. Widyawati and W.P.S. Suprayogi, *IOP Conf. Ser. Earth Environ. Sci.* **724**, 1 (2021)
12. Y. Yusmadi, N. Nahrowi, M. Ridla, *J. Agripet* **8**, 1 (2008)

13. M.Y. Klau, A.F. Pendong, R.A.V. Tuturoong, M.R. Waani, *Zootec* **40**, 2 (2020)
14. Y. Yanti, M. Yayota, *Rev. Agric. Sci.* **5**, 0 p. 65–76. (2017)
15. N.P. Mariani, I.G. Mahardika, S. Putra dan I.B.G. Partama, J. Peternakan Indo. **17**, 1 (2015)
16. Y. Yanti, T.J. Wiweka, S. R. Soegiarto, W. Pawestri, J. Riyanto, R. Dewanti, M. Cahyadi, A. K. Wati, *J. Livest. Anim. Heal.* **5**, 2 (2022)
17. S.C. Karekar, B.K. Ahring, *Biocatal. Agric. Biotechnol.* **47**, November 2022 p. 102526. (2023)
18. F. Parnian-Khajehdizaj, S.J. Noel, M. Johansen, M.R. Weisbjerg, A.L.F. Hellwing, O. Højberg, M.B. Hall, P. Lund, *J. Dairy Sci.* **106**, 6 (2023)
19. E. Rianto, M.Y. Efendi, A. Purnomoadi, *Sains Peternak.* **8**, 1 (2010)
20. C. Matthews, F. Crispie, E. Lewis, M. Reid, P.W. O'Toole, P.D. Cotter, *Gut Microbes* **10**, 2 (2019)

Energy optimization based on project life cycle (PLC): a literature review

Uda Subrata Aditama Kittie Aidon[1*], Frieda[1], Desriantomy[1], *Meilawaty* Okta[1], Supiyan[1], Devia[1], and *Saritha Kittie* Uda[2]

[1]Civil Engineering Department, Faculty of Engineering, University of Palangka Raya, Jl. Yos Sudarso, Palangka Raya, Central Kalimantan, Indonesia
[2]Biology Study Program, University of Palangka Raya, Jl. Yos Sudarso, Palangka Raya, Central Kalimantan, Indonesia

Abstract. The occurrence of global climate change is caused by the rising temperature of the earth as a result of high CO2 emissions due to excessive energy consumption in the construction sector. Energy consumption in construction activities starts from the pre-construction stage, then the construction stage and finally the post-construction stage. This study aims to find assessment indicators in reducing energy consumption based on literature studies within the scope of the project life cycle (PLC). The method applied here is a qualitative approach by identifying and reviewing some of the literature on reducing energy consumption in construction projects. The literature study found that there are 5 parts in the building life cycle that can be used to identify sources of energy consumption in construction projects, namely the building materials production stage, the construction process, the building use stage, the end of the building's life and disassembly stage. We identified that there are 3 stages of energy calculation in buildings, namely the stages of Manifest Energy, Operational Energy and Demolition Energy. We find that optimizing indirect energy at the production and planning stage has a significant impact on reducing energy consumption at a later stage through building policy and design. While direct energy optimization can be carried out at the construction and operation stages through energy efficiency in each activity.

1. Introduction

The construction sector consumes 40% energy and 16% water per year globally through material management and manufacturing, transportation, construction, operation, maintenance, renovation, demolition and disposal. Embodied energy in a building is energy consumed in all construction phases other than the operation phase [1, 2]. Energy used during the phase of building operation via space conditioning, water heating, lighting, other tools of building operations and other operational activities is called operational energy. The total energy of the building is the total of operational and embodied energies [1]. Operational energy consumes an average of 80-90%, while the rest (10-20%) by embodied energy during

* Corresponding author: subrataaditama@jts.upr.ac.id

the building life cycle [3]. Every nation should strive to achieve sustainable development and many countries have been working on preventing the climate change due to global warming and mitigate the effects to the society and environment. The increase in the earth's temperature due to high CO_2 emissions, one of which is caused by energy consumption in construction activities, such as the use of petroleum fuels in heavy equipment, the production of building materials and electricity, will have a major impact on climate change. Therefore, optimizing the use of energy at each stage of construction will directly reduce the amount of CO_2 emissions produced. Hence, reducing carbon emission has been an ultimate aim in every development project including construction projects.

2. Literature review

2.1. Definition of energy

In the field of construction, energy is produced directly and indirectly from fuel to carry out the construction process starting from planning to demolition. These activities include material production (cement, iron, steel, glass, etc.), construction processes (heavy equipment used during construction, material transportation, lighting, etc.), operational processes (water heating, heating and air conditioning, mechanical equipment electrical in the buildings, etc.), the demolition process (heavy equipment used for demolition, transportation of waste materials, the material recycling process, etc.). The dictionary defines embodied energy as "the sum of the energy associated, directly or indirectly, with the delivery of a good or service" [4]. It is very important to manage the use of energy due to its significant environmental and social impacts to the society. Buildings consume large amount of energy at each stage of the building projects starting from design and construction through operation and final demolition [5].

2.2. Life Cycle Energy (LCE)

In the construction sector, Life Cycle Energy Analysis (LCEA) is often used in order to ensure that the sustainable goals are achieved and energy consumption is minimized. This methodology is practically similar to the simplified Life Cycle Analysis (LCA). The difference is, the only parameter used in the LCEA is energy. According to the Dixit stated that buildings, materials and their components consume almost 40% of global energy every day in their life cycle, such as production and procurement of the building materials, construction, operational and demolition [6]. Thus, in order to better understand the use of energy in a building, it is important to differentiate and assess different energy needed in each phase of the life cycle.

2.2.1. Embodied Energy (EE)

The embodied energy is the total energy that relates, directly or indirectly, to the delivery of goods or services starting from mine workings and natural resources processing until manufacturing, transporting and delivering the products [4,7]. Researchers, however, have different point of views on the embodied energy depending on their field of study. There are three views that are mostly adopted: cradle-to-gate, cradle-to-site, and cradle-to-grave [8]. Each building is a complex combination of numerous processed materials which contribute to the total energy of the building. The energy consumed by each material relate closely to the level of processing of the materials. The more complex the materials are, the more processing are needed; and therefore, the more energy is consumed. A high level of energy use implies a higher level of pollution and emission emitted at the end of the production. Concrete, aluminum and steel are the materials which use the highest energy and release high level of CO_2 emission [9].

2.2.2. Operational Energy (OE)

The operational energy in the buildings is the energy imported from outside the system to operate the lighting, lift and escalator, ventilation system, heating and cooling system, water heater, and pumping system in order to maintain the environment inside the building. This energy includes energy from electricity, gas, and fuel combustion, such as oil or coal [3]. The highest operational energy consumption generally takes place in office buildings. Thus, it is important in the designing phase to assess the energy requirement for the building lifetime. As the operational energy is necessary for comfort conditions and stable maintenance of the internal environment; the target of this energy usage will vary depending on the needs and function of the building and the surrounding environment. Some researchers state that the operational energy is the biggest contributor (up to 80%) to the total energy consumption in a building's life cycle [3]. Unsurprisingly, researchers are more concerned on how to reduce the operational energy than the embodied energy and the demolition energy. Some researches mention that energy in the demolition state represents only 1% of the total energy of the building life cycle [10].

2.2.3. Demolition Energy (DE)

The demolition phase as the end of a building life cycle can be divided into 4 sub phases: deconstruction, transport of the waste materials, waste processing and waste disposal [11]. In each of these sub phases, the amount of embodied energy is calculated based on the activities and equipment used of which fuel consumption plays role to the amount of energy produced [12]. Based on samples from ten residential buildings going through the demolition and excavation phases, Moncaster and Symons found that the amount of embodied energy and embodied carbon were 77 MJ/m^2 and 5 $kgCO_2e/m^2$ respectively. The embodied energy during the demolition process includes all energy used on site for the deconstruction of the building parts, sorting of materials resulted from demolition and the transport of the waste materials to the final disposal site and [11]. In the end-of-life phase the energy is needed to operate the machines used for building demolition as well as the energy to level, store and transport the materials from the building site to the landfill site or final processing plant. The removal stage usually consumes about 1-4% of energy during the building's life cycle [13]. At this phase the reutilization of the waste materials or building components will reduce the environmental impact and minimize energy consumption [14].

Research results

3.1. Assessment energy building

The Life Cycle Energy Assessment Method (LCEA) uses a detailed approach to estimate the total amount of input energy, energy that occurs during a building's life cycle, and energy output [15]. Energy assessment is carried out for embodied energy and operational energy. Embodied energy is calculated from the material production stage at the factory (raw material supply, raw material transport, manufacturing), then the building construction process (material supply / transport material and construction installation process), through the building use period (building operational, materials/equipment installed in the building, maintenance, repair, replacement and refurbishment), and finally to the end of building life (deconstruction, transport of waste, waste processing and disposal stage). Embodied energy itself can be divided into two parts: initial embodied energy and recurring embodied energy. The initial embodied energy corresponds to the energy required for the extraction, manufacturing and transportation of the building materials, and it is also the energy necessary for the entire construction process. Meanwhile, the recurring embodied energy is the energy used for the maintenance and renovation activities that are related to replacement of building materials after building construction. Operational energy, on the other hand, is the energy used to maintain the comfortable conditions inside the building through processes such as heating, cooling, ventilation, lightning, hot water and appliances/equipment operation. Finally, demolition energy is the energy required to demolish the building and, transport the waste materials to landfill sites and/or recycling plants. The total energy in a building life cycle is therefore the sum of all energy consume [3,16]. Assessment of energy in construction building based on the building life cycle can be calculated by the following formula (table 1):

Table 1. Assessment of Energy Consumptions in Building [3,17].

No	Category Assessment	Calculation Formula	Explanations
1	**Embodied Energy**	$EE = EE_i + EE_r$	EE = Embodied Energy; EE_i = Initial Embodied Energy of the building; EE_r = Recurring embodied energy of the building
	Initial Embodied Energy	$EE_i = E_M + E_C$	EE_i = initial embodied energy of the building; E_M = Energy for building material manufacturing; E_c= Energy used for building construction
	Energy for building material manufacturing	$E_M = \sum m_i . M_i$	E_M = Energy for building material manufacturing; m_i = Quantity of building material required to produce a building; M_i = Energy content of material per unit quantity.
	Energy used for building construction stage	$E_C = E_{C\_T} + E_{C\_E}$	E_c = Energy used for building construction; E_{C_T} = Energy required to transport building materials from the plant to the construction site; E_{C_E} = Energy required to construction equipment
	Energy required to transport building materials from the plant to the construction site	$E_{C\_T} = 2. \sum_{j=1}^{m} \sum_{i=1}^{n} TD_i^j . Q_i . EC_i^j$	E_{C_T} = Energy required to transport building materials from the plant to the construction site; TD_i^j = Transportation distance of the building material i from the plant to the construction site using vehicle j; Q_i = quantity of material i; EC_i^j = Energy consumption per km of vehicle j for a ton or cubic meter of material i.
	Energy required to construction equipment	$E_{C\_E} = \sum_{k=1}^{l} \sum_{i=1}^{n} D_i^k . C_i^k$	E_{C_E} = Energy required to construction equipment; D_i^k = Duration of equipment k usage;

			C_i^k = Energy consumption per hour of equipment *k*.
	Duration of equipment usage	$D_i^k = Q_i / C_i^k$	D_i^k = Duration of equipment *k* usage; Q_i = Quantity of material *i*; C_i^k = Capacity for work per hour of equipment *k*.
	Recurring Embodied Energy	$EE_r = \sum m_i M_i [(L_b/L_{mi}) - 1]$	EE_r = Recurring embodied energy of the building; m_i = Quantity of building material required to produce a building; M_i = Energy content of material per unit quantity; L_b = Life span of the building; L_{mi} = Life span of the building material.
2	**Operational Energy**	$OE = E_{OA} . L_b$	*OE* = Operational Energy; E_{OA} = annual operating energy; L_b = life span of the building.
3	**Demolition Energy**	$DE = E_T + E_D$	*DE* = demolition energy; E_T = energy incurred for destruction of the building; E_D = energy used for transporting the waste materials.
TOTAL ENERGY		$LCE = EE + OE + DE$	*LCE* = Life Cycle Energy of the building; EE = Embodied Energy of the building; OE = Operating Energy of the building; DE = Demolition Energy of the building.

3.2. Energy optimization base on project life cycle

The amount of research on Embodied Energy (EE) is more than those on the operational energy. This shows that the high interest to optimize and reduce the EE in construction process. However, research on the conventional buildings in the Northern and Central Europe shows that the total of Embodied Energy in the building materials represents about 10 - 20% of the total energy whilst the operational energy covers 80 - 90% and less than 1% are Embodied Energy occurs during the maintenance period and the end of the building's life [18]. The research by Mohammed found that the embodied energy from construction phase until end-of-life phase, it gets the cumulative value of embodied energy for 6.44 GJ/m^2. Meanwhile the operational energy has a cumulative value of 70.28 GJ/m^2 and is more than 85% of total energy for 50 years period [17]. Based on these studies researchers should focus on the efforts to optimize embodied energy (EE) from the beginning so that it will lead to the reduction of the energy during the operational period of the building. The optimization strategy should be implemented since the beginning of the project activity in order to make the energy control and management efforts easier and more measurable. The initiation and design phases are the important part to start the energy reduction efforts. Without a strong commitment to prioritize the construction of environmentally friendly buildings, it will be difficult to reduce impacts of high energy consumption for the global climate. It is possible to produce an effective planning without reducing the strength and beauty of a building structure. Furthermore, it is equally important to reduce the amount of energy during the operation and service life of the buildings.

4. Conclusion

This study shows that the implementation of energy optimization strategy is effective to decrease the amount of energy in the life cycle of a construction project, however, it requires and the commitment of all stakeholders on minimizing the impact of buildings on the environment. Stakeholder commitment should be established in the form of contract documents that legally bind the consultant planners, supervisors and contractors as well as the building managers. As mentioned in several studies, incorporating energy optimization at each stage of construction will affect the on-building operation, especially electrical energy during the life of the building, since operational energy consumes 80 - 90% of total energy. Energy minimizing strategy can be carried out by applying strict rules on the use of energy in each construction phase including the design stage. The building, for example, can be designed to have natural lighting and maximum air circulation to minimize the use of air conditioners, as well as the selection of materials with low-energy s production process, maximizing the use of natural materials and implementing the waste management plan. Besides, optimizing the transportation of materials and construction equipment has a direct impact on reducing the use of fossil fuels. Energy optimization can be performed by creating a model and pattern based on the existing indicators above, then develop the standards and operational evaluations of the construction activities for all stakeholders: consultants, and contractors as the executor in the scope of project life cycle (PLC).

References

[1] F.H. Abanda, L. Byers, An investigation of the impact of building orientation on energy consumption in a domestic building using emerging BIM (Building Information Modelling), *Energy*, **97** (2016) 517–527.

[2] C. Carbonaroa, Y. Casconeb, S. Fantuccib, V. Serrab, M. Perinob, M. Duttoc, Energy Assessment Of A PCM-Embedded Plaster: Embodied Energy Versus Operational Energy, (2016).

[3] C.K. Chau, T.M. Leung, W.Y. Ng, A review on Life Cycle Assessment, Life Cycle Energy Assessment and Life Cycle Carbon Emissions Assessment on buildings, *Appl Energy*, **143** (2015) 395–413.

[4] C. Cleveland, C. Morris, Dictionary of Energy (Expanded Edition, 2009).

[5] J. Davison, D. Densley Tingley, Design for deconstruction and material reuse, *Energy*, **164** (2011) 195–204.

[6] T. Ibn-Mohammed, R. Greenough, S. Taylor, L. Ozawa-Meida, A. Acquaye, Operational vs. embodied emissions in buildings - A review of current trends, *Energy Build*, **66** (2013) 232–245.

[7] B.E. 15978 British Standards Institution, Sustainability of construction works-Assessment of environmental performance of buildings-Calculation method, (2011).

[8] L. Aye, T. Ngo, R.H. Crawford, R. Gammampila, P. Mendis, Life cycle greenhouse gas emissions and energy analysis of prefabricated reusable building modules, *Energy Build*, **47** (2012) 159–168.

[9] L. Guan, M. Walmsely, G. Chen, Life Cycle Energy Analysis of Eight Residential Houses in Brisbane, Australia, *Procedia Eng.*, **121** (2015) 653–661.

[10] A.B. Larriba, A.B. Larriba, Analysis and Evaluation Of 3 Rd Draft Criteria for Buildings and Next Steps, Institute for Prospective Technological Studies. JRC Europe Commission, UE. I (2010) 9–18.

[11] M. Dixit, J. Fernandez-Solis, S. Lavy, C. Culp, Need for an embodied energy measurement protocol for buildings: A review paper, *Renewable and Sustainable Energy Reviews*, **16** (2012) 3730–3743.

[12] A. Moncaster, K. Symons, A method and tool for 'cradle to grave' embodied carbon and energy impacts of UK buildings in compliance with the new TC350 standards, *Energy Build*, **66** (2013) 514–523.

[13] A. Carolina, S. Monteiro, Assessing Initial Embodied Energy in Building Structures using LCA Methodology, (2015).

[14] Kotaji S., Schuurmans S., Edwards S., Life-Cycle Assessment in Building and Construction : A state-of-the-art report, (2003), (n.d.).

[15] S. Sattary, D. Thorpe, Optimizing embodied energy of building construction through bioclimatic principles, (2012).

[16] S. Schimschar, K. Blok, T. Boermans, A. Hermelink, Germany's path towards nearly zero-energy buildings—Enabling the greenhouse gas mitigation potential in the building stock, *Energy Policy*, **39** (2011) 3346–3360.

[17] K. Tarabieh, M. Khorshed, Optimizing Evaluation Methods for the Embodied Energy and Carbon Management of Existing Buildings in Egypt, *Buildings*, **9** (2019).

[18] T. Ramesh, R. Prakash, K.K. Shukla, Life cycle energy analysis of buildings: An overview, *Energy Build*, **42** (2010) 1592–1600.

Spatial optimization of the economic benefits of the potential hydropower locations: study case upper citarum river

Assyifa Rosani[1], *Faizal Immaddudin Wira* Rohmat[3*], *Azman Syah Barran* Roesbianto[2,3], *Nabila Siti* Burnama[2,3], *Hadi* Kardhana[2,3], *Arno Adi* Kuntoro[2,3], and *Mohammad* Farid[2,3]

[1]Master Program of Water Resources Management, Faculty of Civil and Environmental Engineering, Bandung Institute of Technology, Ganesa Street, Number 10, Bandung 40132, West Java, Indonesia
[2]Faculty of Civil and Environmental Engineering, Bandung Institute of Technology, FTSL Building, Ganesa Street, Number 10, Bandung 40132, West Java, Indonesia
[3]Water Resources Development Center, Bandung Institute of Technology, Ganesa Street, Number 10, Bandung 40132, West Java, Indonesia

Abstract. *Hydropower* is a renewable energy source that provides clean energy services and contributes to climate change mitigation by replacing fossil fuel energy production sources. The Upper Citarum Watershed in West Java, Indonesia, is described as a giant basin known as the Bandung Basin, which is type by mountainous relief with an elevation of 750-2300 meters above sea level, a slope of 30-90%, and a radiating river flow pattern. This study focuses on the potential of small hydropower with a run-of-river scheme that acts as a filler for the study of hydropower potential with a pump storage scheme by the Australian National University. Climate data in the study area were analyzed using the SWAT model to obtain Cumulative Discharge Frequency and identify potential discharge by adopting the method of previous studies. The results identified 7 SHPs with P exceeding 500 kW and below 2000 kW. BHUMI ATR BPN can predict land value and land cover in areas designated for hydropower development. The findings suggest improvements in aspects such as proper forecasting, on-the-ground valuation, and calculations relating to economically viable sustainability.

1 Introduction

The IPCC in 2018, reported that the challenge of keeping the Earth's temperature rise below 1.5°C is becoming increasingly serious. At the current rate, warming will continue to increase until 2040 [1], causing both hydrological and other aspects of climate change [2]. These changes are also due to continued population and economic growth, which increases the demand for sustainable renewable energy as an alternative measure in the conventional use of fossil fuel reserves [3]. Renewable energy sources can provide clean energy services and contribute to climate change mitigation by replacing fossil fuel energy production sources. The importance of renewable energy is associated with all future climate change scenarios,

* Corresponding author: faizalrohmat@itb.ac.id

as results the penetration of renewable energy on a significant scale in the energy network [4].

The Upper Citarum Watershed in West Java, Indonesia (Figure 1) is described as a giant basin known as the Bandung Basin, characterized by a mountainous relief with an elevation of 750–2300 m above sea level, an upper slope of 30-90%, and radiating river flow pattern [5]. The Upper Citarum River has potential for small hydropower development, with potential of 700 megawatt. A result of population growth, government continues to receive requests for additional capacity of up to 1,400 megawatt (MW) [5]. The Indonesian government stated in PLN's 2021–2030 RUPTL that hydropower will be one of the backbones of the national net zero strategy, expected to be the main source with 25.6% of the national electricity supply by including mini-hydropower, whose potential is currently estimated at 95 gigawatt (GW) [3]. Small hydropower can contribute significantly to meeting global renewable energy targets [6]. The Australian National University (ANU) has analyzed the potential for large hydropower with pump storage schemes. The purpose of this study is to obtain the potential of small hydropower with a run-of-river (ROR) scheme that acts as a filler for the study of hydropower potential with a pump storage scheme conducted by ANU. Thus, river segments that are not included in pump storage can be known for their ROR potential.

Due to hydropower development there are often complaints regarding inadequate and unfair compensation also the difficult resettlement site [7]. For this reason, this study focuses on land use in selecting potential hydropower sites and economic data as input to estimate the cost of planning potential hydropower sites to avoid potential social and cultural problems and create economic benefits.

2 Materials and method

2.1 Study location

The Upper Citarum River is located between 107° 10' - 108° 00' E and 6° 40' - 7 20' S, covering an area of about 1700 km^2 (Fig.1). Administratively, it covers the area of West Bandung Regency, Bandung City, Cimahi City, and Sumedang Regency [8]. The Citarum River contributes greatly to the economic activities of the people of Bandung Regency with various activities especially land use such as agricultural, tourism, and industrial activities [9]. As part of the Bandung Basin, which has a rainfall of 2,300 mm per year, the Upper Citarum River West Java with a length of 6,614 km^2 and a width of 269 km has a meander relief shape with 4 existing hydropower plants [5,10]. The high human population is feared to affect riverbank land use and add constraints as a potential hydropower development area. Therefore, identifying potential hydropower points in residential areas and rice fields affects the prediction of the research.

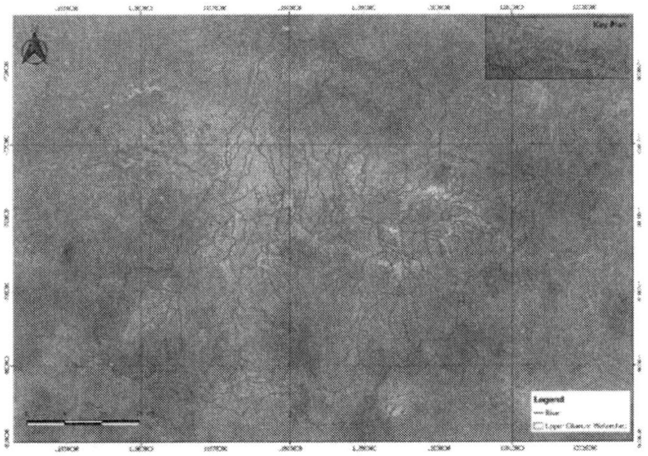

Fig. 1. Upper citarum watershed.

2.2 Data description

2.2.1 Rainfall data

Satellite-based rainfall estimates provide complement measurement over a wide area having few or even no in situ data [11]. In Kardhana et al [12], climate data used Climate Forecast System Analysis (CFSR) for 10 years from January 1st 2004 until December 31st 2011. The CFSR grid data is corrected by observation data in 31 rainfall station from BMKG, PLN, and PUSAIR utilized as data input to predict water balance in a watershed.

2.2.2 Land used data

The land use data utilized in this study was sourced from Peta Rupa Bumi Indonesia by the Geospatial Information Agency (BIG) with scale 1:25.000 in 2011. It is open source, can accessing on http://tanahair.indonesia.go.id (accessed on 25 August 2023). The content and classification of RBI map data types are standardized by scale, thus that uniformity of data and information can be maintained [13].

2.2.3 Topography data

Topographic data is digital data from remote sensing, commonly called Digital Elevation Model (DEM). In this research used as input for hydrological analysis, the DEM used comes from the NASA Shuttle Radar Topographic Mission (SRTM) with a resolution of 90 m at the equator. With data availability of ± 80% of the earth's surface, SRTM DEM is one of the reliable global data and is distributed free of charge [12].

2.2.4 Economic data

The land value of land parcels and information according to the layers listed can be obtained. The information about the value of land in a field from a Web GIS-based land big data, namely www.bhumi.atrbpn.go.id. The information provided is the result of a super impose of two layers, namely land parcels and ownership information, type of rights, NIB, use and land area. Meanwhile, to find out the value of a land parcel, the web provides a data set of the value of a zone [14].

2.3 Multi optimization

Determination of hydropower potential locations calculated by diversion algorithm in Python programming. Adopting Kardhana [12], river grid were sampled every 30 pixel, each sample has one selected pixel spot which has maximum P. The value data land used from ATR BPN helps select feasible choices among potential hydropower locations for land compensation cost. This work proposes a framework to guide the multi-objective optimal design of a small run-of-river hydropower plant, which simultaneously maximizes its economic value and minimizes its impact on hydrological connectivity. Determining variables in the optimization are the maximum power from flow processed by the plant capacity (Q) with reliable discharge and the possible location [6].

2.4 Methodology

This research focuses on the Upper Citarum Watershed to expect the initial representation in planning the potential location of hydropower plants with considering land used and economic benefits analysis. The step of this method can be seen as follows below.

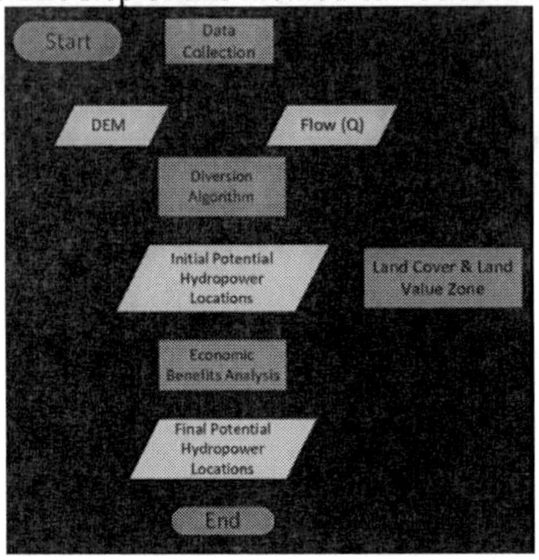

Fig. 2. Workflow.

The schematic workflow follows 4 step: collecting data, utilized diversion algorithm for determining initial potential hydropower location, applied spatial data overlay to land cover and land value zone in BHUMI ATR BPN, and results benefits cost analysis for final potential location (Fig.2). The data obtained were analyzed using the diversion algorithm in Python programming. The analysis results show the location of potential hydropower plants and the potential power that can be generated, the points are then layered into BHUMI ATR BPN to calculate the cost of hydropower development based on land value as shown in Figure 3.a. Furthermore, based on electricity production and electricity selling value, the amount of hydropower profit can be obtained.

3 Results

The diversion algorithm is implemented within a Python program that utilizes collected DEM (Digital Elevation Model) and discharge data. Climate data from the study area is analyzed using the SWAT model to calculate cumulative discharge frequency (CDF) and identify potential firm flow. Following this, the carrier channel tracing begins by generating a grid at the river's intake, employing 90 m DEM SRTM data for each pixel size. The tail race location is determined by selecting the river pixel with the lowest elevation among its neighbors. The horizontal penstock length is the predefined distance between this neighboring river pixel and the head pond. The process entails sequentially picking lower pixels from the head pond to establish the route, halting when no further lower pixel is available. These chosen pixels constitute the headrace. The river pixel's potential head is established by selecting the highest head from the available alternatives in the head pond, with routing disallowing loops. Meanwhile, predicting potential hydropower locations requires resampling procedures, which may yield similar spots but differing probabilities (P) due to reasonable changes in CDF [12]. To enhance feasibility, selected spots with P below 100 kW are discarded. The outcome identifies 7 SHPs with P exceeding 500 kW with nearly half possessing firm power surpassing 1600 kW which is located in large tributaries as shown in Figure 3.b the layer data applied on https://re100.eng.anu.edu.au/global/ accessed on (September 1, 2023).

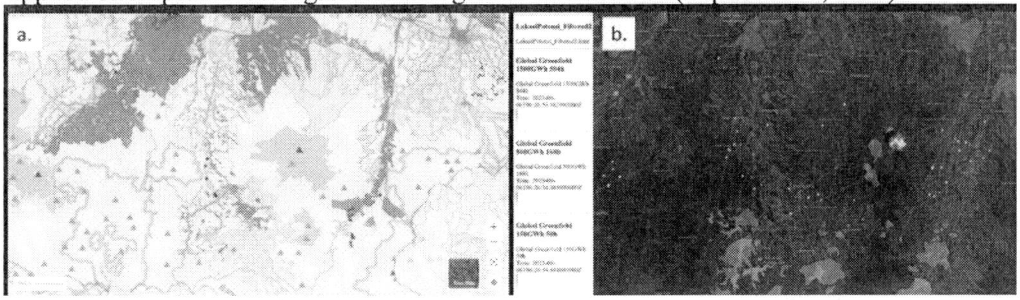

Fig. 3. a. Potential Location Results of Hydropower in BHUMI ATR BPN; b. Hydropower Potential Location ROR Scheme.

The study for diverting flow to generate hydropower from these locations will require significant cost and thus be unlikely feasible also requires additional variables. When the point results have been placed in the BHUMI data, the addition of data catalogues in the form of protected paddy fields, raw paddy fields, land use, and land value zones. During the overlay process, potential locations located in residential areas and rice fields are said to be eliminated. The filtered locations have land values that will be used as economic data for estimating the cost of hydropower development. In addition, it can also be calculated the profit of hydropower from the value of annual production against the cost of buying and selling electric power which refers to the amount of the cost of supplying electricity generation in the decision of the Minister of Energy and Mineral Resources of the Republic of Indonesia Number 169.K / HK.02 / MEM.M in 2021. From both aspects of the calculation, the feasibility level of hydropower potential points can be determined.

4 Conclusions

This research focuses on improving the methodology in finding potential hydropower locations with the appropriate ROR scheme. The diversion algorithm can develop analysis study to predict suitable locations for hydropower potential. The findings from the analysis of potential hydropower sites based on ROR considerations address the deficiency in available locations compared to the ANU global renewable energy map, which focuses on pump storage (shown in Figure 3.b) To get clear results after doing the layer, unfortunately the BHUMI web still does not have a feature to download the image of the layer results. Moreover, the BHUMI ATR BPN has the capability to predict the land value in the designated areas for hydropower development. Evaluations suggest enhancements in aspects like precise forecasting, on-site assessments, and calculations of economically viable sustainability.

References

1. U. N. P. UNEP, *Making Peace with Nature: A Scientific Blueprint to Tackle the Climate, Biodiversity and Pollution Emergencies.* (Norwegian Ministry of Climate and Environment, 2021), pp. 14–15
2. K. Calvin, et al. *IPCC, 2023: Climate Change 2023: Synthesis Report. Contribution of Working Groups I, II and III to the Sixth Assessment Report of the Intergovernmental Panel on Climate Change [Core Writing Team, H. Lee and J. Romero (Eds.)]. IPCC, Geneva, Switzerland.*, First (Intergovernmental Panel on Climate Change (IPCC), 2023)
3. IHA, (2022)

4. C Skoulikaris and K Kasimis, IOP Conference Series: Earth and Environment **899**, 012026 (2021)
5. BBWS Citarum, *Physical and Spatial Condiiton* (2014)
6. S. Basso, G. Lazzaro, M. Bovo, C. Soulsby, and G. Botter, Applied Energy **280**, 115936 (2020)
7. P. H. Ty, A. C. M. Van Westen, and A. Zoomers, Land **2**, 678 (2013)
8. A. A. Kuntoro, M. Cahyono, and E. A. Soentoro, ITB Journal Publisher **50**, 364 (2018)
9. A. S. U. Mudjiardjo, S. S. Moersidik, and L. Darmajanti, IOP Conference Series: Earth and Environmental Science **623**, 012062 (2021)
10. O. P. Purnama, (2019)
11. Nyoman Sugiartha, Nyoman Sugiartha, N. Sugiartha, K. Ogawara, T. Tanaka, and M. S. Mahendra, International Journal of Environment **1**, (2017)
12. H. Kardhana, D. K. Arya, I. K. Hadihardaja, Widyaningtyas, E. Riawan, and A. Lubis, AIP Conference Proceedings **1903**, 100015 (2017)
13. S. D. E. Pinudji, A. I. Savitri, M. Noormasari, D. Wijaya, and A. Kurniawan, Jurnal Dialog Penanggulangan Bencana **10**, 22 (2019)
14. D. F. Ferdiansyah, E. B. Wahyono, and S. Widodo, Tunas Agraria **5**, 47 (2022)

Policy strategies for climate-resilient infrastructure in climate change adaptation : jurisdictional comparison between Chile and Indonesia

Maskun Maskun[1], *Nurul Habaib Al* Mukarramah[2*] , and *Josse Charmario Wara* Angi[2]

[1]International Law Department, Faculty of Law, Hasanuddin University 90245 Perintis Kemerdekaan KM.10, Makassar, Indonesia
[2]Research Assistants, Faculty of Law, Hasanuddin University 90245 Perintis Kemerdekaan KM.10, Makassar, Indonesia

Abstract. The impact of climatic change such as extreme weather events and sea level rise have resulted in a greater burden for countries surrounded by waters including Indonesia as an archipelagic state and Chile as a coastal state to adapt to climate-related disasters and building resilience by developing climate-resilient infrastructure. This article is purposed to analyze legal regulation to strategize preparedness in climate adaptation policy by incorporating climate-resilient infrastructure. This article utilizes jurisdictional-based comparative legal methods as a part of normative legal research, with statutory approach through laws and regulations as its primary data, with country practices obtained through literary sources used as secondary data. The result reveals that there are two strategies needed to be taken by countries for climate adaptation policy: 1) Policy initiation and formulation based on climate change risk mapping by considering potential financial cost, appearing social problems and environmental impact; 2) Policy enactment through legal instruments mandating state to provide climate-resilient infrastructure to adapt to climate change. Chile demonstrates best practice in policy formulation strategies through initiation of ARClim and enacting climate change legislation in 2022, compared to Indonesia with no existing climate change legislation and national climate risk mapping. In conclusion, this research implies that countries prone to climate change impacts including Indonesia shall urgently initiate, formulate and enact policies incorporating climate-resilient infrastructure through the proposed strategies.

1 Introduction

The historical emission from global north countries during its industrial revolution period till current, have become the main driver of rapid changes across climate system and environment due to the warming of atmosphere, ocean and land [1]. Climate change will

* Corresponding author: habaib.edu@gmail.com

continue accelerate sea level rise (SLR) which potentially submerge countries with low-lying coastal ecosystems and encroaching coastal settlements and infrastructure, thus putting threat to the ecosystem from irreversible biodiversity loss [2]. Majority of affected countries located in global south, including Chile and Indonesia [3,4]. Due to the disparity of capacity to adapt to climate change between global north and global south countries, this further widen the gap of inequality which obstructs the pathway in achieving climate justice. Therefore, it is in dire need for the common but differentiated responsibility (CBDR) [5,6] principle to be implemented in the effort of limiting the temperature increase to 1.5°C and simultaneously building resilience from climate change including establishing climate-resilient infrastructures.

Various international legal instruments and policy directives such as the Sendai Framework for Disaster Risk Reduction 2015-2030 (Sendai Framework) have urged states to establish strategies in adapting to disasters. The Sendai Framework adopted by both Chile in 2015 and Indonesia in 2020 mandated states to substantially reduce affected society from disaster through investing in resilient infrastructure and land use planning to minimize vulnerability from danger [7,8]. Simultaneously, the CBDR principle should play a pivotal role to substantially increase financial and technical support to developing countries for capacity building in disaster risk reduction [8]. However, not every country have the ability to interpret this framework to be linked in strategizing adaptation from climate-related disaster. This led to few countries having regulatory framework on climate adaptation, particularly providing access to climate infrastructure. Although most research only identifies climate-resilient infrastructure, this research offers a novel approach by providing a micro-comparison on perspectives of two selected country.

On a national level, climate adaptation policymaking remains a challenge. While climate-related disasters continue accelerating and impacting vulnerable communities which urgently require climate adaptation policies including climate resilient infrastructure [7], several countries do not yet put climate change issues as a priority legislative agenda. Lack of a detailed national-local level climate risk-mapping further affects an evidence-based climate policymaking which potentially can be ineffective after its enactment. Based upon the background, there are two identified problems: lack of regulatory framework putting priority to climate adaptation, and lack of detailed local and national climate risk assessment. This research attempts to explore and answer two questions: 1) What are the necessary policy framework strategies in incorporating climate-resilient infrastructure in climate change adaptation policy? 2) How does Chile and Indonesia strategize in climate-resilient infrastructure and climate adaptation policy?

2 Research Method

The method utilized in this research is a jurisdictional comparative legal method as a part of normative legal research with statutory approach. This paper will not compare different legal systems in Chile and Indonesia, given that both countries are adopting European continental legal systems. Therefore, this paper will compare and contrast practices of each countries' climate law and policymaking, particularly in the area of climate change adaptation upon the policy domain elaborated in figure 1.

 https://doi.org/10.1051/e3sconf/202346705001

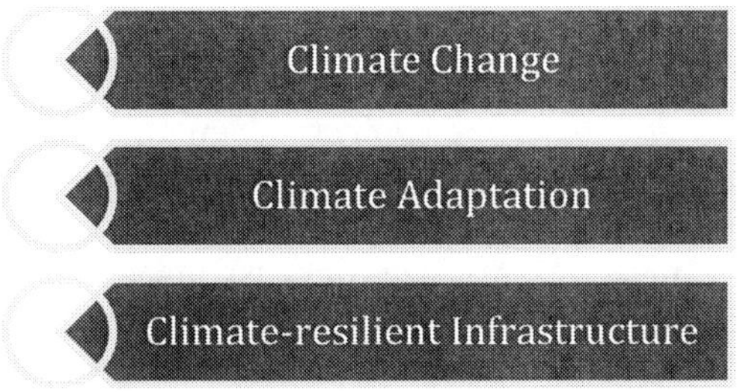

Figure 1. Selected policy domains.

The primary legal materials are obtained from relevant national regulations and authoritative legal materials including policy directives, the secondary legal material are obtained from various related literatures. All collected materials relates to the three policy domains, and will be analyzed qualitatively by describing facts and its linkage to the regulations to construct argumentations as the result of the legal analysis.

3 Results and Discussions

An infrastructure is defined to be climate-resilient if it is planned, designed, built and operated specifically to anticipate, prepares for, adapts and can recover rapidly from the changing climate [7]. Climate resilient infrastructure is a critical component to climate adaptation, it is essential to promote resilience in areas vulnerable to climate disaster [9]. The first necessary step from the three stages as portrayed under figure 2 prior to establishing climate resilient infrastructure is to strategize a climate adaptation policy which incorporates a comprehensive climate resilient infrastructure planning.

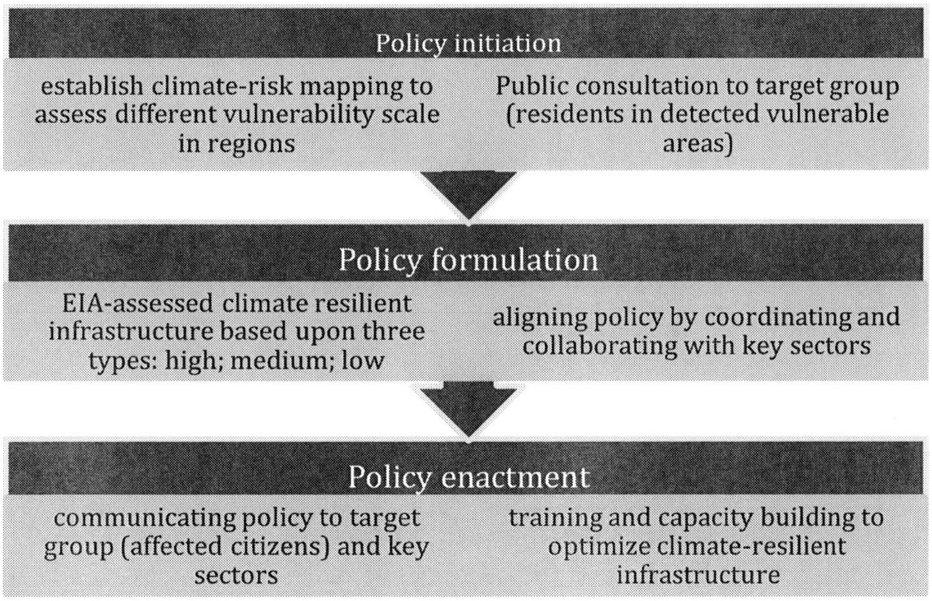

Figure 2. Climate adaptation policy framework strategies.

There are three critical stages in strategizing climate adaptation policy : initiation, formulation, enactment. On the initiation stage, climate disaster risk among cities (including types of climate disaster, frequency of hazard, possible economic losses and socio-economic background) shall be assessed and recorded. This recorded assessments will function as a guideline to provide an evidence-based policy in the formulation stage, given there are possibilities of different extent and types of climate-risk infrastructure in various cities and detecting high risk climate disaster areas. The assessments shall be ensured to be coherent with empirical fact that can be obtained from public consultation to target groups, this includes the residents located in detected vulnerable areas.

The climate-risk mapping shall accommodate information of areas divided into three different risk scales: high; medium; low. In the policy formulation stage, climate risk infrastructure shall be planned to prioritize high risk areas, with considering precautionary principle by only establishing infrastructures with a prior Environmental Impact Assessment (EIA). Furthermore, policymaker should ensure to align targets and strategies by coordinating and collaborating with related key sectors (including public, private, society) through various approaches such as (but not limited to) multi-stakeholder dialogue [10]. This stage is important for early detection of possible issues including fragmenting agencies.

Once the climate adaptation policy with climate-resilient infrastructure planning enacted, it shall provide pathways of communicating policy to targeted groups and key sectors. The policy shall be able to encourage optimizing climate-resilient infrastructure through training and capacity building. On this stage, the CBDR-based substantial increase of Sendai Framework has critical role to provide technical and financial support for infrastructure building and achieving adaptation goals as mandated under the enacted policy. It is also essential to ensure a harmonized cooperation between local governments in targeted areas, with a gradual monitoring and evaluation upon the policy implementation.

3.1 Discussion: Jurisdictional Comparison

This section will compare different approaches between Chile and Indonesia in strategizing climate-resilient infrastructure and climate adaptation policy. The established comparison aspects includes : the existence of a national climate change legislation, local/national climate change adaptation policy, national climate risk mapping and policy directives on climate disaster risk reduction (DRR) / disaster risk management (DRM), and coherence to submitted Nationally Determined Contribution (NDC).

3.1.1. Chile

Chile is highly vulnerable to climate change impacts, particularly from coastal erosion and SLR [11]. Therefore, the Chilean Ministry of Environment established the *Atlas de Riesgos Climáticos* (ARClim), a national atlas of current and projected climate change risk in Chile. With 14 designated working groups including coastal and climate adaptation, those are specifically enacted to strategize suitable pathways for adaptation, DRR and DRM in various areas based upon climate risk mapping.

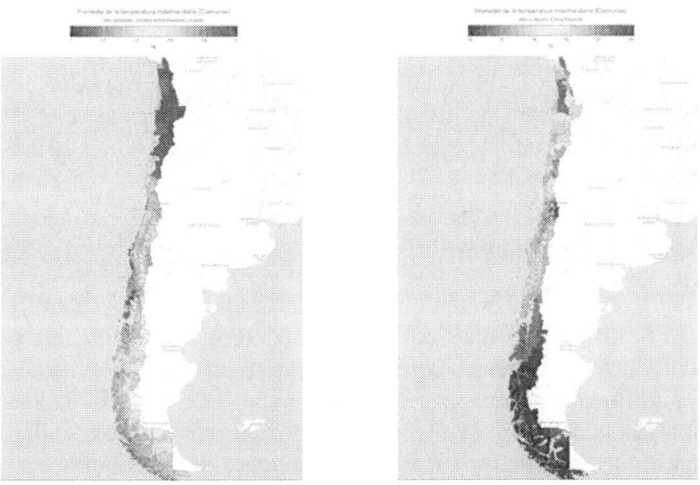

(a) (b)

Figure 3. Maximum Daily Temperature Change ((a) Current/2023; (b) Future)[12].

Chile also puts climate change as a priority legislation agenda, through promulgating the Climate Change Framework Law No.21,455 in 2022. This Law strategizes its provisions based upon the limited amount of time to reverse serious effects of climate change, therefore establishing management instruments at national, regional and local level by assigning responsibilities to specific ministries in addressing climate change. The law mandated to establish financing for mitigation and adaptation projects and actions by environmental protection fund and climate change financial strategy [13].

The climate-risk mapping and laws have strengthened and further optimize the implementation of Chile's National Climate Adaptation Plan 2014 and Adaptation Communication 2022, in coherence to the submitted NDC aimed to fully implement the National Policy for Disaster Risk Reduction 2019-2030 [14]. Several climate DRR examples includes the construction of artificial beaches and beach nourishment in several areas (Iquique, Tocopilla, Antofagasta) to reduce areas with high erosion rates [15]. However, existing implementation barriers crucial to be addressed includes community knowledge and awareness on climate change and DRR, coherent implementation to regional and local level, adequate support on financial and human resources to municipal government [16].

3.1.2. Indonesia

Indonesia is a part of coastal areas that are likely to be highly vulnerable and threatened due to climate change [17]. It is crucial to develop a legal framework that can adequately respond to the challenges posed by climate change [17,18]. However, the country has not yet introduced or implemented comprehensive climate change legislation since the ratification the Paris Agreement by Law No. 16 of 2016. Indonesia is focusing on advancing infrastructure as "economic transportation", with a planned total budget of IDR 392 trillion which increased by 7.8% compared with the infrastructure budget in 2022 that reached Rp363.8 trillion [18,19]. With this budget, Indonesian development is expected to focus on sustainable infrastructure development and climate change mitigation-based infrastructure.

The primary obstacle to advancing resilience-based climate change mitigation strategies in Indonesia is the absence of explicit references to climate change within its legislative framework. Specifically, Law of the Republic of Indonesia Number 24 of 2007, which pertains to Disaster Management, notably omits any mention of the terms 'climate' or 'climate change' within its provisions. Consequently, the disasters regulated by this law encompass a wide array of common events in Indonesia, such as floods, earthquakes, and landslides [20].

Similarly, Law of the Republic of Indonesia Number 27 of 2007, which deals with the Management of Coastal Areas and Small Islands, aligns with the definition and provisions regarding mitigation as outlined in Law Number 24 of 2007 [21].

In this law, the concept of 'mitigation' is exclusively associated with natural disasters, thus neglecting the specific considerations required for addressing climate-related issues. Conversely, when compared to its predecessor legislation, Law No. 6 of 1996, which pertains to the Ratification of the Climate Change Convention, presents a more comprehensive approach. This earlier law enhances the definition of mitigation by incorporating climate change as a crucial element within its climate change mitigation framework [22].

There is a noticeable lack of synchronization between different legislative acts in Indonesia, leading to legal uncertainties. This lack of alignment has significant implications for the formulation of policies and regulations related to climate change [23]. This issue is particularly critical as Indonesia experiences rapid infrastructure development. Despite Indonesia's commitment to addressing climate change, as evident in its existing regulations, these regulations remain insufficient to adequately incorporate the essential aspect of climate change resilience into the country's infrastructure development efforts.

Conclusion

There are three strategies in incorporating a climate-resilient infrastructure in climate adaptation policy: initiation, formulation, enactment. This shall be carried out by assessing climate disaster risk coherent with empirical fact among cities, aligning targets and strategies by coordinating and collaborating with related key sectors, communicating policy to targeted groups and key sectors.

Certain contrasts exists between Chile and Indonesia's strategy in climate-resilient infrastructure and climate adaptation policy. While Chile have been taking substantial measures by policy intervention by enacting Climate Change Framework Act and ARClim, Indonesia still grapple with the absence of a comprehensive climate change legislation. This obstructs the formulation of essential policies and guidelines for enhancing climate resilience, particularly within the context of rapid infrastructure development. It is imperative for the Indonesian government to draft and implement comprehensive climate change legislation that will serve as a guiding framework for subsequent laws and regulations.

References

1. IPCC, Climate Change 2023: Synthesis Report. A Report of the Intergovernmental Panel on Climate Change. Contribution of Working Groups I, II and III to the Sixth Assessment Report of the Intergovernmental Panel on Climate Change [Core Writing Team, H. Lee and J. Romero (Eds.)]. (IPCC, Geneva, 2023)
2. Intergovernmental Panel On Climate Change (Ipcc), Climate Change 2022 – Impacts, Adaptation and Vulnerability: Working Group II Contribution to the Sixth Assessment Report of the Intergovernmental Panel on Climate Change, 1st ed. (Cambridge University Press, 2023)
3. M. Pucheta and A. B. Sánchez, Just Transition in the Global South (ETUI, Brussels, 2022)
4. S. Caney, Midwest Stud Philos 40, 9 (2016)
5. United Nations, Paris Agreement (2015)
6. United Nations Framework on Climate Change, Glasgow Climate Pact (2021)
7. OECD, Climate-Resilient Infrastructure (2018)
8. United Nations General Assembly, Sendai Framework (2015)
9. C40 Cities Climate Leadership Group, Why All Cities Need to Adapt to Climate Change (C40 Knowledge Hub, 2019)

10. The World Bank, Guide to Climate Change Adaptation in Cities (The World Bank Group, Washington DC, 2011)
11. The World Bank, Climate Risk Country Profile: Chile (The World Bank Group, Washington DC, 2021)
12. ARCLim, (2023)
13. Ministerio Del Medio Ambiente, Ley Núm. 21.455. Ley Marco De Cambio Climático (2022)
14. Gobierno de Chile, Chile Nationally Determined Contribution (UNFCCC, Bonn, 2020)
15. C. Martínez, M. Contreras-López, P. Winckler, H. Hidalgo, E. Godoy, and R. Agredano, Ocean & Coastal Management 156, 141 (2018)
16. N. Banwell, A. S. Gesche, O. R. Vilches, and S. Hostettler, International Journal of Disaster Risk Reduction 50, 101703 (2020)
17. B. Neumann, A. T. Vafeidis, J. Zimmermann, and R. J. Nicholls, PLoS ONE 10, e0118571 (2015)
18. Office of Assistant to Deputy Cabinet Secretary for State Documents, Cabinet Secretariat of The Republic of Indonesia (2022)
19. Kementerian Keuangan, Informasi APBN 2023 (2023)
20. Republic of Indonesia, Law No. 24 Year 2007 Concerning Disaster Management (2007)
21. Republic of Indonesia, Law No. 1 Year 2014 Concerning Amendment of Law No. 27 of 2007 of Management of Coastal Zone and Small Islands (2014)
22. Republic of Indonesia, Law No. 6 Year 1994 Concerning United Nations Framework Convention on Climate Change Ratification (1994)
23. D. Hilman, Jurnal Legislasi Indonesia 6, (2009)

The impact of climate change on tobacco farming income in Temanggung Regency, Central Java, Indonesia

Eni Istiyanti[1*], and *Gandhi Singgih* Wicaksono[1]

[1]Departement of Agribusiness, Faculty of Agriculture, Universitas Muhammadiyah Yogyakarta Jln. Brawijaya, Tamantirto, Kasihan Bantul, 55183, Yogyakarta, Indonesia

Abstract. Tobacco is produced extensively in Indonesia, with the most excellent yields discovered in Temanggung Regency, located in Central Java. Climate change and increasing tobacco excise affect farmers' income. The study aims to determine the cost, profit, and feasibility of chopped tobacco farming after climate change and rising excise tax. This study involved 74 respondents, selected using simple random sampling, and utilized both primary and secondary data. The feasibility analysis employed the Revenue Cost Ratio (R/C) and Break Even Point (BEP) criteria. The results unveiled that chopped tobacco production was 773 kg/5,625 m^2 or 1,374 kg/hectare and climate change had no effect on production. Chopped tobacco farming per growing season for five months required a total cost of IDR 37,013,965. a revenue of IDR 38,551,056, and a profit of IDR 1,537,091. Climate change and rising excise tax on chopped tobacco farming could reduce profits. However, following the R/C and BEP criteria, chopped tobacco farming was feasible.

1 Introduction

Tobacco (*Nicotiana tabacum L.*) is a major annual plantation crop in Indonesia. Tobacco leaves are commonly employed for industrial purposes, especially as a raw material for making cigarettes. This commodity is crucial for farmers' livelihoods and the country's economy [1]. *Nicotiana tabacum* and *Nicotiana rustica* are Indonesia's two most common tobacco species [2]. From 2017 to 2021, Indonesia's tobacco production increased by 44.09%, from 181,142 tonnes in 2017 to 261,011 in 2021. The largest tobacco-producing province in Indonesia is East Java, with a total production of 140,283 tonnes, followed by Central Java, with 55,667 tonnes in 2021 [3].

Temanggung Regency is the center of tobacco production in Indonesia. Farmers sell chopped tobacco to Indonesia's well-known cigarette companies, such as PT Gudang Garam Tbk, PT Bentoel Prima, and PT Djarum. Temanggung's local tobacco has a unique flavor, texture, and scent, giving it a competitive and comparative advantage in some years. Tobacco can be a source of income for farmers during the peak harvest season [4].

* Corresponding author: eniistiyanti@umy.ac.id

Production, personal, policy, pest, disease, and financial risks threaten tobacco farming [1]. Like other agricultural commodities, tobacco market prices fluctuate with various internal and external factors [5]. Tobacco quality is an internal factor that significantly affects farmers' selling prices. Climate change causes unpredictable rainy and dry seasons as well as attacks by pests and diseases, increasingly threatening the income of tobacco farmers [6]. Tobacco productivity in Temanggung dropped by half when precipitation levels were above average [7]. The government's new tobacco product excise policy, as outlined in Minister of Finance Regulation No. 156/PMK.010/2018 and has been in force since January 1, 2019, has only worsened the problem [8]. Changes in tobacco product excise accounted for a rise of 8.5% to 12.5% in 2012 and 23% in 2020 [9]. Therefore, it is clear that climate change affects chopped tobacco farming. A question emerged in response to such a condition, is tobacco worth producing?

Research related to income and feasibility of tobacco farming has been conducted by other researchers, namely by [10] in Situbondo, [11] in Jember, [12] and [13] in Limapuluh Kota Regency, West Sumatra [14] in Kendal Regency and [15] in Ngadirejo District, Temanggung Regency. Previous research has not linked climate change to the income and feasibility of tobacco farming.

Based on the above problems, this study aims to analyze the costs, profits and feasibility of chopped tobacco farming due to climate change in Temanggung Regency, Central Java.

2 Research method

Temanggung Regency has generated the greatest tobacco production in Central Java Province, making it one of the tobacco centers in Indonesia. Hence, this study took place in this regency. There has been a suitable tobacco farming area in all districts in Temanggung Regency. Bulu District has the largest area of tobacco growing compared to other sub-districts in Temanggung Regency. The area of tobacco land in Bulu district in 2020 was 2,385.79 hectares with a total of 1,457 farmers [16]. A total of 74 tobacco farmers as the research respondents, representing 5% of the population, gathered through simple random sampling. This study employed both primary and secondary data. A descriptive quantitative analysis was run to accomplish the research objectives. The following equations were applied to determine the cost, profit, Revenue Cost Ratio (R/C), and Break Even Point (BEP) of chopped tobacco farming [17].

$$TC = FC + VC \quad (1)$$

$$\text{Profit} = TR - TC \quad (2)$$

$$R/C = TR/TC \quad (3)$$

$$\text{Production BEP} = FC/1 - (VC/TR) \quad (4)$$

$$\text{Price BEP} = TR/Q \quad (5)$$

Description:
TC = Total Cost
FC = Fixed Cost
VC = Variable Cost
TR = Total Revenue
Q = Quantity of Product
BEP = Break Even Point
R/C = Revenue Cost Ratio

Farming is profitable and feasible if R/C is higher than 1, chopped tobacco production is greater than production BEP, and chopped tobacco price exceeds price BEP.

3 Results and discussion

The age of tobacco farmers in Temanggung ranged from 22 to 57, with more than 50% being 42 to 51. These tobacco farmers were of productive age and physically capable of tending their fields [18]. Tobacco farmers in Temanggung Regency were relatively younger than those in Limapuluh Kota Regency, West Sumatra [12]. Most tobacco farmers in this regency only had an Elementary or Junior High School education. Despite the low level of education, they reported no difficulties in managing their tobacco farming. They could network with one another and obtain advice from agricultural extension workers in farmer groups. Farmer groups serve as a means of learning, production unit, and cooperation [19]. On top of that, farmers in Temanggung Regency have been running tobacco farming for over 20 years on average. This long experience has given them a high level of expertise in the field. Experienced farmers produce better crops since they understand what they do [20].

Temanggung Regency is a mountainous area over 700 m height above sea level. Tobacco was farmed on dry land managed by the farmers, with plot sizes averaging 5,625 m^2, ranging from 2,000 to 21,500 m^2, belonging to the small and medium categories. Since most farmers possessed more than two family members, they typically employed family labor to run their farming successfully. Meanwhile, non-family labor was widely utilized in post-harvest activities, encompassing chopping and drying tobacco. Table 1. displays that the cost of family labor was greater than that of non-family labor. Tobacco farming in Kenya and Zambia also experienced a similar situation [21]. Tobacco farmers incurred the largest cost proportion on labor. It aligns with previous research [15] in Ngadirejo District, Temanggung regency and [14] in Kendal regency.

Table 1. Total Cost of Chopped Tobacco Farming in Temanggung Regency, 2022.

No	Description	Amount (IDR)	Percentage (%)
	Fixed Cost		
1	Land tax	83,727	0.23
2	Depreciation costs	602,168	1.63
3	Own land rent	4,687,500	12.66
	Variable cost		
4	Manure	8,564,231	23,14
5	Chemical fertilizer	4,933,446	13,33
6	Pesticide	649,625	1.76
7	Non-family labor	5,001,351	13.51
8	Transportation	204,820	0.55
9	Basket	1,545,811	4,18
10	Fuel	52,558	0.14
11	Rafination sugar	223,284	0.60
12	Family labor	9,918,919	26.80
13	Own capital interest	546,526	1.48
	Total Cost	37,013,965	100

Source: Primary data analysis

Based on Tabel 1. most costs farmers incurred on production facilities went toward purchasing manure. At the beginning of the growing season, manure was utilized as basic fertilizer, administered simultaneously with land preparation. Adding manure could boost the soil's organic C content, benefiting the growth of tobacco plants [22] and [23].

Tobacco farmers in Bulu District, Temanggung, employed various types of chemical fertilizers: Urea and ZA to meet N element needs, white KNO_3 fertilizer to meet K and N element needs, and Fertila fertilizer containing N, P, and K elements with a ratio of 8:15:19. Fertila fertilizer was widely selected for containing complete macronutrients and was a directive from PT Gudang Garam Tbk, adjusted to the Regulation of the Ministry of Health to reduce cigarette nicotine levels. Fertila fertilizer produced tobacco with a light brown color, less sharp aroma, light suction, and lower nicotine content than tobacco with urea fertilizer. The manufacturer offered a higher price for tobacco with such qualifications.

The total cost of chopped tobacco farming in Temanggung Regency was IDR 37,013,965 per 5,625 m^2 of land area or 65,802,604 per hectare. It was much higher than the total cost incurred by tobacco farmers in Ngadirejo District, Temanggung [15] and Bondowoso Regency [10]. The rising price of production facilities led to quite a large difference in cost.

Profit is the difference between total revenue and total costs. The feasibility of chopping tobacco farming was analyzed using Revenue Cost Ratio (R/C) and Break Even Point (BEP). Profit, R/C and BEP can be seen in table 2.

Table 2. Profit, R/C and BEP of Chopped Tobacco Farming in Temanggung Regency, 2022.

No	Description	Value
1	Fixed cost (IDR)	5,373,396
2	Variable cost (IDR)	31,640,570
3	Revenue (IDR)	38,551,056
4	Profit (IDR)	1,537,091
5	R/C	1.04
6	Production BEP (kg)	601.06
7	Price BEP (IDR/kg)	47,884

Source: Primary data analysis

Table 2 lists that the revenue of chopped tobacco farming in Temanggung Regency was IDR 38,551,056, with a price of IDR 49,872/kg and an average production of 773 kg. Chopped tobacco productivity in Temanggung in 2022 is 1,374 kg/hectare higher than in 2019 which was 648.37 kg/hectare and in 2020 which was 750.97 kg/hectare [24], higher than in Jember Regency was 960 kg/hectare [11] and Limapuluh Kota Regency, West Sumatra Province was 907 kg/hectare [13].

There were rainy days in Temanggung Regency in 2022, with an average of 17 rainy days per month and an average rainfall of 359 mm for the entire year [24]. Since the average month in 2021 had 15 rainy days with 389 mm of heavy rainfall [16]. The existence of high rainfall in 2022, has been anticipated by farmers, by managing agricultural land sustainably. Farmers use more manure and regulate irrigation water according to the needs of tobacco plants [6], resulting in less impact of climate change on tobacco production. It was a common practice for farmers to harvest early to minimize losses [7].

The profit of chopped tobacco farming in Temanggung Regency reached IDR 1,537,091 per 5,625 m^2 or IDR 2,732,606 per hectare. It was lower than that received by tobacco farmers in Pantrang District, Jember Regency was IDR 22,546,992 per hectare [11] and Limapuluh Kota Regency, West Sumatra Province was IDR 6,832,913 per hectare [13]. Based on information from farmers, the price of chopped tobacco in Temanggung in 2022 is close to IDR 50,000 per kg, which is lower than IDR 75,000 in 2017. The decline in tobacco prices was influenced by the quality of tobacco due to climate and weather factors. Farmers in Temanggung Regency usually prepare for planting in April, plant in May and harvest in late July to early August. However, the weather is not supportive with a short dry season, which causes the drying of the tobacco is not optimal and the quality of the tobacco is not good. This situation is exacerbated by the existence of government regulation No. 109 of 2012 concerning Control of Materials Containing Tobacco Addictive Substances for Health and Minister of Finance Regulation No. 156/PMK.010/2018 concerning Changes in Excise Tax on Tobacco Products [8].

An R/C of 1.04 implies that chopped tobacco farming was feasible because, for every IDR 1 incurred, the revenue was 1.04. However, it was lower than simporis tobacco farming in Situbondo Regency, reaching 3.28 [10], Jember Regency, acquiring 2.4 [11], Limapuluh Kota Regency, West Sumatra Province obtaining 2.5 [13] and 1.09 [12].

BEP could be employed for future planning to determine the minimum price or production to make tobacco farming profitable. Following the BEP analysis, farmers must produce chopped tobacco higher than the production BEP of 601.06 kg with a selling price greater than the price BEP of IDR. 47,884 to gain profit and make farming feasible. The average tobacco production in Temanggung was 773 kg, with an average price of IDR 49,872. In other words, chopped tobacco farming was profitable because production was higher than the production BEP, and the price was greater than the price BEP. These findings resembled what happened in Patrang District, Jember Regency with a price BEP value of IDR 3,664,- and a production BEP of 1,056 kg [11].

4 Conclusion

Production costs of chopped tobacco farming in Temanggung Regency per land area 5,625 m^2 amounted to IDR 37,013,965 and the profit obtained was IDR 1,537,091. Climate change affects the quality of tobacco which in turn affects prices and profits. Tobacco farmers encountered a declining profit due to the government's policies regarding the protection of addictive tobacco substances and changes in tobacco excise. Based on the R/C and BEP criteria, tobacco farming in Temanggung Regency was still profitable and feasible.

Reference

1. H. Akcaoz, H. Kizilay, and O. Ozcatalbas, J. Food, Agric. Environ. **8**, 717 (2010)
2. O. Murhawi, P. Benih, T. Ahli, and P. Penanam, Tek. Budid. Tembakau (Nicotiana Tabacum L.) 1 (2014)
3. K. P. R. I. Direktorat Jendral Perkebunan, 1 (2020)
4. D. Setiawan, J. Ketahanan Nas. **21**, 43 (2015)
5. L. Verona, Supriyadi, N. E. Diana, and G. S. A. Fatah, IOP Conf. Ser. Earth Environ. Sci. **974**, (2022)
6. E. Nurjani, R. Harini, A. B. Sekaranom, and A. S. Mutaqqin, IOP Conf. Ser. Earth Environ. Sci. **451**, (2020)
7. A. B. Setiawan, B. R. Pratama, N. Sabtiningtias, and W. Fatriasari, Songklanakarin J. Sci. Technol. **44**, 128 (2022)
8. Menteri Keuangan Republik Indonesia, *Peraturan Meteri Keuangan Republik Indonesia Nomor: 152 /PMK.010/2019 Tentang Tarif Cukai Hasil Tembakau* (2019)
9. A. Suprihanti, N. Harianto, B. M. Sinaga, and R. Kustiari, J. Agro Ekon. **37**, 1 (2019)
10. Puryanto, J. Ilm. Agribios **16**, 55 (2018)
11. A. Nadhirah, T. S. Napitupulu, and D. Saadilah, J. Emak **4**, 167 (2023)
12. S. Sefrimon, J. Appl. Agric. Sci. Technol. **2**, 41 (2018)
13. D. Latifa, I. Ispinimiartriani, S. Soemarsono, and M. Syakib Sidqi, Int. J. Econ. Business, Accounting, Agric. Manag. Sharia Adm. **1**, 221 (2021)
14. R. K. A. Muhamad Farras Nur Islami, Joko Sutrisno, Agrista **7**, 204 (2019)
15. M. Tri Puspita Ganes Agustina, Siswanto Imam Santoso, J. Ekon. Pertan. Dan Agribisnis **5**, 819 (2021)
16. Badan Pusat Statistik Kabupaten Temanggung, *Kabupaten Temanggung Dalam Angka 2022* (Badan Pusat Statistik Kabupaten Temanggung, Temanggung, 2022)
17. Agustina Shinta, *Ilmu Usahatani* (Universitas Brawijaya Press, 2011)
18. S. Aprilyanti, J. Sist. Dan Manaj. Ind. **1**, 68 (2017)
19. D. K. . S. Hermanto, Anal. Kebijak. Pertan. **9**, 371 (2011)
20. M. Mandang, M. F. L. Sondakh, and O. E. H. Laoh, Agri-Sosioekonomi **16**, 105 (2020)

21. R. Lencucha, J. Drope, P. Magati, and G. A. Sahadewo, Tob. Control **31**, 308 (2022)
22. M. D. Sukmasari, Z. Zannah, and U. Dani, J. Ilmu Peternak. **7**, 70 (2019)
23. R. Kartika Putri, J. Tanah, F. Pertanian, U. Brawijaya, and B. Penelitian Tanaman Pemanis dan Serat, J. Tanah Dan Sumberd. Lahan **5**, 2549 (2018)
24. Badan Pusat Statistik Kabupaten Temanggung, *Kabupaten Temanggung Dalam Angka 2023* (Badan Pusat Statistik, Temanggung, 2023)

Justice element in just energy transition partnership decarbonization policy: a conceptual legal analysis

Maskun Maskun[1], *Valeria Erika Sari* Paliling[2*], *Alfiyah Nur Inayah* Hamzah[2], and *Nurul Habaib Al* Mukarramah[2]

[1]International Law Department, Faculty of Law, Hasanuddin University 90245 Perintis Kemerdekaan KM.10, Makassar, Indonesia
[2]Research Assistants, Faculty of Law, Hasanuddin University 90245 Perintis Kemerdekaan KM.10, Makassar, Indonesia

Abstract. Decarbonization is an essential strategy in reducing greenhouse gas emissions. Therefore, Indonesia committed to adhere to the Just Energy Transition Partnership (JETP) to pursue its climate targets by establishing energy transition targets for decarbonization. Nevertheless, the "Just" element in JETP remains unclear. Therefore, this paper aims to examine the justice aspect in JETP through a conceptual legal analysis of justice. This research method is a normative legal analysis conducted by examining legal theories, concepts, principles, laws, and regulations as primary data, with literary sources as secondary data analyzed through a conceptual approach. The result shows that there are several targets and current directives of JETP that may not fulfill the just aspect and are incoherent to the concept of justice. In conclusion, the JETP shall improve its directives and targets prior to its implementation by involving regional government and society as key sectors to ensure a just energy transition.

1 Introduction

To achieve the climate target by limiting temperature to 1.5C° as mandated by the Paris Agreement, it is imperative to reduce emissions through decarbonization by lowering the amount of greenhouse gas (GHG) emissions [1]. This decarbonization shall target sectors associated with the use of fossil fuels, electricity generation, and industrial processes [2]. Therefore, countries should ensure the incorporation of a decarbonization strategy in their Nationally Determined Contribution (NDC).

Indonesia's NDC set its goal to reduce GHG emissions by 32% (unconditional) and 43,20% (conditional) [3]. To reach this goal, Indonesia launched a Just Energy Transition Partnership (JETP) agreement in 2022. JETP was first proposed at the 26th annual meeting of the United Nations Framework Convention on Climate Change (UNFCCC) conference of parties held in Glasgow. It was created for developing countries heavily relying on fossil fuels as energy sources [4]. Therefore, on November 15, President Biden, the current

* Corresponding author: valerika.study@gmail.com

President of Indonesia (Mr. Joko Widodo), and other world leaders agreed to initiate JETP in Indonesia through 20 billion funds over the next three to five years to focus on five targets as the following:

1) Transmission network development,
2) coal power plant early retirement,
3) baseload renewable energy acceleration,
4) Variable renewable acceleration, and
5) renewable energy supply chain development.

The JETP aims to reach peak emission from the electricity sector by 290 million metric tons of CO_2 in 2030, a 34% renewable energy mixed sources in 2030, and achieve net-zero emission in the energy sector by 2050[5]. In reaching this JETP goal, strategic actions from various key sectors including energy industries, public authority, society, labor, and consumers are pivotal. This JETP shall ensure its "just" element is able to resolve the underlying socio-economic issues that have existed and potentially emerged due to this transition. This includes the electricity oversupply concentrated in Java despite the electricity crisis on other islands and potential labor issues due to coal-fired power plant phase-out plans [6]. These issues further challenge the role of the state as the provider of welfare.

Therefore, this research aims to elaborate on the concept of justice, focusing on distributive and procedural justice which will be utilized to examine the "just" element of JETP by answering two research questions: What is the concept of justice? Does JETP coherent with the concept of justice? The result will explore discussions concerning the justice aspect of JETP.

2 Research method

The data used includes primary legal materials obtained from relevant legal instruments (Table 1), while secondary legal materials are collected from related kinds of literature. This data will be analyzed and assessed to interpret just elements in JETP as the result.

Table 1. Primary legal instrument inventory.

Instrument	Year
Government Regulation Number 79 concerning National Energy Policy as the basis for the preparation of the National Energy General Plan	2014
Presidential Regulation Number 112 concerning the Acceleration of Development of Renewable Energy for the Provision of Electricity	2022
Presidential Regulation Number 11 concerning Additional Concurrent Government Affairs in the Energy and Mineral Resources Sector in the New Renewable Energy Sub-sector	2023

The analysis will begin to dissect the JETP policy and analyze the primary legal materials in the legislative inventory. Furthermore, the justice element in JETP will be interpreted through 3 concepts of justice obtained from secondary legal materials mainly sourced from journal articles and policy briefs. The results will reveal the justice element in JETP and its relevancy to the three-justice concepts.

3 Results and discussions

JETP has two important components to support renewable energy. First, it promotes renewable energy and clean energy. Second, it supports decarbonization for economic growth. Apart from encouraging climate change in the form of an energy transition, the JETP model also seeks to promote a green economy and meet the economic needs and social groups of people vulnerable to the impacts of the energy transition. JETP may encounter challenges in overcoming climate change and sustainable development, including defining the concept of justice.

Just energy refers to environmental justice, the implications of social justice, and the transition to low-carbon energy. Sovacool defines energy equity as a global energy system that equitably distributes both the benefits and burdens of energy services and a system that contributes to more representative and inclusive energy decision-making [7]. In context to this research, JETP will start a comprehensive investment and policy plan in August 2023. In its preparation, Indonesia should have the principles of energy justice as a distinctive vision of the Ministry of Energy and Mineral Resources (ESDM) which explicitly includes the principles of just energy in its policies and programs.

The challenge of equitable access to energy has been a matter of concern to the international community for a long time. For example, the United Nations (UN) has included access to electricity as part of the Sustainable Development Goals to ensure access to affordable, reliable, and sustainable energy for all people [7]. The national legal framework for just and fair energy has been stipulated under Law Number 30 of 2007 where energy is managed based on the principles of efficacy, rationality, efficiency with justice, increasing added value, sustainability, community welfare, preservation of environmental functions, national resilience, and integration by prioritizing national capabilities. Government Regulation Number 79 of 2014 concerning National Energy Policy is the basis for the preparation of the National Energy General Plan (RUEN) and the National Electricity General Plan (RUKN), which state that national energy policies must be formulated based on the principles of justice. Finally, an academic draft of the New and Renewable Energy Law also states that to achieve the management of national energy security must be based on management principles that uphold justice and equality.

After observing the high priority of equitable energy development in ensuring the just transition framework, Indonesia needs to consider the basic principles of justice that must be obeyed by all stakeholders including foreign investors providing funds to assist Indonesia in achieving its decarbonization target. However, this research will only limit the philosophical foundations of just energy to procedural justice, distributional justice, and recognitional justice.

i. Procedural justice requires access to decision-making. Procedural justice must be realized as justice by involving all stakeholders in a non-discriminatory way. In addition, procedural fairness must ensure transparency, such as information disclosure and the involvement of all public consultations [5].
ii. Distributional justice, the government is required to provide energy that is affordable and accessible to everyone [7]. Article 7 of the Energy Law 30/2007 requires energy to be determined based on reasonable economic value, and the government is obliged to provide subsidies to the poor who still have difficulty accessing energy.
iii. Recognitional justice, recognitional justice is emphasized on guaranteeing complete and equal political rights of various social groups and identities [8]. An example of a case regarding recognition of justice is that PLN has the responsibility to provide access to affordable electricity for all Indonesian people, but there are still many areas in Indonesia that are considered economically unfit for investment that are still neglected and as a result, do not get access to electricity. In this case, viewing access to electricity only from an economic perspective will only lead to a failure in recognizing energy justice [9].

3.1 Discussion

The Presidential Regulation Number 112 of 2022 concerning the Acceleration of Development of Renewable Energy for the Provision of Electricity (Presidential Regulation 112/2022) and Presidential Regulation Number 11 of 2023 concerning Additional

Concurrent Government Affairs in the Energy and Mineral Resources Sector in the New Renewable Energy Sub-sector (Presidential Regulation 11/2023) as the legal basis for JETP has not been able to precisely define the principle of justice in implementing just transition [5]. There are three concepts to interpret the justice element in JETP: 1) procedural, 2) distributional, and 3) recognitional justice.

Procedural, distributional, and recognitional justice focuses on the ability of the state to accommodate the needs of society by guaranteeing access to energy for underprivileged communities. Regional governments, as authorities that deal directly with the community, should have the authority to implement energy transition policies to guarantee energy access for the community. However, local governments have not been involved in the JETP energy transition policy, considering that the Regional Government currently does not have a regulatory framework implementing Presidential Decree No. 11/2023 and stated that the Presidential Decree did not address the energy transition needs. This certainly impacts the autonomous region's economy given that JETP's main focuses include an early coal-powered plant retirement.

The regional governments have the obligation to ensure an increase and equal distribution of society's income [10,11]. Moreover, the delegation of authority in the JETP policy in Presidential Decree 11/2023 does not grant authority to district/city governments, only to the provincial government. Considering that JETP will directly impact regional income, local government involvement in JETP policy is highly needed. As a result, the Regional Government did not provide material protection guarantees for society after the coal-powered plant closure.

The employment sector which focuses on transportation will be one of the affected areas. This indicates that if there is no anticipation through preparations, including capacity building, this will seriously jeopardize society, particularly the lower-middle class. This cycle will also impact economic circulation, considering that transportation has become a means to meet people's needs. Therefore, in a just energy transition, an emphasis on anticipating affected groups is necessary by establishing regulations in accordance with the legal issues faced by society because of JETP implementation.

Conclusion

The Justice Element in JETP refers to the availability, accessibility, and affordability of clean energy. This means that it must consider the high dependence of society on energy so that public guarantees within the scope of trade unions need to be considered. Moreover, the scope of the energy transition is limited to access to energy and protects all elements subject to replacing coal as an energy source. To activate public participation, the establishment of implementing regulations by regional governments as state organs that interact directly with the public is the primary step. However, until now there is an absence of Regional Regulations regarding anticipation after the implementation of JETP. This will threaten economic sustainability because people's dependence on coal is very high. Here, local governments can design appropriate regulations, especially in anticipating human resource capabilities after the implementation of JETP to get closer to the concept of justice contained in JETP.

References

1. TWI Global. What Is Decarbonisation? *[online] available https://www.twi-global.com/technical-knowledge/faqs/what-is-decarbonisation#:~:text=Decarbonisation%20is%20the%20reduction%20of,greenhouse%20gasses%20into%20the%20atmosphere* (2023)
2. Intergovernmental Panel on Climate Change. *Summary for Policymakers in Climate Change 2021: The Physical Science Basis.* (Cambridge: Cambridge University Press)

(2021)
3. Republic of Indonesia. *Enhanced Nationally Determined Contribution* (Indonesia: Republic of Indonesia) (2022)
4. Andayani A R D, Marthen A A, Imelda H and Setiadi K. *Just Energy Transition Partnership (JETP) Indonesia* (Jakarta: Indonesia Research Institute for Decarbonization (IRID)) (2022)
5. Celios. *Percepatan Transisi Energi Berkeadilan: Tantangan dan Peluang untuk Daerah* (Jakarta: CELIOS) (2023)
6. IESR, Energiewende A and LUT University. *Deep decarbonization of Indonesia's energy system: A pathway to zero emissions by 2050* (Institute for Essential Services Reform (IESR)) (2021)
7. Jenkins K, McCauley D, Heffron R, Stephan H and Rehner R. Energy justice: A conceptual review *Energy Research & Social Science* **11** 174–82 (2016)
8. Pellegrini-Masini G, Pirni A and Maran S. Energy justice revisited: A critical review on the philosophical and political origins of equality *Energy Research & Social Science* **59** 101310 (2020)
9. Setyowati A.B. Mitigating inequality with emissions? Exploring energy justice and financing transitions to low carbon energy in Indonesia *Energy Research & Social Science* **71** 101817 (2021)
10. Republic of Indonesia. *Law Number 23 of 2014 concerning Regional Government* (2014)
11. Republic of Indonesia. *Constitution of Indonesia* (1945)

The role of audit committee expertise and sustainability performance in the Indonesian mining industry

Wika Harisa Putri[1]*, *Rini* Raharti[2], *Clara Wuri* Vitaningsih[1], and Sulistyarini[1]

[1]Accounting Department, Janabadra University, 55231, Yogyakarta, Indonesia
[2]Economics Department, Janabadra University, 55231, Yogyakarta, Indonesia

Abstract. The worldwide concern about climate change has compelled the corporation to revamp its strategy to align with sustainability. The corporate governance structure significantly impacts sustainability performance. The previous study examines the impact of audit committee characteristics on sustainability performance, and the study's findings remain inconclusive. This study analyses the impact of specific attributes of the audit committee to forecast its effect on sustainability performance. Audit committee expertise is crucial in exerting influence on sustainability performance. However, the study's findings indicate that competence within the audit committee has yet to impact the sustainability performance of the Indonesian mining industry.

1 Introduction

Business activities have been entrusted with challenges such as climate change, pollution, greenhouse gas emissions, and other detrimental environmental impact. It is anticipated that corporations will disclose their sustainability performance to relevant parties. Previous studies have established that board governance mechanisms determine the quality of sustainability activities disclosed in a company's sustainability report [1]. Stakeholders and academic scholars commonly exhibit mistrust, lack of confidence, and skepticism towards companies [2], which challenges the underlying assumptions of agency and stakeholder theories [3,4] proposed by Freeman (1984) and Jensen and Meckling (1976). In such situations, companies allegedly utilize Sustainability Reporting (SR) to shape their corporate image by presenting exaggerated sustainability claims that may not align with reality. This practice serves to justify their business activities. Therefore, companies must own a scenario that more authentically adheres to Sustainability Reporting (SR) regulations and standards.

Within the sustainability framework, it is challenging to generalize about the relationship between corporate governance and financial reporting when it comes to sustainability reporting. This situation is primarily due to the unique characteristics of sustainability reporting, such as its extensive scope of subjects, the diverse nature of its contents, and the

* Corresponding author: wikaharisa@janabadra.ac.id

qualitative nature of the information being reported [5]. Implementing sustainability reporting may pose greater complexity and challenges than enforcing financial reporting [6].

Sustainability Corporate Governance (SCG) is essential in influencing management behavior. As a system to control and direct companies sustainably, SCG must be supported by experienced personnel with sufficient vision and competence regarding sustainability practices. The audit committee is one of the parts which holds a significant role within the corporate governance framework. The audit committee is recognized as a representation of the sustainability corporate governance (SCG) structure due to the results found in the initial investigation conducted by researchers, which indicate that only a handful of firms in Indonesia had a dedicated body responsible for the implementation of sustainability initiatives, such as a sustainability committee. The responsibility for incorporating sustainability practices in most corporations in Indonesia is assigned to the audit committee.

Based on that situation, according to previous research, committee audit expertise as one characteristic must be examined to ensure its role in contributing to the implementation of sustainability [7]. However, the contribution of the audit committee expertise to the sustainability performance is still unclear. Therefore, this research aims to examine the contribution of the audit committee expertise to the sustainability performance and focus on mining industries because mining companies face significant sustainable development challenges as extractive industries, particularly concerning environmental and social issues. Hence, mining businesses must be sincerely dedicated to fulfilling their sustainability obligations [8].

2 Literature review and hypothesis development

2.1 Corporate Governance Structure and Sustainability Performance

Corporate sustainability has been crucial since Sustainable Development Goals (SDGs) became a global commitment to a better future [9,10]. The advantages of corporate sustainability programs encompass the capacity to establish a competitive edge, bolster the company's standing within society, and cultivate positive perceptions among consumers and society at large. Finally, corporate sustainability will create value for the organization by integrating sustainability into operational processes, significant cost reduction and improvement [11].

Implementing sustainability practices and activities within the company can be attributed to establishing a sustainable culture, which is deeply ingrained in the organizational approach and driven by top-level leadership. One illustrative instance of the growing significance of sustainability inside the organization may be observed by establishing its three guiding principles. In this context, the critical elements of utmost importance are people, the planet, and economic prosperity [12].

Corporate governance mechanisms are essential to implementing sustainability strategies and achieving sustainability performance. Previous research has indicated that the corporate governance structure plays a crucial role in establishing the necessary corporate mechanisms for effectively implementing sustainability initiatives [13,14]. This study contributes to the current body of literature by examining the influence of specific corporate governance structures on several aspects of sustainability performance, as outlined by the GRI framework, particularly within the context of the mining industry in Indonesia.

To enhance comprehension of the influence of corporate governance (CG) features on corporate sustainability, this study investigates the impact of audit committee characteristics on sustainability performance. The picking of this issue was motivated by the observation that none of the existing empirical studies have comprehensively examined the three dimensions of sustainability performance, particularly in the context of the mining sectors in Indonesia.

Prior research examines the interconnection between board characteristics and sustainability performance [15,16] and uses board size, board independence, CEO duality, women on the board, and board meetings per year. A further investigation examines the features of audit committees (AC) and their impact on transparency in sustainability reports. This study considers factors such as the size of the AC, the independence of AC members, and the frequency of AC meetings [16–19]. The study's findings indicate that the size of the audit committee, the independence of its members, and the frequency of its meetings significantly influence the level of sustainability report disclosure. Due to the scarcity of comparable studies within the mining sector, particularly in Indonesia, the researchers deem it imperative to undertake this research inside the Indonesian setting.

2.2 Hypothesis development

It is imperative to establish the veracity of the assertions about the interconnectedness among various dimensions of sustainability and their respective significance regarding the associations between the characteristics of an audit committee and sustainability dimensions. The previous research examines the audit committee as a component of the company's organizational structure, indicating its significant involvement in supporting the sustainability initiative[17–19]. This study also investigates the audit committee's role within Indonesian corporations as the entity responsible for sustainability performance.

Empirical findings suggest that a larger AC may be more effective due to its potential to foster a broader range of knowledge and experiences. This fact, in turn, can enhance the efficacy of the regulating mechanism that influences corporate social responsibility (CSR) disclosure [20], but contrary to Jizi et al. [21] an AC size does not affect CSR disclosure. This inconclusive finding [20] supports this study to propose the hypothesis as follows :

Hypothesis 1 Audit committee size impacts the sustainability performance of a firm.

H1a Audit committee size impacts economic sustainability performance.

H1b Audit committee size impacts environmental sustainability performance.

H1c Audit committee size impacts social sustainability performance

H1d Audit committee size impacts governance sustainability performance

A study in bank industries in Australia found that gender diversity positively influences the e level of CSR disclosure [20]. Because of the different sector characteristics, we propose the hypothesis in Indonesian mining industries as the following:

Hypothesis 2 Audit committee gender impacts the sustainability performance of a firm.

H1a Audit committee gender diversity impacts economic sustainability performance.

H1b Audit committee gender diversity impacts environmental sustainability performance.

H1c Audit committee gender diversity impacts social sustainability performance.

H1d Audit committee gender diversity impacts governance sustainability performance.

Frequency meetings also found in previous research as one variable that impacts CSR performance [20]. However, another research found no connection between the AC meetings frequency and voluntary ethics disclosure level [22]. The inconclusive findings trigger an examination and propose the hypothesis as the following:

Hypothesis 3 The number of committee audit meetings impacts the sustainable performance of firms.

H3a The number of audit committee meetings impacts the economic SP of a firm.

H3b The number of audit committee meetings impacts the environmental SP of a firm.

H3c The number of audit committee meetings impacts the social SP of a firm.

H3d The number of audit committee meetings impacts the governance SP of a firm.

Combining audit committees with financial skills can enhance the resolution of issues that may challenge management and external auditors regarding financial transparency [23]. However, prior research found a negative association between AC financial expertise and sustainability reporting [18], and in this study, we propose the hypothesis as follows :

Hypothesis 4 Audit committee expertise impacts the sustainable performance of firms.
H4a Audit committee expertise impacts economic sustainability performance.
H4b Audit committee expertise impacts environmental sustainability performance.
H4c Audit committee expertise impacts social sustainability performance.
H4d Audit committee expertise impacts governance sustainability performance

Following prior research, the variable use of company size and age as control variables are positively associated with corporate sustainability reporting [18]; in this study, we use firm size, firm age, and sustainability report age as control variables to re-examine those variables.

2.3 Data and analysis

The study sample comprised 22 Indonesian mining companies listed on the Indonesian Stock Exchange/IDX. The final sample consists of 36 reports issued by selected firms over two years, specifically from 2020 to 2021. It should be noted that the number of sustainability reports has been unbalanced data across the years. Following the approach employed by Plumlee et al. (2015), we utilize a two-stage manual content analysis methodology to quantify the data [24]. According to the GRI information structure, we used the ESGI Airlangga University dataset to measure sustainability performance. All the observed variables are summarized in Table 1.

Table 1. Measurement of dependent, independent, and control variables.

Name of Variable	Mnemonics	Role	Measurement
Economic Sustainability Performance	EcSP	Dependent	ESGI Economic Disclosure Index
Environmental Sustainability Performance	EnSP	Dependent	ESGI Environmental Disclosure Index
Social Sustainability Performance	SSP	Dependent	ESGI Social Disclosure Index
Governance Sustainability Performance	GSP	Dependent	ESGI Governance Disclosure Index
Audit committee size	ACSIZE	Independent	Total number of the audit committee
Audit committee gender diversity	ACGEN	Independent	A binary variable takes value one if there exists a gender diversity and 0 otherwise.
Audit committee meeting	ACMEET	Independent	Number of audit committee meetings per year
Audit committee expertise	ACEXPERT	Independent	Percentage of audit committees which have economics/financial background
Firm Size	SIZE	Control	Log of total assets of the firm
Firm Age	AGE	Control	Age of the firm since its founding
SR Age	SR_AGE	Control	Age of the SR since the 1st SR launched

3 Results and Discussion

The classical assumption test and assessment of data normality were conducted to ascertain the suitability of the study data for subsequent analysis using multiple regression. The multiple regression test results are presented in Table 2.

Table 2. Regression results of AC characteristics and sustainability performance (t-test).

Variables	EcSP	EnSP	SSP	GSP
(Constant)	0.022	0.007	0.001	0.003
ACSIZE	0.690	**0.010**	0.896	0.745
ACGEN	0.873	0.160	0.630	0.712
ACMEET	0.949	0.095	0.819	0.394
ACEXPERT	0.500	0.058	0.537	0.974
SIZE	0.498	0.285	0.223	0.854
AGE	0.528	0.248	0.586	0.629
SR_AGE	**0.004**	**0.014**	0.526	0.334

Empirical results in Table 2 show that no variable is significantly related to economic SP, social SP, and governance SP. Only audit committee size was found to be significant with environment SP. Firm size and firm age were also insignificant as control variables—only sustainability reporting age significantly influenced the economic and environment SP. These results align with the early initiative on sustainability, which prioritizes economic and environmental issues as addressing stakeholder pressure over other bottom lines on sustainability reports [20,25]. This empirical observation indicates that the mining industry needs to prioritize sustainability activities adequately. The detrimental socio-environmental repercussions of the mining sector have led to the formation of anti-mining campaigns, films, and civil society protests, which is a predictable outcome [26].

4 Conclusion

Mining operators have the potential to employ various strategies to promote positive social outcomes for individuals and communities [26]. These strategies encompass grants and donations, directly supporting healthcare, education services, and other social needs. Creating employment opportunities through mining operations can contribute to poverty reduction and offer improved job prospects directly and through engagement with suppliers and contractors.

Furthermore, capacity development initiatives can be undertaken to bolster local businesses and support community-driven projects for local development. These efforts can also focus on strengthening the economic and managerial capacities of the region, facilitating economic diversification. Lastly, establishing partnership relations can ensure stakeholders' involvement and equity participation in local mining projects. Suggestions for future research could be the use of other CG elements. The role of sustainability committees could be an interesting research question for sharpening empirical evidence in Indonesian companies on sustainability implementation.

References

1. E. Moses, A. Che-Ahmad, and S. O. Abdulmalik, Cogent Bus. Manag. **7**, (2020)
2. I. Lock and P. Seele, J. Clean. Prod. **122**, 186 (2016)
3. C. Jensen and H. Meckling, J. Financ. Econ. **3**, 305 (1976)
4. R. E. Freeman, *Stakeholder Theory Handbook* (1984)
5. T. Gerwing, P. Kajüter, and M. Wirth, *The Role of Sustainable Corporate Governance in Mandatory Sustainability Reporting Quality* (Springer Berlin Heidelberg, 2022)
6. D. M. Christensen, Account. Rev. **91**, 377 (2016)
7. A. J. Trotman and K. T. Trotman, Audit. A J. Pract. Theory **34**, 199 (2015)
8. Y. Lazarenko, O. Garafonova, V. Marhasova, and S. Grigashkina, E3S Web Conf. **278**, (2021)
9. M. Naidoo and A. Gasparatos, J. Clean. Prod. **203**, 125 (2018)
10. A. Alberici and F. Querci, Corp. Soc. Responsib. Environ. Manag. **23**, 283 (2016)
11. E. S. Kasim, N. Md Zin, S. Sulaiman, S. F. A. Hamid, and Z. N. Zainuddin, Int. J. Sustain. Soc. **10**, 161 (2018)
12. J. Elkington, *Cannibals with Forks The Triple Bottom Line of 21st Century Business*, 1st ed. (Capstone Publishing Limited, Oxford, United Kingdom, 1997)
13. P. K. Biswas, M. Mansi, and R. Pandey, Pacific Account. Rev. **30**, 517 (2018)
14. N. Hussain, U. Rigoni, and E. Cavezzali, Corp. Soc. Responsib. Environ. Manag. **25**, 1198 (2018)
15. N. Hussain, U. Rigoni, and R. P. Orij, J. Bus. Ethics **149**, 411 (2016)
16. A. Dwekat, E. Seguí-Mas, G. Tormo-Carbó, and P. Carmona, Corp. Soc. Responsib. Environ. Manag. **27**, 2879 (2020)
17. A. M. Buallay and E. S. AlDhaen, *The Relationship between Audit Committee Characteristics and the Level of Sustainability Report Disclosure* (Springer International Publishing, 2018)
18. A. Buallay and J. Al-Ajmi, J. Appl. Account. Res. **21**, 249 (2020)
19. M. Arif, A. Sajjad, S. Farooq, M. Abrar, and A. S. Joyo, Corp. Gov. **21**, 497 (2020)
20. R. Appuhami and S. Tashakor, Aust. Account. Rev. **27**, 400 (2017)
21. M. I. Jizi, A. Salama, R. Dixon, and R. Stratling, J. Bus. Ethics **125**, 601 (2014)
22. R. Othman, I. F. Ishak, S. M. M. Arif, and N. A. Aris, Procedia - Soc. Behav. Sci. **145**, 330 (2014)
23. A. Helfaya and T. Moussa, Bus. Strateg. Environ. **26**, 1061 (2017)
24. M. Plumlee, D. Brown, R. M. Hayes, and R. S. Marshall, J. Account. Public Policy **34**, 336 (2015)
25. A. Helfaya and T. Moussa, Bus. Strateg. Environ. **26**, (2017)
26. A. Fonseca, M. L. McAllister, and P. Fitzpatrick, J. Clean. Prod. **84**, 70 (2014)

Interconnecting Science-Based Target with company supply chain to contribute climate action – lesson learned from Indonesia

Wika Harisa Putri[1*], and *Irfan* Bakhtiar[2]

[1]Janabadra University, Accounting Department, 55231 Yogyakarta, Indonesia
[2]WWF Indonesia, Climate and Market Transformation, 12540 Jakarta, Indonesia

Abstract. Science-Based Target (SBT) has become popular since 2020 as a promising platform for reducing greenhouse gas emissions. Recent studies found that SBT diffusion in low- and medium-income countries remains unresponsive even in emission-intensive sectors. Setting realistic targets for aligning actions across each chain, planning and implementing targeting methods, and developing SBT governance to respond to stakeholder interests can influence the climate change response. This paper describes how the international supply chains influence Indonesian companies to apply SBT through the demand side. This paper will depict early initiatives and evidence with descriptives data that announce the Indonesian company's plan to achieve established SBT. The result shows that Indonesian company still rare to implement SBT and the initiatives begin when triggered by international demand on supply chain.

1 Introduction

Since green and growth have become a widespread initiation to adding value in sustainable development, the linkage between a company's innovation and green business triggers the supply chain management process [1]. Aligned with that initiation, the Paris Agreement greenhouse gas (GHG) emission reduction commitments strengthen the company strategy to transform the Supply Chain Management (SCM) process greener in order to maintain their business. In that context, the Science-Based Target (SBT) platform proposes methodologies to reduce the company's GHG emissions to face the company's climate action[2]. The hope is that SBT and other corporate climate initiatives will become empirical evidence to enforce climate policy [3].

SBT initiative (SBTi) was created to respond to the Paris Agreement commitment and provide tools, guidelines, and methodologies for the company to reduce the GHG emission targets set at the company level [4]. Recently, SBTi become a more expansive platform implemented in global companies, appealing to the academia's worldwide attention [5]. Nevertheless, the empirical evidence shows that low and medium-income countries face many obstacles, such as regulatory pressure, linking the chain in SCM, lack of government

* Corresponding author: wikaharisa@janabadra.ac.id

programs that support this initiative, and lack of assessment tools and methods for combating climate change. These limitations are also relevant in Indonesia and depict the low number of company voluntary movements to join SBTi. The purpose of this review is to observe how why, and how many Indonesian companies participating in SBT initiatives for reducing GHG systematically.

According to the SBTi dashboard https://sciencebasedtargets.org/target-dashboard, the number of Indonesia's companies that joined voluntarily in the initiative is 28, active status and committed the near term is 25 companies, 18 companies in net zero commitment, and only one company dealing with long-term commitment, two companies with approved targets, and only one company removed their commitment. Furthermore, based on the data, most are manufacturing companies (22 from 27); others are services companies (banking, hospitality, services, and real estate). Hence, interconnecting SBT and company supply chain management becomes essential to accelerate cleaner production practices and effectively contribute to climate action.

In order to adopt sustainability practices, manufacturing companies have some motivations, such as (1) stakeholder pressure [6], (2) concern for environmental and social impacts [7,8], and (3) gaining a competitive advantage [9]. They also manage their performance and take appropriate metrics to measure their sustainability action. The metrics also serve as a tool to monitor and evaluate sustainability performance. Measuring the quality and effectiveness of corporate climate disclosure and target setting has become more prominent as these strategies have become more widely adopted [10]. SBT become an appropriate metric Indonesian companies use to meet the Paris Agreement's goals to reduce the GHG emission targets set at the company level. One of the benefits of establishing ambitious environmental objectives is the facilitation of collaboration between prominent companies and governmental bodies. This collaboration aims to foster the advancement of low-carbon technologies through innovation while concurrently assessing and formulating new regulations that promote effective implementation [11].

2 Literature review

2.1 Science-based target

Implementing a science-based target (SBT) setting assists enterprises in establishing greenhouse gas (GHG) reduction objectives that align with a predetermined temperature threshold. According to the SBTi, targets implemented by corporations to mitigate greenhouse gas (GHG) emissions qualify as "science-based" when they align with the necessary level of decarbonization essential to limit the rise in global temperatures to below 1.5°C compared to pre-industrial levels [12]. The SBTi is a joint initiative from the World Wide Fund for Nature (WWF), the World Resources Institute (WRI), the UN Global Compact, and the CDP, first launched in 2014 [13].

The SBTi aims to engage the commercial sector in addressing climate change with an awareness of concern. Adopting SBT approaches can facilitate the implementation of the Paris Agreement's socio-political objectives and ensure its practicality for stakeholders not directly involved in the agreement. This action may be achieved by incorporating the most relevant and up-to-date scientific knowledge into decision-making. Implementing Science-Based Targets (SBTs) should be considered a fundamental aspect of everyday business practice when creating strategies to reduce greenhouse gas (GHG) emissions [4]. SBTi also supports that the private sector is leading in achieving net-zero emissions by implementing the Corporate Net-Zero Standard and the upcoming Net-Zero Standard for Financial Institutions [12]. The incorporation of SBTi into firm supply chain management has the potential to facilitate a just transition towards a low-carbon society.

The SBTi establishes global-level, sector-specific decarbonization pathways compatible with the International Energy Agency (IEA) 2 °C scenario[14]. However, these pathways are only set for 13 aggregated global sectors without consideration of regional differences concerning technical capability, historical responsibility, resource endowment, and political environment. Hence, it is imperative to enhance the applicability of the SBTi framework by incorporating two distinct scientific domains: biophysical planetary boundaries[15], which encompasses a comprehensive understanding of the interrelated dynamics and comprehensive coverage of production and consumption patterns and intermediates demand-driven consumption-based accounting, which focuses on the products utilized by industries in the production of final consumer goods [9]. In the Indonesian context, the appropriate approach between the company supply-chain mechanism and the low-carbon measurement will be a priority to fulfill the global demand for final or intermediate products.

2.2 Company supply chain and greenhouse gas protocol

According to the Greenhouse Gas Protocol, the world's most widely used greenhouse gas accounting standard [16], we find the terms Scope 1, 2, and 3 emissions **(Fig.1)**. That term enables companies to understand their entire value chain emissions and focus their efforts on the most significant reduction opportunities. Scope 1 addresses emissions resulting from sources that an organization possesses or exercises direct control over, such as the combustion of gasoline in its fleet of vehicles. Scope 2 covers the corporation's indirect emissions from producing purchased and utilized energy, such as electricity generation for buildings. Scope 3 concerns emissions that the company itself does not directly cause, nor are they a consequence of actions conducted by assets owned or controlled by the firm. Instead, these emissions are attributable to entities the company bears indirect responsibility along its value chain [17]. The current standard provides obvious definitions for Scope 1 and 2 emissions but lacks detailed quantification and benchmarking methods for indirect emissions /Scope 3 [18]. When assessing specific measures or technologies to mitigate greenhouse gas (GHG) emissions, it is crucial to prioritize actions with the most significant potential to address both direct and indirect emissions [19].

Scope 3 emissions were, for the first time, taken into account in analyzing drivers and trends on a national level in the Intergovernmental Panel on Climate Change (IPCC) [20]. The focus on Scope 3 has only emerged in recent years, despite earlier studies indicating that the most significant emissions increase occurred inside Scope 3 [18]. Why? Because the footprint of a product across its supply chain will form the footprint of global consumption. Moreover, the most emissions increase occurred in developing countries. Therefore, indirect supply chain targets must align with carbon mitigation scenarios to gradually reduce the global upstream Scope 3 emission intensities. Nevertheless, implementation of this agenda has been limited due to the difficulty in obtaining emissions data from vendors and customers [9].

Scope 3 emissions encompass the emissions linked to using manufactured capital to produce energy and products. This scope includes coal mines, vehicles, roads, and buildings emissions. In the green SCM context, emission which does not directly cause by itself must be reduced with the green procurements. Moreover, the company has an opportunity to minimize GHG emissions by electing green manufacturing processes. Within the context of the entire supply chain, a formal request to the logistics supplier, pressing them to utilize vehicles powered by renewable energy sources, will also be a way to minimize GHG emissions.

https://doi.org/10.1051/e3sconf/202346705005

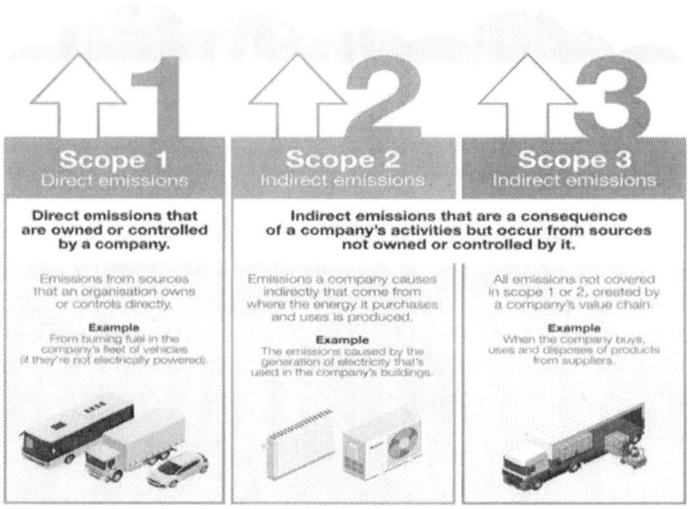

Fig.1. Scope 1,2,3 definition [17].

3 Method

This exploratory was derived from an elaborate analysis of prior studies on SBT, specifically focusing on data about Indonesian enterprises participating in the SBT initiative. Additionally, the study considered the efforts taken by these companies to reduce greenhouse gas emissions, with a particular emphasis on the perspective of supply chain management. The data of members' SBT can be accessible at the following URL: https://sciencebasedtargets.org/target-dashboard.

4 Result and discussion

The manufacturing industry is responsible for substantial carbon emissions released into the environment. As a result, manufacturers face considerable pressure from various stakeholders to address sustainability concerns and actively contribute to achieving the United Nations Sustainable Development Goals. In manufacturing industries, it is crucial to prioritize activities that generate the highest value, and doing a materiality analysis can aid in this process. Backcasting and materiality analysis are valuable methodologies for establishing quantifiable objectives, and these goals must receive scientific validation through the SBTi. Using the backcasting method with the SBT methodology will enhance the company's supply chain by prioritizing commitment to the sustainability platform.

The footprint of a product is the sum of the footprints of the processes along the product's supply chain. These processes may take place in different geographical areas. The operational footprint of a company is the sum of the footprints of its operations. There are always several entities playing a role in causing a footprint: the investors, the suppliers, the recipients, and the regulators. Hence, the responsibility for moving toward a sustainable footprint will be shared among them. Exploring how we can better institutionalize full supply chain responsibility is one of humanity's significant research challenges toward achieving a sustainable future [19].

Upstream activities include operations relating to the initial stages of producing an intermediate good or service, such as material sourcing, material processing, or supplier activities. For example, for a textile manufacturing company, "upstream emission footprint" refers to the emissions of all material extractions, conversions, processing, and delivery of inputs required to produce textiles, including emissions related to services. This footprint measurement does not include the end-use of goods and services the company sells. For the illustration of upstream and downstream activities in different geographical areas, the

https://doi.org/10.1051/e3sconf/202346705005

interconnecting between SCM Indonesian company and the global company depicted in **Fig.2** will be an example.

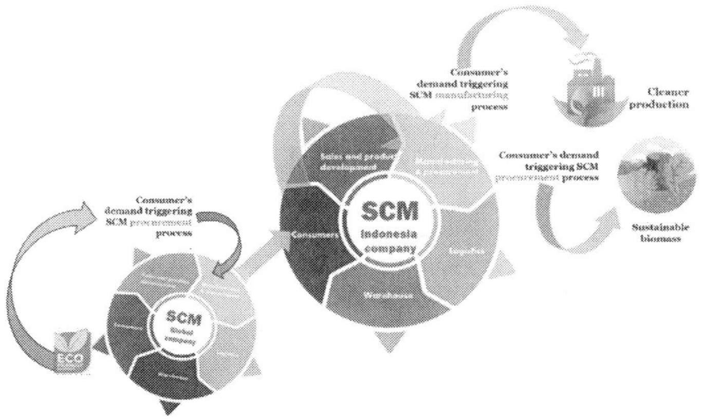

Fig. 2. Interconnecting between SCM Global Company and Indonesia Company.

This paper concerns a case study within the textile industry relevant to the Indonesian context. The analysis of the SBT target dashboard for 2023 reveals that the sectors demonstrating the most proactive initiatives are textiles, clothes, footwear, and luxury goods. The information mentioned above is supported by the observation that 8 of the 27 enterprises examined are actively involved in sustainability initiatives, as illustrated in **Fig.3**, which are in the textile, clothes, footwear, and luxury goods sectors. All processes involved in the extraction, conversion, processing, and transportation of raw materials necessary for textile production, as well as the emissions associated with these activities, were carried out by Indonesian companies. Integrating cleaner production practices and adopting sustainable fuel sources, including sustainable biomass, were mandatory components within the supply chain management framework (SCM). The commencement of the procurement process for green supply chain management (SCM) is motivated by the demand triggered by global firms adopting green SCM methods.

Fig. 3. Indonesian companies taking action in SBTi 2023.

5 Conclusion

Based on early evidence of applying SBTi in SCM at Indonesian companies, we can conclude that we have to raise technical and policy interventions on the supply and demand sides. Energy efficiency, cleaner production, green procurement, using renewable energy in the entire supply chain, and aligning actions across each chain will be essential actions to support the climate change response.

References

1. T. C. Herrador-Alcaide, M. Hernández-Solís, and S. Cortés Rodríguez, Humanit. Soc. Sci. Commun. **10**, (2023)
2. A. N. Ayoub, A. Gaigneux, N. Le Brun, S. Acha, and N. Shah, Build. Environ. **186**, (2020)
3. G. Marland, T. Kowalczyk, and T. L. Cherry, J. Ind. Ecol. **19**, 934 (2015)
4. P. C. S. Faria and N. Labutong, Sustain. Accounting, Manag. Policy J. **11**, 591 (2020)
5. A. Bjørn, J. P. Tilsted, A. Addas, and S. M. Lloyd, Curr. Clim. Chang. Reports **8**, 53 (2022)
6. A. T. Mengistu and R. Panizzolo, Meas. Bus. Excell. **27**, 54 (2023)
7. M. Zarte, A. Pechmann, and I. L. Nunes, Int. J. Sustain. Eng. **12**, 149 (2019)
8. D. Beekaroo, D. S. Callychurn, and D. K. Hurreeram, Ecol. Indic. **96**, 250 (2019)
9. M. Li, T. Wiedmann, and M. Hadjikakou, Environ. Sci. Technol. (2019)
10. Y.-J. Zhang and J.-Y. Liu, Front. Eng. Manag. **7**, 47 (2020)
11. J. Rissman, C. Bataille, E. Masanet, N. Aden, W. R. Morrow, N. Zhou, N. Elliott, R. Dell, N. Heeren, B. Huckestein, J. Cresko, S. A. Miller, J. Roy, P. Fennell, B. Cremmins, T. Koch Blank, D. Hone, E. D. Williams, S. de la Rue du Can, B. Sisson, M. Williams, J. Katzenberger, D. Burtraw, G. Sethi, H. Ping, D. Danielson, H. Lu, T. Lorber, J. Dinkel, and J. Helseth, Appl. Energy **266**, 114848 (2020)
12. Science Based Targets, https://sciencebasedtargets.org/faqs#what (2023)
13. J. Giesekam, J. Norman, A. Garvey, and S. Betts-davies, (2021)
14. J. Giesekam, D. D. Tingley, and I. Cotton, Energy Build. **165**, 106 (2018)
15. A. Bjørn, N. Bey, S. Georg, I. Røpke, and M. Z. Hauschild, J. Clean. Prod. **163**, 106 (2017)
16. P. Bhatia, L. Draucker, H. Lahd, C. Cummis, D. Rich, and A. Brown, World Resources Institute and World Business Council for Sustainable Development. https://ghgprotocol.org/sites/default/files/standa (2011)
17. National Grid, https://www.nationalgrid.com/stories/energy (2023)
18. E. G. Hertwich and R. Wood, Environ. Res. Lett. **13**, (2018)
19. A. Y. Hoekstra and T. O. Wiedmann, Science (80-.). **344**, 1114 (2014)
20. Intergovernmental Panel on Climate Change, in *Clim. Chang. 2014 Mitig. Clim. Chang.*, edited by O. Edenhofer, R. Pichs-Madruga, and Y. Sokona (Cambridge University Press. Cambridge, Cambridge, 2014), p. 35

Antibody titer uniformity in Indonesian Native Chickens against avian pathogens in a tropical environment: a three-time series study

Zainudin Al Wahid[1], *Luthfi Adya* Pradista[1], *Rahayu* Kusumaningrum[2], Lanjar[3], *Nuzul* Widyas[4], *Adi* Ratriyanto[4], and *Sigit* Prastowo[4]*

[1]Master Program of Animal Science, Faculty of Animal Science, Universitas Sebelas Maret, Surakarta, Indonesia
[2]Doctoral Program of Agricultural Science, Faculty of Agriculture, Universitas Sebelas Maret, Surakarta, Indonesia
[3]National Research and Innovation Agency, Jakarta, Indonesia
[4]Department of Animal Science, Faculty of Animal Science, Universitas Sebelas Maret, Surakarta, Indonesia

Abstract. Elevated temperatures in tropical environments detrimentally affect poultry immunity, increasing susceptibility to infectious diseases and compromising overall health. This study investigated the antibody titer uniformity in Indonesian native chickens (Kedu, Arab, and Lignan) against H5N1, H9N2, and ND-G-VII viruses within a tropical climate. The research involved a three-time series monitoring of log2 antibody titers at ages 10, 30, and 45 days. The 90 Indonesian native chicken blood samples were collected to assess their immune system. Results revealed a consistent pattern of antibody titer uniformity for H5N1 and H9N2 throughout the early growth stages, indicating a consistent immune response in the tropical environment. Contrasting with ND-G-VII, the antibody titers increased significantly ($p<0.05$: 4.90 ± 1.20 to 6.20 ± 1.23) at 10 to 45 days, reflecting a progressive immune response in Indonesian native chickens against ND-G-VII. The findings emphasize the importance of considering environmental context when interpreting immune responses in poultry reared in tropical climates, which is influenced by the uniformity of immune responses against different pathogens. Such knowledge could have implications for designing effective vaccination strategies and disease prevention measures tailored to the specific challenges of poultry farming in tropical regions.

1 Introduction

High temperatures in tropical environments can be detrimental to poultry antibody responses, as sustained heat stress has been shown to hinder the development and function of avian immune defenses [1]. In tropical regions such as Indonesia, where the poultry industry has seen significant growth in recent years, the presence of various avian pathogens has posed continuous challenges. Historically, Indonesia has faced outbreaks of avian influenza and

* Corresponding author: prastowo@staff.uns.ac.id

other infectious diseases that have had profound economic and public health implications. The progress made in mitigating these pathogens and the associated losses has been substantial, yet the persistent high temperatures in the region present an ongoing challenge, potentially exacerbating the susceptibility of poultry to these infections. Given this context, it becomes increasingly important to understand the specific impact of heat stress on poultry immunity in tropical environments.

Chronic heat stress (CHS) is a condition that arises when animals are exposed to prolonged periods of elevated temperatures, such as during a hot summer. While this extended exposure to high temperatures may not result in immediate death, it can have adverse effects on the animal's growth, immune system function, and ability to resist diseases [2]. The tropical climate can be characterized as being both warm and humid, and there is extensive documentation regarding how ambient temperature and relative humidity (RH) serve as significant stressors for chickens [3].

Vaccination is a successful method for stimulating the adaptive immune system to prevent the occurrence of infectious diseases caused by harmful bacteria or viruses. This widely employed immunization approach has been applied to industrial animals, including chickens. In the poultry sector, numerous vaccines have been employed, particularly to manage Avian Influenza and Newcastle disease. Nevertheless, when chickens raised for egg production are subjected to chronic heat stress, the levels of antibodies that typically rise in response to these vaccinations do not increase as anticipated [1].

It is important to examine the proper vaccine administration management to minimize the possibility of virus infection in local Indonesian chickens. This research will evaluate antibody titer responses to Avian Influenza and Newcastle Disease, considering the vaccine management that have been implemented on Kedu, Lignan, and Arab chickens at the Breeding Center for Maron Chickens, Temanggung, which are Indonesian Native Chickens breed. This research contributes to understanding the immune development of Indonesian native chickens under tropical conditions, providing insights into their responses to specific viral infections at different stages of growth.

2 Material and methods

2.1 Animal testing and sample collection

This experiment uses 3 breeds of Indonesian native chicken, including Kedu, Lignan, and Arab reared in the breeding center for Maron chickens, Temanggung. Chickens are kept in colony cages for each breed and each age. A total of 3 groups of chickens: Kedu, Lignan and Arab were vaccinated with NDIB killed and NDIB Live Vaccine at the time of hatching, then given a NDIB Live booster at 21 days and NDAI killed at 35 days ages. Data were collected at 10, 35, and 45 days. Basal feed is obtained from a mixture of corn, rice bran, soybean meal, and several other feedstuffs and premixes to meet NRC standards. 1.2 g/kg betaine supplementation (Excential Beta-key, Orffa ©) was given in feed to improve quality [4]. Drinks were given to the chickens ad-libitum.

The Avian influenza and Newcastle Disease vaccine we use medivac NDAI H5N1 Subtype 2.3 and ND G7 Strain. We measured antibody titers using haemagglutination and hemagglutination inhibition test following the OIE standards chapter 3.3.4 for avian influenza [5] and chapter 3.3.14 for Newcastle disease [6] which were carried out at Japfa Comfeed's Agrilab.

2.2 Environment microclimate data

To assess the microclimate conditions in the Temanggung region, we have accessed climatological information from the Meteorology and Geophysics Agency of Indonesia in April 2023 [7], focusing particularly on temperature and humidity data for the Temanggung area. We provide microclimate data in the form of THI (Temperature Humidity Index). THI was calculated based on Zulovich and DeShazer's [8]:

$$THI = db^{o}C - [(0.31 - 0.31\,RH)(db^{o}C - 14.4)]$$

where THI: temperature humidity index, db°C: dry bulb temperature in Celsius, and RH: relative humidity in percentage.

2.3 Experimental design and statistical analysis

A completely randomized design and analysis of variance were used in this experiment to evaluate the antibody titers in every sample collected (10, 35, and 45 days of age). Duncan test was used if there were significant differences in ANOVA test results. Software R version 4.1.2 performs statistical data analysis according to the R Core Team procedure. Package 'agricolae' and 'ggplot2' are used to support the analysis process in R software.

3 Result and discussion

Following our analysis of the weather conditions in the Temanggung region using data from BMKG (Meteorology, Climatology, and Geophysics Agency), it can be deduced that chickens experience constant exposure to heat stress on a daily basis, as indicated by Figure 1.

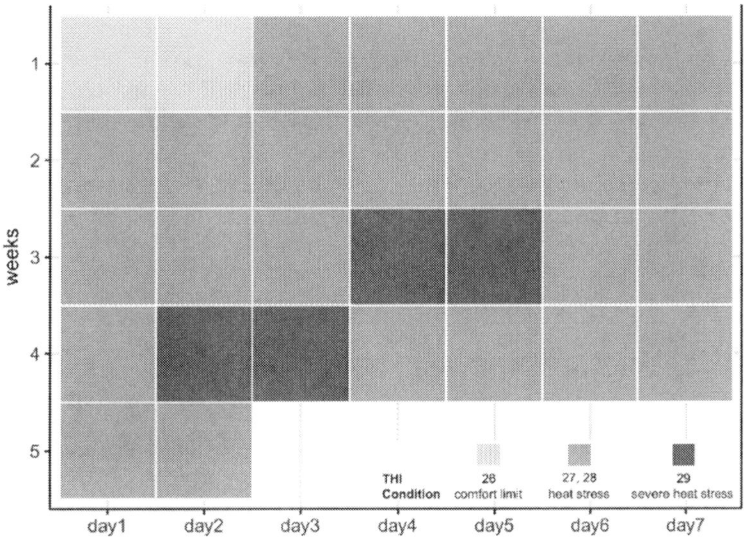

Note: data retrieved from BMKG 2022.

Fig. 1. Heatmap of THI condition.

The values applicable to animals like poultry and pigs that do not have sweat glands are as follows: normal when less than 27.8, moderate between 27.8 and 28.8, severe within the range of 28.9 to 29.9, and very severe (reaching emergency levels) when equal to or greater than 30.0 [9]. Elevated heat stress conditions have been linked to negative effects on the immune system's response [1]. Recent research has revealed that the anticipated increase in antigen-specific antibody response following immunization with a commercial vaccine is not observed in heat-stressed broiler chicks, particularly those with heightened levels of the stress hormone corticosterone (CORT) [10]. Base on figure 2, antibody titers for H5N1, H9N2, and ND G VII were examined in Indonesian native chickens at three different ages: 10 days, 35 days, and 45 days.

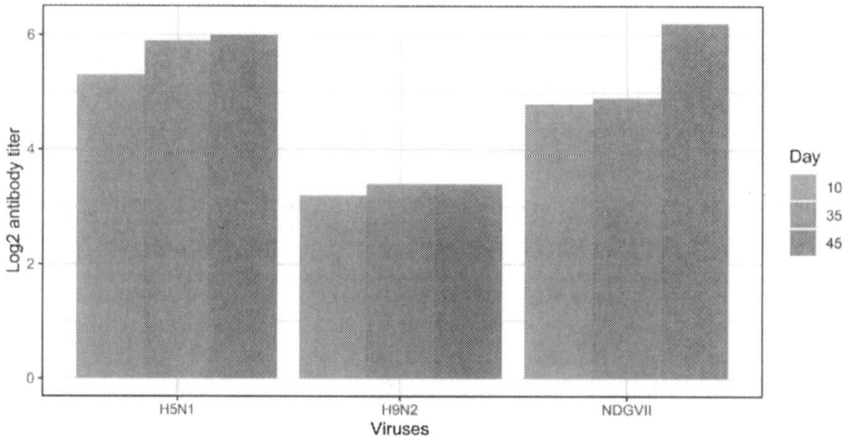

Fig. 2. Boxplot of Log 2 antibody titers in native chicken

The results indicated that there were no significant differences in antibody titers for H5N1 and H9N2 among the age groups. However, for ND G VII, a significant difference was observed, with higher antibody titers at 45 days of age compared to 10 and 35 days. This suggests that age has an impact on the antibody response to ND G VII in these chickens. These findings provide valuable insights into the immune responses of Indonesian native chickens to specific pathogens at different stages of development, which can have implications for poultry health management and vaccination strategies. The OIE (World Organization for Animal Health) has established a threshold for immunity levels against ND (Newcastle Disease) and AI (Avian Influenza) viruses. If the antibody level is greater than 4 HI log 2, it is considered protective. However, if the antibody level is less than 4 HI log 2, it is deemed not protective against the virus [11]. The results of the antibody titer examination at 10 days of age indicated a protective antibody level for the specific antibody against AI strain H5N1, with a value of 5.30±0.95 (table 1).

Table 1. Log 2 antibody titers of H5N1, H9N2, dan ND G VII in Indonesian native chicken.

Age	n	H5N1	H9N2	ND G VII
10-day	10	5.30±0.95	3.20±0.42	$4.80{\pm}1.48^{b}$
35-day	10	5.90±0.74	3.40±0.70	$4.90{\pm}1.20^{b}$
45-day	10	6.00±1.05	3.40±0.52	$6.20{\pm}1.23^{a}$
P-Value		0.205	0.656	0.042

Note: ^{a,b} difference superscript in the same column indicate significant difference ($P<0.05$).

Similarly, the antibody levels against ND strain G7 also showed a protective level at 4.80±1.48. Observations on day 35, following the Live ND vaccine booster administered on day 21, revealed that the levels of AI and ND antibodies remained protective. During the third observation on day 45, following the NDAI vaccine booster given on day 35, protective titer results were still observed. However, after conducting three monitoring sessions, it became evident that the antibody level against H9N2 did not reach a protective threshold. This occurred because the killed vaccine administered did not include a combination with the AI H9N2 strain.

When we consider the influence of heat stress on immunity, it becomes evident that native traditional chickens in Temanggung exhibit an ability to adapt to challenging environmental conditions. Upon closer examination, it is suggested that providing betaine to chickens of all ages may have a positive impact on coping with environmental fluctuations [12–14]. Research conducted on this matter affirms that administering betaine can effectively counteract the detrimental effects of heat stress [14]. Many studies state that betaine plays a role as an osmoprotectant for poultry [15]. In hot environmental conditions, it can reduce excess digestive tract cell fluids. Furthermore, the stable condition of the digestive tract cells will trigger the metabolic system to run optimally, and the more influential the body can absorb the nutrients [12,16,17]. On the other hand, nutrients in the digestive tract are also a significant factor in determining the microbial composition of the digestive tract [13]. Digestive tract microbes directly play a role in many processes. One of the main ones is in the immune system by producing anti-inflammatory products that can reduce pathogens in poultry [13,18]. In addition, betaine has been found to elevate lymphocyte count by up to 22.59% and enhance immunity against the ND virus by up to 103.48% in terms of increased antibody levels [19]. Furthermore, betaine supplementation has been found to decrease the heterophil percentage and increase the lymphocyte percentage, leading to a decrease in the heterophil/lymphocyte ratio [15].

4 Conclusion

In conclusion, this study demonstrates that native chickens exhibit an ability to adapt to heat stress conditions. Additionally, the research highlights the adaptability of native traditional chickens in Temanggung to challenging environmental conditions. It suggests that betaine supplementation, especially for all age groups, may positively impact their ability to cope with environmental fluctuations. Administering betaine was found to mitigate the adverse effects of heat stress and significantly enhance immune responses following vaccination.

These findings contribute to a deeper understanding of the immune development of Indonesian native chickens in tropical conditions, providing valuable insights into their responses to specific viral infections at different growth stages. Furthermore, it underscores the importance of considering the environmental context when interpreting immune responses in poultry reared in tropical climates. This knowledge holds potential implications for designing effective vaccination strategies and disease prevention measures tailored to the unique challenges of poultry farming in tropical regions.

We would like to express our heartfelt gratitude to the National Research and Innovation Agency (BRIN), Indonesia and the Educational Fund Management Institution (LPDP), Ministry of Finance, Indonesia for their invaluable support and funding, which made this research possible. Their commitment to advancing scientific knowledge and education has been instrumental in the success of our project.

References

1. R. Hirakawa, S. Nurjanah, K. Furukawa, A. Murai, M. Kikusato, T. Nochi, and M. Toyomizu, Front. Vet. Sci. **7**, 1 (2020)
2. D. Meng, Y. Hu, C. Xiao, T. Wei, Q. Zou, and M. Wang, Biomed Res. Int. **2013**, (2013)
3. C. Tirawattanawanich, S. Chantakru, W. Nimitsantiwong, and S. Tongyai, J. Appl. Poult. Res. **20**, 409 (2011)
4. NRC, *Nutrient Requirements of Poultry*, Ninth (National Academy of Sciences, Washington, 1994)
5. OIE, in *Man. Diagnostic Tests Vaccines Terr. Anim.* (World Organisation for Animal Health, 2021), p. 282
6. OIE, in *Man. Diagnostic Tests Vaccines Terr. Anim.* (World Organisation for Animal Health, 2021)
7. BMKG, BMKG Stasiun Klimatologi Jawa Teng. (2023)
8. J. M. Zulovich and J. A. DeShazer, *Estimating Egg Production Declines at High Environmental Temperatures and Humidifies* (ASAE, St. Joseph, Mich., 1990)
9. I. F. M. Marai, M. S. Ayyat, and U. M. Abd El-Monem, Trop. Anim. Health Prod. **33**, 451 (2001)

10. R. Jahanian and E. Rasouli, J. Anim. Sci. **93**, 3355 (2015)
11. OIE, World Organ. Anim. Heal. **1**, 1185 (2013)
12. A. Ratriyanto, *BETAIN: Aditif Pakan Ternak Untuk Daerah Tropis*, I (UNS Press, Surakarta, Indonesia, 2022)
13. L. A. Pradista, S. Prastowo, N. Widyas, and A. Ratriyanto, IOP Conf. Ser. Earth Environ. Sci. **1114**, 3 (2022)
14. Z. A. Wahid, A. Ratriyanto, and S. Prastowo, IOP Conf. Ser. Earth Environ. Sci. **902**, 8 (2021)
15. A. Ratriyanto and R. Mosenthin, J. Anim. Physiol. Anim. Nutr. (Berl). **102**, 1634 (2018)
16. A. Ratriyanto, L. A. Pradista, S. R. P. Sari, S. Prastowo, and N. Widyas, IOP Conf. Ser. Earth Environ. Sci. **902**, (2021)
17. M. Eklund, E. Bauer, J. Wamatu, and R. Mosenthin, Nutr. Res. Rev. **18**, 31 (2005)
18. G. Mátis, M. Mackei, B. Boomsma, H. Fébel, K. Nadolna, Ł. Szymański, J. E. Edwards, Z. Neogrády, and K. Kozłowski, Animals **12**, (2022)
19. N. Chand, S. Naz, H. Maris, R. U. Khan, S. Khan, and M. S. Qureshi, Pak. J. Zool. **49**, 1857 (2017)

Exceptions for stop building coal-fired power plants in Indonesia

Andriko Sandria[1,2*], *Kresna Damar* Adji[1], and *Jericho Julius* Prabowo[1]

[1]International Relations Department, Universitas Sebelas Maret, Surakarta, Indonesia
[2]ASEAN Studies Centre Universitas Sebelas Maret, Surakarta, Indonesia

Abstract. Indonesia planned to stop building new Coal-Fired Power Plants after 2023 which are in line with its NDC for Global Climate Change Regimes. But, Joko Widodo, Indonesia President, issued regulations that the government will allow to build new coal-fired power plants until 2030. This paper will explain Indonesia policy in allowing the construction of new coal-fired power plants. By using international regimes in renewable energy approach, Indonesia's role in fulfilling the needs in energy and also commitment to transform to clean energy will be explained. The Results of this paper show that Indonesia used rhetoric to show their commitment to get international support in funding for renewable energy developments. Yet, Indonesia allowing the construction of new coal-fired power plants has been used for bargaining positions for most of its interest in international forums. Indonesia got $20 billion deals from G7 Countries for transitions in The Just Energy Transition Partnership (JETP) scheme but is still getting international attention to get more investments. Indonesia also handled most of Non Governmental Organizations efforts to stop building new coal-fired power plants by making public statements for its reliable plant of transitions. In conclusion, Indonesia is using coal-fired power plants rhetoric to get funding for its transitions coping with the climate changes agenda in energy sectors.

1 Introduction

This paper will explain Indonesia's reasons for continuing to grant permits to build coal-fired power plants after a commitment to stop using coal as a source of electrical energy. Each country agrees to meet the Paris Agreement targets and reduce the impact of climate change. One of the big agendas of this regime is reducing carbon emissions. In the energy sector, the use of coal needs to be replaced with cleaner renewable energy sources to reduce emissions by a large amount. Since Indonesia signed the Paris Agreement, Indonesia has been asked to commit to reducing carbon emissions. One of these commitments is to produce clean energy by utilizing new, renewable energy sources. Thus, Indonesia is planning a transition from using coal for electricity generation to renewable energy sources. This is demonstrated by

* Corresponding author: andrikosandria.ma@staff.uns.ac.id

the State Electricity Company (PLN) which is targeting carbon neutral electrical energy by 2060.

Indonesia is still tug-of-war in the use of coal, especially in PLTUs. Indonesia plans that in 2025 it will start closing Steam Power Plants (PLTU) that use coal. Indonesia also stipulates not to build new PLTUs in Presidential Regulation Number 112 of 2022. However, in the eyes of the public, Indonesia is inconsistent because it is still planning to build new PLTUs at the end of 2022 until 2025. Some researchers say that Indonesia has difficulty releasing coal as an energy source. Indonesia still provides large subsidies with electricity rates that are lower than international standards. Apart from that, the costs used to produce electricity are much lower if you use coal in a PLTU [1]. Indonesia is not fast enough to make the transition because it does not yet have a nuclear power plant like in European countries for renewable energy sources. Indonesia will only build a nuclear-powered power plant in 2049[1]. The Domestic Political Situation is also an obstacle for Indonesia to make the transition. Entrepreneurs in coal mining in Indonesia have an important role in the energy chain in Indonesia. Many entrepreneurs in the regions who have businesses in the coal sector absorb a large workforce, so there is community dependence and emotional attachment to coal mining. There will be changes in economic and social structures that require firm policies to encourage transition [2, 3].

This paper explains further the reasons why Indonesia is still using coal under pressure from the international regime for renewable energy. Several diplomatic and negotiation efforts in various forums and organizations by Indonesia to gain support for coal phase-out are the main object of this paper.

2 Methods

Mitigation efforts in the Paris Agreement are realized by many countries with transition plans to use clean energy. For countries that depend on coal for their energy, it will be difficult to make a transition [4]. Demands from the international public regarding the large carbon emissions produced by burning coal make these countries have to think about ways and strategies to demonstrate their commitment to the Paris Agreement [4], [5]. Indonesia is a country that is under pressure from the international public because of the use of coal as an energy source. This paper will look at public diplomacy efforts in various international forums conducted by Indonesia. Discourse is a tool that can be used in public diplomacy [6]. The discourse behind the tug-of-war over Indonesia's use of coal will be dissected to find reasons for Indonesia to build a reputation for meeting the Paris Agreement targets.

This article analyzes the meaning of the statements made by the Indonesian government in each international forum. The data in this paper was collected through tracking Indonesian government speeches and statements relating to electricity generation and renewable energy. The results of the interpretation of these speeches and statements illustrate the form of Indonesia's diplomatic efforts in carrying out the energy transition and reducing the negative impacts of climate change.

3 Results and discussion

One of the reasons Indonesia is making the transition to New Renewable Energy (NRE) which is considered slow, especially regarding the existence of PLTU or Steam Power Plants, can be seen from an economic perspective. In this case, there are many opportunities where Indonesia, through government figures, has intensified campaigns that Indonesia is a country that has a strong intention to transition to NRE [7]. Indonesia has ratified the Paris Agreement and transferred it to Law Number 16 of 2016, which encourages reduction of greenhouse gas emissions through a Nationally Determined Contribution (NDC) and this has become a point of emphasis in international forums. However, on various occasions, there is one consistent theme that can be found from various actors at various occasions, in front of various audiences – efforts to attract foreign investment. This is done through various opportunities, the majority of which are meetings attended by parties from two or more countries in the world [8].

The construction of PLTU in Indonesia is a form of investment by several development banks in Asia. China is a country that invests a large amount for PLTU in Indonesia. However, in 2019, China began reducing and cutting funds for PLTU construction. Several PLTU development plans with Indonesia are among those that have an impact. In this situation, Indonesia seeks to attract the attention of China which is also committed to reducing the use of coal as an energy source [9]. President Joko Widodo seeks to promote green energy projects along with the Capital City of Nusantara (IKN) project to Chinese entrepreneurs at an event at the Shangri-La Hotel, Chengdu, China, July 28 2023, which resulted in an investment agreement between the two countries.

Efforts to promote green energy were also carried out in Germany, at the Hannover Messe industrial exhibition, April 17 2023. On this occasion, Joko Widodo made a statement regarding the transition to green energy, "Of course this requires investment, requires large financing, at least USD 1 trillion by 2060 Indonesia invites investors from Germany to develop a green economy. Once again, Indonesia is very open to investment and cooperation." Promotional efforts were also carried out in Singapore, on 7 June 2023. On this occasion, Joko Widodo directly stated that "...this meeting was to promote his main investments in the fields of energy transition, green infrastructure and the development of the Indonesian Capital City (IKN)."

In front of 800 participants at the 2023 Ecosperity Conference, Joko Widodo stated "I advise you not to wait too long. This is a very interesting golden opportunity for Indonesia and you can be a part of it." "EV battery production next year, as well as production of 1 million electric cars and 3.2 million electric motorbikes in 2035," said Joko Widodo.

Apart from that, Indonesia is also actively promoting in several international forums. One of them is the International Economic Modeling Forum which is supported by ARISE+ Indonesia, an organization funded by the European Union, and Oxford Economics. On this occasion, the Deputy for Economic Affairs of the National Development Planning Agency, Amalia Adininggar Widyasanti invited the potential for collaboration and cooperation for the use of EBT in Indonesia.

On another occasion, Indonesia also promoted efforts to transition to green energy in Glasgow, Scotland, at the UN Climate Change Conference (COP26). Indonesia said that Its transition should be supported by technology and funding so it won't put an enormous burden on the people, the industry, and the state finances. Indonesia needs US$50 billion for transformation towards new renewable energy and it costs US$37 billion for forestry sector, land use, and sea carbon.

The effect of this vigorous promotion of EBT investment can then be seen through the investment or agreements that appear. It can be seen in several of the meetings above that there is potential or even agreement, both formal and non-formal, such as the meeting with China which immediately resulted in "...an agreement between Indonesia and China to continue to encourage investment in Indonesia", according to Joko Widodo's comments and with Germany which "...will encourage investment among G7 countries into Indonesia of up to €10 billion.", according to a reply speech by Chancellor Olaf Scholz. At the G20 meeting, Indonesia also succeeded in securing an agreement on the Just Energy Transition Partnership (JETP) to reduce dependence on fossil fuels and coal. As a form of follow-up to JETP,

Indonesia also succeeded in agreeing on cooperation in the development of small modular reactor nuclear energy renewable energy technology with the United States through the Indo-Pacific Chamber of Commerce and Industry Business Forum meeting in Nusa Dua, Bali, on March 18 2023, containing technical and economic assistance, including a development fund of USD 1,000,000. Similar things also have an impact on other countries that have not received visits, one of which is Singapore, which on March 16 2023 succeeded in agreeing on a Memorandum of Understanding (MoU) which includes facilitating the development of solar photovoltaics (PV) and Battery Energy Storage Systems (BESS) technology and EBT exports. This MoU is considered to "...strengthen the energy infrastructure, energy transition and energy security for both Singapore and Indonesia, and also support regional initiatives such as the Asean power grid.", said the Prime Minister of Singapore, Lee Hsien Loong. Other countries that have succeeded in agreeing on similar agreements are countries such as South Korea – which focuses on investment cooperation, technology transfer, capacity-building, and accelerating renewable energy development.

4 Conclusion

Indonesia postponed the PLTU moratorium to ensure that domestic energy needs are met to grow the industry. Indonesia seeks to attract investors with rich resource offers to develop green industries. Rhetoric as a developing country is used to gain support in the transition process, starting from technology transfer, infrastructure development and other investments. The target of accelerating the use of nuclear technology in power generation in Indonesia is starting to appear with collaboration with investors to develop small modular reactors. Indonesia's diplomacy to show commitment and efforts to reduce carbon emissions is directed at getting new investors in renewable energy. The sooner Indonesia gets investors, the sooner it will realize an emissions-free Indonesia which will contribute to reducing the impact of climate change.

References

1. A. Soemanto, R.H.T. Koestoer, Scen. Ins. **6** (2023)
2. A.B. Setyowati, J. Quist, Cont. trans. **165** (2022)
3. M. Jakob, J.C. Steckel, *The Political Economy of Coal: Obstacles to Clean Energy Transitions*, (Routledge, New York, 2022)
4. E. Brutschin, F. Schenuit, B.V. Ruijven, K. Riahi, Expl. Enab. **10** (2022)
5. M.R. Davidson, X. Gao, J. Busby, C. Shearer, J. Eisenman (2023)
6. Q. Jiuding, Z. Leilei, Rsch. Rev. (2023)
7. M. Maulidia, P. Dargusch, P. Ashworth, F. Ardiansyah, Reth. ren. **101**, 231-247 (2019)
8. M.M. Islam, K. Sohag, S.O. Mamman, H. Herdhayinta, Resp. Indo. (2023)
9. G. Lim, Chin. Rivl. **46**, 105-132 (2022)

Adaptation and mitigation in agrotourism development

Sugihardjo[1], *Eksa* Rusdiyana[1*], *Eny* Lestari[1], Widiyanto[1], *Retno* Setyowati[1], and *Nur Kholiq Birr* Samsita[2]

[1]Agricultural Extension and Communication Study Program, Faculty of Agriculture, Universitas Sebelas Maret, Jl. Ir Sutami 36 A, Kentingan, Jebres, Surakarta, Jawa Tengah, Indonesia
[2]Dept. Of Undergraduate Agricultural Extension and Communication Study Program, Faculty of Agriculture, Universitas Sebelas Maret, Jl. Ir Sutami 36 A, Kentingan, Jebres, Surakarta, Jawa Tengah, Indonesia

Abstract. Climate change and global warming have an impact on the development of the tourism sector. Tourism village development programs must align with ecological interests and politics to align with tourism and environmental sustainability. This study aims to analyze adaptation and mitigation in efforts to develop agro-tourism. This study uses a qualitative descriptive approach. Data was collected through interviews, involved observation, and FGD. The location was chosen purposively, namely agro-tourism Ngargoyoso, Karanganyar, Indonesia. Research informants were obtained through purposive and snowball sampling of village heads, youth leaders, tourism awareness groups, and community groups. Data validity was determined using source triangulation and method triangulation. Data were analyzed using the interactive Milles and Huberman data analysis model. The analysis results show that the community has adapted to maintain the sustainability of tourism objects by managing irrigation water, selecting plant varieties, planting annual crops to prevent landslides, and marketing and processing crops. Mitigation includes making appropriate infrastructure easy to reach for visitors to tourist attractions.

1 Introduction

The impact of climate change on national development results in an increase and change in temperature and reduces natural resources [1]. A handful of parties need help to solve the problem of climate change. However, all individuals must participate and become actors in development who care for the environment and are sustainable. Communities must create attitudes and behaviors that care for the environment and play an active role in the success of development that maintains environ mental balance and is sustainable. In one of the efforts to deal withclimate change, all parties, including the community, need to have knowledge and skills and be able to support environ mental preservation programs.

Development in the tourism sector is one of the efforts to improve the economy of local communities and globally. One of these developments is carrying out tourism-based village

* Corresponding author:eksarusdiyana@staff.uns.ac.id

 https://doi.org/10.1051/e3sconf/202346705008

development. The development of a tourist village is a form of tourism that creates a relationship between tourists and rural components (norms, values, beliefs, and customs) [2]. The development of tourist villages is an alternative form of tourism that can contribute to the Sustainable Development Goals (SDGs) by facilitating development at the local level. Tourism contributes directly and indirectly to rural development. The management of tourist villages is one of the village development efforts in an integrated manner which aims to encourage the socio-cultural and economic aspects of the village. Tourism villages encourage regional economic development and growth, tourism destinations which are a priority to achieve economic, socio-cultural, and environmental sustainability goals [3]. However, village development must alsopay attention to aspects that can support the sustainability of village development, such as adaptation and mitigation.

Ngargoyoso District is one of the sub-districts in Karanganyar Regency, Central Java, consisting of 7 villages that have very beautiful natural scenery because they are located to the west of Mount Lawu, which makes Ngargoyoso District located on the slopes of Mount Lawu. With this natural landscape, 23 tourist destinations developed in the form of natural and historical tourism in Ngargoyoso District, one of which is a tourism-based village. The existence of this tourism-based village offers an opportunity for tourists to experience the lifestyle of rural communities. Tourism-based villages in Ngargoyoso District are managed by Village-Owned Enterprises (BUMDes) and the private sector. Institutions in tourism are a component to support tourism sustainability because tourism institutions play a role in facilitating and mobilizing community participation in tourism [4].

One of the problems in developing tourist villages in Ngargoyoso District is adaptation and mitigation of climate change and weather. This study aims to analyze the impact of climate change adaptation and mitigation on the development of tourist villages. Climate change is a climate change that is influenced directly or indirectly by human activities that change the atmosphere's composition, increasing the variability of the observed climate over a long period [5]. Climate change can be anticipated with mitigation and adaptation. Mitigation means prevention efforts that need to be done. At the same time, adaptation is adjustment activities that need to be done to live and survive, increase resilience and flexibility, and lead to migration due to different climatic conditions [6]. Based on this background, thisresearchaims to describe adaptation and mitigation to climate change in tourism management, especially in the field of agrotourism in rural areas.

2 Material and method

This study examines community involvement in the tourism-based village development process. The research method used descriptive qualitative approach. Collection techniques with triangulation (combined), data analysis is inductive/qualitative, and the results of qualitative research emphasize meaning rather than generalization [7] with data collection techniques through focus group discussion (FGD), direct interviews, documentation, and field observations, the data collection process took approximately 3 months. Qualitative research methods will be suitable for understanding social processes or interactions [8–10]. Several indicators must beconsidered, including destinations and attractions, facilities and marketing, the potential for partnerships, enthusiasm from the community, and the availability of public facilities. Tourism-based villages also have principles, namely paying attention to environmental characteristics, having a positive impact on the area, materials according to the local environment, environmentally friendly materials, and involving local communities in developing tourism-based villages. The data sources used in this research are books, scientific journals, theses, and other documents relevant to climate change adaptation and mitigation impacts on tourism-based village development in Ngargoyoso District.

3 Results and discussion

Ngargoyoso District, Karanganyar Regency, Central Java Province (Indonesia), has beautiful natural scenery because it is located west of Mount Lawu. With these natural views, 23 tourist destinations have developed in Ngargoyoso District, one of which is a tourism-based village. The existence of this tourism-based village offers an opportunity for tourists to experience the lifestyle of rural communities. Tourism-based villages in Ngargoyoso District are managed by Village-Owned Enterprises (BUMDes) and the private sector. A tourist spot cannot be separated from disasters, whether in the form of natural disasters, artificial disasters, or both. A disaster is a series of events that threaten, disrupt, and damage people's lives and livelihoods caused by natural, non-natural, and human factors [11]. One form of disaster that people need to understandis the climate change disaster. Climate change (*climate change*) is unavoidable due to global warming (*global warming*) and is believed to broadly impact various aspects of life, including the tourism sector. Changes in rainfall patterns, an increase in the frequency of extreme climate events, and increases in air and sea level temperatures are serious impacts of climate change that Indonesia faces [12]. Tourismis one of the sectors experiencing the most serious impacts due to climate change, especially agriculture-based tourism (agro-tourism) and tourism that utilizes natural resources directly. It is necessary to make an effort to preserve nature; the efforts made can be in the form of adaptation and mitigation.

3.1 Adaptation to agrotourism management

Adaptation is an action taken to reduce vulnerability and increase resilience. Emphasis on adaptation is necessary because even though the climate change impact scenario is so worrying, in several places in various parts of the world, the impacts of climate change can be overcome. In the local contest (Indonesia), adaptation is needed immediately to reduce the impact of climate change that is already underway [13]. Adaptation, as stated in Article 57 of Law Number 32 of 2009 mentioned efforts to maintain the environment are carried out in three ways, namely natural resource conservation, natural resource reserves, and or preservation of the function of the atmosphere. Adaptation can also be referred to as an effort to conserve natural resources and prevent changes in natural conditions that can lead to natural disasters. Implementation of adaptation in the development of a tourist village in Ngargoyoso includes the practice of conserving natural resources and protecting the natural resources themselves.

3.1.1 Conservation of flora and fauna

The community conserves natural resources by planting ornamental plants and woody plants that follow the region's geographical conditions, not logging plants indiscriminately, and carrying out plant propagation at tourist village location points. So, apart from being part of the adaptation effort, it also provides aesthetic value in tourism objects. In addition, conservation is also carried out by preserving the fauna in protected tourism areas.

3.1.2 Protecting natural resources

Efforts to protect natural resources include making the best possible use of water resources, utilizing abundant water to store it as water to support the tourism sector, such as bathing pools and for community needs, and channeling the remainder for irrigation purposes on agricultural land, livestock, and fisheries.

3.1.3 Preservation of atmosphericfunctions

The cold Ngargoyoso area encourages tourism managers and accommodation service providers to minimize the use of air conditioning so that apart from reducing electricity consumption costs, it also reduces carbon consumption. Cold regional conditions also encourage people and tourists to adapt to wearing thicker and warmer clothing.

3.2 Mitigation in tourism object management

Disaster Mitigation is an effort to reduce the risks and impacts caused by disasters on communities in disaster-prone areas, whether natural disasters, manufactured disasters, or a combination of the two within a country or community [14]. Mitigation efforts can becarried out in structural mitigation by strengthening buildings and infrastructure potentially exposed to hazards, for example, creating building codes, engineering designs, and construction to delay and strengthen structures or buildings to protect against landslides, coastal walls, and others. In addition, mitigation efforts can be carried out in non-structural forms, including avoiding disaster-prone areas by building away from disaster locations, which can be identified through spatial and regional planning and by empowering communities and local governments.

Tourism in the Ngargoyoso subdistrict is very diverse, including guava-picking agrotourism in Jatirejo Village. In maintaining environmental sustainability, agrotourism managers also carry out several forms of adaptation and mitigation. One form of mitigation to address climate change is related to irrigation; the handling is in the form of irrigation using pumps (wells) when the dry season arrives. In addition, increasing special access to guava picking agro-tourism so that larger vehicles can access the vehicle space when something happens. The form of adaptation carried out is adjusting the conditions of changing weather and climate related to the irrigation of guava plants. Water is the source of life and determines plant growth. The availability of water and the response of plants to water also influence the growth process and crop yields [15]. During the dry season, irrigation canals are opened from above to irrigate guava trees in agro-tourism areas, while during the rainy season, irrigation canals are closed to keep the trees from drying out. This irrigation is very influential on the yield of guava fruit. Irrigation volume affects the growth of guava plants, as it is reported that increasing water availability can increase the area and number of leaves and the weight of plant biomass [16]. The less water, the guava produced will be sweeter but difficult to grow, while the more water, the fruit will quickly grow and mature but lesssweet. Another form of mitigation is protecting the guava fruit, which is still small and young, by wrapping the fruit in a special plastic. This packaging aims to protect the fruit from the influence of external plant conditions such as pests and rain water. Agrotourism managers and farmers in Jatirejo have never received special training related to adaptation and mitigation. However, groups of guava farmers receive counseling at least once every 3 months from the local Agricultural Advisory Office (BPP).

Other tourist attractions in Ngargoyoso District include tea plantations, Pucung river tubing, and paragliding in the village of Kemuning. Kemuning Village is one of the popular tourist villages in Karanganyar Regency, so many visitors spend weekends and holidays in Kemuning Village. Mitigation in Kali pucung tourism is also givenless attention, only paying attention to river flow, especially during the rainy season, when the flow is too heavy (floods). The tubing is temporarily closed, and if the river flow is normal, it will be reopened.

Tourism also greatly impacts the surrounding conditions, especially for people whose homes are close to tourism sites, both positive and negative impacts [17]. The surrounding community tourism spot feelsboth positive and negative impacts. Like wise, the people of Kemuning village, Ngargoyoso. It positively impacts the local economy because many visitors come, but it has other negative impacts, such as poorhygiene. Environmental awareness and concern cannot grow without internal desire and encouragement. Movement with the community is needed to achieve behavior and environmental changes. That is why it is necessary to form an empowerment community where the community is active and empowered to solve social problems that exist in the community [18]. The Kemuning people, aware of this, have formed a community of volunteers to preserve nature in Kemuning Village called the Kemuning Care Community (KPK), founded 3 years ago to preserve nature and tourist attractions. The Kemuning Caring Community consists of members ranging from youth to adults; in addition to keeping the environment clean at the Kemuning tour, this community also works with the Regional Disaster Manajement Agency (BPBD Karanganyar) to learn about natural disaster mitigation. The Kemuning Caring Community conducts cleaning activities in the tourist area in Kemuning every weekend. This activity began by picking up trash from the Bale Rante restaurant, the Kemuning terminal, and the tea garden.

Berjo village is a tourist village with tourist destinations for Jumog Waterfalls and Telaga Madirda. Tourism in Berjo Village is managed directly by BUMDes by appointing managers who are selected through village meetings. All parties involved in the tourist sites in Berjo Village, from managers to parking attendants, are only filled by the people of Berjo Village, and BUMDes also prohibits the involvement of parties outside Berjo Village. The Berjo village youth organization is also directly involved in managing Jumog Waterfall and Madirda Lake tourism to become volunteers when a disaster occurs. Youth in Berjo Village were trained directly by the Indonesian Navy for mitigation at Telaga Madirda. In addition, BUMDes and the surrounding community also work together to preserve Jumog Waterfall. One of the ways is to always replant (reforestation) trees around the cliffs that are getting old and collapsing. Reforestation is an activity to replant bare forests. It is useful for improving the quality of human life by absorbing pollution and dust from the air, rebuilding natural habitats and ecosystems, preventing global warming by capturing carbon dioxide from the air and utilizing the results [19]. Reforestation has had a major impact on reducing abrasion and preventing floods and landslides in the Jumog Waterfall area, surrounded by high cliffs. As well as access to Jumog Waterfall, special road access was developed for persons with disabilities so that access to the waterfall's location can be carried out by anyone, regardless of the physical condition of other people. People with disabilities in their mobility will indirectly experience difficulties in carrying out activities. Compared to physically normal people, people with disabilities experience weaknesses in moving their bodies optimally [20].

Based on field observations, the mitigation carried out by the agro-tourism manager in Ngargoyoso structurally is by strengthening the structure of buildings or tourist objects, including :

1. Building wide roads to facilitate access for vehicle mobility to and from tourist sites to minimize damage,
2. Wise use of water for tubing tourism,
3. Utilizing waste and paying attention to the cleanliness of tourism objects to prevent environmental pollution and
4. Reforestation activities around waterfalls and on cliffs to prevent landslides during the rainy season.

Meanwhile, non-structurally, climate change mitigation activities in supporting agrotourism development are carried out in 2 ways, namely: (1) forming a tourism-oriented community that cares about the environment, (2) Avoiding outside investors by involving the younger generation and the local community as managers of tourism objects. It is feared that foreign investors will not practice adaptation and mitigation values for natural preservation and sustainability (only for profit) and (3) Avoid expanding tourist areas identified as prone to landslides.

4 Conclusion

Climate change affects tourism development regarding the arrangement and development of tourism objects and supporting facilities. Rainfall, water discharge, temperature, and weather require adaptation and mitigation to develop tourist villages to align with climate change, nature conservation, and business sustainability. In order to realizethis goal, adaptations carried out by tourism managers in Ngargoyoso include (1) the conservation of flora and fauna, (2) the preservation of natural resources, and (3) the preservation of the function of the atmosphere. Meanwhile, mitigation is implemented in the form of (1) structural mitigation, (a) building road access for vehicle mobility to and from tourist attractions, (b) wise use of water, (c) keeping waste clean, and (d) reforestation. (2) Non-structural mitigation is carried out by (a) forming a tourism awareness community that cares about the environment, (b) avoiding outside investors, and (c) avoiding expanding tourist attraction areas identified as being prone to natural disasters.

References

1. H. Ni'mah and M. F. Ma'ruf, Jurnal Ilmu Administrasi Negara **7**, 1 (2019)
2. M. S.-U. Rahman, A. Al Muneem, Md. A. R. Avi, and S. Sobhan, Rajshahi University Journal of Business Studies **11**, 131 (2018)
3. M. M. Navarro, Pasos:Revista de Turismo y Patrimonio Cultural **15**, 841 (2017)
4. R. Alfianor, N. A. Safitri, and A. Hidayati, Ruang **6**, 11 (2020)
5. A. Harmoni, Proceeding, Seminar Nasional PESAT (2005)
6. T. Butarbutar, Jurnal Analisis Kebijakan Kehutanan **9**, 1 (2012)
7. Sugiyono, *Metodologi Penelitian Kuantitatif, Kualitatif, Dan R&D*, 2nd ed. (Alfabeta, Bandung, 2013)
8. C. Auerbach and L. B. Silverstein, *Qualitative Data: An Introduction to Coding and Analysis In Qualitative Data: An Introduction to Coding and Analysis* (New York University, New York, 2003)
9. Sugiyono, *Metodologi Penelitian Kualitatif*, 1st ed. (Alfabeta, Bandung, 2018)
10. L. J. Moleong, *Metodologi Penelitian Kualitatif*, 36th ed. (Remaja Rosda karya, Bandung, 2011)
11. H. A. Permanasari and Sunarto, Jurnal Kesehatan Masyarakat Nasional **6**, 42 (2011)
12. E. Surmaini, E. Runtunuwu, and I. Las, Jurnal Litbang Pertanian **30**, 1 (2011)
13. C. Folke, Global Environmental Change **16**, 253 (2006)
14. Barri and Adhitya, *Muhammadiyah Dan Kesiapsiagaan Bencana* (Risalah MDMC, Bandung, 2009)
15. M. Mawardi, *IrigasiAsas Dan Praktek*, 1st ed. (Bursa Ilmu, Yogyakarta, 2016)
16. V. H. Manh and C. H. Wang, APCBEE Procedia **8**, 32 (2014)
17. S. Oktavia, M. C. A. Aziz, W. D. Putri, I. L. Hakim, and Zulbaidah, PROCEEDINGS OF UIN SUNAN GUNUNG DJATI BANDUNG **1**, 52 (2021)
18. L. A. Akhrani, I. Herani, and A. Hany, Journal of Dedicators Community **5**, 159 (2020)
19. D. O. Effendi and D. D. Winandya, SEMNAS TEKNO MEDIA ONLINE **3**, 29 (2015)
20. F. Priscyllia, Lex Crimen **5**, 105 (2016)

Green harvest: consumer attitudes as seeds of change towards a sustainable food revolution

Andika Andika[1]*, *Wika* Harisa Putri[1], *Della* Nanda Luthfiana[1], and *Mohamad* Najmudin[1]

[1]Janabadra University, Jl Tentara Rakyat Mataram No. 55-57, Yogyakarta, Indonesia

Abstract. The breakthrough of organic food plays a crucial role in mitigating climate change. This is based on several research findings stating that organic food can reduce the use of chemicals, improve soil conditions, reduce greenhouse gas emissions, and support biodiversity. However, the high price constraint of organic food leads most people to prefer non-organic food daily. The focus of this research is to determine the level of consumer willingness to pay more (WTPM) for organic food by examining the role of knowledge of organic food (OFK), price consciousness (PC), and consumer attitudes (CA). The 300 respondents were collected from all islands in Indonesia, including Java, Bali, Kalimantan, Sumatra, Nusa Tenggara, Sulawesi, Bangka Belitung, and Papua. This study uses Structural Equation Modeling (SEM) to test the research hypothesis. The research results indicate that OFK and PC directly and significantly influence CA. Furthermore, PC and CA also directly and significantly affect WTPM. Interestingly, OFK does not directly drive consumer WTPM. On the other hand, CA plays a significant mediating role between OFK and PC regarding WTPM. The implications of this research finding are that to increase the adoption of organic food as a climate change mitigation measure; efforts are needed to enhance consumers' knowledge and awareness of the benefits of organic food to form a more positive attitude towards the product. This, in turn, can improve their willingness to pay a slightly higher price to support organic food consumption in Indonesia.

1 Introduction

In the past two decades, global organic food sales have experienced phenomenal growth, rising from $18 billion in 2000 to $132.74 billion in 2021 [1], and is expected to reach $294.54 billion by 2023 at a CAGR of 13.7% [2]. However, this positive growth occurs within a broader global context, where regional factors and specific economic conditions influence consumer consumption patterns. While organic food is becoming a mainstream choice in many European and North American countries [3], social and economic challenges in developing countries such as Indonesia may limit access and adoption of organic food despite the potential benefits.

* Corresponding author: andika@janabadra.ac.id

Organic food, far from being just a contemporary trend, is pivotal in global climate change mitigation and sustainability efforts. Its cultivation bolsters soil health, cuts greenhouse gas emissions, and produces superior-quality food, promoting ecological stability and human health [4,5]. Despite these advantages, the global market share of organic products is still minor, only about 5% [3]. This is particularly evident in developing countries such as Indonesia, where people are aware of organic food's health and environmental benefits but are still hesitant to consume it, mainly due to cost concerns [6]. Higher prices often deter the general public, who are focused on efficient daily spending, causing stagnation in the growth of the organic food market.

While numerous studies have identified factors influencing low organic food consumption, including higher perceived prices and a lack of understanding of its benefits [6,7], more profound challenges related to pricing still need to be solved. The tendency of organic food to command higher prices than conventional food often serves as a primary barrier for consumers despite their awareness of the associated ecological and health advantages [8]. Therefore, this study aims to analyze the impact of knowledge about organic food and price consciousness on consumer attitudes and intentions to consume organic food products.

This study uses Ajzen's Theory of Planned Behavior (TPB) to explore the factors influencing consumers to switch to organic food [9], using the willingness to pay more as a proxy for behavioral intention [10]. We aim to provide insights into pricing strategies that make organic food more accessible to low- and middle-income Indonesians. The findings will guide policymakers in promoting the understanding and consumption of organic food in developing countries, particularly Indonesia, and provide insight into the impact of price on the acceptability of organic food.

2 Materials and Method

2.1 Materials

This study examines the phenomenon of organic food consumption in Indonesia by taking a sample of respondents consisting of regular consumers or those who have at least once consumed organic food, with a minimum age of 17 years. The sample selection includes the islands of Java, Bali, Kalimantan, Sumatra, Nusa Tenggara, Sulawesi, Bangka Belitung, and Papua. Data was collected through an online questionnaire (Google Form) distributed via WhatsApp and Instagram messages, which was explicitly addressed to organic food consumers using a purposive sampling method. The questionnaire instrument consisted of 13 questions answered on a four-level Likert scale ranging from strongly disagree to agree strongly. This questionnaire is based on existing literature [11–13]. From a total of 305 respondents, we processed data from 300 respondents, as some data were incomplete and there were outliers. The collected data were analyzed using Partial Least Squares (PLS) within the Structural Equation Model (SEM) framework.

2.2 Method

2.2.1 Hypotheses Development.

The Theory of Planned Behavior (TPB) has been frequently used to study behavioral intentions, including its application in pro-environmental behavior [9,14]. Several studies have demonstrated its effectiveness in predicting consumers' purchase intentions for green products [15] and their readiness to pay more [16]. At its core, the TPB examines the attitudes individuals perceive and approach behavior [17]. A wealth of evidence underscores a robust correlation between attitudes and the inclination to invest more in eco-friendly products. Kirmani's research [16] found that attitudes toward green products significantly and positively influence the willingness to pay a premium. However, Pileliene and Tamuliene's findings [18] diverged from this perspective, suggesting that consumers in Lithuania maintain a negative attitude towards organic food and are disinclined to pay higher prices for such products. Given the discrepancies in previous research outcomes, this study seeks to revalidate the relationship between attitudes and willingness to pay more, leading to the formulation of the following hypothesis:

H1: Attitudes towards organic food significantly influence the willingness to pay more

Consumer knowledge significantly influences purchasing decisions. When individuals assimilate and analyze product information, their depth of understanding greatly influences their interpretation of product labels, tags, and displays [19]. A study conducted by Peterson et al. [21] indicated that a deep understanding of environmental impact enhances consumers' confidence and appreciation of the benefits associated with organic food. This underscores the pivotal role of consumer awareness and knowledge of organic food in shaping their purchase decisions. Previous studies have consistently supported that knowledge about organic food influences consumers' attitudes toward such products [14], subsequently affecting their willingness to pay more [10,16]. Based on these findings, the following hypothesis is proposed:

H2: Product knowledge significantly affects attitudes toward organic food
H3: Product knowledge substantially affects the willingness to pay more

Price is pivotal in shaping consumer attitudes and intentions regarding product purchases. Despite various factors fostering a cheerful disposition towards organic food products, their typically higher costs than conventional food often serve as a deterrent [20]. This scenario is particularly noticeable among consumers who prioritize affordability, revealing a potential negative correlation between price consciousness and the propensity to purchase organic food [21]. Contrarily, recent research by Matharu et al. [22] indicates that price consciousness does not significantly diminish the enthusiasm or purchase intent for organic food among young Indian university students. This inconsistency in empirical insights necessitates further investigation into the dynamics of the relationship. Therefore, this study seeks to shed light on this aspect by proposing the following hypothesis:

H4: Price consciousness negatively influences consumer attitudes toward organic food
H5: Price consciousness has a negative influence on the willingness to pay more for organic food

There is ample evidence indicating a cause-and-effect relationship among product knowledge (PK), price consciousness (PC), consumer attitude (CA), and willingness to pay more (WTPM) [12,16,22,23]. Several researchers have investigated PK, PC, CA, and WTPM relationships. Sun and Wang [14] and Suprihartini et al. [24] have identified direct relationships among these variables, with PK and PC influencing CA and CA affecting WTPM. However, previous research has yet to explore the potential mediating role of CA in connecting PK, PC, and WTPM. Previous studies have established that PK affects CA, and both PK and PC positively influence CA, with CA having a direct positive impact on WTPM. Furthermore, CA has been recognized as a crucial factor and has been considered a mediator in several studies [14,15]. Based on these findings, this study incorporates attitudes toward organic food as a mediator, formulating the following mediation hypotheses:

H6: CA mediates the negative effect of PC on consumers' WTPM for organic products
H7: CA mediates the effect of PK on consumers' WTPM for organic food products.

3 Results and Discussion

3.1 Descriptive Statistics

The demographic characteristics of the respondents in this study have been categorized based on several factors, namely gender (36% male, 64% female), age (with the majority of respondents falling within the age range of 17-26 years, 55%), family income level (where the majority of respondents have an income between 1-3 million, 46%), and also based on educational level (divided into three categories: high school, bachelor's degree, and postgraduate), with the majority of respondents having a bachelor's degree education background (45%). Additionally, most respondents responded that the reasonable price difference for consuming organic food over non-organic food is approximately 10-20% (47%). This result is in line with the findings of Suharjo et al. [27] in Indonesia and Kranjac et al. [25] in Serbia, that consumers are willing to pay more for organic food with a difference of 10%-20% compared to conventional food.

3.2 Indicator Test

In indicator evaluation, analysis is conducted on the measurement model to evaluate validity and reliability based on specific criteria. These criteria include convergent validity evaluated based on factor loadings, discriminant validity through Average Variance Extracted (AVE), and composite reliability based on Composite Reliability. The findings showed factor loadings for all indicators exceeded 0.6, which confirmed their validity based on significant correlations with their constructs. Furthermore, the discriminant validity of each variable was identified as good, with AVE values exceeding 0.50. Overall, composite reliabilities for all constructs proved firm, with values above 0.70, indicating high consistency and stability. In conclusion, the instrument proved reliable in measuring the variables under study, with integrity that has been exhaustively validated [26].

3.3 Hypothesis Testing

This study proposed seven predictions of causal relationships. Statistical analysis confirmed six relationships while not confirming one relationship, H2. Hypothesis testing was conducted using SmartPLS 3.3, applying the Bootstrapping method, and focusing on the T-statistic and P-value. Critical evaluation criteria included the T-statistic value to be greater than 1.96, the P-value to be less than 0.05 (5%), and the requirement for the beta coefficient to be positive. Full details of the hypothesis testing results are presented in Table 1.

Table 1. Path coefficient results.

Hypothesis	Effect	Original Sample (O)	T Statistics (|O/STDEV|)	P Values	Conclusion
H1 OFK -> CA	Direct	0.384	5.891	0.000	Accepted
H2 OFK -> WTPM	Direct	0.146	1.561	0.12	Rejected
H3 PC -> CA	Direct	0.425	6.079	0.000	Accepted
H4 PC -> WTPM	Direct	0.423	5.177	0.000	Accepted
H5 CA -> WTPM	Direct	0.212	2.576	0.01	Accepted
H6 OFK -> CA -> WTPM	Indirect	0.081	2.152	0.032	Accepted
H7 PC -> CA -> WTPM	Indirect	0.09	2.493	0.013	Accepted

Based on the results of the path coefficient in Table 1, the meal can be explained as follows:

The results of this analysis reaffirm that individuals' attitudes toward organic food significantly impact their willingness to pay more. These attitudes are fundamentally influenced by experiences and personal beliefs, which are constructed through various forms of education related to the health benefits of organic food, as well as support for environmental issues. This finding is consistent with Kirmani's research [16], where people responded positively despite the higher price of organic food.

The analysis also shows that a more profound knowledge of organic food correlates with a more positive attitude. This understanding reduces consumer hesitation and creates a positive emotional response due to its nutritional quality. This knowledge can also influence social perceptions, encourage consumption, and change views on its price. This finding aligns with the study of Peterson et al. [27], who asserted that knowledge influences consumer appreciation and trust in organic food.

However, the results of our analysis show that even if there is knowledge about organic food, this does not necessarily correlate with willingness to pay more. Contrary to what we hypothesized, there is no significant influence between knowledge and willingness to pay more. Factors such as price considerations and socio-cultural aspects may influence purchasing decisions. This is consistent with Khoirunnisa's [28] findings, which indicate that good knowledge about a product does not always result in a desire to buy it.

In addition, the results showed a negative but significant correlation between price awareness and attitude and willingness to pay more for organic food. The more sensitive consumers are to price, the less likely they are to be willing to pay more for organic food. This is consistent with the research of Aschemann-Witzel and Zielke [29], who found that price is a significant barrier to purchasing organic food.

In light of these findings, individual attitudes are critical in linking consumers' knowledge and willingness to incur additional costs, especially in the 10-20% range [30]. Our research revealed that, although organic food is often perceived as exclusive and expensive [6], a 10-20% increase in organic food is still affordable for low- to middle-income consumers in Indonesia. This signifies the unexplored potential to increase organic food penetration in this segment. Through careful pricing strategies and inclusive tactics such as subsidies or discount programs, access to organic food can be expanded, embracing consumers from different economic strata. This approach opens up new opportunities for producers and supports healthier and more sustainable consumption in an increasingly diverse society.

Therefore, policymakers and producers must design inclusive approaches that focus on making pricing more affordable and making organic food accessible to different segments of society, including those with low and middle incomes.

4 Conclusion

This study confirmed the significant impact of attitude towards organic food on willingness to pay more for organic food. Product knowledge shapes positive attitudes, although the effect on willingness to pay more is insignificant. On the other hand, price awareness has a significant influence on attitude as well as willingness to pay more for organic food. In this context, attitude towards organic food is a potential mediator between the effects of product knowledge and price awareness on willingness to pay more. This study clarifies previous findings on the tolerance limit of price increases, which is around 10-20%, which can be an essential guide in formulating tax incentives. This will facilitate access to organic food for various segments of Indonesian society. By utilizing practical literacy and appropriate marketing strategies, organic food consumption can be increased sustainably, thus potentially contributing to climate change mitigation.

References

1. M. Shahbandeh, Statista (2023)
2. The Business Research Company, *Organic Food Global Market Report 2023* (2023)
3. H. Willer, B. Schlatter, and J. Trávníček, *The World of Organic Agriculture*

Statistics and Emerging Trends 2023 (Bonn, 2023)

4. H. Davis, A. Magistrali, G. Butler, and S. Stergiadis, Foods **11**, 1 (2022)
5. A. Haller, Agric. **12**, (2022)
6. M. Najib, U. Sumarwan, S. Septiani, H. Waibel, D. Suhartanto, and F. Fahma, J. Int. Food Agribus. Mark. **34**, 320 (2022)
7. F. J. Torres-Ruiz, M. Vega-Zamora, and M. Parras-Rosa, Sustain. **10**, 1 (2018)
8. C. Barbarossa and A. Pastore, Qual. Mark. Res. An Int. J. **18**, (2015)
9. I. Ajzen, Organ. Behav. Hum. Decis. Process. **50**, 179 (1991)
10. S. Wei, T. Ang, and V. E. Jancenelle, J. Retail. Consum. Serv. **45**, 230 (2018)
11. B. L. Alford and A. Biswas, J. Bus. Res. **55**, 775 (2002)
12. H. V. Nguyen, N. Nguyen, B. K. Nguyen, A. Lobo, and P. A. Vu, Int. J. Environ. Res. Public Health **16**, (2019)
13. Y. H. Shin, H. Moon, S. E. Jung, and K. Severt, J. Hosp. Tour. Manag. **33**, 113 (2017)
14. Y. Sun and S. Wang, Asia Pacific J. Mark. Logist. **32**, 860 (2020)
15. Y. M. Wang, H. M. F. Zaman, and A. K. Alvi, SAGE Open **12**, (2022)
16. M. D. Kirmani and M. N. Khan, South Asian J. Bus. Stud. **7**, 73 (2018)
17. N. A. . Abdelwahed and B. . Soomro, Educ. + Train. **65**, 210 (2023)
18. L. Pilelienė and V. Tamulienė, J. Entrep. Bus. Econ. **17**, 269 (2021)
19. K. Oh and L. Abraham, Int. J. Consum. Stud. **40**, 66 (2016)
20. Y. Joshi and Z. Rahman, *Factors Affecting Green Purchase Behaviour and Future Research Directions* (Holy Spirit University of Kaslik, 2015)
21. W. H. Putri, A. R. Setianan, and D. Puspitasari, Proc. 3rd Int. Conf. Banking, Accounting, Manag. Econ. (ICOBAME 2020) **169**, 207 (2021)
22. G. K. Matharu, T. von der Heidt, G. Sorwar, and A. Sivapalan, J. Int. Consum. Mark. **34**, 497 (2022)
23. L. Hua and S. Wang, Sustain. **11**, (2019)
24. L. Suprihartini, D. Herdiansyah, and M. Fahrizal, Enrich. J. Manag. **12**, 3730 (2022)
25. M. Kranjac, J. Vapa-Tankosic, and M. Knezević, Econ. Agric. **2**, (2017)
26. Hair, Hult, Ringle, and Sarstedt, *A Primer on Partial Least Squares Structural Equation Modeling (PLS-SEM).*, Second (SAGE Publications, Inc, Los Angeles, 2017)
27. H. H. Peterson, G. M. Hustvedt, and Y. J. Chen, Cloth. Text. Res. J. **30**, 35 (2012)
28. D. Khoirunnisa and Albari, Int. J. Res. Bus. Soc. Sci. (2147- 4478) **12**, 80 (2023)
29. J. Aschemann-Witzel and S. Zielke, J. Consum. Aff. **51**, 211 (2017)
30. V. Radulescu, I. Cetina, A. F. Cruceru, and D. Goldbach, Sustain. **13**, 1 (2021)

Supply chain performance of red chili from coastal land farming in the production center of Bantul Indonesia

Susanawati[1*] and *Adnin Jasmin Wafak* Hida[1]

[1]Departement of Agribusiness Faculty of Agriculture Universitas Muhammadiyah Yogyakarta Indonesia

Abstract. Coastal land can be used by farmers to cultivate plants, including red chilies. Global climate change can affect the productivity of coastal land through changes in temperature so that production of red chilies is disrupted as a result, the supply of red chilies in the market is unstable. Therefore, this research was conducted with aim of describing the red chili supply chain and analyzing the performance of red chili supply chain produced from the cultivation of coastal land in terms of product flow, money flow, and information flow. The location of this research is coastalland in Bantul which was chosen deliberately because it is the highest area for red chili production in coastal land. Sampling of farmers in this study was carried out in stages from district to villagelevel totaling 75 farmers from Manunggal farmer group. Sampling after farmers used snowballsampling technique, so that a total sample of 144 respondents consisted of 75 farmers, 1 auction market, 5 collectors, 21 retailers, and 42 household end consumers. Data analysis was carried out by describing the shape of image for red chili supply chain. Red chili supply chain performance was analyzed using scores and categories for product flow, cash flow, andinformation flow. The results of analysis show that supply chain of red chili cultivated on coastal land consists of three chains, namely (1) farmer-Manunggal auction market-collector I-retailer I-end consumer; (2) farmer-collector II-retailer II-end consumer; and (3) farmer-collector III-retailer III-end consumer. The best red chili supply chain performance is chain 2. The results of this study can be used as a source of information to choose the best performing supply chain.

1 Introduction

Coastal land has the characteristics of land that is unable to store water and nutrients, and is low in organic matter. Thus, costs and special treatment are required for cultivation on coastal sand land [1]. However, on the other hand, beach sand land is easy to process because its texture is loose so farmers save time and processing costs, the land is relatively safe from disease so it has quite potential for farming [2]. Bantul Regency is an area with extensive coastal sand land and is widely used by farmers to grow horticultural commodities, such as red chilies.

* Corresponding author: susanawati@umy.ac.id

Red chili is a type of plant horticulture important which cultivated in a way commercially because it has a fairly complete nutritional content and economic value tall [3]. Along with the increasing population in Indonesia, the amount of consumption vegetables as well as fruit – fruit also join in increased also with chilli red. The more varies type of food there is, so consumption chilli red will keep going increase [4]. Besides, chili productivity is very high and time needed for relatively short planting, so the economic value of chili quite high [5].

The red chili harvest season in 2022 will occur in Sanden District, Bantul Regency, did not receive any production results maximum because condition bulk rain which tall. The impact of climate change can have a major influence in influencing the dynamics of OPT, fluctuations in temperature and humidity can stimulate the growth and development of OPT

In operate activity chain that supply lasts from Sanden district start from process production, bottleneck on process delivery, obstacle information for each supply chain actor is indicated to be experiencing problems. Therefore thing the influence performance of supply chain which happen in field, so research is needed to analyze red chili supply chain performance.

The performance of the supply chain encompasses all efforts made by supply chain actors to fulfill consumers' needs and desires. According to [6], supply chain performance can be effectively and systematically managed. Three streams must be handled in the supply chain: the flow of products, the flow of money, and the flow of information. The smoothness of the three flows indicates supply chain success. Multiple factors, including sourcing, inventory, facilities, pricing, and transportation, can impact supply chain effectiveness [7]. Performance measurement is essential for evaluating past performance, establishing goals, and determining subsequent programs. [8] have identified many supply chain performance measures, covering product quality, competitiveness, profit level, and market share. The two key indicators of supply chain performance are efficiency and responsiveness [9]. This research was conducted with aim of describing the red chili supply chain and analyzing the performance of red chili supply chain produced from the cultivation of coastal land in terms of product flow, money flow, and information flow.

2 Methodology

A descriptive analysis was run to examine the robusta coffee supply chain and the activities of supply chain actors. Meanwhile, supply chain performance analysis applied a score between 1 and 5 for product and money flows and 1 to 2 for information flow. The product flow employed five indicators: product quality [10], number of products, availability of storage facilities [11], means of transportation, and delivery time. Four indicators were utilized to determine the flow of money: the transaction system, the transaction method [12], the site of the transaction, and the time of the transaction. Moreover, five indicators were employed to measure the flow of information: production information, demand information, price information, payment information, and quality information.

Sampling of farmers was carried out in stages starting from the sub-district's own market auction. Retrieval sample farmer taken in a way on purpose based on the number of farmers who have succeeded in cultivating to harvest red chilies in year 2022 in land sand as many as 69 people. As for determining the perpetrators of the red chili supply chain after farmer done with use technique Snowball Sampling that consists from 1 market auction, 5 trader collector, 21 trader retailer, And 42 consumer end in Area Special Yogyakarta.

Supply chain analysis chilli red is done descriptively in the form of a chart to follow description perpetrator's activities chain supply. Analyze supply chain performance by looking at whether product flow, money flow and information flow are smooth or not through several indicators. There are six indicators for product flow, namely amount product, cleanliness product, freshness product,time delivery, tool transportation, and availability place storage product. Four_ indicator on money flow, namely the transaction system, method of payment of transactions, place of payment transactions, and transaction payment times. There are 5 information flow indicators, namely production information, information request, information price, information time delivery, and information payment.

The three streams were determined using the Sturges formula, categorized as rough and smooth :

$$Z = \frac{X - Y}{K} \quad (1)$$

Description: Z = Class interval; X = Highest score; Y = Lowest score; K = Number of classes

3 Results and discussion

3.1 Red chili supply chain

Supply chain is a concept where there is a regulatory system related to product flow, information flow and financial flow [13].The supply chain network for red chilies from coastal sand cultivation in Bantul Regency is formed by 3 chains with the following actors: (1) Farmers – Auction Market – Collecting Traders I – Retailer I – End consumer; (2) Farmer – collector traders II – Retailer II – end consumer; (3) Farmers – collector traders III – retailers III – end consumer. In the first chain, the number of farmers who deposit red chilies is as much as 75 farmers, in the second chain there are 4 farmers and in the third chain there are 14 farmer. The three red chili supply chains formed as a whole originate from farmer which combined in group farmer Unity. The flow of red chili products starts from farmers in the Sanden Then through market auction then distributed to trader collectors and retailers to get to the final consumer in Special Region of Yogyakarta. The money flow is the opposite of product flow which where Money will flow from end consumer to going to return to farmer. The flow of information reciprocally or two direction, namely from farmers to consumer or from consumer to farmer. This condition is different from [14] research that the supply chain for cayenne pepper commodities in Kumelembuai Village consists of 5 types of channels that distribute products in the form of fresh cayenne peppers.

3.2 Performance of red chili supply chain

Supply chain performance of red chilies can be measured with see smoothly nope product flow, money flow, and information flow. Performance analysis is based on each chain with the following conditions:

3.2.1 Chain 1: Farmers – Auction Market – Collector trader I – Retailer I – End consumer

The results of the analysis of the six indicators in product flow show that they are in the smooth category for each supply chain actor involved. Money flow in chain 1 is at in category smooth in every supply chain actor. The highest score is among perpetrators of retail I with consumer with score 20.00. Then on score highest second followed by actors that occurs between the collector I and the retailer I with an average score of 19.50. Then the third highest score occurs between auction market with collector I with an average score of 19.00. Last score the lowest occurred between farmers and auction market with an average score of 18.28.

Information flow on chain 1 of red chili farmers to end consumers and vice versa are categorized as smooth at a vulnerable score of 7.50 – 10.0. The information flow that has the highest score exists between collecting traders I and retail traders I with an average of the average score is 9.00 then the second highest score is followed by the actor who occurred between auction market and collecting traders I with an average score of 8.50. Final score Lowest happen between farmer with auction market and retailer I with consumer with score 8.00. Based on results score calculation on product flow, money flow and information flow in chain 1 can be described and their overall performance can be seen through Figure 1.

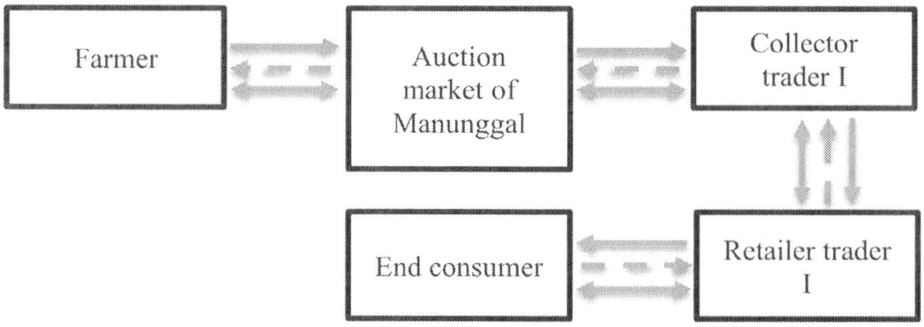

Fig 1. Performance of chain 1

Description:
Product flow smooth
Money flow smooth
information flow smooth

3.2.2 Chain 2: Farmers – Collector trader II – Retailers II – End consumer

The results of analysis of the six product flow indicators in chain 2 show in category smooth in every actor which involved. The product flow which has highest score among farmers with collectors II with score 28.79. Then on score highest second followed by actor which happened between retailer II with end consumers of the Special Region Yogyakarta with an average of 28.67. And finally the lowest score is between collector traders II with retailers retailers II with an average score 28.53 respectively. Each indicator on product flow which happen from farmer until end consumers of the Special Region of Yogyakarta are considered very appropriate and appropriate so that condition product flow enter into the smooth category.

The results of analysis of four money flow indicators in chain 1 shows being in the smooth category in each supply chain actor involved. The money flow that has highest score is among the retailer I with end consumers with an average score of 20.00 then with the second highest score followed by actors who occurred between farmers and collector trader II with score 19.47. Final score l owest happen between collector tII with retailer II with score 19.20.

The information flow on chain 2 of red chili farmers to end consumers and vice versa are categorized as smooth at a vulnerable score of 7.50 – 10.0. The flow of information that has highest score exists between farmers and collector trader II with an average score of 9.00 Then on score highest second followed by actors between collectors II with retailer II with score 8.95.Final score lowest happen between retailer II with end consumer with score 8.00. The results of [15], the flow of information that is carried out by recording the demand and availability of corn, both in terms of quantity and price can provide accurate information so as to reduce high price fluctuations. Based on the results of the calculation of scores and flow categories product flow, money flow, and information flow, the overall performance of chain 2 can be seen in Figure 2.

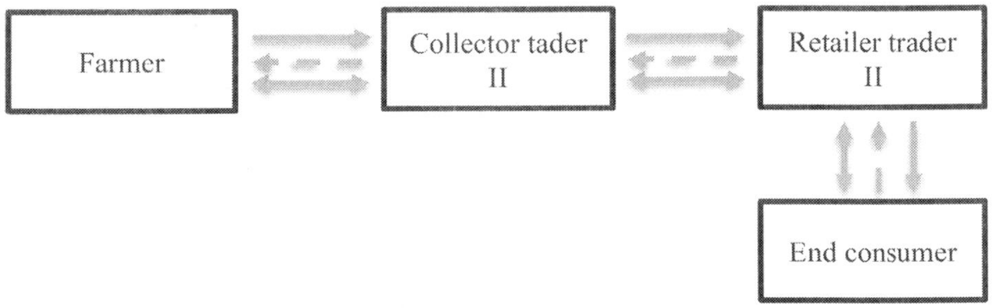

Fig 2. Performance of chain 2

Description:
Product flow smooth
Money flow smooth
Information flow

3.2.3 Chain 3: Farmers – collectors trader III – Retailer III –End consumer

The product flow in chain 3 is located on smooth category in every supply chain actors which involved. The flow of product which own score highest between collector trader III with retailer III with score 30.00 then on the second highest score followed by the actor that occurs between farmer with collector trader III with an average of 29.18. And finally the score lowest between retailers III and end consumers Special Yogyakarta with average score 29.00.

The flow of money on chain 3 is located in category smoothly. The highest score was among retailers I with end consumer with score 20.00 Then on score highest second followed by actors which happen between farmer and collector trader III with average score 19.28. Finally, the lowest score occurred among collecting traders II and retailer II with an average score of 19.25.

The flow of information in chain 3 is categorized as smooth at a score of 7.50 – 10.0. The highest score is among collector trader III and retailer III with score 8.50 Then with the second highest score, followed by actors that occur between farmers and collector trader III and retailer III with end consumer with average score 8.00. Based on the results of calculating the scores for product flow, money flow, and information flow, the performance of chain 3 can be described as follows:

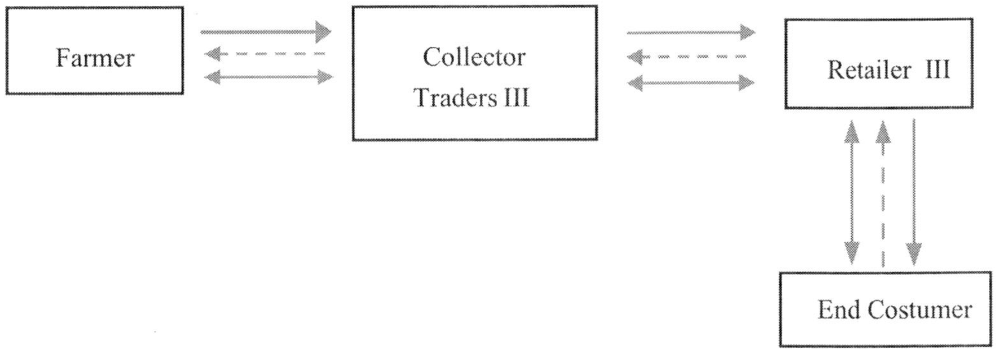

Fig 3. Performance of chain 3

Performance of red chili supply chain from the cultivation of coastal land consists of from 3 chain which own condition which different based on actors and flows analyzed. These flows include product flows, money flows, and information flow. On chain 1 own amount actor which morelLots compared to chains 2 and 3. The fewer number of actors involved in the supply chain can minimize obstacles that impact performance supply chain cannot run smoothly. Product flow, money flow as well information flow on 3 chain supply that already analyzed own condition smoothly. On chain 1 own amount actor which more than chains 2 and 3. Red chili supply chain performance overall whole can seen on picture 4.

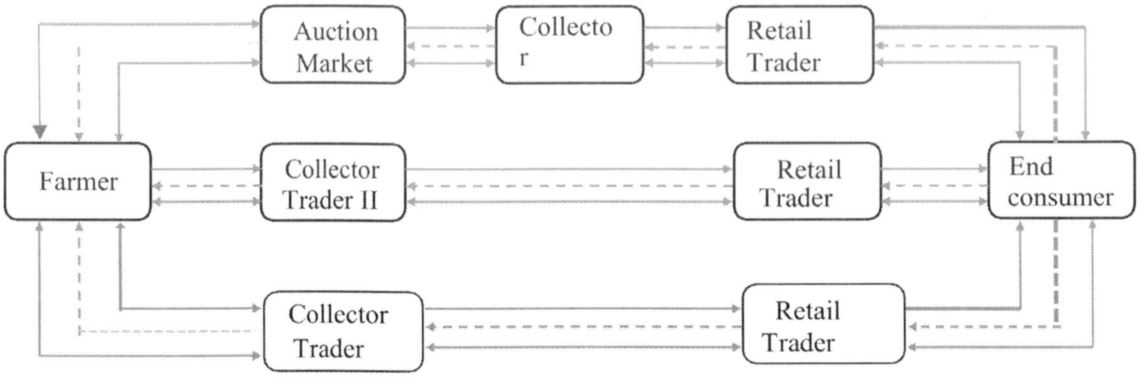

Fig 4. Overall red chili supply chain performance

4 Conclusion

Based on the results of the data analysis performed can obtained conclusion that red chili supply chain network formed three chains and five actors. Of the three chains formed, chain one has more number of actors than chain two and three. The product flow of red chili is the most smooth there is on chain 3 with actors are farmer – collector traders III – retailer III – end consumer. The most money flow conditions smoothly, there is on chain 2 with actors are farmer – collector traders II - retailer II – end consumers. The smoothest flow of information is in chain 2 with actors are farmers – collectors III – retailers III– end consumers. By whole performance of red chili supply chain, the best is in chain 2, namely farmers - collector trader II – retailer II – end consumer.

5 Suggestion

Based on results the study, there is suggestion which can become consideration namely on the flow of information expected by retailers. The end consumer more active in exchanging related information chili request red so that can increase performance of supply chain. On the side other, on money flow is expected time payment which done by auction market to farmer no experience cancel which so long.

Reference

1. Widodo, A. S., *Kajian Usahatani Lahan Pantai di Kabupaten Bantul. Jurnal Faperta Universitas MuhammadiyahYogyakarta*, (2009) 355 -367.
2. Iriani, E., *Prospek pengembangan inovasi teknologi bawang merah di lahan sub optimal (lahan pasir) dalam upaya peningkatan pendapatan petani. Jurnal Litbang Provinsi Jawa Tengah*, **11**, 2 (2013), 231-243.
3. Nurlenawati, N., Jannah, A., *Respon Pertumbuhan dan Hasil Tanaman Cabai Merah (Capsicum annuum L.) Varietas Prabu Terhadap Berbagai Dosis Pupuk Fosfat dan Bokashi Jerami Limbah Jamur Merang,* (2010).
4. Wulansari, F., Rahayu, S., Aulia Qonita, R., *Analisi Rantai Pasok Cabai Merah*

(Capsium Annum L.) di Kabupaten Magelang. Desember, **9**, 4 (2021), 114– 124.

5. Muhajir, B., Dalimunthe, E. L., Panggabean, A., *Pertumbuhan Dan Produksi Tanaman Cabai Merah (Capsicum annum L.) Terhadap PemberianPupuk Organik Pada Berbagai Media Tanam, Agrotekma,* **2**, 1 (2017).
6. Utami, N. P. V. *Analisis Kinerja Rantai Pasok Ikan Lele di Wilayah Istimewa Yogyakarta. Fakultas Teknologi Pertanian. Universitas Gajah Mada Yogyakarta,* (2016).
7. Chopra, S., Meindl, P., Supply Chain Management: Strategy, Planning, & Operations (3rd ed.). New Jersey: Pearson Prentice Hall (2016).
8. L.S. Simchi, P. Kaminsky, E.S. Levi, Designing and Managing The Supply Chain Concept Strategies and Case Studies International Edition, New York: McGraw-Hill (2009).
9. T. W. Bernard, S.R. Russel, Operations Management 7th Edition, New Jersey: John Wiley and Sons, (2011).
10. Susanawati, *Rantai Pasok Komoditas Bawang Merah Di Pulau Jawa (Doctoral dissertation, Universitas Gadjah Mada)*, (2016).
11. Aryadi, N.R., *Rantai Pasok Jamur Tiram Di Kabupaten Sleman. Fakultas pertanian. Universitas Gajah Mada Yogyakarta*, (2017).
12. Saptana, N. Ilham, *Manajemen Rantai Pasok Komoditas Ternak dan Daging Sapi. Jurnal Analisis kebijakan pertanian,* **15**, (2017) 83-98.
13. N.A. Josine., L.R. Pangemanan, C.B. Pakasi, *Analisis Rantai Pasok Komoditi Cabai Rawit di Kota Manado. Agri-Sosio Eonomi Unsrat*, **14**, 1 (2018) 207–214.
14. L.S. Tubagus, M. Mangantar, H. Tawas, *Analisis Rantai Pasokan Komoditas Cabai Rawit di Kelurahan Kumelembuai Kota Tomohon. Jurnal EMBA*. **4**, 2 (2016) 613-621.
15. A.I. Fajar, *Analisis Rantai Pasok Jagung di Provinsi Jawa Barat, Sekolah Pasca Sarjana, Institute Pertanian Bogor*, (2014).

Perception of rainfed farmers on the impact of climate variability in Bengkulu Province, Indonesia

Gita Mulyasari[1*], *Priyono* Prawito[2], *Ira Nurhayati* Djarot[3], *Nugroho Adi* Sasongko[3], *Titin* Handayani[3], *Agusta Samodra* Putra[3], *Netty* Widyastuti[3], *Akhmad* Rifai[3], *Sri Peni* Wijayanti[3], *Hismiaty* Bahua[3], Nuha[3], *Febrian* Isharyadi[3], and *Ari Kabul* Paminto[3]

[1]Department of Agricultural Socio-Economics, University of Bengkulu, Jl. WR. Supratman, Kandang Limun, Bengkulu, 38371, Indonesia
[2]Department of Soil Science, University of Bengkulu, Jl. WR. Supratman, Kandang Limun, Bengkulu, 38371, Indonesia
[3]Research Center for Sustainable Production System and Life Cycle Assessment, National Research and Innovation Agency, PUSPIPTEK Area, Serpong 15314, Indonesia

Abstract. This paper considers how farmers perceive to climate variability impacts, and suggests that perception these impact responses is as important as understanding responses to climate variability impacts. Based on a survey of 215 rainfed farmers in Bengkulu Province, we tested two hypotheses regarding climate variability impacts: (1) that climate variability has negative impacts on rainfed farming based on farmer's perception, (2) that age, formal education, farming experience, land area, and participation in farmer groups will influence farmer's climate variability perceptions. Using a multiple weighted score analysis and linier regression model we found support for the two hypotheses and furthermore showed that most of the farmers agreed that climate variability has a negative impact on rainfed farming. Farmers' formal education and participation in farmer groups influence farmers' perceptions on the impact of climate variability. With farmer groups, farmers can discuss various problems that arise due to climate variability.

1 Introduction

The agricultural sector is very vulnerable to climate change [1] and is experiencing the most serious impacts due to climate change [2]. Climate change affects production [3, 4], production systems [1] and the sustainability of agricultural production [5]. Climate change causes changes in rainfall, evaporation, water runoff, and soil moisture which will affect agricultural productivity. Soil fertility will decrease by 2-8 percent in the long run, which will result in a decrease in annual production [6]. According to [5], the influence of climate change on the agricultural sector is multidimensional, starting from resources, agricultural infrastructure, and agricultural production systems, to aspects of food security and independence, as well as the welfare of farmers and the community.

*Corresponding author: gita.mulyasari@unib.ac.id

 https://doi.org/10.1051/e3sconf/202346706003

Climate change often has a negative impact on rainfed rice farming. Climate change causes the growing season to change so it is difficult for farmers to predict the planting season, increases in pests and diseases, decreases and failures in production, decreases in yield quality, decreases income and increases the risk of crop failure and farming losses. The impact of climate change on farming varies among farmers. The majority of farmers also do not know about climate change [7].

Climate change is felt by farmers through climate indicators. Farmers perceive climate change because: 1) changes in temperature and precipitation [8, 9, 10]; 2) wind changes [11,12,13]; 3) extreme climate events in the form of floods and droughts [11,14]; 4) seasonal shifts [7]. The indicators of climate change that are most felt by farmers are rainfall and temperature. Other climatic elements such as wind are felt unchanged by farmers [7]. Farmers know about climate change through various ways such as feeling rising temperatures, erratic rainfall, warmer and stronger winds over the past few years. [15] stated that farmers perceive climate change more to the conditions felt by farmers such as drought, very hot air, changing rain patterns, strong winds, late start of the rainy season, quick end of the rainy season and very hot sun are indications of climate change. Based on this background, the aims of this study were 1) to analyze perception of rainfed farmers on the impact of climate variability, 2) to analyze the factors that influence farmers' perceptions on the impact of climate variability.

2 Methodology

This research was conducted in Bengkulu Tengah and Seluma Regencies, Bengkulu Province, Indonesia (Figure 1). In this research, the analysis unit is rainfed farmers who have their land as key informants to get information about their perceptions or knowledge on the impact of climate variability. The method used in determining the respondents is purposive random sampling, with a total of 215 farmers selected. The questionnaire was carefully designed to get information on socio-economic characteristics and perceptions on the impacts of climate variability selected by farmers based on their knowledge, understanding, and experience.

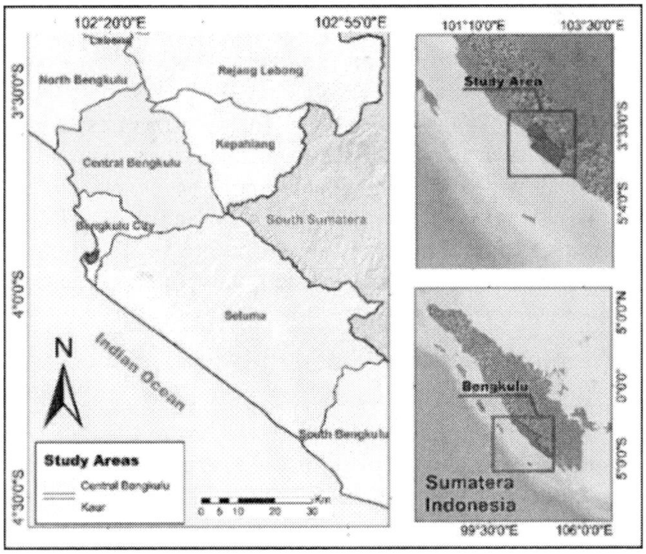

Fig. 1. Research area.

Analysis of climate variability was assesed from a social perspective, namely farmers' perceptions of climate variability that has an impact on rice farming [16]. Data analysis was performed using descriptive statistics and ranking method using multiple weighted for a set of parameters/factors impacting climate variability [17]. Logit analysis is used to determine the factors that influence farmers' perceptions on the impact of climate variability by involving various independent variables based on previous studies.

3 Results and discussion

3.1 Characteristics of rainfed farmers

This research was conducted on rainfed farmers who were vulnerable to affected by climate variability. There are several characteristics of farmers in Bengkulu province which include age, formal education, farming experience, land area and household size. The age of farmers is closely related to the experience and readiness of farmers in farming. The age of a farmer affects his physical abilities and reaction to new things in running. Based on Table 1, the average age of farmers is in the productive category. According to [18] older farmers generally have less physical ability, but have more work experience which is better. Meanwhile, young farmers are willing to take risks, so they are generally more physically fit and open to new things that are recommended.

Table 1. Characteristic of rainfed farmers.

Characteristics	Min.	Max	Mean	Std. Dev.
Age (years)	21.00	76.00	48.34	11.10
Formal education (years)	0.00	18.00	9.37	3.72
Farming experience (years)	3.00	55.00	22.44	11.10
Land area (Hectare)	0.20	1.50	0.50	0.29
Household size (Person)	0.00	8.00	3.47	1.49

Source : Primary data is processed, 2023

In general, education affects the way of thinking of farmers. Farmers who have higher education will more easily accept new things and changes in how to work. Based on the results of the study (Table 1) it shows that the average level of formal education of farmers is in the category of graduating from junior high school. [19] explains that the more educated farmers, the better the knowledge of farmers about the phenomenon of climate change. This is because farmers with higher education (college or high school) have a better ability to understand new information.

The results of the study (Table 1) also show that rice farmers in Bengkulu province are categorized as experienced in conducting farming activities. This means that farmers are good at considering what is done for the development of the success of their farming. Rice farmers also have an average land area that is not too broad. Based on Table 1 shows that rice farmers in Bengkulu Province have an average number of family dependents of 3 people. The more the number of family dependents owned, the more contribution in decision making and also the more diverse knowledge that will be obtained [20].

3.2 Perception on the impact of climate variability

Climate variability is a symptom arising from climate change. Climate variability causes fluctuations in rainfall to be unpredictable and tend to be erratic, thus often causing droughts and floods. The results of the study (Table 2) show that most of the farmers have an obvious perception that hot temperature pressure which causes decreasing rice quality. In addition, farmers also significantly perceive that climate variability increases the risk of crop failure and the risk of farming losses which of course has a direct effect on the sustainability of rainfed rice farming.

Based on the results of the interviews, the farmers said that the impact of climate uncertainty was causing the dry season to get longer. This has a detrimental impact on farming productivity which can increase the risk of crop failure. Reduced productivity continuously makes many farmers must have other jobs to cover losses from decreased production. Many of the farmers are also confused about the sustainability of the farming in the future. This is in line with [16] that most farmers perceive that there are three impacts of climate change on their farming business, namely the higher the risk of crop failure, the effect on farming sustainability and higher farming losses and high temperatures which can cause increases and new pests appear. In addition, research conducted by [21] said that the impact of climate change would be very detrimental to farmers due to crop failure. Most farmers rely on trade and other occupations such as raising cattle to provide for their families.

Table 2. Farmers' perception of climate variability impacts on rainfed farming.

Climate variability impacts	Not obvious		Somewhat obvious		Obvious	
	∑	%	∑	%	∑	%
Earlier harvesting date	98	46.01	32	15.02	83	38.97
Hot temperature pressure that causes decreasing product quality	52	24.41	23	10.80	138	64.79
Increasing cost of irrigation both in long dry season – buying extra water, and in long wet season – adjusting drainage	101	47.42	14	6.57	98	46.01
Changes in cultural practices that tend to demand more intensive use of input – increasing cost of production	170	79.81	6	2.82	37	17.37
Higher air temperature that causes the increase of pests/diseases incidence and the emergence of new pests and diseases	25	11.74	31	14.55	157	73.71
Extremely long dry or wet season that causes yield decrease	35	16.43	42	19.72	136	63.85
Increasing pot-harvest and marketing cost per unit product	162	76.06	14	6.57	37	17.37
Higher risks of crop failure	17	7.98	20	9.39	176	82.63
Increasing loss-risk in farming business that may directly affect farm sustainability	7	3.29	38	17.84	165	77.46

Source : Primary data is processed, 2023

The impacts of climate variability whose significance occupies the three highest orders are (a) Increasing loss-risk in farming business that may directly affect farm sustainability, (b) Higher risks of crop failure, and (c) Higher air temperature that causes the increase of pests/diseases incidence and the emergence of new pests and diseases (Table 3). [22] explained that risks in agriculture include losses and profits where the level of risk is determined before action is taken based on farmers' expectations or assumptions as a decision. The risk of loss and the sustainability of farming in uncertain climatic conditions is an important aspect of farming activities. The dynamics of weather conditions can cause changes in rainfall patterns, temperature patterns, and wind. Farmers perceive these changes to lead to prolonged dry seasons or unforeseen excessive rainfall and are often associated with risks of reduced crop productivity and farm income. This line of thinking tends to be consistent with the results of previous empirical research [23, 24].

Table 3. Rank of importance of climate variability impacts on rainfed farming.

Climate variability impacts	Average of rank value	Rank
Earlier harvesting date	1.9296	VII
Hot temperature pressure that causes decreasing product quality	2.4038	V
Increasing cost of irrigation both in long dry season – buying extra water, and in long wet season – adjusting drainage	1.9812	VI
Changes in cultural practices that tend to demand more intensive use of input – increasing cost of production	1.3756	IX
Higher air temperature that causes the increase of pests/diseases incidence and the emergence of new pests and diseases	2.6197	III
Extremely long dry or wet season that causes yield decrease	2.4742	IV
Increasing pot-harvest and marketing cost per unit product	1.4131	VIII
Higher risks of crop failure	2.7465	II
Increasing loss-risk in farming business that may directly affect farm sustainability	2.7512	I

Source : Primary data is processed, 2023

3.3 Determinants of rainfed farmers' perception

The results of the study (Table 4) show that the level of formal education of farmers and the participation of farmers in farmer groups have a significant and positive effect on farmers' perceptions of the impact of climate variability on rainfed farming. Increasing the understanding of rainfed farmers in Bengkulu province on climate variability can be done by increasing knowledge of climate variability and agriculture through increasing public education. High knowledge can be obtained through formal education and informal education such as counseling and training. Thus, education can increase the empowerment of farmer groups and can open farmers' access to information about climate change [25].

Table 4. Result estimation

Variable	B	t	Sig
Constant	15.718	13.630	0.000
Age	0.043	1.221	0.224
Formal education	0.199	3.376	0.001*
Farming experience	-0.056	-1.481	0.140
Land area	-0.113	-0.167	0.867
Membership of farming group	1.669	3.443	0.001*
F = 9.426			
Sig. = 0.000			
R-square = 0.185			

Source : Primary data is processed, 2023
*significance at $\alpha = 1\%$

Farmer groups have an influence in providing knowledge about climate variability, which is a place for farmers to learn, discuss, and exchange information. Farmers who are active in farmer groups have a positive and real relationship with the level of ability of farmers to manage their farming, management in dealing with climate change for example [26]. [27] also said that the activeness of farmers in farmer groups influences farmers' understanding of climate variability and farmers' decisions in implementing adaptation strategies to deal with climate variability.

4 Conclusion

This research was conducted on rainfed farmers who are susceptible to climate variability. The majority of farmers stated that the impact of climate variability was felt obvious to threaten farming, namely higher risks of crop failure, increasing loss-risk in farming business that may directly affect farm sustainability, and higher air temperature that causes the increase of pests/diseases incidence and the emergence of new pests and diseases. The formal education level of farmers and the participation of farmers in farmer groups have a significant and positive effect on farmers' perceptions of the impact of climate variability on rainfed rice farming. It is important to increase farmers' understanding of the impact of climate variability and one way that can be done is through strengthening farmer groups as a forum for farmers to learn, discuss and exchange information.

References

1. Haryono, I. Las, *Strategi mitigasi dan adaptasi pertanian terhadap dampak perubahan iklim global*, Prosiding Seminar Nasional Era Baru Pembangunan Pertanian: Strategi Mengatasi Masalah Pangan, Bioenergi dan Perubahan Iklim, Pusat Sosial Ekonomi Kementerian Pertanian, Bogor (2011)
2. E. Surmaini, E. Runtunuwu, Upaya sektor pertanian dalam menghadapi perubahan iklim, J. Litbang Pertanian, **30**, 1, 1–7 (2011) Doi:10.21082/JP3.V30N1.2011.P1-7.
3. J. Gornall, R. Betts, E. Burke, R. Clark, J. Camp, K. Willett, A. Wiltshire, Implications of climate change for agricultural productivity in the early twenty-first century, Biological Sciences, **365**, 1554 2973–2989 (2010) Doi: 10.1098/rstb.2010.015
4. K. Mittenzwei, T. Persson, M. Höglind, S. Kværnø, Combined effects of climate change and policy uncertainty on the agricultural sector in Norway, Agricultural Systems, **153** 118–126 (2017) Doi:10.1016/j.agsy.2017.01.016.
5. P. Rejekiningrum, I. Las, I. Amien, E. Surmaini, A. Pramudia, Y. Sarvina, *Adaptasi perubahan iklim sektor pertanian*, Badan Penelitian dan Pengembangan Kementerian Pertanian, Jakarta (2011)
6. P.M. Kopittke, N.W. Menzies, P. Wang, B.A. McKnna, E. Lombi, Soil and the intensification of agriculture for global food security, Enviro. Inter. **132** (2019) https://doi.org/10.1016/j.envint.2019.105078
7. F. Kurniawati, *Pengetahuan dan adaptasi petani sayuran terhadap perubahan iklim*, Thesis, Universitas Padjajaran, Bandung (2012)
8. L. Roco, A. Engler, B.E. Bravo-ureta, R. Jaras-rojas, Farmers ' perception of climate change in mediterranean Chile, Regional Environmental Change, **15** 867-879 (2014) Doi: 10.1007/s10113-014-0669-x.
9. A.M. Limantol, B. E. Keith, B.E., B.A. Azabre, B. Lennartz, Farmers' perception and adaptation practice to climate variability and change: a case study of the Vea catchment in Ghana, SpringerPlus, **5** 1–38 (2016) Doi:10.1186/s40064-016-2433-9.
10. N. Arunrat, C. Wang, N. Pumijumnong, S. Sereenonchai, W. Cai, Farmers' intention and decision to adapt to climate change: A case study in the Yom and Nan basins, Phichit province of Thailand, J. Clean. Prod. **143** 672–685 (2017) Doi:10.1016/j.jclepro.2016.12.058.

11. M. Esham, C. Garforth, Agricultural adaptation to climate change: insights from a farming community in Sri Lanka, Mitigation and Adaptation Strategies for Global Change, **18**, 1 535–549 (2013) Doi:10.1007/s11027-012-9374-6.
12. J.A. Yaro, The perception of and adaptation to climate variability/change in Ghana by small-scale and commercial farmers, Regional Environmental Change, **13** 1259–1272 (2013) Doi:10.1007/s10113-013-0443-5.
13. M.G. Barrucand, C. Giraldo Vieira, P.O Canziani, Climate change and its 197 impacts: perception and adaptation in rural areas of Manizales, Colombia, Climate and Development, **9**, 5 415–427 (2017) DOI:10.1080/17565529.2016.1167661.
14. Md. A. R. Sarker, K. Alam, J. Gow, Assessing the determinants of rice farmers' adaptation strategies to climate change in Bangladesh. Int. J. Clim. Chan. Strat. Manag. **5**, 4 382–403 (2013) Doi:10.1108/IJCCSM-06-2012-0033.
15. K. Murniati, *Adaptasi perubahan iklim dan keterkaitannya dengan produktivitas dan ketahanan pangan rumah tangga petani padi di Kabupaten Tanggamus*, Dissertation, Universitas Gadjah Mada, Yogyakarta (2014).
16. W. Adiyoga, R.S. Basuki, Perception of vegetable farmers on the impact of climate change in South Suawesi. J. Hort. **28**, 1 133-146 (2018)
17. S.H. Rahayu, N. Abdullah, A. Aziz, *Ranking method using multiple weighted score analysis*, School of Science and Technology, University Malaysia (2005).
18. F. Hernanto, *Ilmu Usahatani*, Penebar Swadaya, Jakarta (1996).
19. S. Notoatmodjo, *Pengembangan Sumber Daya Manusia*, PT Rineka Cipta (2003)
20. S. Umyati, S. Tuhpawan, Faktor-faktor yang mempengaruhi petani dalam pengalokasian dengan ganti rugi konversi lahan pertanian, J. Agrib. Sos. Ekon. Pert. **4,** 1 622-631 (2019).
21. G. Nuraisah, R.A.B. Kusumo, Dampak perubahan iklim terhadap usahatani padi di Desa Wanguk Kecamatan Anjatan Kabupaten Indramayu, J. Pemikiran Masy. Ilmiah Berwawasan Agrib. **5,** 1 60-71 (2019).
22. Soekartawi, *Definisi Usahatani*, Universitas Indonesia, Jakarta (2011).
23. Y. Syaukat, The impact of climate change on food production and security and its adaptation programs in Indonesia, J. ISSAAS **17**, 1 40–51(2011)
24. B.S. Choudri, I. Al-Busaid, M. Ahmed, Climate change, vulnerability and adaptation experiences of farmers in Al-Suwayq Wilayat, Sultanate of Oman, Int. J. Clim. Chan. Strat. Manag. **5**, 4 445–454 (2013).
25. Suminah, S. Anantanyu, Strategy for empowering corn farmers due to climate change in Wonoharjo Village Kemusu Boyolali, IOP Conf. Series: Earth and Environmental Science, **1165** (2023) Doi:10.1088/1755-1315/1165/1/012035
26. Erwandi, Ramainas, Tingkat keaktifan anggota kelompok tani di Kecamatan Tanjung Tiram Kabupaten Batu Bara, J. Agrica Ekstensia **10**, 2 45–54 (2016).
27. S. Angles, M. Chinnadurai, A. Sundar, Awareness on impact of climate change on dryland agriculture and coping mechanisms of dryland farmers, India J. Agric. Econ. **66**, 3 365–72 (2011).

Contribution and economic valuation of non-wood forest products from community forest in Plosorejo Village, Karanganyar Regency, Central Java

Aghdiatama Dava Indratna[1*], *Yus Andhini Bhekti* Pertiwi[1], and *Ana* Agustina[1]

[1]Forest Management Study Program, Faculty of Agriculture, Sebelas Maret University, Surakarta, 57126, Indonesia

Abstract. Community forests can help to mitigate and adapt to climate change by storing carbon and providing ecosystem service. Non-Wood Forest Products (NWFPs) can serve as an adaptation strategy for forest-dependent communities in dealing with climate change. In addition, the availability of food and sources of income can be fulfilled by the existence of NWFPs. Given the significant role of NWFPs in climate change adaptation, this study was conducted to identify the types of NWFPs, evaluate their economic value, and determine their contribution to household income in Plosorejo Village. The data collection method used in the present study was field inventory and interviews with 30 respondents. Based on the obtained results in this study, 26 species of NWFPs developed in the community forest of Plosorejo Village, generate an economic value of 923.2 Million Rupiah per year. The commodity that gave the highest contribution to the economic value of NWFPs was durian, with a value of 630 Million Rupiah per year or 68.24% of the total economic value of NWFPs. The average contribution of the economic value of NWFPs to household income was 58.94%. Educational background and the area of community forest owned show a significant effect on the community income from NWFPs.

1 Introduction

The issue of climate change is increasingly being discussed at this time. There were many impacts arising from this issue, one of which is on non-wood forest products (NWFPs). On the other hand, NWFPs can serve as a safety net for forest-dependent communities in dealing with the impacts of climate change on crop production and other livelihoods [1]. They provide alternative sources of income and food security [1].

Several researchers reported the NWFPs contributions to the household income in Indonesia [2][3][4]. It was reported that in Gempolan Village, Karanganyar Regency, NWFPs from community forest contributed 40.5% to the revenue of community [2]. A similar situation occurred in Mandiangin Barat Village, Banjar Regency where NWFPs contributed 42.94% of the community income [3]. The income generated from NWFPs in

* Corresponding author: davaindratna@student.uns.ac.id

Halaban Village, Lima Puluh Kota Regency surpasses that of the previous location [4]. NWFPs from community forests contributed 63% of the total income from community forests in Desa Halaban, Lima Puluh Kota Regency, West Sumatra [4].

Based on the description above, NWFPs has importance role and high potency to further development. In addition, further research is needed to study their contribution to community income in other regions. Therefore, the present study was conducted to obtain information and an overview of the utilization of NWFPs by the villagers of Desa Plosorejo. Furthermore, to find out the extent of its contribution to the community total income.

2 Materials and methods

The research was conducted in Plosorejo Village, Kerjo District, Karanganyar Regency, Central Java during March to May 2023. The research location has an average elevation of 515 meters above sea level [5]. In the present study, respondents were determined at 5% by purposive sampling. Samples were taken from respondents who owned community forests and utilized NWFPs. Based on this method, there were 30 people as respondents in this study.

There were two methods of collecting data that used, namely inventory, and interviews. The inventory was carried out by making plots in each respondent's community forest with an area of 0.1 ha (40×25 m). Then, the interviews were conducted to find out local market prices regarding NWFPs, harvesting intensity, or other matters related to research studies.

To calculate the density per hectare of a NWFPs commodity the following formula was used (1)

$$D = \frac{\Sigma n_j}{A} \quad (1)$$

Where,
D = Density per hectare (individual/hectare)
Σn_j = Total individual per species
A = Plot area (hectare)
j = Type of NWFP

In addition, to determine the economic value of a NWFPs commodity the following formula was used (2).

$$EV_j = V_j \times P_j \quad (2)$$

Where,
EV_j = Economic value of NWFP (j) (Rupiah/year)
V_j = Volume of harvest (unit/year)
P_j = Price of NWFP (j) (Rupiah)

Furthermore, an analysis was carried out to find out the influence of independent variables (age, length of stay, educational background, number of family dependents, the area of community forest owned, and distance from home to community forest) to dependent variable, namely community income in the form of NWFPs from community forests. Independent variables were tested simultaneously by using F-test. Furthermore, each variable was tested by using the t-test to determine its significance level. The regression equation model was used to determine the effect of the independent variables on the dependent variable (3).

$$y = a + b_1x_1 + b_2x_2 + b_3x_3 + b_4x_4 + b_5x_5 + b_6x_6 \quad (3)$$

Where,
y : Income from NWFPs
a : Regression constanta

b_1- b_6 : Regression coefficient
X_1- X_6 : Age (year) ; length of stay (year) ; Educational background (year) ; Number of family dependents ; The area of community forest owned (m^2) ; Distance from home to community forest (km), respectively

3 Results and discussion

3.1 Respondents characteristics

Information on the respondents' characteristics includes sex, age, educational background, length of stay, and the area of community forest owned was showed in Table 1.

Table 1. Respondent Characteristics.

Characteristics	Category	Quantity	Percentage (%)
Sex	Male	27	90.00
	Female	3	10.00
Age	35 - 44 years old	2	6.67
	45 - 54 years old	13	43.33
	55 - 64 years old	11	36.66
	65 - 74 years old	2	6.67
	> 74 years old	2	6.67
Educational Background	No formal education	10	33.33
	Elementary school	8	26.67
	Junior high school	7	23.33
	Senior high school	5	16.67
Length of Stay	< 35 years old	2	6.67
	35 - 44 years old	2	6.67
	45 - 54 years old	12	40.00
	55 - 64 years old	11	36.66
	65 - 74 years old	2	6.67
	> 74 years old	1	3.33
Number of Family Dependents	1 – 2	14	46.67
	3 – 4	12	40.00
	5 – 6	4	13.33
Distance from Home to Community Forest	< 1 km	21	70.00
	1 – 2 km	9	30.00
The Area of Community Forest Owned	≤ 0,25 ha	15	50.00
	0,26 – 0,50 ha	9	30.00
	0,51 – 1,00 ha	4	13.33
	≥ 1,1 ha	2	6.67

Table 1 shows the characteristics of the respondents who own the community forest in Plosorejo Village. There were 27 male respondents (90%) and 3 female respondents (10%). The low participation of women in forest management was influenced by social perceptions and norms in society [6]. The age distribution of farmers was dominated by the 45-54 year age group with 13 people (43.3%). Farmers aged less than 35 years were not found in this study. This condition might occur due to community forests seem to have less appeal to the younger generation, as compared to other industries [7].

As shown in the Table 1 the majority of respondents (33.3%) were not finished their elementary school/non-educated, which may have an effect on their knowledge. The higher the educational background of the farmer, the wider the insight of the farmer [8]. The length of stay will affect psychological ties, the more physically and non-physically bound they are to the area they live in [9]. There were 76.66% respondents who living in Plosorejo Village for 45-64 years, indicating that most of the respondents were natives to Plosorejo Village.

Based on Table 1, the group of 1-2 dependents showed the majority number (46.67%). The number of dependents in a family affects their cost of living and can impact community forest management [10]. According to the distance between houses and community forests, the majority of respondents live in interface areas. The interface area is a forest area that is in direct contact with the settlement (less than 3 km) [11]. Most of the respondents in the present study owned forest area of ≤ 0.25 ha. This phenomenon is normal on Java Island, where the area of community forest ownership per head of household is less than 0.25 ha [12].

3.2 Identification and the economic value of NWFPs

Table 2 shows scientific name, habitus and density per hectare of NWFPs for each species. There were 26 species of NWFPs cultivated in the community forest of Plosorejo Village and classified into 6 groups (fruit based, rhizome based, grass based, tuber based, sap based, and flower based). The highest and lowest density per hectare of NWFPs species was galangal and melinjo.

Table 2. Identification and the Economic Value of NWFPs.

No	Scientific Name	D	Habitus	v	units	p	E
A.	Fruit based						
1	Durian	1,030	Tree	15,750	pieces	45,000	630,000,000
2	Banana*	2,190	Herb	203	bunch	60,000	12,180,000
3	Avocado*	70	Tree	300	kg	15,000	4,500,000
4	Dog fruit	60	Tree	720	kg	15,000	10,800,000
5	Stink bean	90	Tree	390	bundle	120,000	46,800,000
6	Mango*	40	Tree	270	kg	3,000	810,000
7	Jackfruit	50	Tree	18	pieces	40,000	720,000
8	Mangosteen	20	Tree	100	kg	20,000	2,000,000
9	Rambutan	30	Tree	175	kg	3,500	612,500
10	Longan	20	Tree	50	kg	25,000	1,250,000
11	Snake fruit*	20	Palm	48	kg	10,000	480,000
12	Melinjo	10	Tree	70	kg	15,000	1,050,000
13	Guava*	30	Tree	20	kg	5,500	110,000
14	Pineapple *	40	Herb	40	pieces	6,000	240,000
			Total (A)				711,552,500
B	Rhizome based						
1	Ginger*	6,920	Herb	4,350	kg	7,000	30,450,000
2	Galangal*	7,010	Herb	6,560	kg	2,000	13,120,000
3	Turmeric*	6,120	Herb	2,868	kg	2,000	5,736,000
4	Fingerroot *	50	Herb	80	kg	10,000	800,000
5	Cutcherry *	500	Herb	120	kg	10,000	1,200,000
			Total (B)				51,306,000
C.	Grass based						
1	Elephant Grass *	1,500	Herb	390	bundle	10,000	3,900,000
2	Bamboo	60	Bamboo	18	clump	100,000	1,800,000
3	Lemongrass*	50	Herb	50	kg	2,000	100,000
			Total (C)				5,800,000
D.	Tuber based						
1	Taro*	1,220	Herb	140	kg	3,000	420,000
2	Porang*	380	Herb	120	kg	1,500	180,000
			Total (D)				600,000
E.	Sap based						
1	Rubber	4,290	Tree	20,260	kg	6,000	121,560,000
			Total (E)				121,560,000

F.	Flower based						
1	Clove*	220	Tree	1,080	kg	30,000	32,400,000
			Total (F)				32,400,000
			Total (A+B+C+D+E+F)				**923,218,500**

Notes: *= NWFP that are not mentioned in Ministry of Forestry Regulation No. P35/Menhut- II/2007; D= density per hectare; v= the total volume harvested per year; p= price based on market price approach that collect during interview; E= Economic value/year (Rupiah)

The economic value generated from NWFPs that cultivated in Plosorejo Village was showed in Table 2. Thus, the total economic value of NWFPs on community forest was 923.2 Million Rupiah per year. The economic value was influenced by several factors, such as price, volume (amount), frequency, and type of forest products used which can indicate the level of community dependence on non-timber forest products [13]. When discussing each commodity, the economic value of durian in Plosorejo Village showed the highest value compared to the other commodities.

3.3 NWFPs contribution to the household income

Community income contribution from NWFPs can be determined by comparing the total economic value of NWFPs for each respondent and the total household income. The economic contribution of NWFPs to the total income of the Plosorejo village community has been presented in Table 3.

Table 3. Household Income of Community Forest Farmers in Plosorejo Village.

No	Source of Income	Total (Rupiah/year)	Percentage (%)
1	Total income from NWFPs	923,218,500	58.94
2	Total income outside the utilization of NWFPs	643,200,000	41.06
	Total	1,566,418,500	100.00

The total income from NWFPs was higher than total income outside the utilization of NWFPs. The contribution of income from NWFPs obtained in the present study was higher compared to the other locations, namely Jaranguda Village and Merdeka Village in Karo District, North Sumatra [13]. Income contribution from NWFPs in Jaranguda Village and Merdeka Village were 40.41 and 47.47% [13]. The high contribution of income from NWFPs in Plosorejo Village indicating that the community has a high dependency on the forest.

3.4 Analysis of factors influencing community income from NWFPs

In the present study, the factors that influence the community income from NWFPs in Plosorejo Village was analyzed. Variables that used was taken from the characteristics of the respondents, namely age, length of stay, educational background, number of family dependents, the area of community forest owned, and distance from home to community forest. Those variables were analyzed by using multiple linear regression.

$$y = -77,479,334 + 855,374X_1 + 492,220X_2 + 3,008,273X_3 - 1,507,326X_4 + 4,628X_5 + 64,168X_6 \text{ with } (t_a = -2.221), (t_{X1} = 1.489), (t_{X2} = 1.005), (t_{X3} = 2.865), (t_{X4} = -0,537), (t_{X5} = 4.764), (t_{X6} = 0.009), (R^2 = 0.603), (F = 5.832) \quad (4)$$

Based on model summary, the value of R square was 0.603, indicating that there were strong correlation between all independent variables (age, length of stay, educational background, number of family dependents, the area of community forest owned, and distance from home to community forest) and community income from NWFPs. Based on the F-test, it confidently shows a significant simultaneous influence of the independent variable on the dependent variable.

Furthermore, it is worth noting that the outcomes of the t-test was shown in Equation (4). It was known that only variables X_3 (educational background) and X_5 (the area of community forest owned) have a significant effect on community income from NWFPs. The higher a person's educational background, the more receptive they are to knowledge, information, and innovation from others [14]. Similarly, the area of community forest owned also affects income. With more land, farmers can plant more crops, resulting in higher yields.

4 Conclusion

There were 26 species of NWFPs that were classified into 6 groups with a total economic value of 923.2 Million Rupiah per year. Durian showed the highest contribution in Plosorejo village with a value of 630 Million Rupiah per year. The economic value of NWFPs contributes 58.94% of the total community household income, indicating that the community has a high dependency to the forest. Educational background and the area of community forest owned were two factors that showed significant effect on the community income from NWFPs.

References

1. R. N. Tieminie, C. E. Loh, J. C. Tieguhong, M. F. Nghobuoche, P. S. Mandiefe, and M. R. Tieguhong, Environ. Syst. Res. vol. **10**, no. 1 (2021), doi: 10.1186/s40068-020-00215-z.
2. A. D. Indratna, A. P. I. K. S. Putri, E. O. Azzahra, M. S. Oktavia, T. Widodo, A. Agustina, and Y. A. B. Pertiwi, *IOP Conference Series: Earth and Environmental Science*, Institute of Physics, (2023). doi: 10.1088/1755-1315/1220/1/012014.
3. M. Dedi, M. Naparin, and A. Rezekiah, J. Sylva Sci., vol. **04**, no. 6, pp. 1076–1083, (2021)
4. A. W. Safitri, R. N. Yanti, and E. Suwarno, J. Belantara, vol. **6**, no. 1, pp. 69–79, (2023), doi: 10.29303/jbl.v6i1.921.
5. Badan Pusat Statistik Kabupaten Karanganyar, *Kecamatan Kerjo Dalam Angka 2021*. Karanganyar: Badan Pusat Statistik Kabupaten Karanganyar (2021).
6. S. F. Tobing, C. Maharani, L. Napitupulu, and D. Yulindra Affandi, *Partisipasi Perempuan dalam Hutan Adat: Studi Kasus di Sumatra Selatan & Riau*. Jakarta: WRI Indonesia (2021). doi: 10.46830/wripn.19.00133.
7. S. Sanudin and E. Fauziyah, *Prosiding Seminar Nasional Masyarakat Biodiversitas Indonesia* pp. 696–701 (2015), doi: 10.13057/psnmbi/m010402.
8. D. R. Matitaputty, M. A. Langi, and H. D. Walangitan, Agri-Sosioekonomi, vol. **19**, no. 2, pp. 1067–1074, (2023), doi: 10.35791/agrsosek.v19i2.48638.
9. Y. Sadono, J. Pembang. Wil. Kota, vol. **9**, no. 1, p. 53, (2013), doi: 10.14710/pwk.v9i1.6526.
10. B. Achmad, R. H. Purwanto, S. Sabarnurdin, and Sumardi, J. Ilmu Kehutan., vol. **9**, no. 2, pp. 105–116, (2015).
11. H. Simon, *Pengelolaan Hutan Bersama Rakyat (Cooperative Forest Management) Teori dan Aplikasi pada Hutan Jati di Jawa*. Yogyakarta: Pustaka Pelajar, (2008).
12. N. A. Jariyah and N. Wahyuningrum, J. Penelit. Sos. dan Ekon. Kehutan., vol. **5**, no. 1, pp. 43–56, (2008).
13. B. T. Tarigan, *Pemanfaatan Hasil Hutan Bukan Kayu (HHBK) oleh Masyarakat di Kawasan Taman Hutan Raya Bukit Barisan Sumatera Utara*, Institut Pertanian Bogor, (2015).
14. T. Setiyowati, A. Fatchiya, and S. Amanah, J. Penyul., vol. **18**, no. 02, pp. 208–218, (2022).

Emission load analysis of land transportation on urban roads in Jember regency, Indonesia (a case study of Sultan Agung Road)

Santi Dwi Mulia[1], *Cantika Almas* Fildzah[1*], and *Abdur* Rohman[1]

[1]Department of Environmental Engineering, Faculty of Engineering, Universitas Jember, Jl. Kalimantan 37, Jember, Jawa Timur, Indonesia 68121

Abstract. Jember Regency is located in East Java with a population density of 2,536,729 people in 2020. The higher population in Jember will be proportional to the high transportation activity. This will cause problems regarding air pollution due to vehicle exhaust emissions in Jember Regency. The aims of this study are to estimate the annual emission load of air pollutants on a segment of Jember City's urban roads and to recommend emission reduction strategies. The selection of strategies is conducted by considering experts' judgments through the Analytic Hierarchy Process (AHP). Sultan Agung Road in Jember City was the target of this study because it is a national road in the middle of a busy shopping area around it. The results of this study show that the emission burden generated from the land transportation sector on Sultan Agung Road in Jember Regency includes 177.486 tons of CO/year; 48.638 tons of HC/year; 9.481 tons of NO_2/year; 2.534 tons of PM_{10}/year; 1,245.15 tons of CO_2 /year and 0.421 tons of SO_2 /year. Three strategies recommended by the experts to reduce emissions are BRT (Bus Rapid Transit) Operational Development, The use of Solar Cells in Public Street Lighting, and Smart Driving.

1 Introduction

Sultan Agung Road is a national road in Jember Regency. The road is busy because it is included in the urban area, trade routes and urban traffic. According to [1], busy vehicle activities will cause air pollution due to vehicle exhaust emissions. The main pollutants caused by vehicle activity are CO, HC, NO_2, PM_{10}, CO_2 and SO_2. Pollution due to vehicle exhaust gasses can cause respiratory infection. The relationship between PM_{10} concentration and Covid-19 has been evident in [2] of which the authors found SARS-CoV-2 RNA in PM_{10} samples in Bergamo, Italy. In addition, CO_2 is a major greenhouse gas that drives the global warming. To reduce the impact of the global warming, Indonesia in the Enhanced Nationally Determined Contribution (NDC) has pledged emission reductions of 31.89% unconditionally and 43.20% conditionally by 2030 [3].The objective of this study is to estimate the amount of emission load due to vehicle exhaust gasses on Jalan Sultan Agung Jember and to propose the best emission load reduction strategy with

* Corresponding author: cantikaalmasf@unej.ac.id

https://doi.org/10.1051/e3sconf/202346707001

the novelty of this study compared with previous studies is related to the identification whether there is a statistically significant difference among the number of counted motorcycles in weekdays and weekend morning by using repeated-measure ANOVA. We hope the proposed strategy can be considered by the local government as one of the ways to reduce the emission in Jember and further contribute to the fulfilment of the Indonesian NDC.

2 Method

2.1 Data retrieval

This research uses secondary data such as emission factors from Indonesian Ministry regulations and the 2006 IPCC Guidelines [4-6]. The primary data were the numbers and types of motorized vehicles, obtained from CCTV data recorder combined with manual counting.

2.2 Data collection

The data of the number of vehicles were collected at Sultan Agung Road, Jember Regency. Data collection for manual counting was carried out in October and November. Meanwhile, CCTV data in September. The sampling time was conducted every day in the peak hours of morning (06.00-08.00), afternoon (11.30-13.30), evening (15.30-17.30), and night (18.30-20.30) for 2 hours each.

2.3 Data analysis

2.3.1 CO, HC, NO_2, PM_{10} and SO_2 emission calculation

Data on the number of vehicles that have been obtained are analyzed by referring to [4] in which the emission load was calculated based on the formulas in equations 1 and 2:

$$Ea = \sum_{b=1,c=1}^{n,m} \left(VKT_{b,c} \times FE_{a,b,c} \times 10^{-6} \right) \quad (1)$$

$$VKT_{b,c} = f_{b,c} \times VKT_c \quad (2)$$

Where:
Ea = Emission load for a pollutant (ton/years)
$VKT_{b,c}$ = The total length of trips for b category motorized vehicle by using type C fuel (km)
$FE_{a,b,c}$ = Emission factor (g/km)
a = Types of pollutants (1-6 for CO, NO_2, HC, PM_{10}, SO_2, CO_2)
b = The category of motor vehicles
c = Types of fuel (1-2 for gasoline and diesel)
$f_{b,c}$ = The number of vehicles
VKT_c = Road length (km)

2.3.2 CO_2 emission calculation

The calculation of CO_2 emissions refers to [7] for the values of emission factors. Emission calculation is obtained from the sum of CO_2, CH_4 and N_2O emission values. The CH_4 and N_2O emission values are converted into CO_2 by multiplying them by the conversion factor.

$$Emission = Fuel_a \times EF_a \quad (3)$$

Where:
Emissions = Mass of CO_2 emitted
$Fuel_a$ = Mass or volume of fuel combusted
EF_a = CO_2 emission factor per mass or volume unit (kg gas/TJ)
a = Types of fuel (gasoline or diesel) = Emission load for a pollutant (ton/years)

According to [6], CO_2 emission factors per mass or volume unit are 72.600 kg gas/TJ for gasoline and 74.300 kg gas/TJ for diesel. The fuel consumption calculation stage begins with knowing the value of road length, number of vehicles, fuel density, fuel economy and heating value of motor vehicle fuel. The calculation formula for the energy consumption stage is contained in equations 4, 5 and 6 below.

1. Diesel oil Energy Consumption (KES) (Gg/years):

$$KES = Road\ Length\ (km) \times The\ number\ of\ vehicles\ (unit/years) \times Fuel\ density\ (kg/liter) \times Fuel\ Economy\ (liter/km) \times 10^{-6} \quad (4)$$

2. Gasoline Energy Consumption (KEB) (Liter/years):

$$KEB = Road\ Length\ (km) \times The\ number\ of\ vehicles\ (unit/years) \times Fuel\ Economy\ (liter/km) \quad (5)$$

3. Fuel Consumption (KBB) (TJ/years):

$$KBB = Energy\ consumption\ (diesel\ oil/gasoline) \times Fuel\ calorific\ value\ (diesel\ oil/gasoline) \quad (6)$$

2.3.2 Emission reduction strategy using AHP method (Analytic Hierarchy Process)

The Analytical Hierarchy Process (AHP) method was used in prioritizing the predetermined scenarios [8]. The use of questionnaires was carried out with expert respondents from related agencies such as the Jember Regency Transportation Office. The questionnaire data was then processed using R [9] and RStudio [10] to get the results and conclusions on the scenario of reducing land transportation sector emissions on Jalan Sultan Agung Jember. The criteria for emission reduction scenario were Environmental, Technical and Financial aspects. Three strategies, namely Smart Driving, the use of *Solar Cell* in public street lighting and BRT (*Rapid Transit Bus*) Operational Development, were proposed in the AHP questionnaires.

3 Result and Discussion

3.1 Number of Motorized Vehicles

The number of motorized vehicles on Jalan Sultan Agung is reported in Table 2. It is found that the day with the largest number of vehicles was Friday, with motorcycle vehicles as the dominant category. Sunday has the lowest number of vehicles. According to [11], the number of vehicles in the Karees area, Bandung is very high on weekdays, this is related to community activities in the area such as office, school, commercial activities, and so on. Meanwhile on Sunday, the vehicle activity is quite low. This is because some residents prefer to spend their time resting at home and many vehicles with out-of-town plates were found on that day. The number of vehicles tends to be high on Tuesdays. This is because many community activities are carried out on weekdays [12].

https://doi.org/10.1051/e3sconf/202346707001

Table 1. Average number of motorized vehicles per year (units/year).

Motor vehicles category	The average number of motorized vehicles per years (unit/years)						
	Monday	**Tuesday**	**Wednesday**	**Thursday**	**Friday**	**Saturday**	**Sunday**
Motorcycle	1,194,037	1,196,618	1,187,585	1,169,470	1,207,649	1,067,385	698,745
Car (gasoline)	278,969	277,907	268,858	274,488	271,701	287,136	348,806
Car (diesel oil)	61,422	64,437	59,993	68,622	76,634	71,784	66,243
Bus	96	78	177	104	115	188	42
Truck	26,984	26,854	25,821	26,374	26,040	27,427	21,535
Total	1,561,507	1,565,893	1,542,434	1,539,059	1,582,139	1,453,920	1,135,371

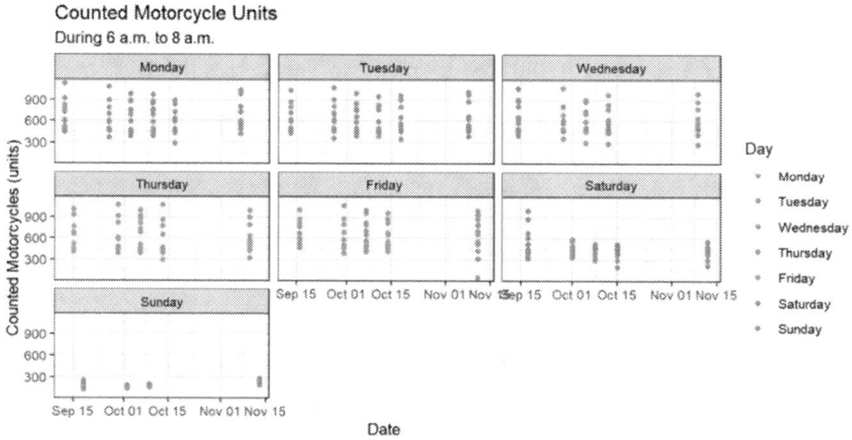

Fig. 1. Number of counted motorcycles during 6 a.m.to 8 a.m. from mid-September to mid-November 2022.

The motorcycle accounts for more than 60% of total vehicle units. Hence, it is worthy of a deeper observation. Figure 1 shows the counted numbers of motorcycles during 6 a.m. to 8 a.m. from mid-September to mid-November 2022. Compared to weekdays, weekends show considerable declines in the number of motorcycles. Based on repeated-measure ANOVA by using *nlme* package in R [13], we found that there is no statistically significant difference among the number of counted motorcycles in weekdays (Monday to Friday) morning from 6 a.m. to 8 a.m. (χ^2 (4) = 8.70, p-value = 0.691). We found a significant difference between weekday mornings and weekend morning in the counted motorcycles (χ^2 (6) = 474.74, p-value < 0.0001).

3.2 CO, HC, NO_2, PM_{10}, CO_2 and SO_2

Emission calculations are based on the CO, HC, NO_2, PM_{10}, CO_2 and SO_2 emission calculation formulas. The calculation of CO_2 emissions is based on equation 3, while other than CO_2 is based on equations 1 and 2. The results of emission calculations are shown in

Table 3. The tables show that the highest emission value varies. The lowest emission value for all emissions is the same on Sunday. CO, HC and PM_{10} emissions are highest on Tuesday. NO_2 and CO_2 emissions are highest on Friday. Meanwhile, SO_2 emissions are highest on Saturday. The difference is caused by the various values of emission factors and the number of vehicles of each motor vehicle. The number of vehicles, fuel consumption and the emission factor influence emission load calculation result [14]. According to [15], the difference in emission rates for each way is caused by a difference between the number of vehicles and the emission factor of the vehicles. PM_{10} concentration is said to be low even though the number of vehicles at one point is greater than at another point because the emission factor of these vehicles is small [16].

Table 2. Calculation results of CO, HC, NO_2, PM_{10}, CO_2 and SO_2 emission loads (ton/year).

Day	Emission (ton/years)					
	CO	HC	NO_2	PM_{10}	CO_2	SO_2
Monday	24.316	7.071	1.374	0.309	181.95	0.057
Tuesday	35.494	9.086	1.380	0.827	182.71	0.058
Wednesday	23.880	7.001	1.334	0.306	178.32	0.055
Thursday	23.880	6.931	1.373	0.307	180.91	0.059
Friday	24.260	7.116	1.397	0.318	185.61	0.062
Saturday	24.929	6.644	1.321	0.263	177.47	0.078
Sunday	20.727	4.790	1.302	0.203	158.18	0.053
Total	177.486	48.638	9.481	2.534	1,245.15	0.421

3.3 Emission Reduction Strategy Analysis Results

Nine respondents were selected in this study to fill AHP questionnaires. The respondents represent stakeholders, namely the Environmental Agency (DLH), the Jember Regency Transportation Agency, lecturers, and the local community. The results of the assessment of each respondent are presented in Table 4.

Table 3. AHP Scores by All Respondents.

Strategy	R1	R2	R3	R4	R5	R6	R7	R8	R9	Normalized Geometric Mean
Smart Driving	0.432	0.077	0.814	0.190	0.143	0.088	0.119	0.131	0.232	0.231
Solar Cells	0.381	0.588	0.093	0.080	0.143	0.440	0.606	0.125	0.302	0.299
Rapid Transit Bus (BRT)	0.187	0.336	0.093	0.730	0.715	0.472	0.276	0.744	0.467	0.470

Respondents Affiliation: R1: University; R2-R4: Department of Transportation; R5-7: Local Environmental Agency; R8-R9: local community members

Table 4 shows that respondents 1, 2, 3 preferred the smart driving strategy most, whereas the respondents 4, 5, 6, 8 and 9 preferred the Rapid Transit Bus (BRT) strategy. Respondent 7 preferred the diesel oil Cell strategy. The Geometric Mean (GM) was used to get the mean values. The results of the Geometric Mean (GM) value are then normalized to make it easier to compare the percentage value and determine the ranking of emission reduction strategies. The results of the normalized value of strategies from all respondents show a value of 0.470 for Rapid Transit Bus (BRT), diesel oil Cell of 0.299 and 0.231 for Smart Driving. The order of priority ranking is Rapid Transit Bus (BRT), Solar Cells and Smart Driving.

4 Conclusion

Based on the results of the research, we conclude that the emission burden generated from the land transportation sector on Jalan Sultan Agung Jember Regency includes 177.486 tons of CO/year; 48.638 tons of HC/year; 9.481 tons of NO_2/year; 2.534 tons of PM_{10}/year; 1,245.15 tons of CO_2/year and 0.421 tons of SO_2 / year. The recommended strategies for emission reduction on Sultan Agung Street in Jember Regency are Smart Driving, Use of Solar Cells on Public Street Lighting, and Development of BRT (Rapid Transit Bus) Operations. Based on the Analytic Hierarchy Process (AHP), BRT Operational Development (Rapid Transit Bus) should be given the highest priority, followed by using Solar Cells in Public Street Lighting and Smart Driving.

The authors are grateful to the Jember District Transportation Office and the Jember District Environmental Office for providing the research data. We also thank students and staffs who assisted us in carrying out the research.

References

[1] S. Rauf, F. A. Aboe, and I. T. Ishak, in *the 17th FSTPT Internasional Symposium* (2014), pp. 1119–1132
[2] L. Setti, F. Passarini, G. De Gennaro, P. Barbieri, M. G. Perrone, M. Borelli, J. Palmisani, A. Di Gilio, V. Torboli, F. Fontana, L. Clemente, A. Pallavicini, M. Ruscio, P. Piscitelli, and A. Miani, Environ Res **188**, (2020)
[3] *Enhanced Nationally Determined Contribution Republic of Indonesia 2022* (2022)
[4] *Peraturan Menteri Negara Lingkungan Hidup Nomor 12 Tahun 2010 tentang Pelaksanaan Pengendalian Pencemaran Udara Menteri Negara Lingkungan Hidup* (2010)
[5] *Pedoman Penyelenggaraan Inventarisasi Gas Rumah Kaca Nasional Buku Ii-Volume 1 Metodologi Penghitungan Tingkat Emisi Gas Rumah Kaca* (2012)
[6] A. Garg, K. Kazunari, and T. Pulles, in *IPCC Guidelines for National Greenhouse Gas Inventories* (2006)
[7] *Peraturan Menteri Negara Lingkungan Hidup Nomor 73 Tahun 2017* (2017)
[8] T. L. Saaty, *How to Make a Decision: The Analytic Hierarchy Process* (1990)
[9] R Core Team, R: A Language and Environment for Statistical Computing (n.d.)
[10] RStudio Team, RStudio: Integrated Development for R (2020)
[11] S. Novianti and D. Driejana, **16**, (2010)
[12] A. L. Ginting and M. Mirwan, **1**, 603 (2022)
[13] J. Pinheiro, D. Bates, and R Core Team. *nlme: Linear and Nonlinear Mixed Effects Models* (2023)
[14] R. N. Rahmasari, D. R. Jati, and J. Jumiati, **21**, 627 (2023)
[15] B. R. Widiatmono, E. Kurniati, and A.T. Imaya, **6**, 40 (2019)
[16] A. Z. Maulana, **15**, 499 (2012)

The adaptation to climate change by *slerek* boat fishermen in Muncar Subdistrict, Banyuwangi Regency

Yuli Hariyati[1]*, *Julfia Rasya* Putri[2], *Kamil* Muhtadi[2], *Ahmad* Zainuddin[1], and *Illia Seldon* Maghfiroh[1]

[1]Department of Agribusiness, Faculty of Agriculture, University of Jember, Indonesia
[2]Department of Undergraduate Program of Agribusiness, Faculty of Agriculture, University of Jember, Indonesia

Abstract. Climate change, the La Nina phenomenon, has various impacts on the fisheries sector, especially capture fisheries. One of the impacts of the La Nina phenomenon is that high rainfall causes fishermen to be unable to go to sea and reduces catches, so that it reduces fishermen's income. Therefore, fishermen need to adapt to climate changes that are occurring. This research aims to determine the form of climate change adaptation, measure the level of adaptation, and income earned by fishermen. This research uses descriptive and analytical methods. The results of the research show that there are 8 forms of adaptation carried out by *Slerek* boat fishermen. The level of adaptation of fishermen to climate change is 74%, of which 53.33% of fishermen are classified as low level of adaptation and 46.67% of fishermen are classified as high level of adaptation. Apart from that, there is a difference in income between fishermen with a high level of adaptation and fishermen with a low level of adaptation. It implied that fishermen with a high level of adaptation are able to combine various forms of adaptation to climate change and earn higher income.

1 Introduction

Climate change is a change in climate conditions that can be identified through temperature, rainfall, air pressure, humidity, wind speed and wind direction, which can last in the long term for decades or more [1]. The cause of global climate change is the El Nino and La Nina phenomena which affect oceans throughout the world. Indonesia, as a maritime country located between the Pacific and Indian oceans, will be affected when changes in temperature and extreme weather occur in the Pacific Ocean region due to El Nino and La Nina, which is increasing wave heights throughout Indonesian seas. This condition can be a challenge for society, especially in the fisheries sector.

The fisheries sector is one of the alternative employment opportunities for Indonesian people to earn income. Communities in coastal areas with livelihoods as fishermen rely on

*Corresponding authors: yuli.faperta@unej.ac.id

https://doi.org/10.1051/e3sconf/202346708001

catching fish or other marine products, including cultivating marine fish [2]. This condition shows that the fisheries sector plays an important role in the life of the Indonesian people to maintain food security [3, 4]. One of the southern coastal areas of Indonesia, precisely on the island of Java as the most numerous and extensive coastal areas, which has the largest production of capture fisheries is Banyuwangi Regency. Based on data from Central Agency on Statistics (BPS) of Banyuwangi Regency, Muncar subdistrict is the largest capture fisheries production center in Banyuwangi Regency with an average production of 35,031.23 tons with a decrease of 39.37% per year (2019–2021). Based on the Performance Accountability Report of Banyuwangi Regency Government Agencies 2020, the decline in capture fisheries production in Banyuwangi Regency was caused by external factors, which was uncertain weather conditions.

The problems that occur in Muncar subdistrict, Banyuwangi Regency are problems caused by climate change. Climate change that occurs is in the form of high waves, high rainfall, wind and changes in currents due to the La Nina phenomenon. Indonesia has experienced La Nina events from mid-2020 to the end of 2022 [5–7]. Based on Climatology Station of East Java, Southern Banyuwangi has an above normal rainy season, which Muncar subdistrict is included in this area. The impact caused by the La Nina phenomenon is an increase in rainfall to reach 50-90% of the average normal rainfall before the onset of the rainy season [7–8]. The occurrence of this phenomenon causes the surface temperature of sea water to change [9], so that it affects the behavior of fish. Changes in sea surface temperature not only affect the fish upwelling zone, but also shift fish species populations to colder or hotter seas and make it difficult for fishermen to catch fish because fish habitats tend to be on the underwater surface [4, 10–12].

Capture fisheries production in Muncar subdistrict comes from traditional fishing units, one of which is the *Slerek* boat. The *Slerek* boat has fishing route that is farther than other types of boats, so it is relatively more vulnerable to the effects of climate change. The occurrence of climate change has made it difficult for *Slerek* boat fishermen to catch fish so that fish catches have decreased. In addition, climate change causes an increase in waves which affect the cost of fishing for fishermen [4, 13]. This has implications for fishermen's income derived from fishing. The fewer fish were caught, the lower income earned will also be. Based on the background that has been described, the purpose of this study is to examine: 1) forms of adaptation to climate change, 2) the level of adaptation of *Slerek* boat fishermen in dealing with climate change, 3) and differences in income of *Slerek* boat fishermen according to the level of adaptation applied.

2 Methodology

The research was carried out purposively (purposive method), which was in Muncar subdistrict as the highest production center and supplier of captured fisheries in Banyuwangi Regency from November to December 2022. The sampling method used the total sampling method with a total of 30 respondents of *Slerek* boat owners. This method uses the population as sampling [14]. The method of data analysis to answer the problem regarding forms and adaptation levels of *Slerek* boat fishermen used descriptive methods and scoring techniques. The problem formulation related to fishermen's income was analyzed using income analysis with the following formulation [15–16]:

$$TR = P \times Q \quad (1)$$

$$TC = TFC + TVC \quad (2)$$

$$\pi = TR - TC \quad (3)$$

Information:
P : Price (Rp)
Q : Quantity (Kg)

TC : Total cost (Rp)
TVC : Total variable costs (Rp)
TFC : Total fixed costs (Rp)
π : Income (Rp)

Furthermore, to determine the difference in income between fishermen with high adaptation levels and fishermen with low adaptation levels, it was analyzed using the average difference test (independent sample t-test) with the following formula [17–18]:

$$t\ value = \frac{\bar{x}_1 - \bar{x}_2}{\sqrt{\left(\frac{S_1{}^2}{n_1}\right) + \left(\frac{S_2{}^2}{n_2}\right)}} \quad (4)$$

Information:
$\bar{x}_1$: Average income of fishermen with a high leverl of adaptation
$\bar{x}_2$: Average income of fishermen with a low level of adaptation
$S_1{}^2$:The standard deviation of fishermen's adaptation level is high
$S_2{}^2$:The standard deviation of fishermen with low level of adaptation
n_1 : Sample of fishermen with a high level of adaptation
n_2 : a sample of fishermen with a low level of adaptation
Decision making criteria: H_0 is rejected if the sig.(2-tailed) value < 0.05

3 Results and discussion

3.1 Forms of adaptation to climate change by *slerek* boat fishermen

Climate change adaptation is an effort by fishermen to minimize the reduction in fish catches due to climate change. There are several forms of adaptation carried out by *Slerek* boat fishermen as an effort to deal with climate change which is explained in Table 1.

Table 1. Forms of adaptation to climate change by *slerek* boat fishermen.

No	Adaptation form	Number of Fishermen	%
1	Moving fishing location	30	100
2	Stretching fishing gear (nets)	30	100
3	Changing fishing times	25	83.33
4	Increasing the amount of oil in the engine	30	100
5	Changing the catchment area	30	100
6	Using plastic sheets as raincoats	24	80
7	Carrying used drums when going to sea	30	100
8	View weather forecast information	30	100

Source: Primary data (2022)

All fishermen (100%) move fishing locations as a form of adaptation to climate change. Location shifts are carried out when fishermen have determined the location points to go to, but are threatened by climate change such as high waves. Fishermen are faced with climate change scenarios while at sea, fishermen will stop fishing and look for alternatives, such as moving locations to a wider area to catch fish more intensively [19–21].

Stretching fishing gear (nets) is carried out by all fishermen (100%) to minimize the risk of reduced catches. Fishermen will stretch their fishing gear to keep catching fish despite climate change while at sea. Stretching the entire fishing gear helps in expanding the reach of fish catches, if the fishing gear is stretched less, fishermen will get results that are not optimal [22–23].

Fishermen (83.33%) made changes to fishing time when bad weather occurred due to climate change before departure. *Slerek* boat fishermen make changes to fishing times by delaying their departure time. The impact of climate variability and change such as extreme events (storms, high waves) and erratic weather can affect fishermen to postpone departure times [24].

The addition of oil to the *Slerek* boat engine is carried out by all fishermen (100%) when climate change occurs such as big waves and also heavy currents. Adding oil to this engine affects the engine's speed when dealing with current. The greater the oil consumption used, the faster the boat will be when sailing against the current [25].

Changes in fishing areas are carried out by all fishermen (100%). If the target fishing area is affected by climate change, such as strong currents, the fishermen tend to change catchment areas that do not have fast currents. Fishermen will tend to change and stay away from fishing areas affected by climate change to avoid decreasing catches [12, 26].

The form of adaptation to the use of plastic sheets as raincoats was carried out by *Slerek* boat fishermen (80%) when it rained while fishing. The owner fishermen and other worker fishermen will enter inside a quite large plastic sheet, usually one sheet of plastic bag is filled by 4-5 people. The use of plastic sheets as raincoats by fishermen is very effective in preventing rainwater [27].

The form of adaptation to carrying used drums is also carried out by all *Slerek* boat fishermen (100%). Fishermen bring used refueling drums to be used as alternative floats when the boat overturns due to big waves, storms, and strong winds. Fishermen usually use plastic drums or bamboo as floats when a cyclone hit [28]. The number of used drums brought depends on the number of the owner and worker fishermen.

The form of adaptation to viewing weather forecast information to arrange departure schedules is carried out by all *Slerek* boat fishermen (100%). The fishermen have an online discussion forum to exchange information regarding weather forecasts through the Meteorological, Climatological, and Geophysical Agency (BMKG) website. This is conducted to anticipate and consider the departure time for fishing, make fishing safer, and preparations against any weather changes [29–30].

Based on the forms of adaptation that have been described, it can also be seen the level of adaptation to climate change based on the results of calculating the score for each form of adaptation presented in Table 2.

Table 2. Level of implementation of climate change adaptation by *slerek* boat fishermen.

Description	Mark
Number of respondents (people)	30
Average total score	18
Average percentage (%)	74
Low level of adaptation (people)	16
High level of adaptation (people)	14

Source: Primary data (2022)

Table 2 shows that the average level of adaptation carried out by *Slerek* boat fishermen in Muncar subdistrict, Banyuwangi Regency is 74%. Fishermen with scores below 74% are included in the category of applying a low level of adaptation as many as 16 fishermen or 53.33% of the total respondents. Meanwhile, fishermen with scores above the average are included in the category of applying a high level of adaptation as many as 14 fishermen or 46.67% of the total respondents.

3.2 Income of *slerek* boat fishermen

The results of the analysis of income of Slerek boat fishermen according to the level of adaptation implementation are presented in Table 3.

Table 3. Income analysis of *slerek* boat fishermen

Component	High Adaptation Rate	Low Adaptation Rate
Total fixed costs	Rp1,265,748	Rp1,337,844
Total variable costs	Rp10,954,643	Rp9,061,875
Total cost	Rp12,220,391	Rp10,339,719
Total revenue	Rp21,085,714	Rp15,562,500
Income	Rp8,865,323	Rp5,162,781

Source: Primary data (2022)

Table 3. shows that the revenue earned and the total costs incurred by fishermen with a high level of adaptation is greater than fishermen with a low level of adaptation in one trip to sea as much as Rp 5,523,214 and Rp1,820,673. This is because fishermen with a high level of adaptation increase the use of oil in the engine, so that the boat is able to move locations or change fishing grounds with longer routes. The results of the income calculation show that the income of fishermen with a high level of adaptation is greater than that of fishermen with a low level of adaptation as much as Rp3,702,542. Based on income analysis, it can be known the income average difference between fishermen with a high level and low level of adaptation as explained in Table 4.

Table 4. Result of S*lerek* boat fishermen's income difference test (independent sample t-test)

		F-stat	Sig.	t-stat	Sig. (2-tailed)
Income	Equal variances assumed	2.003	0.168	-3.986	0,000
	Equal variances not assumed			-3.855	0.001

Source: Primary data (2022)

Table 4 shows the results of the income difference test analysis with sig. (2-tailed) values is 0.000 < 0.05, then the hypothesis H_0 is rejected. This means that there is a significant difference in income between fishermen with a high level of adaptation and fishermen with a low level of adaptation. This shows that fishermen with a high level of adaptation tend to be braver in taking risks and sailing longer routes [19]. Fishermen with a high level of adaptation are able to combine various forms of adaptation. These results also prove that the level of adaptation of fishermen influences the income earned by fishermen.

4 Conclusion

1. There are 8 forms of adaptation carried out by *Slerek* boat fishermen, including moving fishing locations, stretching fishing gear (nets), changing fishing time, increasing the amount of oil in the engine, changing fishing grounds, using plastic sheets as raincoats, bringing used drums, and view weather forecast information.
2. The level of adaptation to climate change of *Slerek* boat fishermen in Muncar subdistrict, Banyuwangi Regency is 74%, as many as 14 fishermen are implementing a high level of adaptation and 16 fishermen are implementing a low level of adaptation.
3. There is a significant difference between the income of fishermen with a high level of adaptation and fishermen with a low level of adaptation.

References

1. LAPAN, *Perubahan Iklim: Penyebab, Dampak, dan Antisipasi* (Pusat Sains dan Teknologi Atmosfer, Bandung, 2017)
2. N. E. N. Laila, S. Amanah, Sodality J. Sosiol. Pedesaan. **3**, 159–168 (2015)
3. S. Sulistiani, J. H. Sinaulan, J. Ekon. **22**, 178–190 (2020)

4. A. Rizal, Z. Anna, World News Nat. Sci. **22**, 119–128 (2019)
5. N. A. Hasan, Y. Chikamoto, M. J. McPhaden, Front Clim. **4**, 01–13 (2022)
6. C. Gao, M. Chen, L. Zhou, L. Feng, R. Zhang, Sci. China Earth Sci. **65**, 2248–2266 (2022)
7. W. N. Harahap, B. Yuniasih, S. Gunawan, Agroista J. Agroteknologi. **7**, 26–32 (2023)
8. S. Risnayah, Bul. GAW Bariri. **2**, 97–104 (2021)
9. T. A. Okey, H. M. Alidina, V. Lo, S. Jessen, Rev. Fish. Biol. Fish. **24**,519–559 (2014)
10. N. Moegni, A. Rizki, G. Prihantono, J. Ekon. dan Stud. Pembang. **15**, 182–189 (2014)
11. G. Rusmayadi, *Agroklimatologi di Era Perubahan Iklim Global* (CV. IRDH, Malang, 2019)
12. R. Patriana, A. Satria, J. Sos. Ekon. Kelaut. dan Perikan. **8**, 11–23 (2013)
13. Lusiani, A. Hendrawan, Wahikun, J. Saintara. **2**, 1–7 (2018)
14. Sugiyono, *Metode Penelitian Kuantitatif, Kualitatif dan R&D* (CV. Alfabeta, Bandung, 2016).
15. S. Sukirno, *Mikroekonomi Teori Pengantar* (Raja Grafindo Persada, Jakarta, 2016)
16. Y. Hariyati, *Ekonomi Mikro (Pendekatan Matematis dan Grafis)* (CSS, Jember, 2007)
17. Santoso, *Statistik Nonparametrik Konsep dan Aplikasi dengan SPSS* (Elexmedia Komputindo, Yogyakarta, 2010)
18. I. Ghozali, *Aplikasi Analisis Multivariate dengan Program IBM SPSS 20* (Universitas Diponegoro, Semarang, 2012)
19. M. Ulfa, J. Pendidik. Geogr. **23**, 41–49 (2018)
20. I. S. Maurizka, S. Adiwibowo, J. Sains Komun. dan Pengemb. Masy. **05**, 496–508 (2021)
21. A. Wahyono, M. Imron, I. Nadzir, J. Kebijak. Sos. Ekon. Kelaut. dan Perikan. **3**, 133–141 (2013)
22. Nurlaili, J. Masy. dan Budaya. **14**, 599–624 (2012)
23. A. A. Samah, H. A. M. Shaffril, A. Hamzah, B. A. Samah, SAGE Open.**9**,1–11 (2019)
24. Sudiyono. J. Masy. dan Budaya. **18**, 263–281 (2016)
25. A. Helmi, A. Satria, Makara Hum. Behav. Stud. Asia. **16**, 68–78 (2012)
26. E. T. Y. Dima, Ekopem J. Ekon Pembang. **5**, 12–22 (2020)
27. U. Wulandari, M. N. Kholis, R. S. Putri, Syafiq, Samakia J. Ilmu Perikan. **12**, 38–46 (2021)
28. M. M. Islam, S. Sallu, K. Hubacek, J. Paavola, Mar Policy. **43**, 208–216 (2014)
29. H. A. M. Shaffril, B. A. Samah, J. L. D'Silva, S. M. Yassin, Int. J. Clim. Chang. Strateg. Manag. **5**, 38–53 (2013)
30. S. Z. Omar, H. A. M. Shaffril, N. Kamaruddin, J. Bolong, J. L. D'Silva, Life Sci. J. **10**, 540–549 (2013)

AUTHOR INDEX